AF247539

Chemical Technology: An Encyclopedic Treatment

VOLUME VIII

Editors

L. W. Codd, M. A.

K. Dijkhoff, Chem. Drs.

J. H. Fearon, B. Sc., C. Eng., M.I.E.E.

C. J. van Oss, Ph. D.

H. G. Roebersen, Chem. Drs. †

E. G. Stanford, M. Sc., Ph. D., C. Eng., M.I.E.E.,

F. Inst. P.

Chemical Technology:
An Encyclopedic Treatment

*The Economic Application
of Modern Technological Developments
Based upon a work originally devised
by the late Dr. J. F. van Oss*

VOLUME VIII

Edible oils and fats

Animal food products

Material resources

General index

Appendix - Recent developments

in materials and technology

NEW YORK

BARNES & NOBLE BOOKS

(a division of Harper & Row Publishers, Inc.)

Published in the U.S.A. 1975 by
HARPER & ROW PUBLISHERS, INC.
BARNES & NOBLE IMPORT DIVISION

Published throughout the world
except United States of America
and the Philippines
by Longman Group Ltd.
and Uitgeverij J. H. de Bussy b.v.
under the title
Materials and Technology

Library of Congress Catalog
Card Number 68-31037

ISBN 06 491109 8

First published 1975

Set in 9 on 11 point Times
and printed in the Netherlands
by De Bussy Ellerman Harms b.v., Amsterdam

Foreword

When I read the first announcement of this new eight-volume encyclopedia I immediately recognized it as a marvellously integrated guide to the World's raw materials unlike any other reference work I know.

The encyclopedia is prepared in such a way that its information is accessible to a large number of readers whose knowledge of science and technology is limited. For the first time, the nonspecialist in need of quick, accurate data about materials and processes can refer to a single reference work that correlates technical facts with economic and financial data in a simple yet highly organised and readily understandable fashion. This encyclopedia is a comprehensive guide, broad enough in scope to include information about all the world's raw materials, the primary manufacturing and agricultural processes involved, world output, prices, and other related information, all presented together in such a way that the unknowledgeable reader can rapidly gain an overall view of a subject or a variety of subjects.

The wealth of economic information presented in these volumes should be of immense value to those engaged in business, manufacturing industry, finance, economics, journalism, public relations, research, analysis, government, and indeed to everyone involved in commercial applications and marketing of raw materials.

The encyclopedia will also be a tremendous asset to librarians and students in science, engineering and economics, who can refer to this information as an authoritative introduction to the whole subject of materials and technology.

Although this work has been planned with the nonspecialist in mind specialists will also find it a useful work of reference. Each chapter includes, in addition to fundamental data about a subject, other information usually found in more technical sources. With this combined approach the technical man now has a source he can use to gain knowledge quickly of the functional importance of materials and processes outside his own special field.

This work may be considered a successor to Dr. J. F. Van Oss's *Warenkennis en Technologie (Systematic Encyclopedia of Technology)*. The present encyclopedia, however, has been thoroughly updated and comprises an entirely new work that is in every way a far more ambitious and elaborate project than its predecessor.

The editors have superbly integrated descriptions of operational processes with descriptions of technical and chemical processes. For example, in the first volume, even though most of the industrial processes describe chemical reactions of one kind or another, the first chapter covers the important physical separation processes involved in the manufacture of some of these materials. The editors intend to carry out this same kind of presentation throughout. Subjects are amply cross-referenced, there are numerous line drawings and photographs, and selected bibliographies at the end of each chapter. Each of the eight volumes is to be indexed separately, with a separate final index to the entire encyclopedia.

We in the United States are used to seeing everything from our point of view and tend to forget the tremendous technological and agricultural world that lies outside our country, a world that includes numerous raw materials that play an important part in our own technology and economy. *Chemical Technology: An Encyclopedic Treatment* deals with them all and will be of inestimable value to the user. I recommend it highly.

John J. McKetta, Jr.
Executive Vice Chancellor for Academic Affairs
The University of Texas System
Austin, Texas.

Contents

Preface

This is the last volume of the Encyclopedia of Materials and Technology and it completes the total survey of materials of commercial interest by dealing with animal food products. As with all the previous volumes, it is written for people who want to know about materials, whether or not they have had advanced training in science and technology. Anyone who has studied physics and chemistry to sixth form level in a secondary school will have little difficulty in following the text.

Because of the growing concern which has been developing since the Encyclopedia was planned about the future shortage of materials for commercial use, the Editors have included in this last volume an additional chapter entitled Material Resources and their Conservation.

The preparation of an encyclopedia of this length goes on over a period of many years and, for all the care taken by the Editorial Board to ensure that the material included would not quickly date, technology and the materials involved in it have continued to develop during this time. Therefore, all authors in the first five volumes were invited to up-date their chapters so that the latest information could be included; and a number of them have taken the opportunity to do so. These short up-dating accounts appear in the Appendix entitled Recent Developments in Materials and Technology.

The volume also includes a consolidated General Index to the whole Encyclopedia.

As with the earlier volumes each chapter has been written by a specialist or group of specialists so as to present a comprehensive and up-to-date account of the subject, which will be of use not only to the general reader but also to those who work with materials in their professional capacity. Thus the encyclopedia will serve the engineer who requires authoritative guidance on materials with which he is not familiar, the business man in commerce and industry, the craftsman who wants to acquire a sound knowledge of the materials used in his trade and, of course, the materials scientist for whom it will provide a work of reference covering the whole field of his subject in a convenient and manageable form.

It is also intended to serve a useful function in education as a link between formal textbooks and the practical applications of academic learning. University and college lecturers and school teachers should find it a valuable work of reference, as also will their students and pupils, in its description of the industrial uses of the materials and processes covered in their courses.

In order to give the reader a measure of the scale of activity in the various fields, quantitative information is given. Every effort has been made to include the most recent figures available but, as there is considerable variation in the speed of publication of such figures from country to country, and from industry to industry, the figures quoted can only be illustrative of trends. Because the encyclopedia can only be revised infrequently, guidance is given wherever possible to the sources consulted so that the most recent information may easily be sought out.

The reader must be warned that some of the information presented is protected by patents and before using it for commercial purposes careful enquiries should be made. Reference in this book to any particular product does not imply approval or recommendation of that product and the use of trade names in reference to any product does not imply that they are generic terms as they may or may not be protected by registered trademarks.

The publishers have made arrangements whereby they hope to be able to answer any enquiries arising out of the use of the encyclopedia. These should be sent to Longman Group Limited, Longman House, Burnt Mill, Harlow, Essex. They should be marked specifically 'For the attention of the Editor, Materials and Technology Encyclopedia'.

The Editors have set out to make *Materials and Technology* both authoritative and approachable: they believe it is unique in presenting so broad a field of technology in so manageable and understandable a form.

THE EDITORS

Authors

R.F. LOONEY,
The Union International Research Centre,
The Union International Company Ltd.   Ch. 1 Oils and Fats

J.M. DAVIES,
The Union International Company Ltd.   Ch. 2 Protein-Containing Foods
                                            in General

T.S. SEALY,   Ch. 3 Meat, Fish and Similar
Martran Ltd.,   Products
E.F. GOLDEN,
E.F.G. Agricultural Advisory Service,
J.M. DAVIES
and
Dr. E.S. KRUDY,
Information Retrieval Ltd.

Dr. J.G. DAVIS,   Ch. 4 Dairy Products
Dr. J.G. DAVIS and Partners (Consultant
Bacteriologists and Chemists)

F. ROBERTS   Ch. 5 Material Resources and
                                            their Conservation

Acknowledgements

We are grateful to the following for their kind permission to reproduce the figures listed below:

Fig. 1.1: by courtesy of Fenno Jacobs and United Africa Company Ltd.
Figs. 1.2, 1.8, 1.9, 1.10, 1.12, 1.13, 1.14, 1.15, 1.16, 1.17, 1.18, 1.19, 1.20: by courtesy of Unilever Ltd.
Fig. 1.4: by courtesy of Pye Unicam Ltd.
Fig. 1.6: by courtesy of Union International Research Centre, St. Albans, through Newport Instruments Ltd.
Fig. 1.11: by courtesy of the United Africa Company Ltd.
Figs. 1.22, 1.23, 1.24: by courtesy of Rose Downs and Thompson Ltd.
Figs. 1.25, 1.26, 1.28: by courtesy of Alfa-Laval Tumba.
Fig. 1.33: by courtesy of The Girdler Company.
Fig. 1.35: by courtesy of Gerstenberg and Agger.
Fig. 1.37: by courtesy of Schweizerische Industrie Gesellschaft.
Fig. 2.2: by courtesy of the University of Wisconsin Press, from *The Physiology and Biochemistry of Muscle as Food, 1* edited by E.J. Briskey, A.G. Cassens and J.C. Trautman.
Fig. 2.4: by courtesy of Courtaulds Ltd.
Fig. 2.5: by courtesy of British Petroleum Company Ltd.
Fig. 2.6: by courtesy of Rank, Hovis, McDougall Ltd.
Figs. 3.1, 3.2, 3.3, 3.4, 3.5, 3.6, 3.11, 3.12, 3.13, 3.14, 3.15, 3.16, 3.17, 3.18, 3.19, 3.29, 3.30, 3.31, 3.32, 3.33: by courtesy of *Farmer and Stockbreeder.*
Figs. 3.7, 3.28, 3.34: by courtesy of *Scottish Farmer.*
Figs. 3.8, 3.22, 3.23: by courtesy of the Ministry of Agriculture, Fisheries and Food.
Fig. 3.9: by courtesy of Accles and Shelvoke Ltd. Birmingham.
Figs. 3.10, 3.20, 3.21, 3.24, 3.25, 3.26: by courtesy of H. Thornton, *Textbook of Meat Inspection* Bailliere Tindall, London.
Fig. 3.27: by courtesy of Her Majesty's Stationery Office, Crown Copyright.
Figs. 3.35, 3.36: by courtesy of Shaver Poultry Breeding Farms Ltd.
Fig. 3.37: by courtesy of Cobb Breeding Co. Ltd.
Figs. 3.39, 3.40, 3.41: by courtesy of *Poultry World.*
Figs. 3.43, 3.44: by courtesy of N. Reich Ltd.
Fig. 3.45: by courtesy of A.B.R. Machinery Co. Ltd.

Fig. 3.46: by courtesy of Barry McKay Industrial Photography Ltd.

Figs. 3.47, 3.48: by courtesy of Union Corporation.

Figs. 3.52, 3.57, 3.58, 3.59, 3.60: by courtesy of Michael Wood.

Fig. 3.61: by courtesy of Findus Ltd.

Fig. 3.62: by courtesy of Hull Fish Meal and Oil Co. Ltd.

Figs. 4.1, 4.3, 4.4, 4.21, 4.25: by courtesy of A.P.V. Company Ltd.

Fig. 4.2: by courtesy of Sordi, Italy.

Fig. 4.5: by courtesy of Dawson Barfos.

Fig. 4.6: by courtesy of Express Dairy Ltd.

Fig. 4.6a: by courtesy of Birmingham Dairies.

Fig. 4.7: by courtesy of Alfa-Laval.

Fig. 4.8: by courtesy of the Milk Marketing Board.

Figs. 4.9, 4.11, 4.13: by courtesy of Unigate Ltd.

Fig. 4.10: by courtesy of Stork, Netherlands.

Fig. 4.12: by courtesy of NIRO, Denmark.

Fig. 4.14: by courtesy of the Condensed Milk Company of Ireland.

Figs. 4.15, 4.17: by courtesy of Professor Demeter.

Fig. 4.16: by courtesy of Soc. Esp. Polenghi, Italy.

Fig. 4.18: by courtesy of Hartingdon Dairies.

Fig. 4.19: by courtesy of Bridel, Retiers, France.

Fig. 4.20: by courtesy of Borden, New York.

Figs. 4.22, 4.23: by courtesy of Lyons Maid.

Fig. 4.24: by courtesy of T. Giusti Ltd.

Figs. 5.1, 5.11, 5.12: by courtesy of F. Roberts, *Future Resources for Engineering.*

Fig. 5.2: by courtesy of B.J. Skinner, *Earth's Resources,* Prentice-Hall Inc.

Fig. 5.4: by courtesy of G.J.S. and M.H. Govett, *Mineral Resource Supplies and the Limits of Economical Growth.*

Figs. 5.5, 5.6: by courtesy of *Resources and Man,* W.H. Freeman and Co.

Fig. 5.7: by courtesy of the Copper Development Association, London.

Fig. 5.8: by courtesy of the Institute of Metals.

Fig. 5.9: by courtesy of the Chartered Mechanical Engineer.

Fig. 5.10: by courtesy of *Understanding the Earth* 2nd edition, Artemis Press.

Symbols and Abbreviations

Symbol	*Name of unit*
*	radioactive
a	are ($= 100$ m²) ($= 119.599$ yd²)
A	ampere
Å	angström ($= 10^{-8}$ m)
	($= 0.003\ 937\ 01\,\mu$ in)
a.c.	alternating current
at	technical atmosphere
atm	standard atmosphere
	($= 101.325$ kN/m²)
	($= 14.6959$ lbf/in²)
b	bar (10^5 N/m²)
	($= 14.5038$ lbf/in²)
Bé	Beaumé's scale
BP	boiling point
Btu	British thermal unit
	($= 1.05506$ kJ)
bu	bushel ($= 36.368\ 7$ dm³)
c	centi ($= 10^{-2}$)
C	coulomb
°C	degree Celsius ($= 5/9$ (°F – 32))
	(temperature value)
cal	calorie (International table)
cd	candela
cg	centigram
Ci	Curie
cm	centimetre ($= 0.393\ 701$ in)
c/s	cycles per second
cwt	hundred weight ($= 50.8023$ kg)
	($= 112$ lb)
d	deci ($= 10^{-1}$)

Symbol	*Name of unit*
da	deca ($= 10^1$)
dag	decagram
d.c.	direct current
degC	degree Celsius (temperature interval)
degF	degree Fahrenheit (temperature interval)
dg	decigram
dm	decimetre
dyn	dyne ($= 10^{-5}$N)
	($= 0.224829 \times 10^{-5}$ lbf)
ϵ	molar extinction coefficient
erg	erg ($= 10^{-7}$J)
	($= 0.737\ 562 \times 10^{-7}$ ft lbf)
	electron-volt ($= 1.60209 \times 10^{-19}$J)
F	Farad
°F	degree Fahrenheit ($= \frac{9}{5}$ °C $+ 32$) (temperature value)
fl oz	Fluid ounce ($= 28.4131$ cm³)
ft	foot ($= 0.3048$ m) ($= 12$ in)
ft H₂O	foot water ($= 2989.07$ N/m²)
g	gram
G	giga ($= 10^9$)
gal	UK gallon ($= 4.596$ litres)
	($= 4.546\ 09$ dm³)
	(cf. US gallon $= 3.78541$ dm³)
gr	grain ($= 64.798\ 9$ mg)
h	hecto ($= 10^2$)
H	henry

Symbol	Name of unit
ha	hectare (= 10000 m²) (= 2.471 05 acres)
hp	horsepower (= 745.700 W)
Hz	hertz (= 1 cycle/sec.)
in	inch (= 2.54 cm)
in Hg	conventional inch of mercury (= 3386.39 N/m²) (= 33.8639 mb)
in H₂O	conventional inch of water (= 249.089 N/m²)
J	joule (= 0.737562 ft lbf)
k	kilo (= 10³)
K	degree kelvin (= °C + 273)
kcal	kilocalorie
kg	kilogram
kgf	kilogram force (= 9.806 65 N) (= 2.204 62 lbf)
kJ	kilojoule
km	kilometre
kp	kilopond (= kgf)
kW	kilowatt
l	litre (= approx 1 dm³) (= 0.220 0 gal) (= 0.24642 US gal)
lb	pound (= 0.45359237 kg)
lbf	pound force (= 4.44822 N)
lm	lumen
lx	lux (= 1 lm per m²)
m	metre (= 1.09361 yd)
m	milli (= 10⁻³)
M	mega (= 10⁶)
mb	millibar (= 100 N/m²)
m.g.d.	million gallons per day
mile	mile (= 1.60934 km)
ml	millilitre
mm	millimetre
mmHg	conventional millimetre of mercury (= 133.322 N/m²) (= 0.0393701 in Hg)
mm H₂O	conventional millimetre of water (= 9.80665 N/m²)
money	£ (= UK pound unless stated to the contrary)

Symbol	Name of unit
$	(= US dollar unless stated to the contrary)
M.P.	melting point
μ	micro (= 10⁻⁶)
μb	microbar (= 0.1 N/m²)
μ in	microinch (= 0.0254 μm) (= 0.000001 in)
μ m	micrometre (micron) (39.3701 m in)
μ mHg	micron of mercury (= 0.133322 N/m²)
n	nano (= 10⁻⁹)
N	newton (= 0.224809 lbf)
n mile	international nautical mile (= 1852 m) (cf. UK nautical mile = 1853.18 m)
oz	ounce (= 28.3495 g)
Oz apoth	apothecaries' ounce (= 31.1035 g) (= oz tr)
ozf	ounce force (= 0.278014 N)
oz tr	troy ounce (= 31.1035 g) (= oz apoth)
Ω	ohm
p	pico (= 10⁻¹²)
P	poise (= 0.1 N s/m²) (= 2.08854 x 10⁻³ lbf s/ft²)
Pl	poiseville (= N m²/s)
pH value	measure of acidity/alkalinity
pK_A	measure of strength of acid
pK_B	measure of strength of alkali
p.p.m.	parts per million
p.s.i.	poundweight per square inch (= 6894.76 N/m²) (= 68.9476 mb)
pt	pint (= 0.568261 dm³)
pz	pieze (= 10³ N/m²)
PVC	polyvinylchloride
q	quintal (= 100 kg)
qt	Imperial quart (1.13652 dm³)
°R	degree Rankine (°F + 459.67)
rad	radian
s	second

Symbol	Name of unit	Symbol	Name of unit
St	stokes ($= 10^{-4}$ m²/s) ($= 558.001$ in²/h)	tonf	ton force ($= 9964.02$ N)
		V	volt
t	metric ton ($=$ tonne) ($= 1000$ kg) ($= 0.984207$ tons) ($= 2204.6$ lb)	W	Watt ($=$ J/s)
T	tera ($= 10^{12}$)	Wb	Weber
ton	Imperial ton ($= 1016.05$ kg) ($= 2240$ lb) ($=$ long ton) (*cf*. US ton $= 2000$ lb $=$ short ton)	wt	weight
		w/w	weight for weight
		yd	yard ($= 0.9144$ m)

Table of Chemical Elements

*These elements, like the transuranic elements (see below), have been produced by artificial means and do not occur naturally (at least, not in any appreciable amount).

Element	Symbol	Atomic Number	Atomic Weight	Element	Symbol	Atomic Number	Atomic Weight
Actinium	Ac	89	227	Gallium	Ga	31	69.72
Aluminium	Al	13	26.98	Germanium	Ge	32	72.60
Antimony	Sb	51	121.76	Gold	Au	79	197
Argon	A	18	39.944				
Arsenic	As	33	74.91	Hafnium	Hf	72	178.5
*Astatine	At	85	(210)	Helium	He	2	4.003
				Holmium	Ho	67	164.94
Barium	Ba	56	137.36	Hydrogen	H	1	1.008
Beryllium	Be	4	9.013				
(Glucinium)	(Gl)			Indium	In	49	114.76
Bismuth	Bi	83	209	Iodine	I	53	126.91
Boron	B	5	10.82	Iridium	Ir	77	192.2
Bromine	Br	35	79.916	Iron	Fe	26	55.85
Cadmium	Cd	48	112.41	Krypton	Kr	36	83.80
Caesium	Cs	55	132.91				
Calcium	Ca	20	40.08	Lanthanum	La	57	138.92
Carbon	C	6	12.01	Lead	Pb	82	207.21
Cerium	Ce	58	140.13	Lithium	Li	3	6.940
Chlorine	Cl	17	35.457	Lutetium	Lu	71	174.99
Chromium	Cr	24	52.01	(Cassiopeium)	(Cp)		
Cobalt	Co	27	58.94				
Copper	Cu	29	63.54	Magnesium	Mg	12	24.32
				Manganese	Mn	25	54.94
Dysprosium	Dy	66	162.46	Mercury	Hg	80	200.61
				Molybdenum	Mo	42	95.95
Erbium	Er	68	167.3				
Europium	Eu	63	152	Neodymium	Nd	60	144.27
				Neon	Ne	10	20.183
Fluorine	F	9	19	Nickel	Ni	28	58.7
Francium	Fr	87	223	Niobium	Nb	41	92.91
				(Columbium)	(Cb)		
Gadolinium	Gd	64	156.9	Nitrogen	N	7	14.008

Element	Symbol	Atomic Number	Atomic Weight	Element	Symbol	Atomic Number	Atomic Weight
Osmium	Os	76	190.2	Thallium	Tl	81	204.39
Oxygen	O	8	16	Thorium	Th	90	232.1
				Thulium	Tm	69	168.94
Palladium	Pd	46	106.7	Tin	Sn	50	118.70
Phosphorus	P	15	30.974	Titanium	Ti	22	47.90
Platinum	Pt	78	195.23	Tungsten	W	74	183.92
Polonium	Po	84	210	(Wolfram)			
(Radium F)							
Potassium	K	19	39.100	Uranium	U	92	238.07
Praseodymium	Pr	59	140.92				
*Prometheum	Pm	61	(145)	Vanadium	V	23	50.95
Protactinium	Pa	91	231				
				Xenon	Xe	54	131.3
Radium	Ra	88	226.05				
Radon	Rn	86	222	Ytterbium	Yb	70	173
(Niton)	(Nt)			Yttrium	Y	39	89
Rhenium	Re	75	186.31				
Rhodium	Rh	45	102.91	Zinc	Zn	30	65.38
Rubidium	Rb	37	85.48	Zirconium	Zr	40	91.22
Ruthenium	Ru	44	101.1				

TRANSURANIC ELEMENTS

Element	Symbol	Atomic Number	Atomic Weight
Samarium	Sm	62	150.43
Scandium	Sc	21	44.96
Selenium	Se	34	78.96
Silicon	Si	14	28.09
Silver	Ag	47	107.873
Sodium	Na	11	22.997
Strontium	Sr	38	87.63
Sulphur	S	16	32.066
Tantalum	Ta	73	180.88
*Technetium	Tc	43	(98.91)
Tellurium	Te	52	127.61
Terbium	Tb	65	158.9

Atomic Number	Element	Symbol
93	Neptunium	Np
94	Plutonium	Pu
95	Americium	Am
96	Curium	Cm
97	Berkelium	Bk
98	Californium	Cf
99	Einsteinium	E
100	Fermium	Fm
101	Mendelevium	Mv
102	Nobellium	No

Units Conversion Table

This table has been compiled with reference to units and magnitudes appropriate to the technical and commercial operations with which these volumes are concerned, and in the light of the present transitional situation. It therefore includes many units which, though widely employed in imperial or metric usage, fall outside the strict Système International des Unités.

IMPERIAL/US AND METRIC/SI

Length

1 mile	=	1.6093 km		1 km	=	0.6214	mile
1 yd	=	0.9144 m		1 m	=	1.0936	yd
1 ft	=	0.3048 m		1 cm	=	0.3938	in
1 in	=	2.54 cm		1 mm	=	39.37	'thou'
1 mil or 'thou'	=	0.0254 mm					
(1/1000 in)							

Area

1 mile^2	=	2.590 km^2		1 ha	=	2.471	acres
	or	259 ha			*or*	0.386	mile^2
1 acre	=	4047 m^2		1 km^2	=	247.1	acres
	or	0.4047 ha		1 m^2	=	1.196	yd^2
1 yd^2	=	0.8361 m^2		1 cm^2	=	0.1550	in^2
1 ft^2	=	0.0930 m^2					
1 in^2	=	645.2 mm^2					

Volume, Capacity

1 yd^3	=	0.7646 m^3		1 m^3	=	1.3079	yd^3
1 ft^3	=	0.02832 m^3			*or*	35.315	ft^3
1 in^3	=	16.387 cm^3		1 dm^3	=	0.0353	ft^3
1 gal	=	4.546 l		1 cm^3	=	0.0610	in^3
1 US gal	=	3.785 l			*or*	0.0351	fluid oz
1 pint	=	0.5682 l		1 l	=	0.220	gal
1 fluid oz	=	28.413 cm^3			*or*	1.760	pints
					or	0.2642	US gal

Velocity

1 mile/h	=	1.6093 km/h		1 km/h	=	0.6214 mile/h
1 ft/min	=	0.00508 m/s		1 m/s	=	3.2808 ft/s
1 ft/s	=	0.3048 m/s		1 mm/s	=	0.0394 in/s
1 in/s	=	25.40 mm/s				

IMPERIAL/US AND METRIC/SI

Mass

1 ton	=	1016	kg
(2240 lb)			
1 short ton	=	907.19	kg
(2000 lb)			
1 cwt	=	50.802	kg
(112 lb)			
1 stone	=	6.350	kg
(14 lb)			
1 lb	=	0.4536	kg
1 oz	=	28.349	g

1 tonne	=	2204.7	lb
(1000 kg)	*or*	0.9842	ton
	or	1.1023	short ton
1 quintal	=	220.47	lb
(100 kg)			
1 kg	=	2.2047	lb
1 g	=	0.0353	oz

Mass per Unit Length

1 ton/mile	=	631.3	kg/km
1 lb/yd	=	0.4961	kg/m
1 lb/ft	=	1.488	kg/m

1 tonne/km	=	1.584	ton/mile
	or	1.774	short ton/mile
1 kg/m	=	2.016	lb/yd
1 tonne/m	=	0.900	ton/yd

Length per Unit Mass

1 yd/lb	=	2.016	m/kg
1 in/oz	=	0.7005	cm/g

1 m/kg	=	0.490	yd/lb
1 cm/g	=	0.0721	in/oz

Mass per Unit Area

1 ton/mile^2	=	392.3	kg/km^2
1 ton/acre	=	0.2511	kg/m^2
1 lb/ft^2	=	4.882	kg/m^2
1 lb/in^2	=	70.31	g/cm^2

1 tonne/ha	=	892.2	lb/acre
1 kg/m^2	=	0.2048	lb/ft^2
1 kg/cm^2	=	14.22	lb/in^2

Area per Unit Mass (Specific Surface)

1 mile^2/ton	=	2549	m^2/kg
1 yd^2/ton	=	0.823	m^2/tonne
1 ft^2/lb	=	0.205	m^2/kg

1 ha/tonne	=	2.511	acre/ton
1 m^2/kg	=	0.542	yd^2/lb

Volume per Unit Mass (Specific Volume)

1 ft^3/ton	=	0.0279	l/kg
1 ft^3/lb	=	62.428	l/kg
1 gal/lb	=	10.022	l/kg
1 in^3/lb	=	36.127	cm^3/kg

1 m^3/tonne	=	1.332	yd^3/ton
1 l/kg	=	0.995	gal/lb

Mass Rate of Flow

1 ton/h	=	1016	kg/h
1 lb/h	=	0.454	kg/h
1 lb/s	=	0.454	kg/s

1 tonne/h	=	0.984	ton/h
1 kg/s	=	2.2047	lb/s

Volume Rate of Flow

1 ft^3/s	=	28.32	l/s
(1 cusec)	*or*	1019	m^3/h
1 gal/min	=	272.76	l/h
1 US gal/min	=	227.10	l/h

1 l/s	=	0.353	ft^3/s
1 l/h	=	0.220	gal/h
	or	0.264	US gal/h

Density

1 ton/yd³	=	1.329 tonnes/m³	1 tonne/m³	=	0.7525 ton/yd³
1 ib/ft³	=	16.018 kg/m³	1 kg/m³	=	0.0624 lb/ft³
1 lb/in³	=	27.680 g/cm³	1 g/cm³	=	0.0361 lb/in³
1 lb/gal	=	0.10 g/cm³			
1 oz/gal	=	6.236 g/l	1 g/l	=	0.1600 oz/gal

Force

1 pound force (lbf)	= or	0.454 kgf 4.449 N	1 kgf (kp)	=	2.205 lbf
1 poundal (pdl)	=	0.1383 N	1 N	= or	0.225 lbf 7.233 pdl
1 ton force (tonf)	=	1.016 tf	1 tonne force (tf)	=	0.984 tonf
1 tonf	=	9.964 kN	1 kN	=	0.1004 tonf

Pressure

1 lbf/in²	= or	0.0703 kgf/cm² 6.8947 kN/m²	1 kgf/cm² 1 kN/m²	= =	14.22 lbf/in² 0.145 lbf/in²
1 lbf/ft²	=	47.880 N/m²			
1 tonf/ft²	= or	10.936 tf/m² 0.1072 N/mm²	1 tf/m²	=	0.914 tonf/ft²
1 atm (760 mm Hg)	=	101.325 kN/m²	1 mb	=	2.088 lbf/ft²
1 in Hg	–	3386.4 N/m²			0.019 lbf/in²
1 in H₂O	=	249.1 N/m²			

Energy, Work, Heat

1 Btu (0.252 kcal)	=	1055 J	1 kcal	=	3.9683 Btu
1 therm (100000 Btu)	=	105.51 x 10⁶ J	1 thermie (10⁶cal₁₅)	=	0.309 therm
1 ft lbf	= or	1.3558 J 3.766 x 10⁻⁷ kWh	1 J 1 kWh	= =	0.7375 ft lbf 2.6552 ft lbf
1 ft pdl	=	0.0421 J			
1 hph	=	2.6845 x 10⁶ J	1 J	=	37.25 x 10⁻ hph

Power

1 hp	=	1.0139 met hp	1 met hp	=	0.9863 hp
1 hp	=	0.7457 kW (1 cv)	1 kW	=	1.3410 hp
1 ft lbf/s	=	1.3558 W	1 kW	=	737.56 ft lbf/s

Various Heat Factors

1 Btu/h	=	0.293 W	1 W	=	3.4121 Btu/h
1 Btu/lb	=	2326 J/kg	1 J/kg	=	0.4299 x 10⁻³ Btu/lb
1 therm/gal	=	2.321 x 10⁴ J/cm³	1 J/cm³	=	4.3089 therm/gal
1 Btu/ft³	=	0.0372 J/cm³	1 J/cm³	=	26.839 Btu/ft³
1 Btu/lb deg F	=	4.187 x 10³ J/kg deg C	1 J/kg deg C	=	0.239 x 10³ Btu/lb deg F
1 Btu/ft²h	=	3.1546 x 10⁻⁴ W/cm²	1 W/cm²	=	3160 Btu/ft² h
1 Btu/ft² h °F	=	5.6783 x 10⁻⁴ W/cm² °C	1 W/cm² °C	=	1760 Btu/ft² h °F
1 Btu/ft h °F	=	1.7307 x 10⁻² W/n °C	1 W/cm °C	=	57.78 Btu/ft h °F

General references

Although a list of references is included in most chapters, additional information may be found in the following major reference works.

I. ARNON. *Crop production in dry regions*,
London, Leonard Hill, 1972 (two volumes)

J. ASHTON and S. J. ROGERS (Eds.). *Economic Change and Agriculture*,
Edinburgh and London, Oliver and Boyd, 1967

F. K. BEILSTEIN. *Handbuch der organischen Chemie*,
Berlin, Springer Verlag

A. R. CLAPHAM, T. G. TUTIN and E. F. WARBURG. *Flora of the British Isles*,
Cambridge, Cambridge University Press, 1962

Comprehensive Inorganic Chemistry. M. C. SNEED, J. L. MAYNARD and R. C. BRACTED,
New York, Van Nostrand, 1958

G. W. COOKE. *The control of soil fertility*,
Frogmore, Crosly Lockwood and Son Ltd., 1967

L. F. and M. FIESER. *Advanced organic chemistry*,
New York, Reinhold, 1961

L. I. FINAR. *Organic Chemistry*, Vol. 1,
London, Longman, 1973 (6th edition)

Gmelins Handbuch der anorganischen Chemie. Gmelins-Institut für anorganische Chemie und Grenzgebiete in der Max-Planck-Gesellschaft zur Forderung der Wissenschaften,
Weinheim, Verlag Chemie, 1946-1968 (8th edition)

R. F. GOLDSTEIN and A. L. WADDAMS. *The Petroleum Chemicals Industry,*
London, E. & F. N. Spon, 1967 (3rd edition)

G. W. C. KAYE and T. H. LABY. *Tables of Physical and Chemical Constants,*
London, Longman, 1973 (14th edition)

R. E. KIRK and D. F. OTHMER. *Encyclopedia of Chemical Technology,* New York,
Interscience, 1947-1960 (1st edition), 1963-1968 (2nd edition)

C. L. MANTELL. *Engineering Materials Handbook,*
New York, McGraw-Hill Book Company Inc., 1958

P. McCONNELL. *The Agricultural Notebook,*
London, Iliffe, 1968

J. W. MELLOR. *Comprehensive Treatise on Inorganic and Theoretical Chemistry,*
London, Longman, 1956-1973 (14th edition)

S. A. MILLER. *Acetylene, its properties, uses and manufacture,* Vol. II,
London, Benn, 1966

S. A. MILLER (Ed.). *Ethylene and its industrial derivatives,*
London, Benn, 1969

D. H. ROBINSON (Ed.).*Fream's Elements of Agriculture,*
London, John Murray, 1959

E. H. RODD (Ed.). *Chemistry of carbon compounds.* A modern comprehensive treatise,
5 vol. in 10 parts,
Amsterdam, Elsevier, 1951-62

N. I. SAX. *Dangerous properties of industrial materials,*
New York, Reinhold, 1968 (3rd edition)

THORPE's *Dictionary of Applied Chemistry,*
London, Longmans, 1937-1954 (4th edition)

Ullmanns Encyklopädie der technischen Chemie,
München-Berlin, Urban und Schwarzenberg, 1951-1968 (3rd edition)

WINNACKER-KUCHLER. *Chemische Technologie,*
Hamburg, Carl Hanser Verlag, 1958-1962

G. WRIGLEY. *Tropical Agriculture,*
London, Faber and Faber, 1969

Foodstuffs in general
N. W. DEROSIER. *Technology of Food Preservation,*
Westport, Conn., Avi, 1959

Food Industries Manual,
London, Leonard Hill, 1940 (20th edition)

J. HAWTHORN (Ed.). *Recent Advances in Food Science,* Vols. I-III,
London, Butterworth, 1962

HOTCHINSON's *Food and the Principles of Dietetics,* revised by V. H. Mottram,
London, Arnold, 1956 (11th edition)

M. A. JOSLYN and J. L. HEID. *Food Processing Operations,* Vols. I-III,
Westport, Conn., Avi, 1963

LILIAN H. MEYER. *Food Chemistry,*
New York, Reinhold, 1960

Statistical References

Most of statistical data in this work has been taken from the following references. The most recent editions of these references should be consulted for more up to date statistics. In addition in most countries there is a government bureau of statistics which publishes annual volumes containing data on production, consumption and trade.

Periodicals (annual) of international production and trade statistics:

Industrie and Handwerk Fachserie D.
Reihe 8 Industrie des Auslandes
      II Verarbeitende Industrie
Stuttgart, W. Kohlhammer.

Statistical Yearbook of the United Nations.
New York.

Commodity Yearbook.
New York, Commodity research bureau Inc.

Periodicals (weekly and monthly) including international production and trade statistics.

Chemical Age (Incorporating *Chemical Trade Journal*).
London, Benn Brothers.

Chemische Industrie.
Düsseldorf, Verlag Handelsblatt G.m.b.H.

Periodicals (annual) of production and trade statistics:

United Kingdom:

Annual statement of the trade of the United Kingdom. Vol II and III.
Imports and exports by commodity.
London, Her Majesty's Stationery Office.

Census of production of the Board of trade of the U.K. (a four year periodical).
London, H.M.S.O.

Annual Abstracts of Statistics.
London, H.M.S.O.

Accounts relating to Trade and Navigation of the U.K. (December number).
London, H.M.S.O.

U.S.A.:

Statistical Abstracts of the United States.
U.S. imports of merchandise for consumption.
U.S. exports of domestic merchandise.
U.S. Department of commerce.
Washington, Bureau of the census.

Chemical and Engineering News (the first September number). An American Chemical
Society publication.
New York, Reinhold Corp.

CHAPTER 1

Oils and Fats

1.1 OILS AND FATS IN GENERAL

1.1.1 Definition and classification

a. Definition. There is no scientific differentiation between edible oils and fats and the two terms may be used interchangeably. The common distinction between them is to consider solid products as fats and liquid ones as oils, but this difference is temperature dependent. At high temperatures all the edible oils and fats are liquid and at low temperatures they appear to be solid. Even this latter statement is only apparently true, as most natural fats, while appearing solid at ambient temperature are strictly a blend of solid and liquid components. The products which form the subject of this chapter could equally well be referred to either as 'fatty oils', or simply fats, irrespective of consistency at ambient temperature. No special significance should therefore be attached to the use of the alternative terms 'oil' or 'fat'. As far as edible products are concerned, usage is largely a matter of convenience and common practice.

b. Classification: oils. The term 'oil' has been applied in general usage to describe certain physical characteristics of various substances rather than their chemical nature or composition. This common use has led to confusion over the nature of oils and it is essential at the outset to understand their true classification. Oils and fats can be divided into groups according to both their origin and their chemical nature.

1. *Fixed oils and fats*

(i) Animal fats derived from the milk and body tissues of animals, including marine animals.

(ii) Vegetable oils and fats obtained from the fruit and seeds of a wide range of plants.

Fixed oils and fats are esters of fatty acids (specifically, glycerides) and are non-volatile.

2. *Mineral oils* which are distilled from petroleum and shale deposits. This group of materials, which includes paraffin oil, fuel oil and most lubricating oils, is considered in volume 4, chapters 1 and 2.

3. *Volatile or essential oils* such as oil of lemon, oil of cloves and cinnamon oil are obtained mainly from plant sources although some are derived from animals. These oils are complex mixtures of aldehydes, ketones, hydrocarbons, alcohols, acids and short chain esters. They are used for flavours, perfumery and pharmaceutical purposes and are discussed in volume 5.

In this chapter we are concerned only with the first group–fixed oils and fats.

c. Classification: fats. The distinction between edible and inedible fats can depend either on the composition or on the purity. Those fats and oils, such as wool fat and castor oil, which are inedible because of their composition, cannot be converted into edible fats by processing, whereas those which are inedible because of the presence of unpleasant impurities can usually be made into edible products by purification. These statements will be clarified later.

Fats (and oils) are further classified as crude, natural, modified, or synthetic. These terms must be defined at the outset in order to establish the meaning used throughout this chapter.

Crude fats are completely untreated, as isolated from the oil-bearing tissue. With a few exceptions, which include olive oil and lard, crude oils and fats are unpalatable, and hence inedible in this state.

Natural fats may have been treated to remove the impurities which make them unpalatable, but otherwise have not been altered.

Modified fats are sometimes erroneously called synthetic. A modified fat is one which has been changed by chemical treatment or physical separation to produce a fat which is different from the original material. Fats subjected to hydrogenation, trans-esterification and fractionation all come under this heading.

Synthetic fats are those which have been chemically made from various source materials. Some have been produced by oxidation of hydrocarbons to fatty acids which are then esterified with glycerol. This process was operated on a large scale in Germany during the Second World War (1939–45); however, not only was it costly, but the fats produced contained unnatural fatty acids which, it was reported, gave rise to toxic effects.

Another process which is partly chemical, partly biological, consists of the synthesis of fatty acids from carbohydrate by micro-organisms, and the esterification of the fatty acids with synthetic glycerol.

Neither of these processes is likely to contribute to the supply of edible fats in the near future, so we will not consider them any further.

1.1.2 History of fats

Man was using fats as part of his diet before recorded history began, and references to their non-edible use are found in early documents. There can be no doubt that the first fat products used by man were obtained by 'rendering' the carcases of animals killed by hunting. The subsequent domestication of animals which were suitable for food made available supplies of fat when they were killed. Marine fats have always played an important part in the diet of the Eskimo and other littoral peoples. Olive oil has been a dietary fat since Biblical times as well as an illuminant.

In more recent times the development of an edible fat industry was brought about by the need to establish manufactured fat products to replace the natural animal fats, lard, dripping and butter.

As the demand for fats grew, it began to exceed the availability of lard and soft tallow. This stimulated efforts to find lard substitutes, and led to the introduction of 'compound' cooking fats which were simulated lard, produced by blending (or compounding) soft vegetable fats with hard tallows. Cooking fat was developed in the United States as a lard substitute, and attempts at fractionation of fats were reported as long ago as 1767.

Methods of fractionation of tallows were soon established and the production of oleo oil and oleo stearin developed. A stimulus to the manufacture of fat products came with the invention of margarine. In 1869, Napoleon III called for a search for a satisfactory cheap substitute for butter, for which he offered a prize. This was won by Mège-Mouriés, who promptly patented his process and installed a factory. This French chemist deduced that the cow must make milk by drawing on the fat depots in its body and converting these tallows into milk fat. He also considered that the difference in melting point between butter fat and tallow was the result of fractionation of the tallow taking place in the body of the animal. His experiments comprised heating tallow to animal body temperature with various enzyme extracts and allowing the mixture to separate. He isolated two fat fractions, one soft and yellow like butter fat, and the other hard. The soft fat was emulsified in skim milk and an extract of cow's udder, to make a synthetic cream which was then churned in the conventional butter-making manner to produce a butter substitute. The process was established, but before long the enzymes and udder extract were found to be unnecessary, and their use was discontinued. The oleo oil produced in the American fractionation process was an ideal soft fat for this product, which was called margarine.

Other workers were taking an interest in the new product, introducing various modifications to the process and adding other oils and fats. Solidification of the emulsion by chilling was soon introduced. The chilling was achieved by pouring the emulsion into cold water, or iced water, to produce a granular mass. When the excess water was drained away the mass of margarine had to be worked into a homogenous product in the same way that butter was worked.

As the available oleo oil was insufficient to meet the needs of the growing industry the use of other fats was introduced. Coconut oil and palm kernel oil became established as suitable solid fats to include in the blend for margarine manufacture. The supply of animal fats was inadequate to meet the demands of the rising industry and soon the limited supplies of coconut and palm kernel oils were not enough to make good the deficiency. The industry received a great fillip when in 1902 Normann succeeded in hydrogenating a liquid oil to produce a solid fat. Catalytic hydrogenation made it possible to obtain sufficient quantities of solid fats for margarine production, and the industry has continued to develop.

Whaling had been taking place for centuries, and while the animals caught provided meat and whalebone, the fat was of limited use. Outlets included oil for illuminants, tanning, and low grade soap production but the quantity used was very small. The development of hydrogenation made whale oil a potential raw material for edible fat production. Expansion of the whaling industry rapidly took place, especially during the 1920–39 period.

Meanwhile, satisfying the demand for oleo oil to meet the needs of the margarine market led to the production of a large quantity of oleo-stearin which, in America, was blended with cotton seed oil to produce a lard substitute. So, while margarine–the butter substitute–developed in Europe, compound fats or shortenings–the lard substitute–developed in USA. The advantages of hydrogenation were soon appreciated in America also, and it was not long before hydrogenation plants were established and producing hardened fats for shortenings. The two branches of the edible fat industry then developed together, improvements to plant and processes being to a great extent of equal value to both.

At first compound fats (shortenings) were regarded as inferior substitutes but in recent years this stigma has been successfully overcome; to a great extent, margarine also is now regarded as a product in its own right. This change in attitude to margarine is

Fig 1.1 Harvesting fruit from a palm tree plantation.

partly due to the development of speciality products in the bakery field and promotion of soft or 'tub' margarine on the domestic market. The advertising of products which will spread straight from the refrigerator, which are different from the hard, difficult-to-spread butter, and are claimed to have medical advantages, is making tub margarine more acceptable to the domestic user.

Another recent development in fat processing is the introduction of inter-esterification, a process which modifies the structure of a fat without chemical alteration.

Developments in butter processing have also taken place, but since butter is inseparable from the dairy industry it is discussed later in this volume.

As the world population grows, and more people adopt the eating habits of industrialized countries, the demand for edible fats also rises. Several schemes to increase the

Table 1.1 Characteristics of common fatty acids

Common name	Chain length (carbon atoms)	Degree of unsaturation	Abbreviated structural formula	Melting point °C
Lauric acid	12	Saturated	$CH_3-(CH_2)_{10}-COOH$	44.2
Myristic acid	14	Saturated	$CH_3-(CH_2)_{12}-COOH$	54.4
Palmitic acid	16	Saturated	$CH_3-(CH_2)_{14}-COOH$	62.9
Palmitoleic acid	16	Mono-unsaturated	$CH_3-(CH_2)_5-CH=CH-(CH_2)_7COOH$	− 0.5
Stearic acid	18	Saturated	$CH_3-(CH_2)_{16}-COOH$	69.6
Oleic acid	18	Mono-unsaturated	$CH_3-(CH_2)_7-CH=CH-(CH_2)_7-COOH$	16.0
Linoleic acid	18	Di-unsaturated	$CH_3-(CH_2)_4CH=CH-(CH_2)_7COOH$	− 5.0
Linolenic acid	18	Tri-unsaturated	$CH_3-(CH_2)_2-CH=CH-CH_2-CH=CH-$ $CH_2-CH=CH-(CH_2)_6-COOH$	− 11.0
Arachidic acid	20	Saturated	$CH_3-(CH_2)_{18}-COOH$	75.4
Erucic acid	22	Mono-unsaturated	$CH_3-(CH_2)_7-CH=CH-(CH_2)_{11}COOH$	34.0

amount of fats available have been proposed, including the growing of groundnuts in East Africa, the replacement of rubber trees in Malayan plantations with oil palms, and the development of sunflower seed and rape seed as alternative crops in the wheat belt of Canada. Schemes which will increase fat supplies are of great value, but we must also look to increasing efficiency in fat processing to give better yields of edible grade oils from the raw materials made available.

1.1.3 General structure and composition

A very high proportion of all oils and fats is composed of a group of closely related compounds, the triglycerides. Minor components together rarely comprise more than 5% of the total material, and are frequently in even lower concentration. The major components will be considered first, and the minor ones later, in section 1.1.7.

a. The triglycerides. These compounds are formed by the reaction of the trihydric alcohol glycerol (commonly known as glycerine) and fatty acids. The predominant acids in natural fats occur as straight chains containing 12 to 18 carbon atoms. A few shorter chain acids (present in small proportions) are characteristic of milk fats, while longer chains (20 and 22 carbon atoms) are found in certain types of material, especially in fish oils. In addition to the chain length of the acids another important variable characterises the fats, namely the degree of unsaturation. While the number of unsaturated acids which could occur is very large, the number found in natural fats is limited. The common fatty acids are listed in table 1.1, together with information regarding their structure and melting points.

The structure of the fatty acids is relatively simple and follows that of the simplest of organic compounds, the normal paraffins. The molecule of a saturated acid consists of a chain of carbon atoms, each with its requisite number of hydrogen atoms attached, except for the terminal carbon which carries a carboxyl, the organic acid grouping. The following example shows the structure of butyric acid, $CH_3.CH_2.CH_2.COOH$:

$$
\begin{array}{ccccc}
 & H & H & H & \\
 & | & | & | & \\
H- & C- & C- & C- & C=O \\
 & | & | & | & | \\
 & H & H & H & OH
\end{array}
$$

The unsaturated acids have a deficiency of hydrogen atoms so that the link between two unsatisfied carbon atoms becomes a double bond, as shown for oleic acid in table 1.1. Conventional methods of writing the fatty acid formulae are both time and space consuming so various short-hand methods have been devised. The one adopted here is simple and easily remembered. The letter C is used to denote carbon, followed by the number of carbons in the chain, and then the number of double bonds in the chain, the two numbers being separated by a colon. This is demonstrated in table 1.2.

It will be noted from table 1.1. that the unsaturated acids have a very low melting point, so that at ambient temperature they are liquid, whereas the saturated acids are solids with fairly high melting points. The different acids have an influence on the melting point of the corresponding triglycerides; thus a triglyceride with three unsaturated acids will be liquid, and one with three saturated, a hard solid. It follows that when the

Table 1.2 Abbreviated nomenclature for fatty acids

Butyric acid	C4:0
Lauric acid	C12:0
Palmitic acid	C16:0
Palmitoleic acid	C16:1
Stearic acid	C18:0
Oleic acid	C18:1
Linoleic acid	C18:2
Linolenic acid	C18:3

unsaturated fatty acids are combined with additional hydrogen, as in the process of hydrogenation, they are converted to saturated acids of a higher melting point – hence the use of the alternative term 'hardening'.

A further complication in the study of unsaturated acids is that they can exhibit different three-dimensional structures, notably a structural difference called *cis-trans* isomerism. (Isomerism is a term used to describe compounds which have the same composition but different structure). Natural fatty acids are the *cis* form but during hydrogenation changes take place resulting in some *trans* acid formation and a consequent shift in melting point (see section 1.6.5).

The most abundant fatty acid is oleic acid, C18:1. It is present in all known fats in amounts ranging from less than 10% in coconut oil up to over 80% in some olive oils. Among the saturated fatty acids, palmitic acid, C16:0, is the most common, while stearic acid, C18:0, is far less abundant than is generally realized, although it is a major component of many animal fats.

It is interesting to note that the natural fats are composed of even-numbered fatty acids. There have been reports of odd-numbered fatty acids being found, but, with the exception of iso-valeric acid in one or two marine oils the quantities involved are only trace amounts. This peculiarity is explained later, in sections dealing with the biochemistry and metabolism of fats and fatty acids.

It is the combination of the different acids and glycerol which gives the wide range of triglycerides, and hence influences the physical characteristics of the various oils and fats which occur in nature. As the fatty acid portion of the fat is about 95% of the total weight of the molecule the chemical performance is principally dependent on the fatty acids present in the triglycerides. The interplay of triglyceride structure and fatty acid composition gives rise to the variations in fat characteristics, and is responsible for wide physical differences between fats having very similar chemical compositions. A good example is the comparison of cocoa butter and tallow. The first of this pair is the major fat component of chocolate. It has a short melting range, is hard and brittle, and melts completely in the mouth. Tallow (a mutton fat) is greasy and soft, has a wide melting range, and never fully melts on the palate. The fatty acid composition of these two fats is almost the same, but the way in which the fatty acids are combined with glycerol to form triglycerides is very different. In the cocoa butter, the triglycerides contain two saturated acids and one unsaturated, and as these triglycerides have very similar melting points the fat has the characteristics described. The tallow triglycerides, by contrast, are a very mixed collection, including some which are liquid at ambient temperatures and others which melt well above body temperature, hence the different behaviour.

The structure of triglycerides can be illustrated with the aid of a typical member of the group, tristearin.

$$
\begin{array}{lll}
1 \quad H_2C\!-\!OH & HOOC\cdot C_{17}H_{35} & H_2C\!-\!OOC\cdot C_{17}H_{35} \quad 1 \\
\quad\quad | & & \quad\quad | \\
2 \quad H\,C\!-\!OH \; + & HOOC\cdot C_{17}H_{35} \;\longrightarrow\; & H\,C\!-\!OOC\cdot C_{17}H_{35} \quad 2 \\
\quad\quad | & & \quad\quad | \\
3 \quad H_2C\!-\!OH & HOOC\cdot C_{17}H_{35} & H_2C\!-\!OOC\cdot C_{17}H_{35} \quad 3 \\
\quad\quad glycerol & three\ molecules\ of & tristearin \\
& stearic\ acid
\end{array}
$$

Tristearin is a simple triglyceride with the same fatty acid attached to each carbon of the glycerol chain. The systematic description of this molecule would be 1,2,3, tristearin, but as there is no distinction between the three molecules of stearic acid, the numbering is dropped. If in our example we replace one of the stearic acid molecules with one of palmitic acid, we can form three different molecules according to the position occupied by the new acid. They would be, 1 palmito–2,3,distearin; 2 palmito–1,3,distearin; and 3 palmito–1,2,distearin. Although these three compounds all contain the same fatty acids, their physical properties are not identical, the melting points in particular differing. Clearly the complexity of the glycerides will increase with the number of fatty acids. For example, three different fatty acids will give rise to six different mixed glycerides. The mixtures which arise when eight, nine, ten or even more fatty acids are present, and in unequal quantities are manifold, whilst the distribution of both glyceride types and fatty acids within those glycerides becomes quite complex. Although some fats may show a random distribution of the fatty acids, there are rules of distribution which appear to hold true for many of the natural fats. The most important one is the 'rule of even distribution', which requires that each fatty acid should be distributed in as many triglycerides as possible. If this was fully obeyed, no fatty acid would occur more than once in a triglyceride molecule unless the proportion of that fatty acid was more than one third of the total. As one would expect in living organisms, the rule is not strictly obeyed but serves as a guiding principle when examining the structures of different fats.

A second generalization, with regard to seed fats in particular, is the tendency for an unsaturated acid, oleic acid, to be in the 2 position in the triglyceride. While this rule applies to some extent to other vegetable fats, animal fats appear to behave differently. Here the presence of palmitic acid in the 2 position is more common, especially in lard.

1.1.4 General properties

The properties of oils and fats are generally the properties of the component glycerides. These in turn are greatly influenced by the fatty acid components.

Fats have a lower density than water and are almost completely insoluble in it, although small amounts of water will dissolve in the fats. They will dissolve about 0.1% of water at room temperature and the solubility increases with temperature, and with the presence of free fatty acids.

In non-aqueous solvents, the solubility of fats is influenced by the chain length of the component acids and their degree of unsaturation. Increased chain length of the saturated acids decreases solubility, but increased unsaturation increases it. The solubility is also influenced by temperature. Advantage can be taken of these factors in the separation of glycerides by fractionation at different temperatures. As the solution of a fat in acetone is cooled, the more saturated triglycerides crystallize out and the solution becomes progressively richer in unsaturated glycerides. This has been used as the basis for fractional crystallization of the component glycerides of fats when attempting to investigate their structure. Unfortunately, the process is complicated by the effect of the more soluble glycerides, which increase the solubility of the less soluble ones. Most of the common organic solvents such as acetone, diethyl ether, petroleum spirit, carbon tetrachloride and chloroform, will dissolve fats in the liquid state. If the fat is solid, the solvent will slowly take up the fat.

Some of the more important properties of fats are the melting and solidification characteristics and the influence on them of polymorphism. It has been observed that when fats are melted and then solidified, the melting point is dependent on the method of solidification. The greater the temperature shock during cooling, the lower the melting point becomes. It is therefore very important that the method of cooling is specified when comparing the melting points of fat samples. The manner of cooling affects not only the melting point of the fat but also other properties such as plasticity, appearance and uniformity. Slow cooling can lead to fractional crystallization, producing large crystals of the higher melting glycerides with separation from the lower melting ones. Rapid cooling produces crystals which hold the liquid glycerides within the structure, so producing a plastic solid. The solidification of margarine and shortening takes advantage of these properties to produce a plastic material with the performance required.

When exposed to the oxygen of the air, fats absorb it at a rate dependent on the degree of unsaturation. As oxygen absorption takes place, the rate of oxidation increases, and the fat reaches a state of rancidity. The rate of oxidation is also influenced by temperature and light. As oxidation proceeds, peroxides form and these then break down to form aldehydes, ketones, and other shorter chain compounds, which together give the unpleasant smell and taste which we associate with rancid fats. The formation of peroxides can be followed by the determination of peroxide values (see section 1.2.4).

The susceptibility of fats to oxidation and rancidity is accentuated by traces of metals in the fats. The common metals, iron and copper, are used in chemical plant construction, but the latter metal must never be allowed to come into contact with fats. It has been shown that one part of copper in twenty million parts of lard will halve the time required for the development of rancidity. Iron also has an accelerating effect on oxidation, but only about one tenth as serious as that of copper. Thus it is essential that copper and its alloys should be avoided in the construction of equipment to handle fats. Even the use of brass valves and gun-metal pump impellers should be avoided.

Oxidative deterioration is a self-accelerating process – a chain reaction. The progress of oxidation starts slowly and continues gradually for a time. After a period, known as the induction period, the rate of oxidation accelerates suddenly and the fat rapidly becomes rancid. Although heat will increase the rate of oxidation, the peroxides formed decompose rapidly at high temperatures. It may be observed, therefore, that a fat which is being heated continuously does not develop a high peroxide value, but will be come oxidized.

The chain reaction can be broken by the addition of anti-oxidants. These compounds, which are effective at a level of a few parts per million, will often postpone the development of rancidity for a long time.

Highly unsaturated fats, when heated in air, undergo polymerization. The exact mechanism of this process is not fully understood, but it is sufficient for our purposes to accept that cross-linking takes place between adjacent molecules. The double bonds of the fatty acids are involved in this cross-linking, which is of great importance to the paint industry. In the edible oil field it can give rise to the deterioration of frying fats owing to polymerization. As the rate of oxidation and polymerization is related to the degree of unsaturation, it is this factor that limits the use of highly unsaturated fats. Soyabean oil, when used as a frying fat, soon polymerizes on the side of the cooking vessel and builds up a yellow 'varnish' of polymerized material. Polymerization is accompanied by a rise in viscosity, a fall in iodine value (a measure of unsaturation) and deterioration of flavour.

Another property of fats which leads to deterioration is the susceptibility to hydrolysis. Since fats are a combination of glycerol and fatty acids they are chemically classed as esters. Like all esters, the linkage between the alcohol and acid can be broken by hydrolysis, giving the free alcohol and the free acid. Since in fats the alcohol is trihydric, each fatty acid can, under suitable conditions, be separately removed, leaving progressively, diglycerides, monoglycerides and finally, glycerol. This type of hydrolysis is mainly due to enzymes called lipases which occur not only in the digestive juices of man and animals, but are secreted by bacteria and moulds and are present in many plant tissues. Chemical hydrolysis is carried out by water under high pressure. Hydrolysis brought about with alkali will yield glycerol and the alkaline salt of the fatty acid (i.e. a soap and the process is saponification). The amount of alkali required to perform this reaction is directly proportional to the equivalent weight of the fatty acids involved. Thus the reaction can be used to determine the mean molecular weight of the fatty acids and is the basis of the saponification value determination.

The glycerides which contain unsaturated acids will, under certain conditions, absorb hydrogen until the unsaturated acids are reduced in unsaturation and can become saturated. This property is used industrially in the process of hydrogenation. Under these conditions liquid fats can be converted into solid ones.

Fig. 1.2 Coconut kernels ready for drying to make copra.

In the preceding section on the general structure of triglycerides, we discussed the way in which the fatty acids were attached to the glycerol to give an even distribution. If a fat, or mixture of fats, is treated with a catalyst, the fatty acids can be made to change position in the glycerides to produce a random structure. This process of inter-esterification is now being used to make fats for specific applications. Details are given in the section devoted to interesterification.

1.1.5 Biochemistry

The biochemistry of fats is a very complex subject, and this account must be restricted to a brief outline covering the principles. The sub-sections on nutrition and metabolism cannot be isolated completely, consequently there is some overlap. The subject must be considered from two aspects: the synthesis of fats in plant and animal tissues, and the utilization of fat as a part of human diet.

Fat is an important constituent of body tissues. One of its functions is to serve as a store of food which is laid down when the supply of calories exceeds demand, to be utilised when other sources of energy are not available. Subcutaneous fat also serves as an insulation against heat loss. The thick layer of fat which lies below the skin of the whale is essential for the conservation of heat which is so necessary for a warm-blooded animal in a cold sea. A third function of fatty tissue is to support and cushion internal organs such as the kidneys.

A growing plant contains very low concentrations of fat in the leaves, but high concentrations are found in the seed and in a few cases, in the fruit (oil palm and olive). Research has led most workers to accept that the fat is formed in the seed itself, and not formed in the leaf and transported to the seed. It has been shown that, under suitable conditions, olives will produce fat in the fruit after picking. This work is supported by investigations which have demonstrated that the composition of the leaf fat may often be distinct from that of the seed; an example is the oil palm where the fruit fat is very different from the seed fat. Most of the work has been concentrated on the development of fat in the seed, where there is considerable evidence that the plant forms the seed and fills it with carbohydrate. This carbohydrate is then converted to fat. The biochemical process is not simple since it is almost certain that the formation of carbohydrate and its conversion to fat occur simultaneously. Synthesis of fatty acids seems to be temperature dependent in that when the same plants grow at different latitudes the oils formed have a higher iodine value (are more unsaturated), when grown at lower temperatures. The key to the mechanism of conversion undoubtedly lies in the observation that almost 90% of the total weight of vegetable oil fatty acids produced are found as the C18 unsaturated acids; oleic, C18:1; linoleic, C18:2; linolenic, C18:3; and eleostearic, C18:3. The conversion of carbohydrate to fat is very complex process and the specialist works given in the literature (section 1.10) should be consulted for further information.

It was believed at one time that the depot fat of animals was derived directly from the diet. While it is true that part of the fat laid down in the tissues arises in this way, it is also true that some of the fat is synthesized by the animal, from carbohydrate sources. More than a hundred years have passed since feeding experiments were used to show that pigs fed a controlled diet produced more fat in their bodies than had been fed in the diet. It was also shown that the protein in the diet was insufficient to provide a base material for the fat produced. Evidently at least part of the depot fat in the experimental animals had been formed from ingested carbohydrate. Subsequent work has shown that the composition of the animal fat was also different from that of the diet, as well as being in excess of that provided. The fatty acids formed appeared to be only in the C16 and C18 groups, with oleic acid predominating. The route of synthesis has been established; it consists of building up fatty acids from glucose breakdown products and then esterifying the fatty acids with glycerol. Fats in the food supply may be hydrolysed to monoglycerides and these built up to triglycerides again, adding two fatty acid groups to the free hydroxyls in the monoglyceride. Other fats are absorbed directly without breakdown.

Some research workers have shown that the fat synthesized by the body does not remain static, but is utilized as an energy source together with dietary fat, and is partially replaced by dietary fat as well as synthesized material.

1.1.6 The role in nutrition

Fat in the diet provides a concentrated source of energy amounting to about 9.5 kilocalories per gram, more than twice as much as the energy supplied by an equal weight of dry carbohydrate. The function of fat in the food is not just as a high energy source; it also increases the palatability of the food and makes it more satisfying. Fatty foods have a longer residence time in the stomach, increasing the time interval before the stomach is empty and the sensation of hunger recurs.

Digestibility is important in assessing the nutritional value of any food. In the case of fats, melting point is of marked importance. If the fats are liquid at body temperature they are much more readily digested and absorbed than solid ones. The critical melting point appears to be between 104°F (40°C) and 122°F (50°C). The presence of fats of different melting points has a modifying effect, the lower melting components reducing the overall melting point of the mixture. The digestibility depends to a certain extent, therefore, on the circumstances, but it is claimed that vegetable oils and the soft animal fats have a digestibility of 95–99%. The harder fats such as oleo-stearin and mutton tallow have lower coefficients of 80% and 88% respectively.

The nutritional value of a fat also depends on the presence of certain fatty acids which were at first called 'vitamin F', and later 'essential fatty acids'. It has been shown that certain unsaturated fatty acids must be present in the diet if young animals, especially rats, are to grow well and remain healthy. Other experimental animals show deficiency symptoms if they are fed a diet free from unsaturated fatty acids. There is evidence that these animals are unable to convert saturated acids into the required unsaturated ones and that the deficiency symptoms disappear when the missing acids are given. The most effective appear to be arachidonic and linoleic acid, of which the latter is quite widely distributed in natural fats, especially those of vegetable and marine origin. It is more difficult to demonstrate the need for these acids in human health. Most people in the developed countries have access to such a wide range of food fats that it is improbable that they risk essential fatty acid deficiency. However, it has been observed that in societies where large quantities of animal fats are consumed, fatty deposits tend to form in the arteries (atherosclerosis). This type of diet also leads to high levels of fat and cholesterol in the blood. These observations have given rise to the hypothesis that the consumption of large quantities of animal fats leads to atherosclerosis and heart disease. There is evidence that experimental animals develop atherosclerosis if they are fed with a diet high in saturated fat. It has also been shown that the inclusion of poly-unsaturated acids in the diet in place of animal fats results in a lowering of the blood cholesterol. Attempts are being made to persuade people to change their dietary habits in an effort to reduce the incidence of these conditions. It is also stated that the ratio of poly-unsaturated to saturated acids should be at least $2:1$. Support for this hypothesis is forthcoming in the successful treatment of an abnormal condition in which the afflicted individuals are found to have a high amount of lipoprotein (a fatty acid protein complex) in the blood plasma. The treatment consists of providing a diet high in poly-unsaturated acids, together with a drug which lowers blood cholesterol. The situation is not quite as simple as it may appear, however; other evidence shows that high fat levels in the blood have been found after taking a meal that is rich in sugar (sucrose).

There has been growing concern in recent years over the possibility of toxic effects of the long chain fatty acids such as erucic acid (C22:1). This acid is present at a high level in oils of the rape seed and mustard seed group. It has been observed that if laboratory animals are fed with rape seed oil, fat accumulates in the heart muscle during the first few days of the test. As the feeding continues the animals apparently develop a means of meeting the situation, and the level of fat in the heart muscle drops to normal. Much more work will be needed before a positive decision can be made, but already manufacturers are replacing rape seed oil where possible and new plant varieties, which give low levels of erucic acid in the rape seed oils, are being bred.

An indirect contribution of fats to the nutritional wellbeing of the population is the presence in fats of the minor components, vitamins A, D, and E, either in the form of the ester of the vitamin or as a precursor such as β-carotene, which is converted in the body into vitamin A. Palm oil contains a large quantity of carotenes, but other vegetable oils only contribute to the supply of tocopherol (vitamin E). Animal fats are a much richer source, butter supplying vitamins A, D and E, and the fish oils A and D. Fish liver oils, which are not normally regarded as edible fats, supply a substantial quantity of vitamins A and D, for use either in pharmaceutical preparations or as a raw material for the preparation of vitamin concentrates to be used, for example, in fortifying margarine.

1.1.7 Metabolism

As stated above, the primary role of fat in the diet is as a source of energy. Before the fat can be utilized in any form it must first be absorbed from the intestine. Pancreatic secretions which are injected into the intestine contain the lipases which attack the fat, producing hydrolysis to monoglycerides, diglycerides and free fatty acids. The hypothesis that all the fat must be hydrolysed to free fatty acids before absorption is no longer accepted. It is now believed that while some hydrolysis occurs, there is also direct absorption of triglycerides, which is influenced by the solubilizing effect of the bile. There is also evidence that, when absorption of fatty acids takes place, esterification takes place in the intestinal wall.

When the body needs to use its fat supplies as a source of energy the fat is broken down to simpler units by a chain of enzyme reactions. The first step in this process is through the action of a lipase which hydrolyses the glycerides, liberating fatty acids and glycerol. The glycerol is converted to glycerol phosphate which is then oxidized to dihydroxyacetone phosphate and thus enters the same sequence of reactions by which glucose in oxidized to carbon dioxide and water. These oxidation processes generate energy. The fatty acids are processed by a different pathway which also links up eventually with the carbohydrate one.

A fatty acid degradation is started by combining it with a co-enzyme known as co-enzyme A. A series of reactions follows in which the fatty acid is oxidized at the β-position (two places away from the carboxyl group). First, a double bond is formed which is then converted to a hydroxyl group. Another enzyme converts the hydroxyl to a keto group, and finally the fatty acid splits at the β-position. This results in a fatty acid which is two carbon atoms shorter than the original one, and a fragment, consisting of two carbons from the fatty acid (an acetyl group) still attached to the enzyme, forming a complex known as acetyl co-enzyme A.

The fragment broken off from the fatty acid is fed into a metabolic cycle called the citric acid cycle. In this process the acetyl co-enzyme A adds the acetyl group to oxaloacetic acid to form citric acid which then goes through a series of reactions which converts the acetyl group to carbon dioxide and regenerates the oxaloacetic acid. Thus the two carbons which were split off the fatty acid are oxidized to carbon dioxide, with production of more energy.

We have outlined the fate of the fragment of fatty acid which was split off, leaving a shorter chain fatty acid. Successive reduction of the chain length by two carbon atoms continues in the same manner until all the fatty acid has been oxidized to carbon dioxide and the energy produced by the oxidation made available to the body.

Theoretically, the oxaloacetic acid is continuously regenerated, but this does not happen in practice. The level of this component is maintained by conversion of pyruvic acid to oxaloacetic acid. The importance of this route lies in the source of pyruvic acid; it is formed during the metabolism of glucose.

The conversion of acetyl co-enzyme A to carbon dioxide and water as described is not the only way in which the complex is metabolized. Two molecules can condense to form acetoacetic-co-enzyme A which may be transported to various tissues and utilized in several ways which include the synthesis of cholesterol and the building of fats. This is the common nexus between the various metabolic pathways. The utilization of fats, the synthesis of fats not only from dietary fat, but also from carbohydrate and protein, all use this link.

1.1.8 Minor constituents

Although fats are predominantly triglycerides there are minor components present in crude and refined fats and oils. The important ones are phospholipids (or phosphatides), sterols, vitamins and their precursors, and impurities of various sorts.

Phosphatides are a group of compounds which, like the fats, contain fatty acids; they also contain a phosphate group, usually a nitrogenous base (such as choline or serine or ethanolamine) and glycerol. In general, phosphatides may be considered as triglycerides in which one of the fatty acid groups has been replaced by a phosphoric acid derivative. The only well-known member of the group is lecithin, consequently this term is frequently used, incorrectly, as a synonym for phosphatides. Phosphatides are widely distributed in various tissues of both animal and vegetable origin, but it is their presence in crude oils which is of importance in edible oil technology. 'Lecithin' is present in crude soyabean oil to the extent of 2 or 3%. It is also present in cotton seed, groundnut, corn, and rapeseed oils in varying concentrations. Vegetable oil 'lecithin' is usually a mixture of phosphatides containing true lecithin (phosphatidyl choline) and cephalin (phosphatidyl ethanolamine). Both these groups of compounds consists of a diglyceride with the third hydroxyl esterified with phosphoric acid and a base. In the lecithins, the base is choline, and for cephalins, ethanolamine. As in the triglycerides, there are a number of different lecithins and cephalins each with a different fatty acid combination.

Animal fats usually contain only very small amounts of phosphatides, but fish oils, which are made from all the tissue, frequently contain large amounts of these compounds.

Sterols are a group of complex, high melting point crystalline alcohols. They play an important part in biochemical activity, being the basic structure from which animals manufacture bile acids and hormones, as well as serving as cell constituents. They are present at low concentrations in all fats. Although only minor components, and completely different in structure from the fats themselves, they are an important group of compounds, having a marked physiological effect. The sterols are classified into three groups according to their source; zoosterols from animal sources, phytosterols from plant sources and a third group, obtained from fungi, which does not concern us here. The important zoosterol is cholesterol which is present in animal fats and in most animal cells. The phytosterols differ slightly in structure from cholesterol and this fact has been used in a process for examination of the sterol acetate esters by isolation and measurement of melting point to show the presence of vegetable oils in animal fats. The sterols form a constituent of the unsaponifiable matter.

Fatty alcohols are present in waxes and in a few vegetable oils. The waxy coat of maize, sunflower seed and one or two other seeds frequently passes into the refined oil producing a haze, or even a cloud of crystallized wax particles when the oil is chilled. Although the concentration is less than 0.01%, the waxes have to be removed to make the product acceptable to the consumer. The fatty alcohols form part of the unsaponifiable matter.

Hydrocarbons are also present to a small extent in oils. These materials, which are isolated as part of the unsaponifiable matter, rarely exceed 0.1%, except for squalene in olive oil in which amounts up to 0.7% have been reported. Fish liver oils also contain quite substantial quantities of squalene together with similar hydrocarbons. They are not regarded as important.

Tocopherols are an important minor constituent of fats. They constitute the essential nutrient vitamin E and have a marked effect on the stability of the fats. These compounds – there are four in the group, designated by the first four letters of the Greek alphabet – are natural antioxidants. The natural oils and fats were known to have a much greater resistance to oxidation than pure glycerides; it was also observed that vegetable fats were more resistant than animal ones. This stabilizing effect was eventually traced to the tocopherols. The total tocopherol content of oils varies between 0.0005% in a lard, up to nearly 0.5% in wheat germ oil. Animal fats have a content similar to that of lard, vegetable fats averaging around 0.05%.

The flavour of many oils is due to a mixture of trace quantities of a number of compounds. These include ketones, and terpenoid hydrocarbons, both of these groups having very strong flavours.

Colouring matter and oil-soluble pigments are among the natural components of oils and fats. Edible animal fats generally contain very little colour, although inedible tallows can be green or dark brown because the fat dissolves colours from various sources during rendering. Coconut oil also contains very little pigment, but palm oil is very highly coloured with a mixture of carotenoids, and cottonseed oil contains a different type of pigment called gossypol. Other pigments are present in cotton seed oil but gossypol predominates. Most other oils contain small amounts of carotenoids and some, especially soyabean oil and olive oil, contain green pigments, the chlorophylls.

Brown colours in fats are usually an indication of deterioration caused by oxidation or overheating. These brown colours are difficult to remove and the bleached oils are rarely satisfactory.

The carotenoid pigments give a yellow solution when present in low concentrations.

The colour deepens with concentration through amber to orange and deep red. Palm oil contains three carotenes designated α, β and γ. The β isomer is the most important as it is pro-vitamin A. Its structure shows that it is made up of two molecules of vitamin A which are joined together at the ends of the straight chain portion of the molecule. Theoretically, it should be possible to split the β-carotene molecule to form the two vitamin A ones. Of the other isomers, α-carotene will only give one vitamin A molecule and the third one has no vitamin activity. Unfortunately, the metabolic process of the human body is unable to convert the β-carotene to vitamin A at theoretical efficiency.

Gossypol, which occurs in large quantities in cottonseed, is soluble in oil and water. It is also soluble in some organic solvents, including those commonly used in extraction processes. Removal of gossypol from cotton seed oil with alkali is effective because the pigment is a polyphenolic compound which behaves as an acid. The cooking process used to prepare the seed before extracting the oil influences the extraction of the pigments. Under correct processing most of the pigment can be retained in the residual meal, improving the oil at the expense of the meal. As the latter is important to the economics of the oil mill, it is necessary to compromise to achieve the optimum overall result.

Impurities are present in many natural products. It is usual for the crude material to be contaminated with substances which have accidentally, carelessly, or unavoidably become admixed with the fats and oils. Some of these impurities are derived from the fat-bearing tissues of the animal or plant. They include cell debris, connective tissues, husks, mucillagenous material ('gums') and moisture. When a crude oil is left to stand in a tank, many of these impurities settle out to form a layer at the bottom or foot of the tank. This layer is known as 'foots', and when a sample of oil is being examined due allowance must be made for the 'foots' which have already separated. Other impurities are introduced during the handling of the fat or tissues.

Mineral matter such as sand is often present in crude vegetable oils owing to contamination of the seed or fruit.

Solvent residues may be present if the oil has been obtained by solvent extraction methods.

Fuel oil is sometimes found in crude edible fats if it has been carried or stored in tanks which had previously been used for fuel oil. This was a hazard of the whale oil industry where crude oil was carried in the tanks that had been used to transport the fuel oil to the whaling fleet.

Water is a common contaminant of crude oils. Although the solubility of water in pure oil is only about 0.1%, the presence of free fatty acids, phosphatides, monoglycerides and other impurities can markedly increase the water-holding properties of the fat. It is necessary to make a careful check on the amount of water present, as this can be quite high.

1.2 ANALYSIS AND TESTING

The analysis and testing of oils and fats is needed for an assessment of quality and purity as well as for their identification. A number of physical and chemical 'constants' have been established for these purposes. Although many of them are empirical, others are quite specific measurements of the characteristics of the fats. Those most commonly used to establish identity are: saponification value, iodine value, refractive index, and the

Reichert–Polenske–Kirschner values. Other constants have special significance for certain oils such as the hydroxyl value for castor oil, and Evers test for groundnut oil. Colour reactions can also give useful guidance in identifying an oil. Measurements of melting point are of debatable value.

Oils and fats are natural products and are therefore subject to some variation in composition so that the 'constants' fall within a range. Even so, from time to time oils become available which have constants outside the normal range.

The variations in composition are dependent on a number of factors. Vegetable oils vary according to the botanical variety, climatic conditions, soil composition, rainfall and temperature. Similarly, animal fats vary according to the breed of the animal, climatic conditions and feed. According to reports, some of the fatty acids in animal fats are derived from fat in the diet, especially if an abnormal fat is fed in large quantities.

Other data are determined on the oils and fats in order to assess quality. Of these the free fatty acid content, peroxide value, benzidine or anisidine value, A.O.M. test or one of its variants (described later), moisture, impurities, and unsaponifiable matter, all help to give an adequate measure of the quality of the material.

If an oil or fat is required for a special application, additional tests may be demanded in order to check suitability.

1.2.1 The 'constants' and their significance

Each of the constants used in examining the oils and fats is chosen to measure one of the characteristics of the glycerides or fatty acids. An assessment of the types and quantities of the various fatty acids present is related to the composition and therefore the identity of the fats being examined. First the information derived from each determination will be discussed, then the actual methods of examination.

a. Saponification value is a measure of the mean molecular weight of the fatty acids present in the fat. The process of saponification is the hydrolysis of triglycerides into glycerol and the potassium salt of the fatty acids, using a solution of potassium hydroxide in alcohol. This process measures the amount of alkali which is required to combine with the fatty acids liberated by the hydrolysis of the fat, and from this the equivalent weight and molecular weight of the fatty acids can be determined. For the purpose of characterizing the fat in terms of a constant the actual molecular weight is not calculated, the saponification value being expressed as the number of milligrams of potassium hydroxide required to saponify one gram of fat. As each fat has, within the limits of biological variation, a constant fatty acid composition, the determination of the saponification value is a reasonably reliable means of characterizing the fat.

b. Iodine value is a measure of the proportion of unsaturated acids present. There is no iodine present in oils and fats, but the test measures the amount of iodine which can be absorbed by the unsaturated acids. One of the properties of unsaturated organic compounds is the reactivity of the double bonds, especially their ability to form addition compounds with halogens. As the addition takes place at the double bond, measurement of the quantity absorbed is a measure of the number of double bonds present. As the concentration and types of unsaturated acids present in the fat are fairly constant, the iodine value will give a figure for the total degree of unsaturation, expressed as the percentage of iodine absorbed by the fat. Because there are a number of unsaturated acids which could be present in different fats, the iodine value alone will not provide a measure of a specific unsaturated acid, only the total unsaturation. If additional information is needed, the thiocyanogen value may also be determined. Thiocyanogen behaves like halogens in adding to unsaturated acids at the double bond. One molecule of the reagent adds on to a mono-unsaturated acid, one to a di-unsaturated, and two to a tri-unsaturated. Unfortunately, this relationship does not precisely hold, and the result depends on conditions under which the test is made. For these reasons and the development of more reliable methods of determination of the fatty acid composition, the thiocyanogen value is no longer widely used in identification of fats and oils.

c. Reichert, Polenske and Kirschner values are all determined at the same time. They measure the amount of steam-volatile fatty acids which can be recovered from fats under standard conditions. Low molecular weight acids are steam volatile under the test conditions, and are separated from the higher acids. The acids concerned are lauric (C12:0), capric (C10:0), caprylic (C8:0), caproic (C6:0), and butyric (C4:0). They are further divided into those which are water-soluble (chiefly the butyric and caproic) and the water-insoluble ones, caprylic, capric and lauric. The Reichert value measures the water-soluble acids, the Polenske value the insoluble acids, and the Kirschner value butyric acid by separation from caproic by precipitating the latter as a silver salt. The test is empirical, and the procedure must be followed meticulously if the results are to have any value. The greatest significance of these constants is in checking butter, coconut oil and palm kernel oil for adulteration, or in examining mixtures containing these oils. In Europe, only these three fats have high values in this test although in America babassu oil must be allowed for.

d. Unsaponifiable matter is a measurement of the water-insoluble components produced after heating the fat with potassium hydroxide. During this process all the triglycerides are broken down to form glycerol and potassium salts of the fatty acids. Both of these products are water soluble, and the insoluble hydrocarbons, sterols and fatty alcohols, can be extracted from the aqueous solution and their proportion measured. The amount of unsaponifiable matter found in edible fats is usually small, and high figures may indicate contamination or adulteration.

The test measures most of the minor components of the fat, with the exception of the phosphatides which are saponified together with the triglycerides. Sterols and vitamins appear in this portion of the fat so that preparation of unsaponifiable matter is a preliminary step to their examination. One of the purposes of measuring the unsaponifiable matter is to check for the presence of fuel oil or lubricating oil in the fat.

e. The refractive index is a physical attribute of triglycerides, measured by the angle through which a beam of light is bent when passing through a thin film of melted fat. The index of each fat falls within a narrow range and can be used as a characteristic of the fat in checking purity or searching for components of a mixture. It is temperature dependent and is usually measured at 104°F (40°C), a temperature at which most fats are liquid. A correction factor can be used if it is not possible to work at a selected temperature for which reference data are available.

f. Melting and solidification temperature of fat is not easy to measure. Although pure triglycerides are often stated to have a definite melting point, it is well known that triglycerides exhibit polymorphism; that is, a variation of crystal structure and therefore of melting point, according to the thermal history to which the fat has been exposed. As fats are mixtures of a number of different triglycerides, the melting point becomes even more difficult to measure. It is therefore necessary to impose an empirical test procedure which will allow reproducible results to be achieved, but which cannot be used to identify components of a mixture. The two common measurements are 'slip point' and 'titre'.

The 'slip point' measures the temperature at which a carefully prepared sample will move or slip in a capillary tube when heated slowly in a water bath.

The 'titre' is not really a characteristic of a fat itself, but is the solidifying point of the fatty acids prepared from the fat. It has some significance in soap-making technology, but is of little importance in the examination of edible fats.

g. Impurities. Determination of the impurities present is not as simple as it may appear. In principle it is the measurement of solid matter which is not soluble in fat or fat solvents. This is complicated by the fact that some materials may be dispersed in the fat, but insoluble in fat solvents. When the solvent is added during the estimation of impurities these materials precipitate, and, according to the degree of dilution, may not all be measured with the impurities. It is again essential to use a standardized method of measurement.

h. Acid value is a measure of the amount of free fatty acid present in a fat. Some of the deterioration that takes place during storage of either the raw material from which the fat is obtained, or in the fat itself after isolation, results in hydrolysis of triglycerides to yield free fatty acids. As these free fatty acids must be removed in the preparation of edible fats, the commercial yield will be reduced by an amount proportional to (but greater than) the amount of free fatty acid present. The reasons for this will become clear in the section on processing (especially section 1.6.2).

i. The peroxide value is usually used as an indicator of deterioration of fats. As oxidation takes place, the double bonds in the unsaturated fatty acids are attacked, forming peroxides. These break down to produce secondary oxidation products which indicate a condition known as rancidity. The peroxide value can therefore be used to estimate oxidation, but as the compound formed is unstable and oxidation proceeds further, it is not a complete measure of oxidation, and if taken in isolation may give rise to misleading conclusions.

1.2.2 Determination of constants and other characteristics

Standard methods of examination have been worked out for all constants. Various national societies have adopted their own variations of the standard methods, and as all these variations can give slightly different results it is customary to declare the exact technique used. The British Standards Institution (B.S.I.), The American Oil Chemists Society (A.O.C.S.), and the German Society for Fat Research (D.G.F.), all publish their own methods. Great efforts have been made to draw these methods together and produce an internationally accepted collection, but the A.O.C.S. and B.S.I. methods are still the most widely used.

The methods given in the following pages are intended to be examples of general methods of analysis; standard methods are available in the official publications mentioned above. There are also a number of text books available which not only give details of various methods, but also critically compare them.

a. Saponification value. The details of individual standard methods vary, but the general technique is similar to the B.S.I. method. It is necessary to carry out a 'blank' simultaneously with the samples. In the examination of most edible oils the technique given will achieve the required condition that the sample titration is approximately half the blank titration.

About 2 g of fat are accurately weighed into a flask. To this, 25 ml of 0.5 N alcoholic potassium hydroxide are added. A 'blank' is also prepared, by putting 25 ml of the alcoholic potassium hydroxide in a similar flask. Air condensers are fitted to both flasks and the contents are boiled for one hour, swirling the flasks from time to time. The flasks are then allowed to cool a little and the condenser is washed down with a little distilled water. The excess potassium hydroxide is titrated with 0.5 N hydrochloric acid using phenolphthalein indicator. The saponification value is calculated from the difference between the blank and the sample titrations.

$$\text{Saponification value} = \frac{28.05 \times (\text{blank titration–sample titration})}{\text{weight of sample}}$$

where 28.05 is the number of milligrams of potassium hydroxide in 1 ml of 0.5 N potassium hydroxide.

Efforts to improve the method have included the use of different indicators, double indicators and potentiometric titrations.

b. Iodine value. The determination of iodine value is based on the absorption of halogen by the unsaturated fatty acids. It consists of the addition of a substantial excess of halogen, allowing time to react, and then measuring the excess. A number of techniques have been suggested by various workers, each using a different solvent system and halogen combination. The variation in method has arisen partly because no one method is fully satisfactory under all conditions. Probably the one most widely used is that devised by Wijs, and described here. This has been adopted by most organisations who approve standard methods, although the Hanus and Kaufmann methods are also recommended.

The reagent used is iodine monochloride which may be either purchased as such in sealed ampoules or prepared from iodine and chlorine. The A.O.C.S. recommends making the reagent from iodine and chlorine, but the B.S.I. utilize iodine trichloride and iodine.

Preparation of reagent. 10 g of iodine trichloride are dissolved in a mixture of 300 ml of carbon tetrachloride and 700 ml of glacial acetic acid. To this is added 0.55 g of iodine. A check must be made on the halogen content and the solution adjusted.

Sodium thiosulphate solution is prepared by dissolving 24.8 g of sodium thiosulphate, $Na_2S_2O_3 \cdot 5H_2O$, in freshly boiled distilled water and diluting to 1 litre, to give a 0.1 N solution.

Method. It is necessary either to have some idea of the iodine value, or to do a preliminary estimation. (An excess of iodine monochloride must be present in the solution when the reaction is complete.) The weight of sample to be taken is calculated from the anticipated iodine value and the need to have 100–150% excess of the reagent. The appropriate amount of fat is weighed out, and placed in a dry flask. It is essential that the sample and the flask are really dry as the presence of water interferes with the reaction. The fat is dissolved in 20 ml of carbon tetrachloride, and 25 ml of Wijs solution is added from a pipette. A blank must be done alongside the sample flask, using the same pipette.

The stoppered flasks are swirled to mix the contents and stood in a dark cupboard for 30 minutes at normal room temperature (approx. 77° F. (25°C)). They are then removed and to each is added 20 ml of a 15% potassium iodide solution, followed by 100 ml of distilled water. The liberated iodine is slowly titrated with the 0.1 N thiosulphate solution, until the yellow colour just disappears. At this stage 2 ml of starch solution are added and the blue colour which then appears is discharged by further slow additions of thiosulphate.

The iodine value is calculated by the formula:

$$\text{Iodine value} = \frac{(B\text{-}S) \times N \times 126.9}{\text{weight of sample}}$$

where B = blank titration
S = sample titration
N = normality of the sodium thiosulphate solution
126.9 = atomic weight of iodine

The method is capable of great accuracy. Duplicate determinations should always be carried out and agreement should be very close.

Some fats, such as tung oil, contain conjugated double bonds and these do not react quantitatively with Wijs reagent under these conditions. As these oils are not edible, no further comment need be made.

c. Reichert, Polenske and Kirschner values. Special apparatus, the dimensions of which are critical, is required for this determination. A saponifying reagent is needed and this is made by dissolving 20 ml of 50% sodium hydroxide solution in 180 ml of pure glycerol. A quantity of fat (5 ± 0.1 g) is weighed into a 300 ml flask and 20 ml of the glycerol and alkali reagent is added. The mixture is heated over a flame, whilst swirling the contents. Within a few minutes the mixture should become uniform and clear as the fat saponifies; there should be no evidence of unsaponified fat on the flask. The flask is allowed to cool for about 5 minutes and 135 ± 1 ml of boiled water is added to dissolve the saponified mass, warming if necessary. Then 6 ml of 20% sulphuric acid is added and some pieces of porous pot. After fitting the still head and condenser, 110 ml of distillate is distilled in 28 to 32 minutes. The receiving flask is changed for a small cylinder and the flame removed from the distillation flask. The still head is disconnected and the receiving flask immersed in water at 15°C for 15 minutes. After mixing the contents are filtered through a dry 9 cm filter paper. 100 ml of filtrate is collected and titrated with 0.1 N baryta solution (barium hydroxide) using phenolphthalein solution as indicator. The solution contains the water-soluble volatile acids which have distilled. A blank should be run. Reichert value = 1.1 × (sample titration–blank titration). The rest of the distillate collected in the cylinder is run through the filter and the cylinder replaced under the condenser. 15 ml distilled water is used to rinse the condenser, the cylinder and the 110 ml flask before being passed through the filter paper. The washing is repeated twice more and the washings discarded. The washing procedure is repeated using three 15 ml portions of neutral alcohol, the washings being carefully collected as they contain the steam-volatile, water-insoluble fatty acids. The alcohol solution is titrated with 0.1 N sodium hydroxide, using phenolphthalein indicator. A blank determination is carried out.

Polenske value = sample titration–blank titration

The titrated solution after determination of the Reichert value is treated with 0.3 g of finely divided silver sulphate and allowed to stand for an hour, shaking at intervals. The

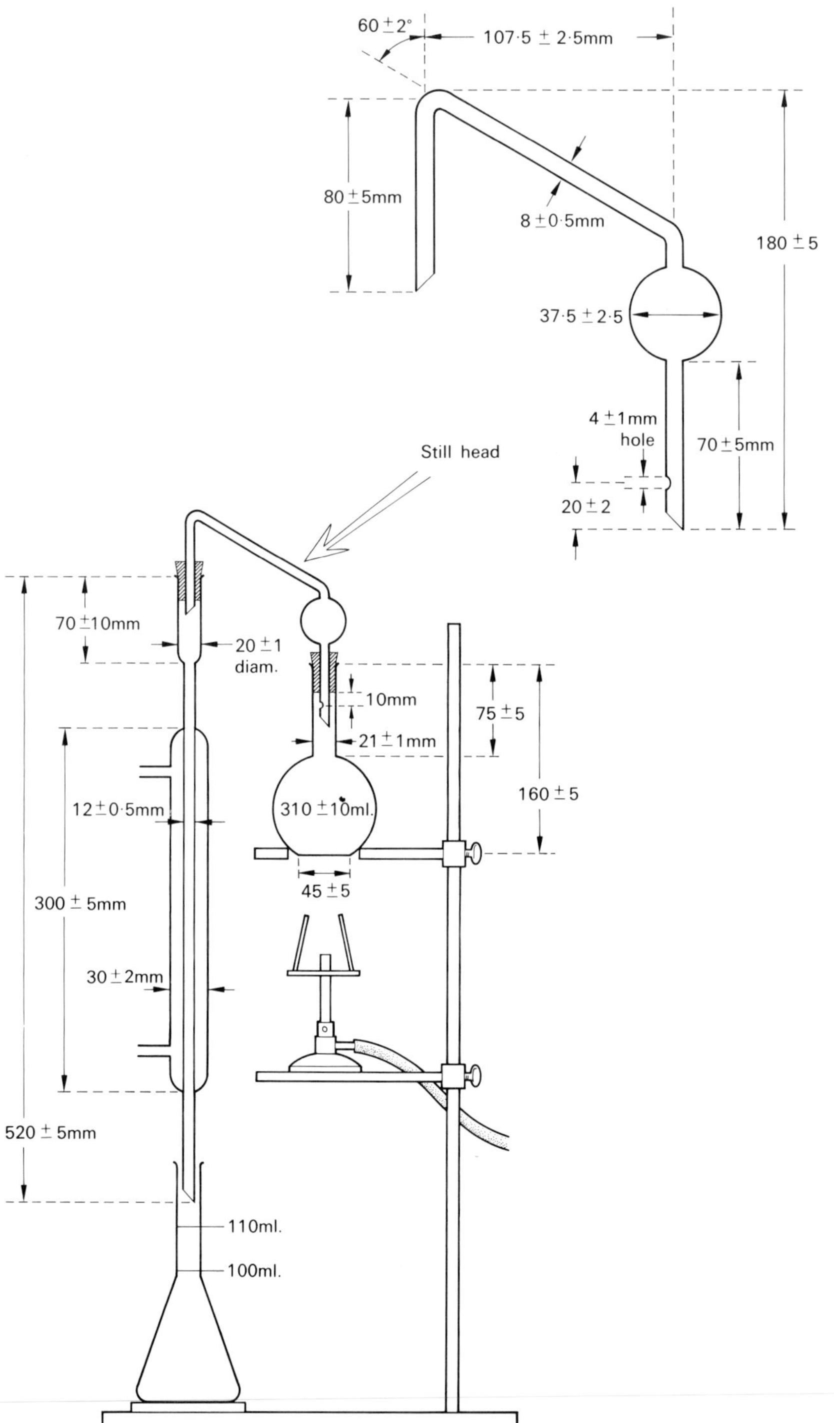

Fig. 1.3 Reichert–Polenske–Kirschner apparatus.

solution is filtered and 100 ml of the filtrate is transferred into the 300 ml distillation flask. 10 ml of $2\frac{1}{2}\%$ sulphuric acid, 35 ml water, and powdered pumice are added. The distillation apparatus is set up and 100 ml is distilled in 20 minutes. 100 ml of the distillate is titrated with 0.1 N barium hydroxide solution using phenolphthalein as indicator. A blank determination is run. The distillate contains any butyric acid distilled in the first (Reichert) distillation.

$$\text{Kirschner value} = \frac{121(100 + T_1)(T_2 - T_3)}{10\,000}$$

where T_1 = sample titration for Reichert
T_2 = sample titration for Kirschner
T_3 = blank titration for Kirschner

d. Unsaponifiable matter. About 2 g of fat, weighed accurately into a conical flask, are required for this test, unless there is good reason to expect a very high result, when less may be taken. 25 ml of a 0.5 N alcoholic solution of potassium hydroxide is added, and, after fitting an air condenser, the mixture is heated on a water bath for one hour, adjusting the boiling rate so that reflux of the alcohol takes place in the lower half of the condenser. When saponification is complete the contents of the flask are transferred to a separating funnel and the flask washed out with about 50 ml distilled water, the washings being added to the solution in the separating funnel. The flask is rinsed with 50 ml diethyl ether which is then carefully poured into the separator. The separator is shaken and the mixture allowed to stand and the two layers to separate. The lower soap layer is run into a second separating funnel, retaining the ether solution in the first. 20 ml water is added to the ether and a further 50 ml of ether to the soap solution. The shaking of the ether and soap solution is repeated. After settling, the soap layer is run off into the original flask. The ether layer is transferred to the first ether extract and the soap solution re-extracted a third time. All three ether extracts are now combined. The ether solution is washed successively with 20 ml portions of water, 0.5 N aqueous potassium hydroxide, water, potassium hydroxide, and then with water until free from alkali. The washed ether solution is poured into a weighed flask and the solvent driven off. When almost dry, a few millilitres of acetone are added and evaporated off. If there is a residue of water present, more acetone is added and again evaporated off. The flask is dried at 80°C till no further loss occurs, and weighed.

$$\text{Unsaponifiable matter} = \frac{\text{weight of residue} \times 100}{\text{weight of sample}}$$

The most important variation in method is the replacement of ethyl ether with petroleum spirit.

If the amount of unsaponifiable matter appears to be unusually high there may be failure to completely saponify all the triglycerides; the analysis should be repeated using the isolated material as recovered, instead of the fat.

e. Refractive index. As the refractive index changes with temperature, it is important to specify the temperature at which the determinations are made. Fats are generally liquid at 40°C so this is the preferred temperature for most tests, but reference tables may give the readings at 20°C. It is possible to correct readings for temperature variation, but as the readings must be taken under controlled conditions it is simpler to work to the reference tables.

A standard refractometer is used and the water jacket round the prisms is connected to a temperature-controlled water circulation unit. The water circulation is started and the instrument left to stabilise, readings being taken on the thermometer in the instrument. The sample of fat must be dry and clear before the measurement is made. A few drops of the melted fat are placed on the face of the prism which is then closed and tightened. Time must be given for the fat to come to temperature and then, after adjusting the light, the instrument is set so that the field dividing line is coincident with the cross wires. The best results are obtained if a sodium vapour lamp is used as a light source.

f. Melting point. Pure glycerides can be carefully crystallised in a form which will have a fairly constant melting point. Even so, different methods of preparation can give crystal forms with some variation in melting. Fats, being a mixture of numerous glycerides, do not have a sharp melting point, and the method of preparation of the sample has a marked effect on the result. A variety of methods have been devised for the determination of the melting point. They include closed capillary, open capillary, complete fusion, Wiley melting point, Ubbelohde melting point, flow point, drop point, and others.

Open capillary. For practical purposes the open capillary method is the most useful. Standard methods for this measurement are known as the slip-point (B.S.I.), softening point (A.O.C.S.), and rise melting point (German standard method). Preparation of the sample is extremely important. The fat is melted at a temperature not more than 10 deg. C above its melting point. If it is not clear it may be necessary to add a little anhydrous sodium sulphate to dry it, and then filter. A standard open-ended capillary tube is lowered into the fat so that a column about 1 cm long is drawn into it. The tube is cooled to 15°C and held near that temperature (15–17°C) for at least 16 hours. The tube is then suspended in a beaker of boiled distilled water alongside a thermometer, so that the lower end of the tube is about 3 cm below the surface of the water and the thermometer bulb is level with the column of fat. The water is then heated at a rate of about 2 deg. C per minute until the fat begins to clear and then slip up the tube. This temperature is recorded as the slip point. Several replicates are done simultaneously and the mean reading taken.

Titre. The first step in determination of the titre is to prepare the fatty acids. Older methods used alcoholic sodium hydroxide to saponify the fat. This has been replaced by the use of glycerol and potassium hydroxide. One reason for this change was the difficulty of removing all the alcohol from the recovered fatty acids.

The saponifying reagent is prepared by dissolving 20 g of potassium hydroxide in 100 g of glycerol. To 110 g of this solution 50 g of the melted fat is added and heated together until the mass becomes homogenous and transparent. Any cloudiness indicates that saponification is not complete. The temperature during saponification rises to about 150°C. The mixture is cooled for a few minutes and then about 300 ml of distilled water is added and stirred till uniform. The soap solution should be clear. Excess dilute sulphuric acid (about 50 ml of 30%) is added, stirring well all the time. The mixture is heated until the fatty acids rise to the top and form a clear layer, without boiling, as some of the lower molecular weight acids may otherwise be lost by steam distillation. The lower (aqueous) layer is removed with a syphon tube and the fatty acids are washed by stirring with hot water. After allowing the mixture to settle again the aqueous layer is removed. The washing is repeated several times if necessary, until the water is free from acid. The fatty acids are dried by running through a dry filter paper and cooling to about 20 deg. C above the expected titre. A titre tube is filled to the mark and inserted into a glass jar through an insulating cork. This arrangement prevents rapid loss of heat and guards against uneven cooling. A thermometer, graduated in tenths of a degree is suspended in the centre of the titre tube. The fatty acids are stirred either with a small wire loop, or, more commonly, with the thermometer, taking care not to touch the side of the tube with the thermometer bulb. As cooling takes place the fatty acids will become cloudy due to the formation of crystals. Stirring is stopped and the thermometer left to hang in the centre of the tube. The temperature is observed carefully: it will fall slowly, hesitate, and then rise quite rapidly for perhaps a degree before falling again. The titre is recorded as the highest temperature reached in the rise. Under some circumstances the rise in temperature may be small, or only a temperature arrest is observed. In this case the slight rise, or the temperature at which the hesitation took place, is recorded as the titre. It should be noted that incomplete saponification will lead to a low titre as will any fault in technique.

g. Acid value and free fatty acid. This is one of the easiest tests to carry out. Alcohol is boiled on a water bath for a few minutes to remove dissolved gases, and neutralized by adding a few drops of phenolphthalein and dilute sodium hydroxide until a pale pink colour is obtained. A quantity of fat, between 2 g and 10 g, is weighed into a flask, about 50 ml of the hot neutralized alcohol is added and the mixture boiled on a water bath. While still hot, the solution is titrated with 0.1 N sodium hydroxide solution until the pink colour returns. The amount of sodium hydroxide used is recorded; it should be between 10 and 20 ml. If it is very much more or less than this the test should be repeated with a more suitable sample size.

Free fatty acid is expressed either as oleic acid or as the commonest acid present in the fat, or on the basis of mean molecular weight of the fatty acids in the oil. For lauric acid oils the free fatty acid should be expressed as lauric acid, palm oil as palmitic acid, and most other fats as oleic acid, e.g.

$$\text{Free fatty acid (as oleic acid)} = \frac{\text{Titration}}{1\,000} \times \frac{282}{10} \times \frac{100}{\text{weight of sample}}$$

For expression as other acids the molecular weight of oleic acid (282) should be replaced by the appropriate figure for lauric acid (200) or palmitic acid (256).

Acid value is a simpler expression as the calculation is independent of the molecular weight of the free fatty acid. It is defined as the number of milligrams of potassium hydroxide required to neutralise the free fatty acids present in 1 g of fat, and is calculated by:

$$\text{Acid value} = 5.61 \times \frac{\text{titration}}{\text{weight of sample}}$$

h. Peroxide value. There are probably more variations in the methods of determining peroxide value than in the measurement of any other fat characteristic. The method given here is widely used and does not require any special apparatus.

Five grams of the melted fat are weighed into a conical flask, and dissolved in 30 ml of a solvent mixture containing 12 ml chloroform and 18 ml glacial acetic acid. To the solution 0.5 ml of a saturated aqueous solution of potassium iodide is added; the flask is stoppered and allowed to stand for 1 minute; 30 ml water is added and the solution titrates with 0.1 N sodium thiosulphate solution until the yellow colour has almost gone. About 0.5 ml of starch solution is introduced and the titration continued adding the reagent drop by drop until the blue colour disappears. During the titration the flask must be shaken vigorously enough to transfer the liberated iodine from the chloroform layer to the aqueous layer. At the end of the titration both layers should be colourless. On standing, the blue colour will return, but this must be ignored; the endpoint is the first disappearance of the colour, provided that shaking during the first part of the titration was adequate.

i. Colour. There are a number of scales for expressing the colour of fats, the scale used in the Lovibond tintometer being one of the most convenient. The tintometer contains a series of calibrated coloured glass slides in the three primary colours, red, yellow and blue. A melted sample of the fat is poured into a glass cell and stood in the tintometer. Inside the instrument, light passes in two parallel beams through the fat in the cell and through an aperture into which the coloured glass slides can be inserted. Both beams of light are brought together into a periscope. Glass slides are put in place until the colour produced by the slides matches the colour of the fat. The colour reading of the fat is then expressed as the number of reds, yellows and (if necessary) blues, which are required. The size of cell ($\frac{1}{2}$ inch, 1 inch or $5\frac{1}{4}$ inch) must also be given; the choice depends on the depth of colour in the sample.

1.2.3 Methods for determination of composition

The methods given for determination of the constants make it possible to examine a single fat and identify it. They also give some idea of the fatty acids present and the total amount of glycerides. A detailed determination of the fatty acids present demands other techniques. The classical methods formerly used for carrying out this investigation were tedious and time consuming. They consisted of preparing from the fat a quantity of fatty acids by careful saponification, using potassium hydroxide and alcohol. The soaps so produced were dissolved in water and the fatty acids precipitated with sulphuric acid, dissolved in ether and separated from the acid layer. After removal of the ether, the fatty acids can then be separated into groups either by lead salt fractionation or by low temperature crystallization. The weight, saponification value and iodine value of each fraction are then determined, and the fatty acid composition of each fraction deduced. Full details of the method are given in Hilditch and Williams' book (see section 1.10, Literature). Ester fractionation can be used as an alternative to fractional crystallization. In this method the fatty acids are converted into methyl esters by dissolving in methanol and adding a little sulphuric acid as a catalyst. The solution is boiled under reflux for several hours, cooled and washed with sodium carbonate solution after adding ether. The ether is then removed and the esters distilled under high vacuum. The isolated fractions are again subjected to examination to determine equivalent weight, iodine value and type of unsaturation.

Various other techniques for separation of fatty acids have been introduced and used in the investigation of fats, the most valuable being gas-liquid chromatography (James and Martin, 1950). The principles of this technique are fairly simple. A high boiling liquid is absorbed onto a porous inert support which is then packed into a glass column and placed in a controlled temperature oven. An inert gas such as nitrogen, argon or helium is passed through the column into a detector which will produce an electrical signal when there is a change in the composition of the vapour passing into it. If a small amount of volatile material is added to the gas stream, it passes through the column into the detector. As the vapour passes over the surface of the liquid in the porous support, it dissolves in it and then re-vaporizes into the gas stream. This process of solution and subsequent vaporization takes place at different rates for the various components of the

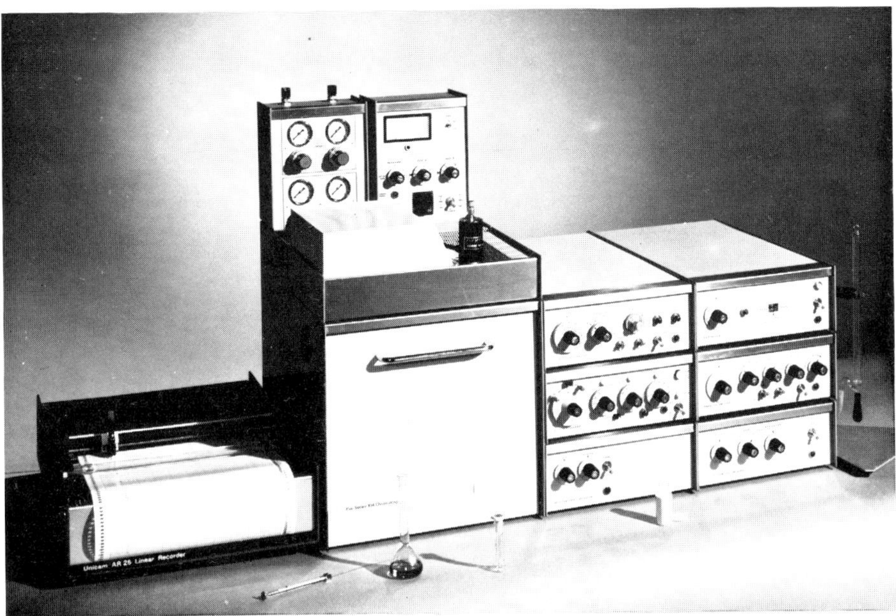

Fig. 1.4 Pye Series 104 gas chromatograph showing recording chart on left.

vapour, so that they slowly separate from each other. This allows each component to pass individually into the detector which then produces an electrical impulse which is recorded on a chart. When a mixture is passed through the column under these conditions a chromatogram (a graph consisting of several peaks) is produced. If pure substances are used to calibrate the equipment, then the position and size of the peaks can be used to identify and measure the component.

The fatty acids are prepared by mixing the fat with methanol and potassium hydroxide. The mixture is heated under a reflux condenser for five or ten minutes and then cooled. A few millilitres of a solution of boron-trifluoride in methanol are added and the solution again heated under a reflux condenser. The reaction mixture now contains the fatty acids as methyl esters. The flask is cooled and the condenser washed down with hexane. Brine is added to the flask to separate the hexane, and the methanol and water layers; the methyl esters enter the hexane layer. There are other methods of making the methyl esters but this method is convenient and gives satisfactory results; it has the advantage over conventional saponification and esterification methods in that the low molecular weight acids are not lost. The esters are used for the estimation of the fatty acids. The column used for methyl fatty acid esters is usually a polyester such as diethylene glycol adipate absorbed onto a celite support. A small quantity of the methyl esters is injected onto the column with a microsyringe and as the component esters emerge into the detector the recorder produces a series of peaks on the graph. The length of time taken for the peak to appear under fixed conditions of temperature and gas flow rate gives the identity of the fatty acid, and the area of the peak is a measure of the amount present. Careful calibration of the equipment using pure samples of the known fatty acids is necessary but under controlled conditions the same column will perform uniformly. Full details of the method and interpretation of the results are given in the American Oil Chemists' Society Methods Book (see section 1-10).

This method of determination of the fatty acid composition must always be checked against the constants determined. Iodine value and saponification value can be calculated from the fatty acid composition and when these are compared with the determined constants, agreement should be good if the proposed fatty acid composition is correct. A final check of the accuracy of the determination can be carried out by making a mixture of fatty acids to match the determined composition, and then establishing the composition of this known mixture by the technique used for the sample. A very full survey of the gas-liquid chromatography of fatty acids is given in volume 12 of *Progress in the Chemistry of Fats* (see section 1.10).

The foregoing methods will give the fatty acid composition of the fat, but not their manner of combination to give the glyceride structure which determines many of its properties. The glyceride structure is very complex, requiring the application of several techniques in order to arrive at a complete picture.

For some applications it may not be necessary to know the full details of the glyceride composition – the distribution of solid and liquid glycerides at critical temperatures may be sufficient. Two techniques are available for this purpose; 'dilatation' measurements giving the 'solid fat index', and the application of nuclear magnetic resonance (n.m.r.). 'Dilatation' measurements depend on the observation that as a glyceride melts it expands without any rise in temperature. If the temperature is plotted on a graph against the measured volume of glyceride, a slight change in volume, owing to thermal expansion, will be observed, followed by a sharp rise in volume as the glyceride melts. After melting, the graph will continue parallel to the original temperature/volume curve. It has been shown that the expansion resulting from the melting of a unit weight of solid glyceride can be regarded as nearly equal for all solid glycerides in a mixture. As most fats are a complex mixture of glycerides, any errors produced by assuming equality tend to cancel out. The method of measurement makes use of a dilatometer consisting of a glass bulb closed at one end with a stopper, and connected at the other end to a graduated capillary tube, the whole apparatus forming a U-tube with unequal sides. A number of variations in design are in use employing different sizes and shapes of bulb, and different methods of closure. A solution containing 1% potassium dichromate is used to fill the capillary and about 2 ml is introduced into the bottom of the bulb. The dilatometer is weighed, with the stopper, before filling the bulb with molten, dry, de-aerated fat. The stopper is carefully inserted, ensuring that no air is trapped in the bulb, and made tight. Excess fat is removed from the outer surface of the bulb with solvent, and the dilatometer is again weighed in order to ascertain the weight of fat enclosed in the bulb. An arbitrary solidification procedure is then closely followed; this is one of the weaknesses of the test. The fat is solidified in a specified manner to give reproducible results which apply only if the method is followed closely. The effect of solidification conditions on the polymorphic form of fat solids makes standardization essential. After solidification the dilatometer is placed in a successive series of controlled temperature baths, starting at the lowest temperature and working up the series until at least two readings have been taken after the sample has completely melted. The volume indicated on the capillary is recorded at each temperature. Half an hour is allowed for the dilatometer and its contents to reach temperature equilibrium before recording the volume. In order to estimate the solid glycerides present at any temperature it is necessary to calculate the thermal expansion from the two readings after completion of melting, and the total expansion between the given temperature and complete melting. The difference between total expansion and thermal expansion is the expansion due to melting. It is assumed that the expansion due to melting is

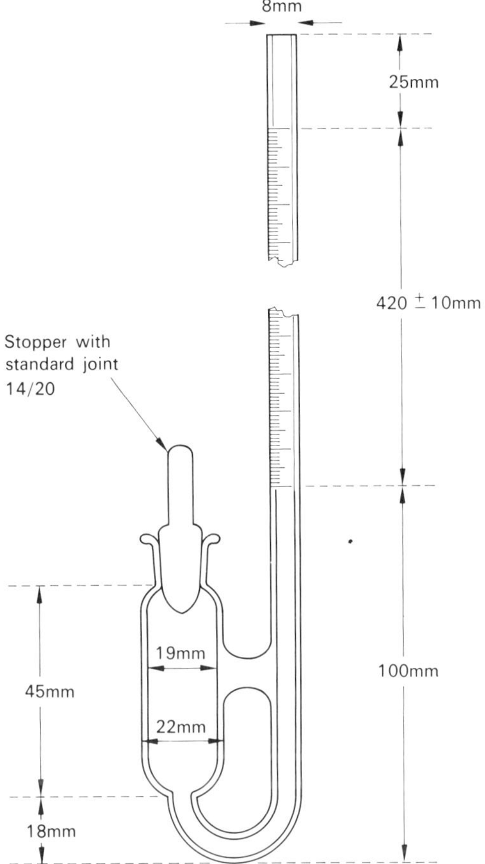

Fig. 1.5   Dilatometer–American Oil Chemists' Society pattern.

0.1 ml per gram. Precise details of the method can be found in *Official and Tentative Methods of the American Oil Chemists Society* (see section 1.10).

In recent years low resolution nuclear magnetic resonance has been applied to the measurement of solid glycerides in a fat. When a mixture of solid and liquid components is subjected to high frequency radio impulses while in a magnetic field, the free hydrogen atoms in the liquid components absorb the energy. This absorption can be measured and the proportion of solid and liquid components calculated. The instrument commonly used for this purpose is the Newport Analyser made by the Newport Instrument Company.

Nuclear magnetic resonance has a great advantage over dilatometry – the sample can be measured in the condition in which it is received. It will be recalled that before dilatometry can be applied it is necessary to melt, dry and de-aerate the sample. For n.m.r. it is only necessary to fill the cylindrical measuring cell with a solid or liquid sample; the adjustment of weight to allow for moisture and non-fatty solids and a correction for the contribution of the moisture to the n.m.r. signal, can all be made after completion of the measurements. Thus n.m.r. enables the actual solid glycerides to be measured as compared with estimating glycerides solidified under arbitrary conditions. It

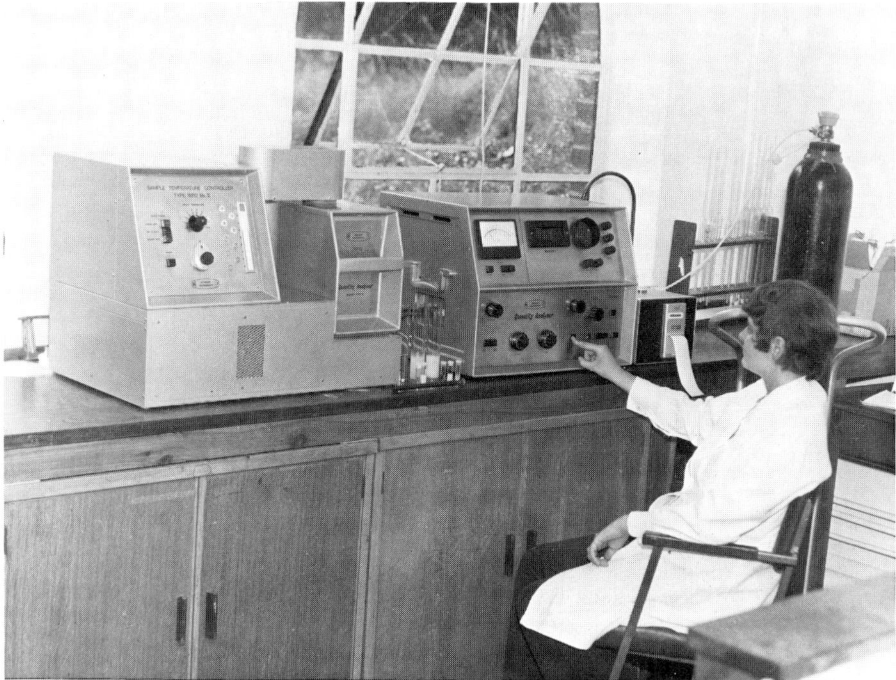

Fig. 1.6 Low resolution nuclear magnetic resonance (n.m.r.) apparatus for measurement of solid glycerides in a fat.

also permits the study of changes in solid/liquid ratios which take place in fats and fat products both during manufacture and storage. Use of the Newport Analyser will permit the measurement of the liquid components of the fat as distinct from dilatometry which measures the solid glycerides. In both cases the solid and liquid components can be derived, but only then can comparisons be made. The two techniques will give comparable results only if the fats have been prepared in the same manner, and this would be desirable only when comparing samples of unprocessed fats, since preparation of a sample for dilatometry neccessarily entails destruction of the fat structure.

The traditional methods of investigating glycerol structure have produced some very useful results but their accuracy is debatable. The proportion of saturated triglycerides can be assessed by a controlled oxidation of the unsaturated acids in acetone solution with potassium permanganate. The glycerides containing unsaturated acids are attacked at the double bonds, producing an acidic triglyceride. Oleic acid, for example, splits at the double bond leaving a C9:0 di-basic acid, azaleic acid, which is still esterified by one of its acid groups, to the glycerol chain. This azaleo-glyceride behaves differently from the fully saturated triglycerides, which are unaffected by the reaction. Separation is achieved by extraction of the oxidized acids by careful washing with alkali. This method still merely reveals the quantity of saturated triglyceride present. More detailed information can be obtained by the use of fractional crystallization from solvents. For this technique the fat is dissolved in acetone and cooled to $-94°F$ ($-70°C$). The solids which crystallize out are re-dissolved and re-crystallized at a higher temperature such as $-58°F$ ($-50°C$) and the solids are again re-crystallized at a higher temperature, and so on. The soluble glycerides at each step are recovered to give a series of fractions which can then be individually examined by conventional tests such as saponification value and iodine value and fatty acid identification. Armed with these data, and making assumptions, it is possible to calculate the triglycerides present in each fraction.

It is perhaps relevant at this juncture to discuss the use of spectrophotometry in resolving the composition of mixtures of unsaturated fatty acids. Oleic, linoleic and linolenic acids as they occur in natural fats do not show very much absorption of ultra-violet light when solutions of these acids are examined in a spectrophotometer, but their conjugated isomers do. (A conjugated acid is one in which there is more than one double bond and the double bonds present alternate with single bonds.) The unsaturated acids of edible oils do not have this structure, but suitable treatment will cause the double bonds to 'wander' and line up in a conjugated sequence. The treatment which produces this change, alkali isomerization, consists of heating the fatty acids at a temperature of 338°F (170°C) or 356°F (180°C)–the optimum conditions–in an alkaline solution of ethylene glycol or glycerol, for a specified time. There are several variations of the method but whichever one is used, the details must be adhered to. Various workers have conducted this test on pure samples of linoleic and linolenic acids and prepared calibration data for the estimation of the acids. After isomerisation the fatty acids are dissolved in purified solvent and the absorption of ultra-violet light is determined at several different wave-lengths. From these readings the concentrations of individual poly-unsaturated acids can be calculated. Having estimated the amounts of these acids present and knowing the iodine value of the mixture, the amount of mono-unsaturated acids can be calculated, and then, by difference, the total saturated acids. This examination can be applied to fractions of glycerides in order to determine their composition.

An alternative solvent fractionation technique uses counter-current distribution between two immiscible solvents. The individual glycerides of a fat may be soluble in two different solvents and if they are shaken together the glyceride dissolved in one will be distributed between them with a greater concentration in one than the other. The ratio between the concentrations, the distribution coefficient, is different for the various glycerides. If a fat is dissolved in one solvent and shaken with the second, the concentration of glycerides in the two solvents will be different from each other. If the two solutions are then separated and each mixed with fresh solvent of the other kind, further redistribution of glycerides will take place. A semi-automatic system has been developed in which the mixing of the solvents, their separation, and mixing again with fresh solvent is carried on a large number of times. In this way a number of fractions of the fat, each containing a different pattern of glycerides, is obtained, and can be further examined.

Recent developments have greatly improved the ease and accuracy with which the examination of glyceride structure can be carried out. A combination of gas-liquid chromatography, thin-layer chromatography and the use of specific enzymes, can give a great deal of information.

The fatty acids which form the glyceride differ in chain length, in unsaturation and in the position occupied on the glycerol part of the molecule. The first of these, chain length distribution, can be determined by gas-liquid chromatography. As already described for fatty acid analysis, a solution of the fat, this time without any pre-treatment, is put onto a suitable chromatographic column and the chromatogram developed. The peaks recorded on the chromatogram represent a series ascending from the shortest chain length to the highest, in steps of two carbon atoms. Thus tripalmitin would have a carbon number of

(3 × 16) + 3 but because the 3 of the glycerol is common to all triglycerides this is usually ignored, leaving this glyceride as 48. Dipalmito-stearin would be 50, distearo-palmitin 52, tristearin 54. As the unsaturated triglycerides do not separate from the saturated ones, the picture is not so clear cut; for example, a peak with carbon number 54 could contain one palmitic, one oleic, and one stearic acid; or 1 palmitic, 2 oleic; or 1 palmitic, 1 oleic and 1 of any other C:18 acid. The correct interpretation of the peaks therefore, is to identify 54 as being three of C:18 acids, or a C:16, C:18, C:20 grouping, or any other group adding up to 54. To some extent the complexities are resolved by working through the chromatogram from the ends, summating the fatty acids positively known to be present from fatty acid analysis previously obtained. If, in our example, all the C:20 acids have been accounted for in the higher carbon number peaks, then the resolution of the 54 carbon peak is much simpler.

In thin film chromatography a flat glass plate or rigid plastic sheet is coated with a thin layer of an adsorbent such as silica gel. A spot of test solution is placed near one edge of the plate, which is then placed on this edge in a trough of solvent in a closed container. The solvent slowly flows up the film of adsorbent, just as paraffin oil climbs the wick of an oil lamp. As the solvent passes the spot of test solution, the latter starts to follow the solvent flow, but the components of the solution may flow at different speeds and so form separate spots. When the solvent has nearly reached the top edge of the plate, the process is stopped by removing the plate and allowing it to dry. If the test solution was coloured, the spots may be readily visible, but if not, the plate must be sprayed with a suitable reagent to make them visible. This technique can be applied to fats in order to separate the triglycerides into groups of different degrees of unsaturation, in a sequence starting with fully saturated triglycerides and then those containing one, two, three, four etc., double bonds. The separated groups can then be scraped off the plate, extracted from the adsorbent, and after conversion to methyl esters, passed through a gas-liquid chromatograph to determine the fatty acid composition.

These methods can provide the data to show the composition of a fat in terms of triglycerides, but another technique is needed if we wish to find the position of the individual acids within the triglyceride molecule. It will be recalled that the fatty acids can be in one of three different positions, designated 1, 2, and 3, on the glycerol part of the molecule. No chemical method is capable of distinguishing between these positions but enzymes are often specific in their activity. A lipolytic enzyme present in pig pancreas will release the fatty acids from the 1 and 3 positions, but not from the 2 position. Both the separated fatty acids and the residual one on the 2 position can be identified. The information collected in these various examinations can then be used to deduce the overall glyceride composition. The calculations are complex, and a certain number of assumptions are made.

In this section so far, we have only concerned ourselves with the composition of the major component of the fat – the triglycerides. The composition of the minor components may also be of interest.

Phosphatides are frequently present in fats, especially some of the vegetable ones. The composition of the individual phosphatides is almost as complex as that of the triglycerides, and similar methods of resolution have been applied to them. The determination of the phosphatide content of fats is not precise. Generally the amount of total phosphorus present is measured, or alternatively, the quantity of material insoluble in acetone. As phosphatides contain a phosphorus atom in each molecule of phosphatide, the amount of phosphorus present would seem to be an easy way to determine phosphatides, and so it would if the phosphatides themselves had a fixed composition. Because the fatty acid components of the phosphatides vary, the proportion of phosphorus in the molecule is not fixed, so that only the amount of phosphorus present can be found. Even this cannot be assumed with certainty to be entirely due to phosphatide. The alternative procedure is to measure the acetone-insoluble matter. Although phosphatides are said to be insoluble in acetone, they have a very low solubility which must be allowed for when estimating the amount present. This is carried out by dispersing lecithin in a quantity of acetone and allowing it to come to equilibrium at the temperature at which the determination will be carried out. Just before use the acetone, now saturated with phosphatide is filtered to remove insolubles. The sample of fat (2–3 g) is dispersed in 30 to 40 ml of the acetone, chilled in ice-water for 10 minutes, and centrifuged in a weighed centrifuge tube. The supernatant liquid is discarded and the precipitate is re-suspended in fresh chilled saturated acetone. Again the suspension is chilled and centrifuged. The washing of the precipitate is carried out three times and the residue dried and weighed. Thin-film chromatographic methods are being developed for the separation and estimation of phosphatides.

Tocopherols (vitamin E) are determined by their reaction with ferric chloride and α,α-dipyridyl, the colour so produced being measured in a spectrophotometer and calibrated against pure samples of tocopherols.

1.2.4 Methods for determination of stability

Stability of a fat is an expression of its resistance to oxidative deterioration, with the changes in odour, flavour and appearance which usually accompany it. There are a number of tests which can be applied to a fat in order to determine its present state of oxidation, from which, with experience, it may be possible to hazard a forecast at maximum shelf life (the term used to indicate the period of time before deterioration becomes detectable by taste or smell). The most widely used measurement for this purpose is the peroxide value, details of which have already been given in an earlier section. Other tests measure some of the breakdown products from the peroxide. The most useful one in this field is based on benzidine, but as the reagent has been found to be carcinogenic it has been replaced by anisidine. It can be shown that as peroxides are broken down by heating, the peroxide value falls and the anisidine value rises, so that the oxidative damage which may be hidden by reduction of the peroxide figure is revealed by the anisidine value. A suggested combination of these values (twice the peroxide value plus the anisidine value) gives a guide to the condition of the fat, and allows a judgement to be made of its possible stability.

The 'Issoglio value' is dependent on the amount of water-soluble substances present in the fat. These materials are extracted from the fat and oxidized with potassium permanganate solution. The amount of oxidizable material is thus assessed, and the higher the figure obtained, the less stable the fat is likely to be. This test was developed for beef tallows in particular and is not widely used.

Accelerated tests used to forecast stability are more severe than the above and are generally based on the Schaal test or the Swift test.

The Schaal test in various modifications is a simple test in which the sample is incubated in a covered vessel at 145°F (63°C) until it develops a peroxide value of a pre-determined level, or until it smells rancid. The number of hours of incubation can be equated to the number of days, weeks, or months of shelf life, according to the

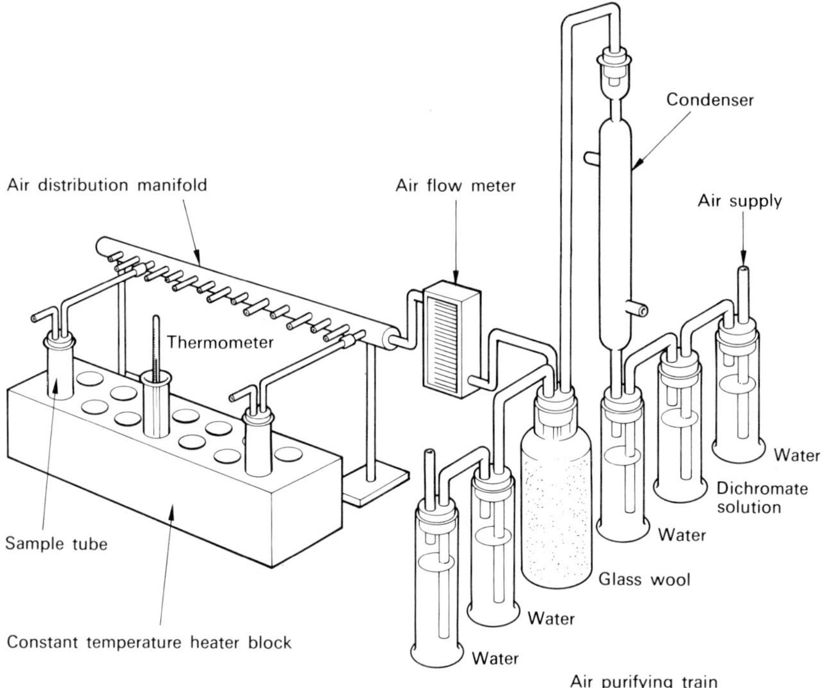

Fig. 1.7 Swift apparatus for assessing stability of fats. American Oil Chemists' Society pattern.

experience of the operator. The relationship between hours of test and stability depends very much on the conditions of the test and the standards set for the product.

The Swift test was originally devised for measuring the stability of lard, but has been modified in a number of ways and is now used for forecasting the stability of most fats. It is performed by passing purified air into a sample tube containing the fat under test at a fixed rate of 2.33 ml/second, while the fat is held at a temperature of 208°F (97.8°C). Peroxide values are determined at intervals, the end point of the test being the time taken for the value to reach a pre-determined level, which for lard was fixed at 20 meq/kg (meq = milli-equivalents of oxygen). The actual endpoint will vary from one fat to another, values of 50 being suggested for groundnut oil, and 150 for soyabean oil. A more satisfactory method is to plot a graph of time against peroxide value over the whole period. The point at which the slope of the curve suddenly increases marks the time at which the first signs of rancidity are detectable.

In the Sylvester method the above procedure is mechanized. About 5 g of the sample are placed in a special flask together with a magnetic stirring rod. The flask is filled with oxygen, connected to a manometer, and placed in a temperature-controlled water bath. A rotating magnet beneath the water bath stirs the oil in the flask, so aerating it. As oxidation proceeds, the oil absorbs oxygen, causing the manometer level to fall, the movement being recorded on a chart.

In both of these accelerated methods it is usually accepted that each hour of test time is equivalent to a certain period of storage stability, but this relationship is subject to assessment by experience with the individual apparatus and conditions.

1.3 VEGETABLE OILS AND FATS

Vegetable oils and fats are obtained from the fruit or seeds of various plants. They may be grouped according to the botanical classification of the plants yielding the oils, but although this method may be systematic it does not readily reveal the compositional relationships of the oils. Another method of classification divides them into drying, semi-drying and non-drying oils. This method also has its weaknesses and is more suited to classification for industrial use than for edibility. The alternative is to group according to the dominant fatty acids present. As this scheme helps to show the interchangeability of the oils in a group, it has been used here. The oils associated with the dominant fatty acids are given below; the chief ones are described in sections 1.3.1 to 1.3.5, the others in section 1.3.6.

Lauric acid oils	Coconut, palm kernel, babassu.
High saturated acid oils	
1. Vegetable butters	Cocoa-butter, shea nut.
2. Palm oil	Palm oil.
Oleic/linoleic oils	Olive, groundnut, maize (corn), sesame, cotton seed, sunflower, safflower.
Linoleic/linolenic oils	Soyabean, linseed.
Erucic acid oils	Rape seed, mustard seed.

1.3.1 Lauric acid oils

There are only three commercially important oils in this group in which the lauric acid content approaches 50% and the unsaturated acid content is low. Coconut oil and palm kernel oil are the most important, babassu being restricted to South America.

a. Coconut oil is obtained from the nut of the coconut palm, *Cocos nudifera*. This palm is widely distributed in tropical areas where it grows wild; especially on the coast and on the islands of the Pacific. Malaysia, Indonesia, the Philippines, the South Sea islands and Ceylon are the principal growing areas, and in these parts large estates are devoted to the cultivation of the coconut palm.

In common with other palms, the coconut palm is classified as a monocotyledon. The stem, which is devoid of branches, grows to a thickness of 18 in (0.5 m) diameter, and reaches 80 to 90 ft (24–30 m) in height. It terminates in a crown of fronds, each of which can reach a length of 15 ft (5 m) or more. Best growth is obtained in areas with rainfall well distributed and amounting to between 50 and 90 in (130–230 cm). Along with this moisture requirement, the palm thrives in a warm sunny climate in which there is little temperature fluctuation, ideal conditions being about 85°F (29°C) in the daytime, with a night temperature of not less than 73°F (23°C) and an overall mean temperature not below 70°F (21°C). New palms are grown by planting ripe coconuts in nursery beds, the nuts being well bedded down but not covered. Development is slow and seedlings may not appear for about two months. Favourable growth conditions will still not produce more than three leaves in four to six months. As soon as the seedlings reach this stage of development they can be replanted in the plantation proper, at intervals of about 30 ft (10 m). Cultivated palms will start to flower and produce fruit after six or seven years, whereas the wild ones will not usually start to bear until they are ten years old. It takes a further nine or ten years before they come into full production, but they may go on

Fig. 1.8 Coconut palm plantation.

bearing for up to 60 years. The fruits, which develop in the crown, can only be harvested by hand cutting, and as there are no branches on the stem, the harvesters have to climb the tree in order to reach the crown and cut the bunches of nuts.

During the first five or six months of the development of the nuts, they are full of liquid, and as they approach maturity the 'flesh' or 'meat' begins to form inside the shell. As the process continues the thickness of meat increases, at the expense of the volume of the liquid. At maturity, nearly a year after flowering, there is very little liquid left in the nut and the meat is about half to three-quarters of an inch (12 to 20 mm) thick.

Each palm will yield about 50 to 70 fruits per year. The nut itself is enclosed in a thick covering of fibre called coir, which is removed before splitting the nut and pouring out the liquid. The meat of the nut at this stage contains about 35 to 37% fat. As it dries, the oil-bearing part shrinks away from the shell and is easily separated. This dried material is known as copra. In order to prevent deterioration of the copra owing to enzyme action or mould growth, the moisture content is reduced by the drying action to about 5 or 6% and the oil content rises accordingly to between 65 and 70%. The copra can be dried by exposure to the sun, or over fires, or in modern hot-air driers. The last-mentioned method gives the best quality copra.

The oil is obtained from the copra by pressing. The residue can be broken up again and re-pressed or milled into flakes and extracted with hexane.

Coconut oil obtained from good quality copra is almost colourless and is very low in free fatty acid. The quality varies with the type of copra used, and it can be quite dark

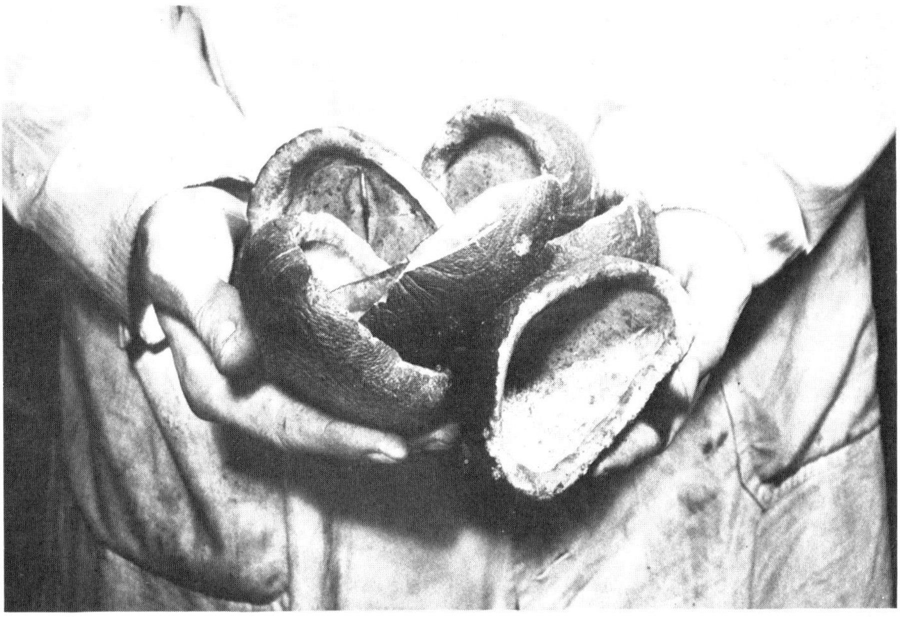

Fig. 1.9 Copra, the dried flesh of the coconut.

yellow/brown in colour. Although the oil is liquid in the countries of origin, it solidifies to a semi-crystalline fat with a fairly sharp melting range of 23 to 26°C. This characteristic, which is similar to that of the harder vegetable butters, is due to the high concentration of saturated acids present and hence the narrow range of melting points of the glycerides. The composition of coconut oil is given in table 1.3.

The outstanding features of coconut oil are the high saponification value and the high Reichert–Polenski–Kirschner values. These constants have been used to identify the presence of coconut oil or palm kernel oil in a mixture, and also to estimate the quantity present. It will be noted from table 1.3 that coconut oil and palm kernel oil have very similar compositions and it is not easy to positively identify which of the two is present in a mixture with other fats, although in the pure state the difference in iodine value readily distinguishes them.

If properly processed, the fresh oil is almost tasteless, and because of its high proportion of saturated fatty acids is resistant to rancidity. Deterioration of coconut oil is usually accompanied by the development of free fatty acids which are very readily detected and rather unpleasant to taste. Care must be taken to avoid incorporating coconut oil in products which may contain active lipases since these enzymes will liberate free fatty acids. Large quantities of coconut oil are used in the manufacture of margarine and confectionery. One of the unusual characteristics of the oil is the short melting range. Most fats pass through a protracted series of changes in consistency as the temperature is increased, but coconut oil liquifies rather suddenly. This behaviour makes the fat brittle, not plastic, and therefore of limited use in food fats, such as shortenings, which are required to have a wider plastic range. In hard table margarines the short melting range of coconut oil has the advantage that it increases the rate of crystallization of the product and the brittleness it gives to the margarine makes its behaviour similar to that of butter.

Fractionation of the fat can be used to produce a 'stearin' which is used in making chocolate couvertures for bakery use, while hydrogenation will produce a hard, stable fat used in biscuit manufacture.

    b. Palm kernel oil. The oil palm (*Elaeis guineensis*) produces a fruit which provides two very different oils. The pulp contains palm oil which is highly coloured, is composed mainly of C16 and C18 fatty acids and behaves like most other vegetable oils apart from its higher melting point. Full details of this oil are given in section 1.3.3. Within the pulp is a kernel which provides the source material for palm kernel oil.

    The kernels are separated from the pulp and fibre and are usually shipped to Europe in this state. They are then cracked to release the nut from the shell, and the nuts are either solvent extracted or subjected to pressure in order to win the oil from them. This oil, although obtained from the same fruit as palm oil, is a very different product. Its composition and behaviour are very similar to those of coconut oil which, for many uses, is freely interchangeable with it. The composition and characteristics are given, together with those for coconut oil, in table 1.3.

Fig. 1.10 Palm kernels.

Table 1.3 Composition and characteristics of coconut and palm kernel oil

Fatty acid composition		Coconut oil %	Palm kernel oil %
Caprylic	C8:0	6	3
Capric	C10:0	6	4
Lauric	C12:0	44	51
Myristic	C14:0	18	17
Palmitic	C16:0	11	8
Stearic	C18:0	6	2
Oleic	C18:1	7	13
Linoleic	C18:2	2	2
Constants			
Refractive index at 40°C		1.448–1.450	1.449–1.452
Iodine value		7.5–10.5	14–22
Saponification value		250–264	245–255
Unsaponifiable matter		0.5	1.0
Reichert value		6–8	4–7
Polenski value		15–18	9–11
Kirschner value		1.6–1.9	0.8–1.2
Melting point		23–26	24–26
Titre °C		20–24	20–28

The oil is frequently darker in colour than coconut oil and does not have such a pleasant smell or taste, but after processing it gives an almost white, pleasant flavoured nut oil which can be used in place of coconut oil in almost any application. It does not have quite as stable a flavour as coconut oil and is prone to give rise to distinctive 'off' flavour when it deteriorates. Advantage can be taken of the higher unsaturated acid content to obtain a higher melting point product by hydrogenation, without losing the advantages of the low molecular weight acids. Hydrogenated palm kernel oil is widely used in the confectionery industry as a replacement for cocoa-butter.

1.3.2 High saturated acid oils

a. Cocoa-butter

The cocoa bean which is the source of both fat and cocoa, is the seed of the tropical plant *Theobroma cacao*. Its natural habitat is the tropical forests of South America and although it has been largely cultivated in West Africa, which is now the main production area, it is still grown in Ceylon, the West Indies and Central America.

The trees are started in a nursery and planted out, about 12 ft (4 m) apart, when they are about a year old. It is necessary to provide some shade, either by special planting or by leaving existing trees, as the cacao cannot stand direct sunlight. In order to thrive and produce well the trees need plenty of rain throughout the year. When about three to five years old, flowering commences with the flowers developing directly on the trunk and larger branches. The fertilized flowers produce large fruits, or pods, which grow to about 8 or 9 in (20–22·5 cm) in length. There is no specific flowering and fruiting season, fruit and flowers can usually be found on the trees all the year round, during the 20 years that they continue to bear. When ripe, the pods are harvested and opened. The pulp and seeds inside are scraped out into heaps and left to ferment and the pulp to liquefy. This process not only simplifies the separation of the seeds from the pulp, but also permits the

Fig. 1.11 Cocoa bean pods growing directly on the trunk.

development of desirable characteristics in the seed itself. The separated seeds are sun dried and are then ready for transport to the mills for fat and cocoa production.

The beans, which contain 50 to 55% fat, are used for cocoa and chocolate manufacture, the production of the fat being a by-product. After harvesting, the beans are shelled and roasted before being pressed to force out the fat. The residual cake is ground to give cocoa powder, but it can also be extracted with solvent to recover more fat.

The fat is a pale-yellow solid with an exceptional melting characteristic and a pleasant flavour. It is very resistant to oxidation and will keep for a long time without developing rancidity. At temperatures below 28°C it is hard and brittle, but at 34 to 35°C it melts quite sharply. It is this almost unique melting behaviour that makes the production of chocolate possible. This characteristic is due to the simplicity of its glyceride structure with one triglyceride providing over 50% of the composition and three di-saturated mono-unsaturated glycerides providing 77%. The composition and constants of cocoa-butter are given in table 1.4.

The special value of cocoa-butter for confectionery and the limited supplies available restrict the use of the material to the confectionery trade. Small amounts are used in the preparation of pharmaceutical products for which the melting characteristics make it ideal.

In the manufacture of chocolate, large quantities of cocoa-butter are used. The conditions of solidification of the fat, cocoa, and sugar mixture are of great importance as the correct crystal form of the fat is needed to give the correct characteristics to the chocolate.

Attempts have been made to produce a suitable cocoa-butter substitute, by hydrogenation or by fractionation of palm kernel oil and coconut oil. These two fats can be used to make a product which has *some* of the characteristics of cocoa-butter, but the critical crystallizing and handling qualities do not suggest that a satisfactory substitute for use in chocolate can be found by these methods. A true synthetic cocoa-butter has been produced by a combination of hydrogenation, interesterification and fractionation, and this material behaves in a very similar manner to the natural fat. The ordinary substitutes can be used satisfactorily in the preparation of confectionery fats for couvertures.

b. Palm oil. Reference has already been made to the oil palm, *Elaeis guineensis,* and the two oils which are obtained from it. The oil from the pulp is so highly coloured that it is frequently referred to as red palm oil. It is much more important than the kernel oil, and is one of the most valuable edible oils available to the United Kingdom market. This is partly due to the introduction of the palm to achieve diversification of crops in areas where the existing crops are no longer economically viable. This applies particu-

Table 1.4 Composition and constants of cocoa-butter

Fatty acid composition		%
Palmitic	C16:0	25
Stearic	C18:0	35
Oleic	C18:1	38
Linoleic	C18:2	2

Constants	
Refractive index at 40°C	1.453–1.458
Iodine value	35–40
Saponification value	190–200
Unsaponifiable matter	not more than 1.0
Melting point	32–35
Titre °C	45–50

larly to rubber growing areas of Malaya where the fall in demand for natural rubber has encouraged the replacement of rubber trees with oil palms.

The oil palm is classified as a monocotyledon. Its leaves are borne directly on the stem and there are no branches. It grows wild in the forests of West Africa and formerly much of the palm oil was from these wild trees. Plantations have been established, and the oil of commerce comes almost exclusively from plantation sources. The palms are grown from seed and planted out about 30 ft (~ 10 m) apart, in order to encourage a dwarf growth, which not only gives better yields, but allows easier harvesting compared with the wild trees. The seed used is the palm kernel referred to earlier. It is not easy to grow the young plants; germination is uncertain and very slow. Several techniques have been tried to improve the situation. Planting in 'hot beds' has been used, the heat being provided by filling a trench with green vegetation and covering it with soil. The heat generated by fermentation beneath and the sun above, provides the warm environment that the seeds need. It has been recommended that this 'natural' system be replaced by a more controlled method where the seeds are heat-treated at temperatures ranging from 95°F (35°C) to 104°F (40°C), according to the shell thickness. The seeds are planted in a prepared nursery bed at 4 to 6 in (100–150 cm) apart. Even after the pre-treatment it is three or four months before evidence of germination appears, and then perhaps only 30% will be growing. In the succeeding months more will germinate and start to grow. When the seedlings have developed three leaves they can be transplanted with a greater spacing, at about 20 in (50 cm). Shade must be provided to protect the young plants and ground cover between the rows is necessary to conserve the moisture in the ground; the earth must not be left bare. When the plants have developed about ten leaves they are ready to be set out in the plantation. Four years elapse before the palms begin to bear fruit, and a further eight or nine years before they reach full yields, which they can maintain for fifty years. The pinnate fronds grow directly from the crown of the palm, which in the wild can grow to a height of 60 to 70 ft (18–21 m). Being a native of tropical rain forests, it requires a great deal of rain, over eighty inches (200 cm) in order to grow well, but at the same time needs plenty of sunshine. The optimum temperature for fruiting is reported to be 75 to 80°F (24–27°C), although development will take place outside this range.

Flowering starts in the dry season, the fruit develops during the following wet season, and ripening comes with the arrival of the dry weather again. The flowers develop in clusters and the bunches of fruit which follow are made up of individual fruits, each the size of a very large plum. Harvesting consists of removing the bunches by hand-cutting, and transporting them as quickly as possible to the factory. The older way of treating the fruit was to drop the bunches into vats of boiling water. This simple process arrested enzyme action and melted the oil out of the fruit to form a separate layer on top of the water. Modern systems heat the bunches of fruit with low pressure steam to destroy the enzymes which are responsible for the deterioration of the oil. During the steaming process the fruits are loosened from the stalks and then the mass is mechanically disintegrated. The oil is recovered by centrifugal separation. Speed of handling is necessary if good quality oil is to be obtained. The fruit contains active lipolytic enzymes which rapidly split the oil forming free fatty acids. The oil which was recovered from wild palm always had a high free fatty acid content, as high as 50% or even 75%, owing to enzyme action being allowed to continue unchecked between collection and delivery to the oil mill. It is obvious that with such an oil it would not be possible to ship the palm fruit any distance, and the establishment of plantations was dependent on the building of an oil mill no matter how primitive.

Fig. 1.12   Artificial pollination of palm flowers by spraying pollen on them.

The oils are classified commercially according to their origin; Malayan, Sumatran, Nigerian etc., the quality of the oil, free fatty acid, colour and ease of bleaching, being related to origin.

The intense red/orange colour of the oil is due to high concentrations of carotenoid pigments which are present to the extent of up to 0.2%. Before β-carotene was synthesized and the synthetic material made available for industrial use, small amounts of palm oil were used in the margarine industry, as a source of β-carotene, a pro-vitamin A.

Fig. 1.13 A collection point on a plantation for bunches of palm fruit.

The pigments can be removed from the oil by adsorption bleaching or by high temperature processing. When the oil is heated under vacuum, to temperatures over 220°C, the pigments are de-colourized and a pale yellow oil results. This oil can be processed to give a bland, white fat which is very stable to oxidation. Its constants and composition are given in table 1.5.

The oil is used extensively for the production of margarine and shortening, and is also used as a fat in biscuit and cake making, in ice-cream and frying fats.

Table 1.5 Composition and constants of palm oil

Fatty acid composition	%
Myristic	1
Palmitic	48
Stearic	4
Oleic	38
Linoleic	9
Constants	
Refractive index	1.453–1.456
Iodine value	44–48
Saponification value	195–205
Unsaponifiable matter	not more than 0·8
Titre °C	40–47

The characteristic of palm oil fatty acids is the high palmitic acid concentration. There is a variation in composition of palm oils drawn from different growing areas, and from different varieties of palm, but the level of saturated acids is always high. Although the fatty acid composition is similar to that of cocoa-butter it is a very different fat. Slow cooling of palm oil will lead to the formation of a mixture of coarse crystals and liquid glycerides. Advantage has been taken of this separation on cooling, to develop fractionation procedures yielding a hard fat and a liquid fraction which is similar to cottonseed oil.

1.3.3 Oleic/linoleic oils

a. Olive oil. Olive oil is undoubtedly the most important oil in the countries bordering the Mediterranean Sea. It is obtained from the ripe fruit of the olive tree, *Olea europaea*, which thrives in the Mediterranean climate; hot dry summers with rain in the winter. It is grown extensively in Algeria, Tunisia, Israel, Syria, Greece, Italy, Southern France and Spain and has been introduced into other countries with a suitable climate, but the main source of olive oil is still the Mediterranean area.

It is raised from cuttings, to produce a gnarled and twisted tree with small leaves, green on the upper surface and silver on the under-side, giving an all-over grey appearance to the foliage. Under suitable conditions the tree may start to flower at two or three years old, but it is eight or nine years before it is producing fruit in commercial quantities. Once established, it will go on bearing for many years, hundreds or even longer, according to some claims. During the summer months the fruits slowly ripen, and in the autumn start to fall. Harvesting should be done just before full ripeness in order to obtain the oil at its best. Deterioration starts when the fruit becomes over-ripe. There is no really satisfactory short cut to harvesting. The best oil is obtained from fruit which has been hand-picked, otherwise fruit is made to fall from the trees onto cloths laid beneath them. This procedure is quicker than hand picking, but the bruising of the fruit which is caused by falling sets free enzymes which start hydrolysis of the oil, with consequent formation of free fatty acid in the oil while the fruit is waiting to be processed.

The fruit is first crushed in a mill, which breaks down the pulp but not the kernels, and then it is pressed to squeeze the oil out of the pulped material. Oil obtained by this simple gentle pressing is rated as the finest available. Further pressing at higher pressures and temperatures leads to more oil recovery. Additional oil can be obtained by treating the pulp with hot water which will displace the oil, when the wetted pulp is again pressed. The kernels can be separated and the oil recovered from them also. Olive kernel oil does not differ significantly from the fruit oil (in contrast with the marked difference in composition of palm fruit oil and its kernel oil) and the oils are frequently blended. Olive fruit residues after pressing can be extracted to recover more oil. In this case carbon disulphide is used as a solvent, producing an inedible oil which is known as sulphur olive oil.

The yield of oil and its composition are dependent to some extent on the variety of the tree and the condition of growth and cultivation. It should be pale yellow; a green colour, due to chlorophyll, often indicates an inferior oil. Its flavour should be pleasant and bland.

The better quality oil from the first pressing (virgin oil) and often from the second, are good enough to be used as a salad and cooking oil without any treatment other than filtration to remove small amounts of pulp. The high price of the oil prevents it being generally available for use in margarine manufacture, although the inedible grades are used for the manufacture of soap.

As a consequence of demand for the oil and the high price, adulteration has been widely practised. Other liquid oils may be detected by careful measurement of the constants and testing by colour reactions, although teaseed oil is so very similar to olive oil that it is difficult to prove admixtures. Another means of extending the edible olive oil yield is to esterify high free fatty acid oils with glycerol, so converting an inedible oil into an acceptable one.

The oil is quite stable to oxidation because of the small amount of linoleic acid present. Its composition and constants are given in table 1.6.

Olive oil remains clear at lower temperatures than most other liquid oils. This is due not only to the low concentration of saturated acids in the oil, but also to the distribution of these solid acids in monosaturated di-unsaturated glycerides.

In most of the countries where olive trees grow, the oil is used to the exclusion of other fats for food preparation, as a salad oil, and as a frying medium, but at times political pressure or economic expediency dictates that the valuable olive oil is exported, and replaced by imports of other liquid oils.

b. Other oleic/linoleic oils. Olive oil is generally used without any refining or other treatment, but the other liquid oils of commerce require some purification before use.

The oil-bearing plants considered so far are perennial and develop into large trees. The oils described below are derived from annuals and hence their cultivation follows a different pattern.

Corn (or maize oil). This oil is a by-product of the industries which use Indian corn (or maize) as a source of starch, either as corn starch, for glucose manufacture, or for the production of food cereals.

The maize plant, *Zea mays,* is classified botanically as a monocotyledon and a grass. It is indigenous to America but is now cultivated in many parts of the world. A great deal of work has gone into the breeding of strains which give high yields of grain. Little is known of the intermediate varieties which have given rise to the large corn cobs with their packed seeds which are familiar today.

The fields for the growth of maize are prepared in the conventional way for annual crops. The seed is sown thinly in rows which are about 3 ft (1 m) apart, as the plant requires plenty of room for growth. As the crop develops it is usually sprayed to control disease, and generally cultivated like any other cereal crop. When fully ripe it is harvested mechanically. The grain, when separated from the cob, is used for many purposes, mainly as an animal feed as well as for human consumption.

Table 1.6   Composition and constants of olive oil

Fatty acid composition	%
Palmitic	7–15
Oleic	70–85
Linoleic	4–12
Constants	
Refractive index at 20°C	1.4690–1.4700
Iodine value	80–88
Saponification value	188–196
Unsaponifiable matter	not more than 1.8
Titre °C	17–26

The oil is only found in the germ of the seed, which is separated from the endosperm by steaming and mechanical action. The separated germ is then pressed to recover the oil.

Maize oil is a dark golden-yellow colour, which, on standing and cooling, clouds owing to the crystallization of flakes of waxy material derived from the seed coat. The refined oil is a bland yellow product much in demand as a salad oil, for mayonnaise, and as a cooking oil. It is also used to a limited extent as a raw material for margarine. In recent years it has been advocated as a health food. Because of its relatively high essential fatty acid content it is considered that the use of maize oil as the main fat in the diet is a means of controlling the blood cholesterol levels. In spite of the high concentration of unsaturated acids present, it is a fairly stable oil due to the presence of tocopherols (vitamin E) which are antioxidants.

The composition of maize oil is given in table 1.7.

Cotton seed oil. The cotton industry is based upon the growth of varieties of *Gossypium* species which produce a seed coated with fibres. The plant is cultivated as a source of fibres for the textile industry, the cotton seeds being a useful by-product. The availability of cotton seed oil depends on the state of the cotton market for textiles.

Cotton is cultivated in many parts of the world, especially USA, India, Sudan, Egypt, the Soviet Union and Brazil. It has been developed from wild species which were native to sub-tropical areas of America and Asia. The seed is sown in rows about 3 ft (1 m) apart, on well prepared ground. It grows to a height of about 3 ft (1 m) in the next two months and then flowers. The seed pods which follow the flowers burst when ripe, revealing the mass of cotton fibres containing the black seeds. When the bolls are ready they are collected, either by hand or by boll-picking machines. The bolls are fed into machines which remove the husk and then separate the seeds from the fibres. The latter go for textile production, and the seeds to the oil mill. At the mill the oil is separated from the seed by pressure or solvent extraction. A very dark oil containing large amounts of gums and pigment, gossypol, is yielded by the seed. Because of the high level of impurities it is usual for a first refining to be carried out with strong alkali, producing a grade of oil referred to as 'washed' cotton seed oil. The oil is generally sold in this condition and is further refined and deodorized for use as a salad oil, cooking oil, or for incorporation into margarine and shortenings. The high linoleic acid content is responsible for the relatively poor resistance to oxidation so manufacturers of

Table 1.7 Composition and constants of maize oil

Fatty acid composition	%
Palmitic	8–12
Stearic	2–5
Oleic	19–49
Linoleic	34–62
Linolenic	trace
Constants	
Refractive index at 25°C	1.470–1.474
Iodine value	103–128
Saponification value	187–193
Unsaponifiable matter %	up to 2.0
Titre °C	14–20

Fig. 1.14 Cottonseed.

shortenings and margarines have turned to the use of hydrogenation not only to increase the solids content of the oil, but also to increase the stability.

Liquid cotton seed oil will become cloudy at low temperatures especially if stored in a refrigerator. In order to avoid this precipitation, which is regarded as undesirable, the oil is 'winterized'. This is achieved by cooling it to a low temperature, allowing crystallization to take place and then filtering off the solids. Oil processed in this way is filled into bottles for retail sale.

Cotton seed oil is readily detected by a colour reaction, developed by Halphen. The test is carried out by dissolving some of the oil in an equal volume of carbon disulphide containing 1% of sulphur. The mixture is placed in a test-tube and tightly sealed. It is then heated for half an hour in a boiling water bath and during this period cotton seed oil develops a crimson colour. This colour reaction is also given by one or two rarer oils including kapok oil, but other confirmatory tests would establish the difference if necessary. Modern refining techniques reduce the amount of material which is responsible for the colour reaction and well-refined and deodorized oils may fail to give a positive result.

Table 1.8 Composition and constants of cotton seed oil

Fatty acid composition	%
Myristic	0.5–1.5
Palmitic	20–23
Stearic	1–3
Oleic	23–35
Linoleic	42–54
Constants	
Refractive index at 25°C	1.468–1.472
Iodine value	99–113
Saponification value	189–198
Unsaponifiable matter	less than 1.5
Titre °C	30–37

Table 1.8 lists the composition and constants of cotton seed oil.

Peanut oil. Peanut oil, groundnut oil and arachis oil are synonyms for the oil obtained from the leguminous plant *Arachis hypogaea*. Its natural habitat is sub-tropical parts of South America, but it is now grown in many parts of the world–Africa, India, North America and China.

There are two types of this low-growing annual, one characterized by a spreading or running growth, the other a more bushy type. Both types will grow in tropical and subtropical climatic conditions, and even in the warmer parts of temperate regions provided that the growing season includes a minimum of four months at a temperature of at least 80°F (27°C). Ideal conditions are those of the humid tropical or sub-tropical grasslands where there is an annual rainfall of at least 40 in (100 cm), but with a dry period approaching and during harvest time. It can also be grown under irrigation. Growth is favoured by a light, loamy, sandy soil in which the seed, decorticated just before sowing, is drilled. Some types of seed are subject to a dormant period and will not germinate if sown less than three months after harvesting. The plants require four to six months to mature according to variety and type. The first flowers appear about three weeks after sowing and the plant continues to flower for up to ten weeks. The seeds take about six weeks to fully ripen before they are ready to harvest.

An unusual growth habit characterizes the last stages of development of the plant. When flowering has been completed and the seed pods begin to develop, the seed-bearing part of the plant bends over and forces the pods underground to mature and ripen. The pods develop as the familiar peanut, a thin brittle shell containing, usually, two brown-skinned seeds. Most of the nuts were shipped from the growing areas for processing in the user countries, but now it is more common for the oil to be produced in the country where the nuts are grown and then shipped in bulk. The oil is obtained from the nuts by expelling or pressing, and the residual cake may be solvent extracted.

After obtaining the oil it is necessary to remove gums and phosphatides before refining. This operation is frequently carried out at the oil mill. Hot water is added to the oil in order to hydrate the phosphatides which can then be removed with the water by centrifuging. Recovery of the phosphatides is achieved by evaporation of the water to produce 'groundnut lecithin'. The de-gummed oil varies in colour from a clear golden-yellow to a dark brownish-yellow. The colour is greatly improved after refining and bleaching. The refined and deodorized oil is a very good edible oil, which, because of

Fig. 1.15 Groundnuts.

the absence of linolenic acid and a low linoleic acid content, is fairly resistant to oxidation and the development of reversion flavours.

Groundnut oil is used widely in every application where edible fats are required. As a liquid oil it is used as a salad oil, for mayonnaise, frying, and as an ingredient in margarines, especially the soft product. It is hydrogenated to various degrees and as such can be used as part of the hard fat in margarines, shortening, and many other applications. The liquid oil suffers from the disadvantage that at low temperatures it clouds and gels. Attempts to remove the constituents that cause the clouding have not been successful, because the gel structure that also forms at low temperatures prevents filtration. Cold oil in bottles can therefore be unattractive to the housewife although the cloudiness and gel will disappear in a warm kitchen.

The fatty acid composition and constants are given in table 1.9. One of the distinctive characteristics of groundnut oil is the small quantity of the high molecular weight saturated acid, arachidic acid; a qualitative test (Belliers test) and a quantititive one due to Evers, depend on the insolubility of the acid in 70% alcohol. In both tests the oil is

Table 1.9 Composition and constants of groundnut oil

Fatty acid composition		%
Palmitic	C16:0	6–9
Stearic	C18:0	3–6
Oleic	C18:1	53–71
Linoleic	C18:2	13–27
Arachidic	C20:0	2–4
Behenic	C22:0	1–3
Lignoceric	C24:0	1–3
Constants		
Refractive index at 25°C		1.467–1.470
Iodine value		84–100
Saponification value		188–195
Unsaponifiable matter		less than 1.0
Titre °C		26–32

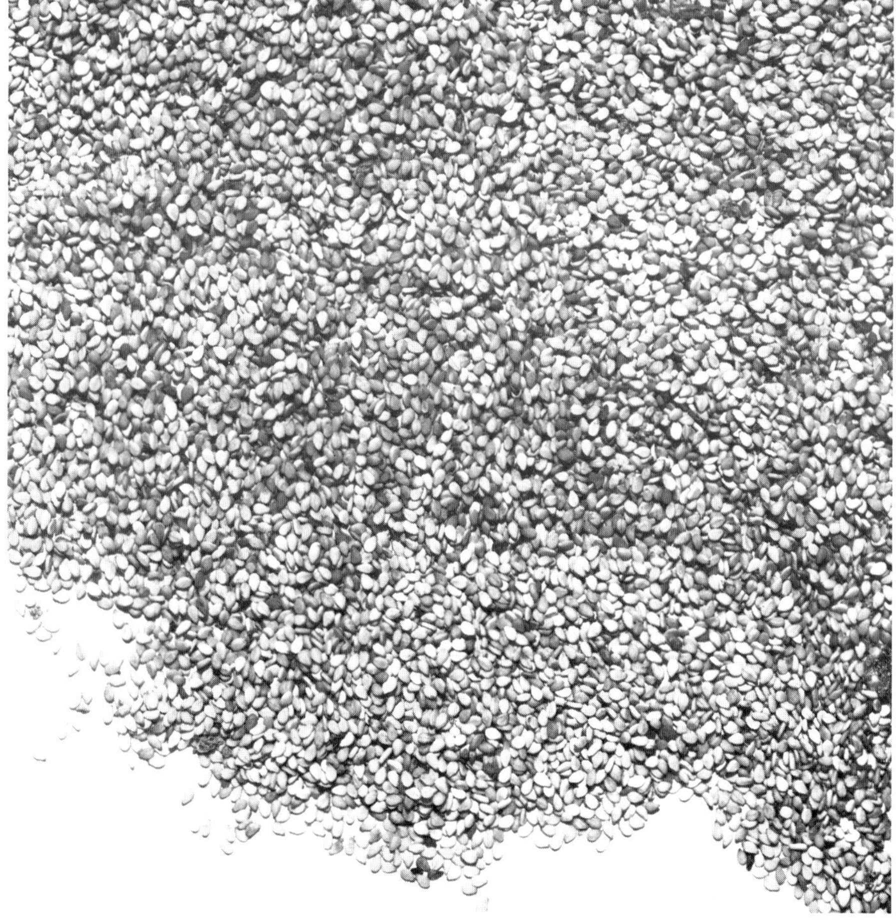

Fig. 1.16 Sesame seed.

saponified with caustic soda, in alcoholic solution, and the soap split with acid. The fatty acids are dissolved in alcohol of fixed strength and temperature in order to allow the arachidic acid to precipitate while retaining the other acids in solution.

Sesame oil. A plant of unknown origin, *Sesamum indicum* which grows well in the warm climates of China, India, Turkey, and many other countries and is the source of sesame oil.

It is an annual with an erect habit. The stems are grooved along their length, and grow to a height between 1 and 7 ft (30–215 cm). Growing best in tropical and sub-tropical climates it requires an equable temperature of about 80°F (27°C). High rainfall and high humidity are detrimental to the growth of this crop, although it needs either about 20 in (50 cm) of rain or irrigation. While some varieties are unaffected by the length of day, others will only grow and flower successfully if the day is less than twelve hours.

The seed, which is sown either by hand or drilling machine, will germinate and start to grow only when the soil temperature is above 70°F (21°C). It normally takes three to four months to reach maturity. The crop is cut, either by hand or with binders, when the lowest pods on the stems are just ready to open. After the seed is harvested, the oil is obtained by pressing or expelling, and the residual cake is sometimes extracted with solvent.

Refining produces a stable yellow oil which is acceptable for all edible purposes, including use in margarine, both as a liquid oil and hydrogenated. Sesame oil contains phenolic substances sesamin and sesamoline, which are not removed from the oil by refining. These permit identification of the oil by the Baudouin test, and its modification, the Villavecchie test, in which the oil is shaken with hydrochloric acid containing 1% of sucrose. On standing, sesame oil develops an intense crimson colour. This distinctive test led a number of authorities to stipulate that all margarines must contain at least 10% sesame oil as a marker so as to make it easy to detect adulteration of butter with margarine. These regulations have been withdrawn in most countries.

The composition and constants of sesame oil are similar to those of cottonseed oil and are given in table 1.10.

Sunflower seed oil. The flowering plant, *Helianthus annuus*, which is indigenous to Mexico, has been extensively cultivated in Eastern Europe, Turkey, Argentina and South Africa. It requires a hot summer and fertile soil to produce a good crop, and thrives in the tropics at higher altitudes.

There are a number of different varieties of sunflower and the growth habit,

Table 1.10 Composition and constants of sesame oil

Fatty acid composition	%
Palmitic	7–9
Stearic	4–5
Oleic	48–49
Linoleic	35–47
Constants	
Refractive index at 25°C	1.470–1.474
Iodine value	103–116
Saponification value	188–195
Unsaponifiable matter	less than 1.8
Titre °C	20–25

particularly the height, varies considerably. The common large variety grows to between 5 and 8 ft (150–240 cm) and bears a flower head of up to 14 in diameter (35 cm); the dwarf strains only grow to about 4 ft (120 cm). The best results are given if the annual rainfall is at least 30 in (75 cm), although the dwarf varieties can grow with as little as 10 in (25 cm) but the rain must be spread over the growing period.

Seed may be broadcast by hand, or with a maize planter or similar suitable drilling machine. If the seed is sown mechanically about 12 in (30 cm) are left between rows.

It takes about four months for the plant to grow from germination to maturity; dwarf varieties mature rather faster and may be ready to harvest in as little as two and a half months. As the head develops, each floret in the composite flower produces one seed. When the seeds are ripe the head is cut down either by hand or by a reaper/binder. The heads are then threshed to obtain the seeds. At this stage of ripeness the seeds still have a high moisture content and if the seed is to be stored the moisture must be reduced to under 10%.

Oil is obtained from the seed either by pressing, or by solvent extraction. As the seeds may contain 20 to 40% oil it may be necessary for profitable operation, to press the richer seeds to bring the oil content down before solvent extraction.

The composition and constants of the oil are given in table 1.11. It should be noted that the composition of sunflower seed oil is affected, more than most oil seeds, by conditions of growth. There is no test which is specific for this oil.

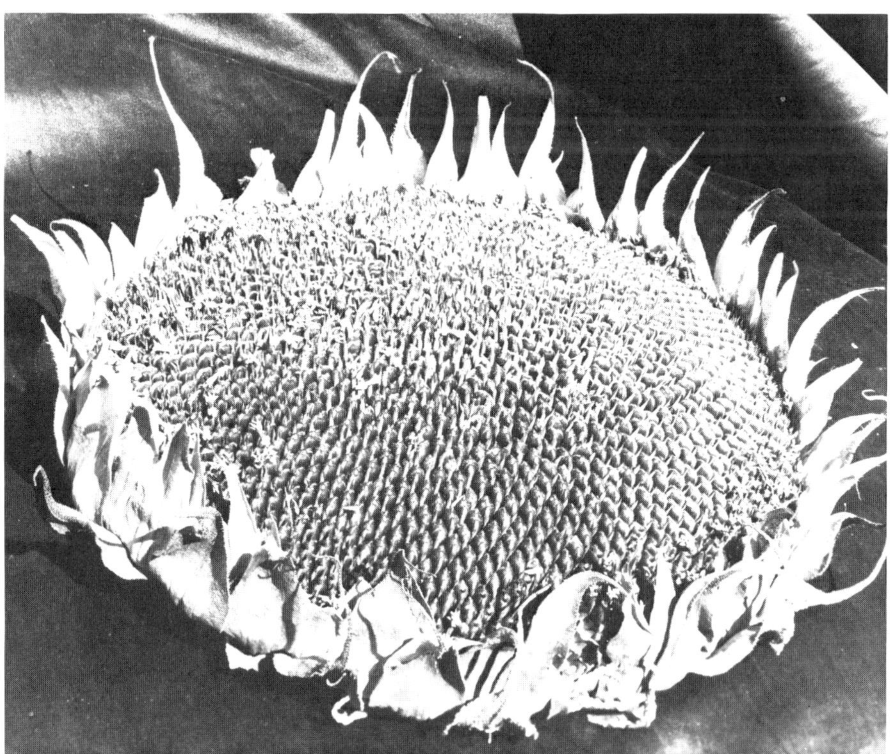

Fig. 1.17 Head of a sunflower showing the arrangement of seeds. The head can reach 14 in diameter.

Fig. 1.18 Sunflower seeds.

Table 1.11   Composition and constants of sunflower seed oil

Fatty acid composition	
Palmitic	3–6
Stearic	1–3
Oleic	14–43
Linoleic	44–75
Arachidic	0.5–1.0
Constants	
Refractive index at 25°C	1.472–1.474
Iodine value	125–136
Saponification value	188–194
Unsaponifiable matter	not more than 1.5
Titre °C	16–20

Refined sunflower seed oil is a pale yellow colour with a pleasant flavour and with a fairly good resistance to oxidation. It is used as a salad oil, for frying, and as a component of margarines and shortenings.

Sunflower seed oil has a rather higher linoleic acid content than corn oil and is therefore, like corn oil, advocated as a suitable dietary fat in conditions of heart disease and high blood cholesterol levels.

1.3.4 Linoleic/linoleic oils

Soyabean oil. The oil variously referred to as soyabean oil, soy bean oil, bean oil or soya oil, is obtained from the seeds of *Glycine max*. This leguminous plant is a native of China, but is now cultivated in many countries, especially the United States and Canada.

Although it is essentially a sub-tropical annual, it is cultivated in tropical regions and the area of cultivation extends as far north as latitude 52°. The limitations on area of growth are imposed by the susceptibility of seedlings, not only to frost damage, but also to poor and slow growth if night temperatures fall below about 55°F (13°C). The seed can be sown broadcast or in drills. Being leguminous, it will benefit by sowing in soil which contains nitrogen fixation bacteria. The plants will grow to 2 to 6 ft (60–180 cm) to form an erect and branched plant if the growing conditions are satisfactory. It is sensitive to length of day, each variety having an optimum day length for blooming and seed setting.

Variety and conditions of climate also affect the growing period, which may range from 11 to 15 weeks. After the flowers have set, narrow hairy pods begin to develop and grow till they are about 2 to 4 in (5–10 cm) long. As the seeds approach maturity, the leaves begin to yellow and fall so that if left till all the pods have matured most of the

Fig. 1.19 Soya bean plants ready for harvesting.

Fig. 1.20   Soyabeans.

leaves will have fallen. Harvesting is carried out either by hand or combine harvesters. The beans are separated from the plant in the harvester and then, if necessary, dried to below 12% moisture, to ensure that they remain sound in storage. The oil is won from the seed by mechanical means (presses or expellers), or by solvent extraction, the latter being the method of choice if maximum yields of oil are sought. Soyabeans are particularly suited for solvent extraction as in comparison with most of the oil seeds considered so far, the soyabean has a low oil content (about 16–18%), which makes pre-extraction preparation easy.

Crude soyabean oil contains relatively high concentrations of phosphatides; samples with 3% have been recorded. In order to separate the oil and phosphatides, the crude oil is mixed with hot water to hydrate the phosphatides and the mixture is centrifuged to separate the clear oil and the phosphatide sludge. Commercial soya lecithin is prepared by drying the sludge under vacuum. This product contains 50–60% phosphatides and the balance is soyabean oil. It is used as an additive in margarine to impart anti-spattering properties, and in chocolate manufacture to make a smoother product.

The oil itself is yellow, often with a greenish cast due to the presence of chlorophyll. It is a difficult oil to refine well and to completely eliminate the typical beany taste. The difficulties are probably associated with the linolenic acid which is present in the oil to the extent of about 10%. Linolenic acid is very susceptible to oxidation, and the oxidation products give rise to 'off' flavours variously described as 'painty', 'beany', or 'fishy'. Careful refining and handling can produce a good bland oil with the minimum development of reversion flavours.

Refined and deodorized soyabean oil is used for all edible purposes but it does not give as good results as the other liquid oils. The explanation is again that resistance to oxidation is poor owing to the linolenic acid present. The addition of antioxidants helps to improve the stability of the oil. Stabilization is best achieved by hydrogenation. Unfortunately, if soyabean oil is hydrogenated by conventional techniques it is converted from a liquid product to a solid one by the time the linolenic acid has all reacted. There are methods available, in the development stage, which enable hydrogenation to be controlled so that no saturated acids are formed but all the linolenic acid is converted to either linoleic or oleic acid. When this technique is made commercially viable, the production of stabilized liquid soyabean oil will have been achieved. Meanwhile, large quantities of soyabean oil are hydrogenated, to a varying extent, to form the basis of margarine, shortening, and various other cooking fats. The composition and constants of soyabean oil are given in table 1.12.

1.3.5 Erucic acid oils

a. Rape-seed oil. Alternatively known as colza oil, true rape-seed oil is isolated from the seeds of the cultivated Cruciferous plants, *Brassica campestris* and *Brassica napus*. Varieties of *Brassica campestris* which grow wild in the Black Sea region are the source of a very similar oil known as ravison. Two other related species, *B. alba* and *B. nigra* yield an oil seed from which is extracted white and black mustard seed oil, a product similar again to rape-seed oil. In table 1.13 the composition and constants of these members of the erucic acid group of oils are given for comparison.

Rape-seed plants attain a height between 18 and 30 in (45–75 cm) and can be grown either as annuals or biennials. The two types are grown according to the prevailing climate which can range between northern temperate regions and sub-tropical areas (as a cool season crop). The annual varieties of *B. campestris* are cultivated in those areas

Table 1.12   Composition and constants of soyabean oil

Fatty acid composition	%
Palmitic	7–11
Stearic	2–6
Oleic	15–33
Linoleic	43–56
Linolenic	5–11
Constants	
Refractive index at 25°C	1.470–1.476
Iodine value	120–141
Saponification value	189–195
Unsaponifiable matter	less than 1.5
Titre °C	21–23

where there is only a short frost-free period. The seed is spring-sown by a grain drilling machine in light soils, or broadcast followed by harrowing if the soils are heavy, especially in high rainfall areas. Biennial varieties of *B. napus* can be autumn sown in the milder climates.

The time taken to grow from germination to maturity is very variable and reports indicate that maturity can be reached from 85 days up to 160 days.

As ripening takes place the leaves fall and the stems and seed pods start to change colour. As might be expected of a crop that can be grown under many different climatic conditions rape-seed and associate species are grown in many different areas: Scandinavia, Canada, India, Poland, the Black Sea area, China and France.

After harvesting, the seed, which though small, contains 40–45% of oil, is pressed in expellers to obtain the oil. The oil has an unpleasant taste and odour due to the presence of sulphur compounds. The dark-yellow crude oil can be refined, bleached and deodorized to give a pale yellow bland oil which is used a a salad oil, cooking oil, and in margarine and shortenings. It is also hydrogenated to give solid fats for blending into margarines and shortenings. Poorer quality oils are used industrially as lubricants and formerly were extensively used as a fuel in lamps for illumination.

There has been some concern in nutrition circles over possible effects on health of the high molecular weight erucic acid (see section 1.16). The Canadian agricultural authorities have taken this concern seriously and have endeavoured to produce a mutant of rape-seed which contains very little erucic acid. This research had reached such a stage in 1971 that the whole of the Canadian 1972 crop was sown with low erucic acid rape-seed (L.E.A.R.). France and other countries are also carrying out similar research. It must be realized that the production of L.E.A.R. oil is of interest only for edible purposes; most industrial uses of rape-seed oil depend on the erucic acid content.

1.3.6 Unusual oils of commercial importance

a. Babassu oil has been briefly mentioned in the section dealing with lauric acid oils. It is produced almost exclusively in Brazil from the wild palms of two

Table 1.13 Composition and constants of rape-seed, ravison and mustard seed oil

Fatty acid composition		Rape-seed %	Ravison %	Mustard seed %
Palmitic	C16:0	1.9–2.8	4.3	1.5
Stearic	C18:0	0.4–3.5	2.1	0.4
Oleic	C18:1	12.0–24.0	15.5	22.0
Linoleic	C18:2	12.0–16.0	20.9	14.2
Linolenic	C18:3	5.0–10.0	9.9	6.8
Eicosenoic	C20:1	3.0–12.0	4.1	7.0
Erucic	C22:1	35.0–55.0	42.2	44.2
Docosadienoic	C22:2	0.6–2.0	1.0	—
Constants				
Refractive index at 25°C		1.470–1.474	1.470–1.474	1.4678–1.4735
Iodine value		97–108	108.5	102
Saponification value		170–180	178.0	173
Unsaponifiable matter		not more than 1.5	not more than 1	0.7–1.5
Titre °C		11–15	11–15	6–10

species – *Orbignya martiana*, which grows in the rain forest areas of Brazil, and *Orbignya oleifeba*, which is found in the drier forest areas of the same country. Like the oil palm, they are monocotyledons and are similar in growth habit, reaching a height of 60 to 70 ft (18–21 m). The slender trunk is crowned with typical palm leaves, each of which reaches a length of about 20 ft (6 m). There is no record of the palms being cultivated in plantations; the trees are self-seeding and there is not much information available concerning the conditions for their development.

It is recorded that the palms are probably about ten years old before they start to flower and produce fruit. Once fruiting starts, they appear in large bunches with several hundred fruits in each bunch. They develop rather like the coconut, with a very thick, hard outer shell. The proportion of kernel is rather less than 40% of the overall weight of the fruit, but when it has been separated it has a high oil content round about 65–70%, the oil being similar to palm kernel oil. As distinct from the oil palm, there is no oily fruit pulp surrounding the kernel and therefore no fruit oil. The fruits ripen between July and November and are harvested by collecting the fallen fruit. Under these conditions the collection and processing of the nuts requires much labour, and the cracking of the very hard, thick shell requires a lot of force. The production is mainly local and although the potential may be good, no large scale development of the product has yet taken place. The oil is used mainly in South America as a replacement of palm kernel oil. Table 1.14 gives the constants and composition of the oil.

b. Safflower oil is obtained from the seeds of *Carthamus tinctorius*, a plant of the same family as the sunflower. It is therefore not surprising that the seed and the oil from it are similar to those obtained from sunflowers. The safflower is not known in the wild state, but is cultivated in areas of India, Egypt, and North America. The plant is an annual which is grown as a winter crop. It is sown very much like cereals, developing well in the cold, short days of the winter. Its growing period is about four months and it reaches a height of about 4 ft (1–1.5 m). The leaves are spiny and the developed plant is perhaps more like a thistle than a sunflower.

Table 1.14   Composition and constants of babassu oil

Fatty acid composition	%
Caprylic	4–5
Capric	6–7
Lauric	44–45
Myristic	15–16
Palmitic	6–8
Stearic	3–5
Oleic	12–16
Linoleic	2–3
Constants	
Refractive index at 40°C	1.4495–1.4506
Iodine value	14–16
Saponification value	247–251
Unsaponifiable matter	not more than 0.5
Titre °C	23
Reichert value	6.1
Polenske value	11.4

Fig. 1.21 Safflower.

Although most of the oil is used locally, its composition suggests that it may become an important health food material. It has a very high linoleic acid content, probably the highest of any commercial oil. With the emphasis that is at present being placed on poly-unsaturated acids, safflower oil is obviously an oil which will have an increased attraction for use in manufactured foods where this high essential fatty acid content would be valuable. The constants and composition of safflower oil are given in table 1.15.

c. Sheanut oil, or as it is sometimes called, Sheanut butter, has some interesting properties, but limited application. It is obtained from the seed of *Butyrospermum parkii,* a deciduous tree which grows to about 40 ft (12 m). Its natural habitat is the drier parts of the equatorial belt of Central Africa where the weather conditions are

Table 1.15   Composition and constants of safflower oil

Fatty acid composition	%
Palmitic	2–5
Stearic	2–5
Oleic	13–35
Linoleic	60–75
Traces of myristic, arachidic and linolenic acids have been reported.	
Constants	
Refractive index at 40°C	1.4679–1.4698
Iodine value	140–150
Saponification value	186–195
Unsaponifiable matter	not more than 1.5
Titre°C	15–18

unfavourable for the growth of the oil palm. It is cultivated on farms but is very slow in yielding fruit, which do not usually appear for at least ten years. Because of the root structure of the tree, the seed is generally planted in the field where it is to remain, rather than in nurseries for transplanting. The fruits which develop grow to about 2 in long (5 cm), with a fleshy pulp covering the seed. As the fruits ripen they fall from the tree and after collection are left for the pulp to decompose, releasing the kernels, which are then dried. After removing the shell, the kernel is extracted to yield about 60% of a greenish-brown fat. As with most other fats, the colour can be removed during processing, but even after bleaching it often has a faint greenish-grey appearance. The crude oil is difficult to refine owing to the presence of 4 to 9% of an unsaponifiable material, which tends to produce emulsions. When refined, it gives a tough, plastic fat, whose unusual behaviour is of considerable interest in the production of special margarines for puff pastry, especially Danish pastry types. Table 1.16 gives the constants and composition of this fat.

d. Tea-seed oil. There are a number of varieties of *Camellia* which are cultivated in order to obtain leaves for tea production. Although the seeds of these varieties contain oil, they are not an important source of the oil which is mainly obtained from the seeds of a variety *Camellia sasanqua* (also called *Thea sasanqua*). The trees or bushes of the tea plant require sub-tropical conditions for growth. They prefer fairly uniform temperatures with no risk of frost, and plenty of rainfall without any dry season. The principal growing region of *Camellia sasanqua* is in China, although some is grown in Japan.

The seed contains a kernel which yields about 60% of a yellowish-brown oil. Like many vegetable oils, tea-seed oil is not really edible in the crude state, but conventional refining produces a bland yellow oil with many of the characteristics of olive oil. Its constants are so similar to those of olive oil that distinction, on these alone, is impossible. The titre is one characteristic which shows a difference, but in mixtures of tea-seed and olive oils this too becomes doubtful. Tea-seed oil has been commonly used for the adulteration of olive oil, although it is a good edible oil in its own right. It can be used satisfactorily in salad oils and frying oils, and would be suitable for margarine production were it more readily available, but the supply is unreliable and restricted. The constants and composition are given in table 1.17.

Table 1.16   Composition and constants of sheanut oil

Fatty acid composition	%
Palmitic	6–8
Stearic	36–41
Oleic	49–50
Linoleic	4–5
Constants.	
Refractive index at 40°C	1.4630–1.4670
Iodine value	52–66
Saponification value	178–198
Unsaponifiable matter	2–11
Titre °C	48–54

The high unsaponifiable matter (usually around 7%) is a distinctive feature of this oil.

Table 1.17   Composition and constants of tea-seed oil

Fatty acid composition	%
Palmitic	7–16
Stearic	1
Oleic	69–84
Linoleic	8–14
Constant	
Refractive index at 40°C	1.4600–1.4640
Iodine value	80–90
Saponification value	188–196
Unsaponifiable matter	not more than 1.0
Titre °C	13–18

The most useful test to distinguish tea-seed oil is due to Fitelson. Details are as follows.

A mixture of 0.8 ml of acetic anhydride, 1.5 ml of chloroform and 0.2 ml of sulphuric acid are cooled to 5°C. To this mixture seven drops of oil are added and the contents of the tube mixed. If the mixture does not clear add drops of acetic anhydride until it does. Stand at 5°C for five minutes, add 10 ml ether and mix. Tea-seed oil gives a red colour which reaches maximum intensity in about a minute and then fades. Some samples of olive oil may give a slight reaction and this makes it impossible to positively identify mixtures containing less than 15 to 20% tea-seed oil.

1.4 ANIMAL AND FISH OILS AND FATS

It is probable that the use of animal fat in food preceeds the use of any of the vegetable oils. Fats from land animal tissues are characterized by having a high concentration of the solids acids, palmitic and stearic. To a great extent also they are by-product fats, made available as a result of the preparation of meat either for sale as meat, or for the manufacture of processed meat products. Provided that the animal fats are well produced they can be used without any further processing. Milk fats are mainly consumed as butter and cream. These are considered later in chapter 4. The other part of the group, the marine oils which include whale oil and fish oils, is very different. The fatty acids in these oils are generally very unsaturated and contain a high proportion of long chain acids. These oils must be hydrogenated before they can be used in edible products. The production of oil is sometimes the major purpose in this section of the fishing industry, fish meal being the by-product; at other times the position is reversed, according to the demands of the world market.

1.4.1 Lard

A very high proportion of the animal fats produced is in the form of lard, a fat obtained by rendering the fatty tissues of pigs (hogs).

Pigs are raised in most countries and are an important source of meat products. As the processing of meat concentrated in large packing factories, the large-scale production of lard also became established. One of the largest producers of lard is the USA, but

Sweden, Poland, France and many other European countries have now become established in this field.

After slaughter of the animals the fat is removed from various edible tissues, especially round the kidneys and from the back and under the skin. The fatty tissue is chilled immediately, usually in iced water, in order to remove body heat. If the tissues are left to cool slowly they are subject to breakdown by enzymes and attack by bacteria. Both of these forms of deterioration will reduce the quality of the fat obtained. After the carcase has been chilled and is cut up for use in various ways, the surplus fat is removed and this also is used for lard production.

The terms used to describe various types of lard offered on the market may have a different meaning from the common usage of the words, and can refer either to the method of isolation of the fat, the source of the fatty tissue, or the country of origin. Some of the more important terms will be explained.

Leaf lard is obtained by dry-rendering solely the internal 'leaf fat' of the animal. This is the fat which lies above the kidneys. It is a firm lard with a distinctive flavour resulting from the dry-rendering process.

Neutral lard is produced by low temperature wet-rendering of selected tissues. It has a very mild, neutral flavour, hence its description.

Prime steam lard is the description given to lard obtained by a wet-rendering process in which the tissue is heated with steam in closed vessels.

Some official distinction has been made between lard and rendered pork fat. The United States Department of Agriculture has defined lard as 'the fat rendered from fresh, clean, sound fatty tissue from hogs in good health etc.', but specifies certain tissues which must not be used for lard production. Rendered pork fat is prepared from most of the tissues which are excluded from use for making lard, but can of course also include those tissues which are permitted for lard.

Details of methods of rendering are given in section 1.5.2 so here it is sufficient to say that there are two main methods: dry rendering, in which the tissues are heated in open pans until the fat melts and separates from the dehydrated tissue, and wet rendering in which the tissues are steam heated, usually under pressure in a closed vessel.

The composition and characteristics of lard are very variable. The breed of animal, the feed, and, where appropriate, the region of the body used for rendering, all play their part in the variations. Leaf lard is harder than the lard produced from the other tissues and it is not uncommon for the iodine value of leaf lard to be as much as 10 units lower than that of the lard from other tissues of the same animal. Controlled feeding tests have shown that some of the fat present in the feed is directly deposited in the body tissue of the animal. Reports show that if the diet contains less than 5% of fat, the pig produces fat of constant composition, while with increased fat in the diet the lard will contain quantities of dietary fat. This makes the interpretation of analytical data extremely difficult. One way round these difficulties is to set a specification which fits the requirements of the purchaser and advise producers that only lard fulfilling the specification will be accepted. Table 1.18 lists the characteristics and composition of a number of lards, and table 1.19 gives the A.O.C.S. recommended standard.

Adulteration of lard with vegetable oils is unlikely because the price of lard is usually lower than that of the oil. Specific tests for oils such as cottonseed oil or sesame oil may give a slight positive reaction owing to the effects of feed. The most common adulterant is

Table 1.18 Composition and constants of some lards

Fatty acids present	%	%	%	%	%
Myristic	1.3	2	1	trace	1
Palmitic	28.3	26	17	16	14
Palmitoleic	2.7	–	–	–	–
Stearic	11.9	12	10	7	26
Oleic	47.5	59	40	57	32
Linoleic	6.0	1	32	20	27
Linolenic	–	–	trace	–	–
Constants					
Iodine value	59.8	52.6	90.7	84.1	77.4
Saponification value	196				
Unsaponifiable matter	0.2				
Titre °C	34–40				

Table 1.19 A.O.C.S. recommended standard for lard

Refractive index at 40°C	1.4559–1.4607
Iodine value	46–70
Saponification value	195–202
Unsaponifiable matter	not over 1.0
Titre °C	36–42

tallow which normally is cheaper than lard. A number of tests have been used to try to detect the presence of beef fat in lard, the best known being the Bömer test. This depends on the difference between the melting points of the glycerides and the fatty acids prepared from them. Some workers claim that this test can give a good measurement of the degree of adulteration, but others regard the test with suspicion. The most reliable method is one made possible with the general introduction of gas chromatography. It has been observed that ruminant fats contain small quantities of branched chain fatty acids which are only present at very much lower levels in genuine lard. Using the gas chromatographic methods these differences can be detected and measured.

Lard is used, without any treatment, as a frying and cooking fat. It can be processed to produce a tasteless fat for use in margarines, shortenings and bakery fats.

The stability of lard is rather poor so that much attention has been paid to improving the keeping properties by the addition of antioxidants. The addition of a few parts of antioxidant per million parts of lard is successful in improving the stability.

The use of lard as a bakery fat is restricted by its very poor creaming properties. To some extent this weakness can be overcome by changing the glyceride structure through a process of interesterification (see section 1.6.6.)

1.4.2 Tallow

The fat obtained from the body tissues of cattle and sheep is given the general name of tallow and is then further identified by the animal from which it was obtained. Thus, from the domestic cattle we obtain beef tallow, and from sheep, mutton tallow. The depot fat of other domesticated ruminants such as the buffalo is also referred to as tallow, but these sources are only of local interest. Much of the tallow produced is now used for non-edible purposes, although beef-producing countries like Australia use tallows in the production of margarines and bakery fats.

Isolation of the fatty tissue from which tallows are to be produced follows a similar pattern to that for lard. Immediately after slaughter of the animal the fatty tissues are removed from the kidneys, caul, and round the heart. They are quickly cooled and then rendered. Additional quantities of fat are removed from the carcase when it is being cut for meat preparation. This fat also is rendered to produce tallows. As for lard production, only the fat taken from edible parts of the animal is used to prepare edible tallow. The various grades of beef tallow are distinguished as described below.

a. Premier jus is considered to be the top quality beef fat. It is prepared from the kidney fat by rendering at a low temperature–about 120°F (50°C). It should be clear and free from suspended matter and it is ready immediately for use in making cakes, biscuits and puff pastry, to which it contributes a distinctive flavour.

b. Oleo oil and oleo stearin are two fractions produced from premier jus or beef tallow. The molten tallow is allowed to cool slowly to produce a sloppy, semi-granular mass. This partly crystallized material is held at a pre-determined temperature for several days in order to stabilize the crystals and then it is transferred to bags and pressed in a hydraulic press. The liquid phase, the oleo oil, is pressed out, and the solid portion, the stearin, is compacted in the bags. Both the yield of liquid and its characteristics are dependent on the temperature at which the mixture is allowed to crystallize. The control of temperature is dictated by the standards set for the products, the ambient temperature, and economic factors.

c. Beef dripping is the tallow derived from rendering unspecified edible fatty tissue. It is usually softer than premier jus.

The characteristics of tallows given in table 1.20 are average figures which are subject to considerable variation due to breed and feed of the animals. It can be seen that the constants of beef and mutton tallow are so close that it is impossible to differentiate using these figures. Often, the odour gives some indication of the presence of mutton tallow. If the fatty acids are examined by gas chromatography, mutton tallow can be identified by the higher linoleic acid level.

Table 1.20   Characteristics of various tallows

	Mutton tallow	Beef tallow	Premier jus	Oleo oil	Oleo stearin
Refractive index at 40°C	1.4536–1.4581	1.4573–1.4586	1.4573–1.4580	1.4576–1.4580	1.4569–1.4576
Iodine value	35–41	35–48	38–45	44–48	18–25
Saponification value	192–197	193–202	195–200	198–202	192–197
Melting point, °C	44–50	47–50	47–49	32–34	50–54
Titre, °C	43–49	40–43	42–44	40–42	48–50

When the fatty tissues are properly handled the free fatty acid of the tallow is very low and the fat can be used for many purposes without any further treatment other than chilling to solidify it. In some areas of the UK dripping is used extensively for commercial frying operations as well as for domestic use. Untreated tallow has a distinctive flavour and this is carried through into the cooked products. Its usefulness can be extended by refining and deodorizing, after which a bland, firm fat is produced. Refined tallows have a wider application; some of the cheap fats sold under proprietary names are nothing but tallow.

Premier jus is used in some countries as the fat of choice in producing biscuits and cakes. Australia in particular uses a high proportion of tallow in margarine production. The desirable properties of the product are achieved by blending the tallow with either oleo oil or oleo stearin to give the correct melting structure for the particular use. High levels of oleo stearin are used to produce puff pastry fats which need a high melting fraction to give the open texture.

Tallows are distinguished by the fatty acid composition in which the saturated acid content is high, palmitic acid accounting for about 30% of the total and stearic acid 20 to 30%. Almost all of the balance is oleic acid, these three acids accounting for about 90% of the total fatty acid composition. Table 1.21 gives the composition of beef and mutton tallows.

1.4.3 Whale oil

A few years ago whale oil was the most important marine oil, but this is no longer so. The decline in availability of whale oil is due to over-fishing, with consequent reduction in the numbers of whales.

The history of whaling dates back many years; records exist referring to whaling more than 1000 years ago. While hunting at that time was for general food supplies, later, especially in the Middle Ages, whale hunters sought such species as the Right whale (*Balaena mysticetus*) for their whalebone (the fat being used for non-edible purposes), also the Sperm whale (*Physeter macrocephalus*) from which a non-edible wax was obtained for illuminants and lubricants. A later development was the hunting of whales for their fat. A fillip was given to whaling when it was discovered that whale oil could be hydrogenated to make a solid and more stable fat. Once hydrogenation processes were established, whale-catching for the oil became a major industry. At first the whales were caught by harpooners in small boats and the carcases transported to land stations for processing. The disadvantages of this method were a limitation on the fishing areas, and a poor quality oil due to the time which elapsed between killing and processing. It was quite common for several days to be spent transporting the whale to the land station, during which period the oil deteriorated. The development of the harpoon gun and the factory

Table 1.21   Composition of beef and mutton tallows

Fatty acid composition	Beef tallow %	Mutton tallow %
Myristic	3–7	5
Palmitic	27–30	24–28
Stearic	14–24	25–32
Oleic	38–49	26–47
Linoleic	1–3	3–5

ships changed the position drastically. A fast hunting boat searches out whales which have been sighted by helicopter or radar and uses an efficient harpoon gun, discharging a harpoon with an explosive head, to catch the animal. When the whale is dead, air is pumped into the carcase to prevent it from sinking, and it is towed to the factory ship. The modern floating factory has a sloping slipway, open at the stern. Whale carcases are towed to the opening, hooked onto a winch cable, and dragged up the slipway onto the 'flensing' deck. The whale has a thick layer of subcutaneous fat, the blubber, and it is this layer which is removed from the carcase in the flensing procedure. As the blubber is removed it drops through hatches into the processing equipment on the deck below. When the blubber has been removed, the rest of the carcase is moved on along the deck to an array of saws which cut up the meat and bones. These again pass through hatches to the cookers below where the bones and meat are processed to recover fat and a meat meal. If whale meat is required either for animal or human food it is selected and frozen before the rest of the carcase is broken up for fat extraction. The time interval between catching the whale and completion of the dismembering process is under 24 hours.

The number of whales which can be caught in the Arctic regions and northern waters generally is very small, consequently for the last eighty or so years almost all the whaling has been carried out in Antarctic regions. A number of species have been fished into extinction. The remaining species which are of commercial importance, four in number, are described below.

The Blue whale (*Balaenoptera musculus*), which is the largest mammal known, commonly grows to over 70 ft (22 m) and can attain 100 ft (30 m) and weigh 140 tons.

The Finback whale (*B. physalus*), will grow to 65 to 70 ft (20–22 m).

The Humpback whale (*Megaptera nodosa*), grows to 60 ft (19 m).

The Sei whale (*B. borealis*), also attains a length of about 60 ft (19 m).

Various international conferences have been convened to try to control the catching of whales, putting limits on total catches. The limits were reduced several times in order to prevent the creatures being brought to extinction, but the reduction in catches and numbers available has resulted in several of the major operators withdrawing. The amount of whale oil used now in the UK is very small.

The methods of handling the catch, quick processing and modern techniques of extraction have produced a pale-coloured whale oil with low free fatty acid content and low impurities. It is not used in the liquid state even after refining, but produces satisfactory fats after hydrogenation. The figures in table 1.22 show that whale oil as such is not satisfactory because of the high concentration of unsaturated fatty acids. These acids oxidize very readily and give rise to compounds which have unacceptable flavours.

The degree of unsaturation of the longer chain C20 and C22 acids is very high, averaging four to five double bonds. These highly unsaturated acids must be hydrogenated if the oil is to be made stable. The presence of the longer chain unsaturated acids is used in analysis as an indication of the presence of marine oils.

After hardening whale oil is used in margarines and bakery fats, but it is not satisfactory as a frying fat.

1.4.4 Fish oils

The fish oils can be divided into two groups, the larger one in quantity being fish body oils and the smaller, fish liver oils. Among edible fats, fish liver oils play a very small part, but the body oils have to a great extent taken over the role formerly filled by whale oil.

Table 1.22 Composition and constants of whale oil

Fatty acids present		%
Myristic		6–9
Unsaturated	C14	1–4
Palmitic		15–22
Unsaturated	C16	12–15
Stearic		2–4
Unsaturated	C18	35–45
Unsaturated	C20	11–15
Unsaturated	C22	1–8
Constants		
Refractive index at 40°C		1.4644–1.4714
Iodine value		100–125
Saponification value		185–202
Unsaponifiable matter		1.2–2.0

The species of fish used for fish oil production depends on the area of operation.

In Europe the most important fish for oil production is the herring (*Clupea harengus*). Large numbers of this fish are caught, for the Norwegian fish oil industry, in the North Atlantic and the North Sea; Iceland also has a large herring oil industry. Other species which may be included are mackerel (*Scomber scompor*), sand eel (*Ammochytes* species), and capelin (*Mullatus villosus*). Usually no attempt is made to segregate the individual species and the whole catch is processed, it is therefore uncommon to specify individual oils and they are marketed as fish oil.

At present probably the biggest producer of fish oil is Peru, where the anchovy (*Clupea*) is the principal source of oil. The development of the Peruvian fish meal and oil industry is comparatively recent. In 1958 only 1 600 tonnes were exported and by 1963 this had increased to over 125 000 tonnes.

The fish are prepared for oil extraction by cooking in boiling water for about 15 minutes. This treatment coagulates the protein and breaks down the cell walls, liberating the oil. When the cooked fish is pressed, a mixture of water and oil is forced out as an emulsion which is further heated and centrifuged to separate the components. Oil and 'stick' water are recovered. The latter material contains dissolved protein, and in order to recover the valuable material the liquor is evaporated and the residue is added to the press cake. The oil is clarified to remove suspended solids, ready for sale or further processing. A great deal of this oil is exported to Europe.

In North America a number of other fish – pilchard, sardine and menhaden – are caught and processed as outlined, for recovery of oil. Some of the characteristics are given in table 1.23. There is a substantial production of fish oil in Japan.

Like whale oil, fish oils are not considered acceptable as simple refined oils but are hydrogenated before being used in margarine, cooking fats and bakery products. The hardened fish oils give to the products a good plastic range.

The foregoing description is a guide to the pattern of fish oil usage for edible purposes. The year-to-year pattern changes as the fish population changes and technical knowledge advances. At one period, the two major oils in use were herring oil and menhaden oil, but as already indicated, the Peruvian anchovy oil is now the most important. This latter oil has a very high iodine value.

Table 1.23 Composition and characteristics of fish oils

Type of fatty acid	Herring %	Peruvian anchovy oil %	Menhaden %	Pilchard %
C14:0	7.0	12.4	6.8	5.1
C16:0	11.7	20.5	15.5	14.4
C16:unsat.	11.8	11.1	14.9	11.7
C18:0	0.8	4.1	3.1	3.2
C18:unsat.	19.6	18.0	23.7	17.7
C20:unsat.	25.9	21.7	17.5	17.9
C22:unsat.	21.6	9.7	10.8	13.8
C24:unsat.			4.0	15.2
Constants				
Refractive index at 25°C	1.473–1.478	1.4807–1.4819	1.4774	1.4785–1.4802
Iodine value	140	184	150–185	170–188
Saponification value	180–192	192	191	188–199
Unsaponifiable matter	0.5–1.7	1.0	1.0	0.5–1.25

The high concentration of poly-unsaturated acids in fish oils make them difficult to convert to stable, bland oils. The ease with which these acids oxidize, leading, to 'fishy' and 'painty' tastes, make them unsuitable for use in edible products without modification. Hydrogenation of the liquid oils to harder ones with a consequent reduction in iodine value and increased stability is described in section 1.6.5. The production of suitable edible fats from fish oil is not quite so simple. The process of hydrogenation not only adds hydrogen to the double bonds of the unsaturated acids, but also causes isomerisation, converting liquid acids into solid ones without change of iodine value. Hydrogenation increases the melting point of the oil, and with the highly unsaturated fish oils great care has to be taken if a satisfactory reduction of iodine value is to be achieved without making the melting point too high to be useful. The exact conditions to succeed in this field are closely guarded secrets of those establishments which have produced good, stable fish oils of low melting point. These oils are used to manufacture margarine, shortenings and bakery fats.

1.5 EXTRACTION

In the sections on various sources of edible oils and fats, the conditions of growth of plants, harvesting of the seed and the characteristics of the oils have been discussed, but the extraction of the oils from the source material has been only briefly mentioned. Although there are differences in the handling of different oil seeds, the general principles are the same throughout. Harvesting of the crop at the correct time is necessary if the maximum yield of oil is to be obtained, but the manner in which the materials are then stored and transported before extraction can make all the difference between a good quality oil and one which is unacceptable as a crude edible product.

There are certain crops such as the fruit of the oil palm which must be processed to extract the oil as soon after harvesting as possible. Similarly, with the animal fats speed in chilling the fatty tissue and then extracting as soon as possible is essential if edible quality fats are to be obtained. On the other hand, many of the oil seeds can be stored for a long period if conditions are correct.

Oil seeds are still a living, if dormant, part of a plant and as such they contain a large number of enzymes which will become active if the seed is damaged, or storage conditions are suitable for germination. Another cause of deterioration of the seeds is bacterial or fungal attack. The action of these micro-organisms can cause severe damage to stored material if the storage conditions favour their development. The deterioration which can take place during storage is manifest in two ways:

1. Lipase activity either of the enzymes in the seed, or enzymes secreted by the micro-organisms will hydrolyse the triglycerides of the fat, producing free fatty acids.

2. Breakdown products of non-fatty constituents of the seed may be extracted with the oil and may cause difficulties during refining. The colour also is liable to deteriorate from the same cause and it is often difficult to bleach an oil which has been damaged in this way.

Similarly, fresh animal tissues contain enzymes which will remain active until the tissues are either chilled to low temperatures or heated to destroy them. Details of the handling techniques are given section 1.5.2. As animal fats will not need refining if treated correctly when being isolated, it is of great economic importance that no deterioration should be allowed to take place.

Moisture levels are very important when storing oil seeds. Not only do moulds and bacteria require a high moisture content, but enzymes also only act in an aqueous environment.

We are all familiar with the 'spontaneous combustion' of damp haystacks which is caused by the heat generated by growing bacteria raising the temperature of the hay to ignition point. Similar 'spontaneous heating' of stored oil seeds has been recorded, with severe damage and charring of the seed. Even if the rise in temperature is not so great and no visible damage occurs, heated seeds will give an oil with a high free fatty acid content. Dry storage of dried seed is therefore of paramount importance.

Some seeds are harvested with a fairly low moisture content and under modern conditions suitable driers are available to bring the moisture content of all seeds down to a suitable figure. There is no general level which can be quoted as suitable for the storage of all types of seed. The moisture which must be controlled is that which is present in the non-fatty portion of the seed. This means that a seed with a low oil content can tolerate a higher total moisture than a seed with a high oil content. The variations quoted range from 13% for soyabean to 8% for palm kernels and 6% for copra.

1.5.1 Extraction of vegetable oils and fats

Fats and oils are separated from the cellular material which forms the rest of the seed either by pressure, or by dissolving the oil in solvent, or by the two techniques used in series. The selection of the method is partly dictated by the oil content of the seed and mechanical difficulties in handling. As the seed is enclosed in a shell or husk, it is necessary to break this open and preferable to remove it as not only does it reduce the capacity of the plant for handling the actual oil-bearing part of the seed, but it also absorbs some of the oil and reduces the value of the residual oil-cake. Some form of pre-treatment must therefore be carried out.

has metal sides. Even when the closed box type of press is used to obviate the need for cloths, the amount of labour needed for operating the process is high. Hydraulic presses are therefore almost obsolete, and have been replaced by expellers.

An expeller operates on similar principles to the domestic mincing machine. It comprises a steel screw rotating inside a closed barrel, the material fed in at one end being transported along the barrel and compressed at the other end. The actual design is not so simple. The pressure required to expel the oil from the seed must be applied gradually if the best results are to be obtained. This is achieved by gradually reducing the internal diameter of the barrel, or increasing the thickness of the screw. A further increase in pressure can be achieved by reducing the pitch of the screw, which results in the seed feeding along the beginning of the barrel faster than the latter part. The barrel is made of bars separated by narrow spacers producing narrow slots through which the oil can escape, but which retain all but the finest particles of meal. The preparation of the seed for expeller processing is similar to that for hydraulic pressing except that the particle size of the meal need not be as small. The conditions of cooking may also vary according to the type of seed and type of expeller, the optimum conditions being applied to each case. The meal from the cookers drops into a hopper above the feed worms of the expeller. As the meal feeds along the expeller barrel the oil is forced out and, escaping through the slots, it collects in a trough below the barrel and can then be pumped to filters for removing meal particles, before passing on to storage tanks. The residual solids are forced out at the end of the barrel where they can be collected on a conveyor. The reduction in the labour content in the expeller operation as contrasted with hydraulic presses is obvious, but this is not achieved without cost; the power requirements are higher and so are the maintenance costs. This extra cost is off set not only by saving in labour, but also by the extra yields which are obtained from expellers

c. Extraction. The extraction technique for the recovery of oil from oil seeds is a development of a laboratory method. The standard method of estimating fat in various materials, including oil seeds, is to wash the material continuously with a fat solvent, until no more oil can be recovered. The well-known Soxhlet extractor uses a slightly different method in which the material to be extracted is flooded with solvent, allowed to soak for a time (during which oil is dissolved), and then the solvent is removed and replaced with fresh. The objection to the direct scaling up of either of these methods to production scale is cost. The amount of solvent required for recovery of a fairly small amount of oil is very high. The development of economic systems depends on obtaining higher oil levels in the solvent. When the oil has been dissolved in the solvent and separated from the residual meal, the solvent and oil have to be separated, the solvent for re-use and the oil for sale. Separation is achieved by distilling the solvent and recovering the vapour in a condenser. The amount of energy required to distil the solvent is very great and hence the cost is high. If more oil is present in the solution going to the still, the cost of solvent recovery per ton of oil obtained is lower. In the interests of economy this high oil level in the solution had to be achieved before the process could become profitable, so that some sort of counter-current process was needed.

The seed preparation for extraction processing is similar to preparation for pressing or expelling, although the details vary with the type of seed and operating conditions. Purification, removal of stones, rubbish and metal, remains the same for extraction processes as for expelling. The removal of lint and shells is also necessary so the dehulling equipment is not changed, but milling and cooking conditions are different.

As the rate of solvent penetration into the particles of meal is dependent upon the thickness of the particles, it is desirable to prepare the seed in the form of flakes rather than irregular small fragments. Also, flakes permit better percolation of the solvent than a finer, powdery material. The moisture content of the flake is more critical in solvent work than when pressure is used to squeeze the oil out. If the flake is too damp, the solvent will be unable to penetrate it, so that it is necessary to introduce a drying process, called crisping. This part of the process not only makes the flakes ready for extraction, but also helps to prevent them from compacting.

Cake from expellers or hydraulic presses can also be further de-fatted by solvent extraction. Before this can be done, the cake must be broken and milled to make the extraction more thorough.

Extraction plants can be divided into groups according to whether they are used for batch or continuous processing. There are many different designs which are either in commercial operation or have been proposed. Space does not permit a full survey of them all, and only the more important types will be described.

One batch process consists of a series of cylindrical vessels filled with prepared oil seed material which is then washed with solvent. The solvent, sprayed in at the top of the vessel, percolates down to the bottom, becoming enriched with oil in the process. If this solution is pumped back to the top of the tower and allowed to percolate again the oil content rises until an equilibrium is reached. At this stage no more oil can be extracted from the seed material, but by pumping the solution to the top of a second vessel full of fresh seed material, the process can be repeated and the concentration of oil in the solution improved. Meanwhile, the first vessel is re-washed with fresh solvent, and the process continued (using several vessels) so as to reach a maximum oil concentration in the solvent and a minimum in the seed material residues. This type of multiple batch system has been developed with an optimum number of vessels in the sequence. As soon as the residual seed material in each vessel has yielded all the oil which could be extracted, the solvent residues in the meal are recovered by steaming. The steam flows through the meal, converting the solvent to vapour which passes, with the steam, into a condenser. The total vapours condense and the solvent is recovered and returned to the system. The meal should be solvent free. Similarly, the solution of oil is transferred to a still in which the solvent is vaporized, condensed, and returned to storage; the oil left in the still should be solvent free. Theoretically, the solvent can be recovered for indefinite use, but this degree of efficiency is never reached and some solvent losses are experienced.

A simpler batch process consists of a cylindrical drum rotating with its long axis horizontal. It is divided horizontally into two compartments, a larger one holding the meal and a smaller one the solvent. The dividing screen is covered with a filter cloth to hold back the particles of meal. As the extractor slowly rotates, the solvent enters the main compartment and extracts the fat. After a number of revolutions the cylinder is stopped with the smaller compartment underneath, so that the solution can filter through and be removed. Fresh solvent can be pumped into the solvent chamber and the process repeated until the optimum oil extraction has been achieved. As in the case of the vertical extractors, rotary extractors can be linked in series and operated in a similar fashion.

Following the batch processes using a number of vessels in series, continuous extractors, such as the Hansa-Mülhe, were developed. The Hansa-Mühle extractor is built inside a tower. It consists of a series of perforated boxes or baskets, fixed to a continuous vertical chain. At the top of the tower the first basket is filled with flaked seed, after which it commences to descend. Almost at once the basket passes under a spray of solvent which washes through the flakes, and dripping through the perforated base, passes into the next lower basket. As the baskets descend the tower the contents are continuously washed with solvent from the supply above, passing through all the baskets to the bottom of the tower. The solvent has then dissolved as much oil as possible in its passage through the flakes and is pumped to the still for recovery. The flakes reaching the bottom of the tower still contain an appreciable oil content so the washing is continued on the other side of the tower. Solvent is sprayed into the top basket again, but this time the baskets are rising up the tower, meeting solvent with less oil in it and therefore giving up more oil until at the top position the almost exhausted flakes are being washed with oil-free solvent. They are then tipped mechanically into a hopper, leaving an empty basket to pass to the top of the descending side of the tower for refilling. The solvent-soaked flakes in the hopper are conveyed to a drying unit in which the solvent is distilled off and recovered, leaving a dry, solvent-free meal ready for use as an animal feed. At the bottom of the ascending side of the tower is collecting tank which receives the solvent which has passed through the partially extracted flakes, and although it contains some oil it is by no means saturated. this solution, called half-miscella, is used to wash the flakes in the descending side which are richer in oil. While the second stage (ascending side) of the extractor is a counter-current process, extracting as efficiently as possible, the first stage (descending side) is a co-current system and, perhaps surprisingly, results in as strong a miscella as a complete counter-current system would give. It is also claimed that the fines which wash out at the top basket are filtered out of the miscella in succeeding baskets so that the final miscella is cleaner than that from other types of extractor.

De Smet designed a continuous extractor in which the meal was conveyed horizontally instead of vertically as in the Hansa-Mühle system. One advantage of this approach is the achievement of a complete counter-current process with fresh solvent entering one end of the unit and the oil-laden meal entering at the other end. A wire-mesh band is used for the conveyor, beneath which are a series of collecting troughs. Above the belt is a series of sprays, each fed from the trough below it. As the meal passes slowly beneath the spray, it is washed with miscella which slowly increases in strength. At the same time fresh solvent is fed into the first trough, and overflows into the second trough and so on through the extractor, in a cascade system, producing a steady flow of solvent with increasing oil content. As in previous extractors, the spent meal is freed from solvent and the miscella distilled to separate the oil and solvent.

The Rotocel system is a departure from the linear flow of the Hansa-Muhle and De Smet processes, the whole process being housed in a cylindrical unit with a rotating series of cells. The bottom of each cell is perforated to allow solvent to drain through, and hinged ready for discharge when extraction is complete. A continuous flow of meal is mixed with miscella to form a slurry which flows into the first empty cell. The miscella drains through the perforations into a receiving trough while the cell slowly moves round to the first set of sprays. The rest of the process is similar to the De Smet system giving counter-current extraction with maximum miscella strength. When the circuit is almost completed the last solvent drains away and the spent meal is discharged by dropping the hinged bottom.

Other designs of extractor use basket conveyor systems like the Hansa-Mühle but with different feed and washing methods. Still others use screw conveyors, paddle

Fig. 1.24 'Rotocel' continuous solvent extractor, installed in Spain, for processing 200 tons/day of oil seed (sunflower, soya, safflower).

wheels or rotating sweeps to move the meal through the plant against the flow of the solvent.

After the solution has been removed from the extractor it normally goes to the recovery plant in which the fat and solvent are separated. During the extraction process the solution has picked up fine particles of meal which have remained suspended. These particles are removed by filters. The filtered solution then passes to a solvent recovery plant. Distillation of the solvent is accomplished in two stages, the first to remove the bulk of the solvent, the second to strip out the last traces. In the main distillation column the miscella is heated by steam coils, boiling the solvent away to the condensers. There are numerous designs of still, the best operating with either climbing or falling films which give minimum heating time to avoid damage to the fats. Evaporation may be assisted by application of vacuum. The last traces of solvent are removed by running the oil through a stripping column in thin films counter-current to a steam flow which boils off the solvent and carries it away to the condensers. These units must be efficient as little solvent loss can be tolerated if the process is to be economical. Solvent recovery can be improved by incorporating solvent scrubbers on all vapour outlets from the plant. Activated charcoal in the scrubbers adsorbs any solvent vapour which has passed the condensers. The solvent can be recovered from the charcoal by steaming and passing the steam and solvent vapours back to the condensers.

Recovery of solvent from the extracted meal is also important. The meal leaving the extractor is saturated with solvent. It is pushed through heated tubes by means of a screw conveyor so that the solvent is vaporized and passes to condensers. Residual solvent is removed from the meal by treating it with open steam.

The choice of solvent for extraction processes is important. Not only must the solvent have a high dissolving power for fats, but it should not dissolve any other components of the seed. Carbon disulphide has been used for the extraction of oil from the press cake residues of olive fruit. The product was inedible and sold as sulphur olive oil; moreover, the solvent is unpleasant to use and has a high fire risk. Chlorinated hydrocarbons such as trichlorethylene are attractive as solvents because they are non-flammable, but not only does trichlorethylene extract more colour from the seed than other solvents, but it has a toxic vapour, causes plant corrosion from acidic breakdown products, and can produce an extracted meal which is toxic to animals. In spite of the high fire risk, hexane and related petroleum fractions have been found to be the solvents for the purpose. They are sufficiently volatile to ensure complete removal from the oil and residual meal and yet can be recovered economically.

An alternative procedure has been introduced into some oil mills. Instead of sending the miscella directly to the stills for solvent recovery, the solution is first subjected to miscella refining. The principles of normal refining apply to a great extent, and it is claimed that losses of neutral oil are low because the presence of solvent reduces the entrainment of neutral oil in the soapstock resulting from alkali refining (section 1.6.2).

1.5.2 Extraction of chemical fats and oils

Fat deposits in the animal body are held in a connective tissue matrix containing also water and protein. They are concentrated to some extent in specific areas, such as along the back, and round the kidneys and other organs. According to the source of the tissue, it may contain as much as 95% fat or as little as 15%, although the low-fat tissues are not commonly used to produce edible fats. While the production of fats from vegetable sources is an industry in its own right, the vegetable fats usually being the primary product, animal fats are largely a by-product of the meat industry, giving rise to a different attitude to the processing plant installed, and a different scale of operations. Oil mills handling vegetable oils tend to be fairly large, and use plant designed and purchased for the purpose, but in many places the scale of animal fat production does not justify or encourage expenditure on modern equipment until the processor has built up a large meat handling plant. Consequently, there are many small units producing small amounts of animal fats, often for local sale, still using the traditional methods of preparation. Animal fats are extracted by application of heat in various forms, collectively called rendering processes, and covering all stages of process development from primitive to automated continuous processes.

a. Pre-treatment. It is important that the fatty tissues should be isolated and chilled or rendered as soon as possible after slaughter of the animals. If it is not possible to render the fat immediately after removal from the carcase, the tissue is washed in iced water and left to lose its body heat. The chilled material is then stored under refrigeration until it can be processed. Before chilling, the fatty tissue is trimmed as necessary, to remove unwanted non-fatty material and pieces of bone. It has also to be reduced in size in order to increase the rate of processing. Fat is not a good conductor of heat, consequently large pieces may take a long time to reach processing temperatures so that the first fat to separate may get overheated and damaged. Breakdown methods depend on the scale of operations. In small-scale plants the fat may be chopped by hand, but on the larger scale, machines which chop or mince the fat are used.

Fig. 1.25 Preparation of fatty tissues by mincing before extraction in a centriflow plant.

   b. Rendering. Batch processes for rendering of animal fats are divided into two types: dry rendering and wet rendering.

   Dry rendering is the simplest process. The chopped or minced tissue is tipped into an open top vessel, often called a kettle, and heated until all the moisture has been driven off. During the heating, the fat cells are disintegrated, releasing the melted fat, the protein part of the tissue becoming dehydrated and shrivelled. Heating is usually by a steam jacket around the kettle, and heat transfer may be assisted by a slowly moving agitator. The temperature of the fat may reach 220°F (105°C) during the melting, giving conditions which are favourable for the oxidation of the fat. Dry rendered fats are often darker coloured than those obtained by other methods, and have a cooked flavour which is preferred by some users. When the water has all evaporated the fat is allowed to settle, and is then pumped out of the kettle. A certain amount of fines are carried over with the fat. These are removed by straining through a cloth; sometimes salt is sprinkled on to the fat in order to hasten the settling of the fine particles of protein. After removal of the fat from the melter some residual fat is left in the 'crackling'. In order to recover as much of the fat as possible the crackling is transferred to a small screw or hydraulic press, and the fat squeezed out. In some areas there is a demand for the crackling as cocktail snacks so that it is important that they are not pressed too severely, or the loss of value of the crackling may be greater than the gain of fat. The efficiency of recovery of the fat is partly dependent on the degree of tissue breakdown in the pre-treatment. The finer the mincing (and hence the smaller the fat fragments) the better the yield of fat, but this leads to a lower sales value for the crackling. Dry rendering in open tanks is slow and not very efficient. The small amount of protein present not only affects the flavour, but if allowed to become too dehydrated it will break up to disperse colloidal particles in the oil. These are very difficult to remove and give a poor colour and poor clarity to the fat.

In some plants a modification of the dry-rendering process is operated in which the open kettle has been replaced by a closed one which can be evacuated. Removal of air helps to prevent oxidation and the vacuum helps to remove water at a lower temperature.

One of the advantages of dry-rendered fats is the low free fatty acid content of the product. There is little moisture present by the time the fat reaches a high temperature so that hydrolysis is limited.

Wet rendering processes use water as a heat exchange medium. If the process is carried out in open tanks, high temperatures are not reached and this has a beneficial effect in minimizing flavour and colour development. The tanks for wet rendering are equipped with stirrers and are heated either by steam coils or a steam jacket, the jacket being preferred. Fatty tissue is chopped or minced and thrown into the tank, together with water. The temperature is raised while the mass is slowly stirred. Under these conditions the fat melts and escapes from the tissue, rising to the top of the tank. At the low temperature employed the proteins of the fatty tissue are not fully denatured, consequently those fat cells which have not been ruptured mechanically will not liberate the fat. Efficient mincing of the tissue is necessary for good yields. The presence of protein matter will promote emulsion formation, so that a clear fat is not obtained during the rendering and it is necessary to sprinkle the surface with salt during the subsequent settling period. Salt breaks the emulsion and helps the water and protein to separate from the fat and settle to the lower layer. The fat obtained by this process has a good colour and neutral flavour. Not all the fat is recovered under these mild conditions; the residues are usually transferred to a steam rendering plant.

A closed pressure vessel is required for the steam rendering process. It is cylindrical, with a domed top and dished bottom. After charging it with fatty tissue and water, including any residues from a low temperature process, steam is blown into the vessel, displacing the air and raising the temperature of the water to boiling point. The air-vent is closed and the pressure raised to 40 or 50 p.s.i. (about 3 kg/cm^2). These conditions are maintained for several hours before cooling is allowed to take place. During cooling, the fat rises to the top and after the pressure has been relieved the fat is run off and settled, to separate it from the water and suspended solids.

Further developments of this batch-pressure process have resulted in plants like the Titan lard plant which will be described in detail.

The Titan wet rendering batch process is suitable for a small producing unit. It consists of a mincer, an autoclave and centrifuge. Fatty tissue is broken down by mincing before being charged into the autoclave, which is made of stainless steel and has provision for admitting live steam. When the full weight is prepared, the lid is bolted on the vessel and steam admitted; moisture in the tissues, together with condensed steam, provides all the water necessary for extraction. At first, the relief valve on top of the autoclave is left open to release the air. It is important that all the air is purged from the vessel before it is allowed to build up pressure; when a vessel is filled with steam only, the pressure and temperature are related, but if air is present the temperature reached at a pre-determined pressure is much lower. Any air left in the autoclave will interfere with the process in this way, as well as damaging the fat by oxidation. When purging is complete the vent is closed, and the pressure allowed to rise to 45 p.s.i. (3 kg/cm^2), at which it remains for $1\frac{1}{2}$ hours. During this period the protein matter is coagulated and the walls of the fat cells are ruptured, releasing the fat. At the end of the period under pressure, the steam is turned off, and the vessel allowed to cool until the pressure has fallen to atmospheric so that the lid can be opened. If there is evidence of emulsion in the liquid, a little salt is sprinkled on the surface to break the emulsion. After standing for five minutes a skimmer pipe is used to pump the fat layer into the centrifuge and separate the two phases. Any dispersed moisture is removed at this stage and the fat runs away to storage. The aqueous phase and the solids contain very little fat.

Wet-rendering processes at atmospheric pressure, and under increased pressure, all give a neutral tasting product, but the pressure systems tend to produce fats with a higher free fatty acid. The pressure and humidity cause some hydrolysis of the fat, the degree of hydrolysis being dependent on time and temperature. It is therefore advantageous to restrict the time cycle to a minimum. This can be assisted by mincing the fatty tissues finely, and, as in the Titan process, restricting the amount of added water. These processes are more efficient than dry rendering, especially when the fat content of the tissues is low. A low fat content is accompanied by a high moisture and protein, which causes a prolongation of the drying (rendering) period.

A continuous fat rendering plant installed in a large slaughter-house has been designed to deal with the fat as soon as it becomes available, without the necessity of chilling. As the tissues are removed they are dropped into chutes and within a very short interval they are being broken up in the large mincer at the beginning of the plant. The mincer is heated so that the tissues are raised to 140° F (60° C), a temperature at which the proteins are coagulated. The fluid mass passes into a centrifugal separator, a 'Super D-cantor'. This type of separator is of the horizontal solid bowl type with an internal scroll. As the bowl rotates at about 4000 rev/min the mixture of fat, water and protein is fed in. The liquids separate from the mass and are ejected from nozzles, while the solids are separately discharged by the scroll which runs at a lower speed than the bowl. No separation of the fat and water takes place at this stage, but the mixture, which still contains some solids, is passed through a high-speed disintegrator to break down any remaining solid matter. It can now be heated in a plate heat exchanger to a temperature of about 195° F (90° C) before passing into an 'autojector' centrifuge. In this unit the solids are thrown to the wall of the bowl while the fat and water are separated and flow out of the unit through separate nozzles. The solids are ejected at intervals through ports in the wall of the bowl. These ports open and close automatically while the centrifuge is still running, so making it possible to keep the process operating without having to stop at intervals for bowl cleaning. Fat is flowing from the plant in about 10 minutes, after the fatty tissue was removed from the carcase, which in a well organized plant can mean that the fat has been isolated in about half an hour from the animal being slaughtered. Under these conditions the free fatty acid content of the fat is kept low, and should be not more than 0.2%. The flavour and colour also are very good since the fat has been isolated quickly, out of contact with air and at a fairly low temperature. The protein matter also is a good quality by-product for the same reasons. When the carcases have been chilled and are presented for cutting up, more fatty tissue becomes available. This can be put through the same plant using a modified procedure. The fatty tissue is coarsely minced, about 5% extra water is added and the mixture heated to about 115°F (45°C) before putting through a disintegrator and feeding to the Super D-cantor. A good recovery of fat is achieved, leaving only 4% of fat in the protein. Very good yields are reported for this plant. The fat obtained is of good quality and can be used as an edible fat without any further treatment.

The 'Centriflow' plant operates on similar principles to the plant already described.

1.6 PROCESSING OF OILS AND FATS

A few oils and fats, such as olive oil and lard, are suitable for use in the freshly produced crude state without and treatment other than filtration to remove debris produced from the tissues during the oil-milling or rendering processes. Most of the others contain various impurities which interfere with their use as edible products or which give rise to an unacceptable taste. The removal of these oil-soluble impurities is accomplished in the process of de-gumming, refining, bleaching and deodorization. Each

Fig. 1.26 'Centriflow' plant.

of these will be defined as we come to discuss them individually. There are a number of other processes which are used to change the characteristics of the fats either by modifying the structure or composition of the fatty acids, by interesterification, hydrogenation, or fractionation. These processes will be examined, but before any can be applied to the fat, the impurities must be removed, and in the right sequence.

The processing of fats is carried out in two ways: batch-wise and continuous. The batch processes are flexible, they can produce one batch of treated oil and then change to a completely different one, but they require a great deal of skill on the part of the operator and the yields are poorer than with the continuous methods. Continuous processing can be highly automated, requires less labour than batch methods, and is not so dependent on operator skill. It is still true, however, that even in a fully automated plant, the ability of the operator to run the plant is reflected in plant efficiency.

In each section, both aspects of processing will be considered as applicable, for although continuous processing has been highly developed in some sections of oil and fat processing, it is not always the best method in other areas.

Most of the suspended mealy matter should have been removed by filtration before storage at the oil mill. There are a large number of different designs of filter but they are all fundamentally the same – a filter medium which is woven cloth made from various fibres or wire, and a pressure unit which forces the oil through the cloth leaving the solids behind. One of the commonest types of filter is the plate and frame press. This consists of a heavy iron frame with two long side arms, on which the plates and frames rest; the plates, which are corrugated on both sides to help oil drainage; the frames, which are just open like a picture frame; and a closing device. The frames are covered with cloth and then the plates and frames, which are hung alternately, are pushed close together and tightened with a screw. Oil enters through channels in the frame aperture and drains from the surface of the plate through passages in the bottom. The oil is fed to the press by means of a pump which forces it through the passages into the hollow formed between the plates and frames. The oil passes through the filter cloth, leaving the solids in the frame space, and then flows over the surface of the plates down to the outlet pipes and thence into a run-off trough. The suspended solids are left in the press and the clarified fat is ready for dispatch to the refinery.

1.6.1 De-gumming

Some of the vegetable oils contain a quantity of natural gums and phosphatides. These gums are good emulsifying agents, and if left in the oil would give rise to serious losses of oil at the next stage. A number of patents have been taken out covering the de-gumming of oils. These include the use of citric acid, sulphuric acid, hydrochloric acid, boric acid, salicylic acid, tartaric acid, sodium chloride, inorganic phosphates and sequestering agents. Sodium silicate has also been used with some success, especially in de-gumming soyabean oil. Some of these methods are used for oils intended for industrial use but for edible purposes the most widely used methods are hydration and phosphoric acid treatment.

In the batch process, oil is pumped into a neutralizing vessel comprising a large steel cylinder with a cone shaped bottom, fitted with an agitator and heating coils. Steam heating is used in the coils to raise the oil temperature to about 203°F (95°C). Hot water is then sprayed into the oil while it is being agitated. The water hydrates the phosphatides and gums to produce a flocculant material of higher specific gravity than that of the oil. When about 4 to 5% of water has been added, the agitator is stopped and the mass allowed to come to rest and settle. If the operation has been performed correctly the hydrated gums will settle to the bottom of the vessel and they can then be run off. The settling process is sometimes helped by scattering salt on the surface of the oil, in order to help to split emulsions and increase the specific gravity of the aqueous phase.

Continuous methods utilize centrifuges to remove the sludge formed on hydration. In some designs of plant the oil is fed continuously into a mixer into which water is also being metered. The two liquids mix and are passed through a small holding tank before entering the separator where the sludge is removed from the oil. Instead of adding water, in some processes strong phosphoric acid is added, which effectively coagulates protein matter and precipitates the dispersed phosphatides. The treated fat is either centrifuged or filtered immediately as in the Zenith process, or passed directly to the neutralizing stage as in the Laval process. Soyabean oil contains a high concentration of phosphatides which, after separation from the oil, can be processed further to produce commercial soya lecithin, a valuable by-product. It is used in a number of manufactured foods including chocolate and margarine. Lecithin is recovered by evaporation of the water used for hydration to yield a dark, sticky product containing free soyabean oil as well as the lecithin itself. When the lecithin is to be recovered it is essential to employ aqueous hydration (using water only) and to separate the phosphatides before passing to the refining stage.

Animal fats do not contain appreciable quantities of gums and phosphatides, especially when compared with the amount present in vegetable fats; so little in fact that it is often stated that none are present and de-gumming is not needed. Practical experience shows that putting an animal fat through a water de-gumming process will remove impurities present and make the subsequent refining easier and give better yields. The impurities removed include proteins and protein breakdown products, which, like the phosphatides, will act as emulsifying agents causing losses in the alkali refining stage.

1.6.2 Alkali refining

The alkali refining step is sometimes referred to as neutralization because the main objective is to remove the free fatty acids which have developed in the oil.

The selection of type and strength of alkali will influence the efficiency of neutralization and the ability of the process to accomplish other desirable improvements, such as removal of pigment and residual amounts of phosphatides and gums not removed by de-gumming.

Neutralization with caustic soda is the most important and widely-used method as it gives not only the maximum reduction in free fatty acid content, but is also effective in making some of the pigments soluble in the aqueous phase. Choice of strength of alkaline solution, temperature of treatment and duration of treatment, varies from oil to oil, and is designed to achieve a balance between maximum refining effectiveness and minimum oil loss. Unfortunately, the treatment with caustic soda in order to remove free fatty acid is not so specific; not only does the alkali react with the free acid, but it also attacks the neutral glycerides, splitting them to produce soap and free glycerol. Moreover, the reaction between the free fatty acid and the sodium hydroxide would not go to completion if only the theoretical amount of alkali was used. A calculated excess of caustic soda is necessary to ensure the removal of as much free fatty acid as possible, but as the excess alkali will attack neutral glycerides, the amount must be kept under control. Dark coloured fats, such as crude cotton seed oil, will require a much more vigorous treatment than that required by light coloured ones like lard in order to reduce the amount

phosphoric acid to precipitate the phosphatides which are then separated centrifugally to give a de-gummed oil for refining. The neutralizing vessel is filled with caustic lye of appropriate strength and temperature, and the fat, after de-gumming, is dispersed as droplets in the bottom of the neutralizer. As the drops of fat rise through the caustic lye, reaction takes place between the alkali and the free fatty acid giving a fat droplet which should now be free from fatty acids and other impurities. The fat droplets collect in the vessel as an upper layer which is automatically decanted continuously to maintain a constant level in the neutralizer. Instead of washing the fat to remove traces of soap, a small quantity of citric acid is added, and the fat is then passed into a vacuum drier. Soap is thereby converted into sodium citrate and free fatty acid, the former being filtered out either immediately, or with the bleaching earth if the fat is to be bleached.

The Laval short-mix process is a truly continuous one which relies on a centrifugal separation of the fat and aqueous phases. The fat to be processed is heated in a plate heat exchanger as it is being pumped into a small reaction mixer. Just before the fat enters the mixer a small amount of concentrated phosphoric acid is added to it, and thoroughly dispersed during its passage through the unit. The precipitated gums and phosphatides

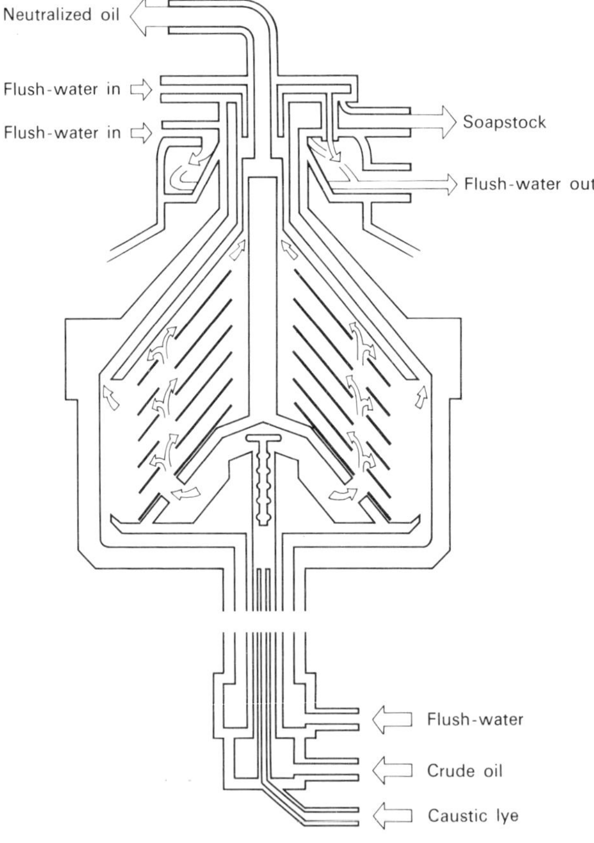

Hermetic Alfa-Laval separator rotor with inlets and outlets.

Fig. 1.29 Cross section of hermetic separator.

are left in the fat which is then continuously mixed with a metered amount of caustic lye and passed into a hermetic separator. Here the fat and soap phases separate into two layers and are separately discharged, the soap (which contains the gums and phosphatides) for by-product recovery, the fat for further processing. The neutralized fat is passed through another heat exchanger to adjust the temperature, and then mixed with a metered amount of hot water and passed into another centrifuge where the fat and water are separated. The fat, which should now contain less than fifty parts of soap in one million parts of fat (50 p.p.m.), discharges from the separator into a vacuum chamber where the water, which is dissolved in the oil boils off, leaving a dry, refined oil. The whole of the process, from the crude oil storage tank to the vacuum drier, is conducted under hermetic conditions so that the fat does not come into contact with the air during process. Consequently, vigorous mixing and high processing temperatures can be used to give the best refining conditions without risk of oxidation of the fat.

Alkali refining can be carried out using alkaline solutions other than sodium hydroxide. Sodium carbonate has been used successfully; it has the advantage over caustic soda that it does not attack neutral glycerides, and therefore, theoretically should give a lower refining loss. It has disadvantages as well. A special technique must be used to prevent the liberation of carbon dioxide gas which would cause frothing and make separation of the soapstock very difficult. Its less vigorous action results in a reduced cleaning effect and a weaker attack upon the pigments in the oil. Overall, the disadvantages of sodium carbonate refining outweigh the advantages to such an extent that the majority of refiners do not use it.

Some oils do not give a completely satisfactory product after caustic soda neutralization so it is necessary to subject them to a re-refining stage. Various refiners have their own ideas on re-refining, using a variety of alkalis to achieve improved quality in the oil. Weak caustic soda, sodium carbonate, sodium silicate, and alkaline phosphates have all been recommended for effecting this. When batch processing is practised, the re-refining stage can follow directly after separation of the refining soapstock, without washing out all the soap from the oil. In a continuous process such as the Laval short-mix process which has been described, an additional re-refining stage is normally introduced between the neutralizing and washing stages. Any type of re-refining lye to suit the requirements of the refiner can be used for this purpose.

All of the alkali refining techniques produce, as a by-product, an alkaline soapstock which contains, in the form of soap, the free fatty acid and saponified neutral oil together with emulsified neutral oil and the gums and phosphatides from the de-gumming stage. The total weight of fatty matter removed in this way can amount to a high proportion of the original crude oil, depending on the free fatty acid content of the starting material. Refining efficiency is often expressed as a refining factor rather than as a percentage yield. The theoretical maximum recovery of refined oil is dependent on the amount of free fatty acid and impurities present in the crude oil, so that the theoretical yield from an oil containing 2% free fatty acid could not be better than 98%, whereas a similar oil containing 4% free fatty acid, refined equally efficiently, could not give better than 96%. In practice these yields would never be achieved as there is always some loss of neutral oil, either by saponification or by emulsification. Frequently, the amount of additional loss is related to the amount of free fatty acid present as well as small independent losses due to entrainment in the various water separation stages. This additional loss can equal or even exceed the loss due to free fatty acid removed, giving, in our two examples, yields of 96% and 92% respectively. While the yield expressed in this way will serve several purposes it does not show how efficiently the refining has been performed. In order to measure this efficiency, the total loss is related to the free fatty acid of the crude oil, giving the refining efficiency as a ratio of actual divided by the theoretical losses, which in our examples would be 2 in both cases. This shows that although the yields are different, the refining operation has been carried out with equal efficiency, the yield discrepancy being due entirely to the poorer starting material in the second example. In the running of a refinery, refining factor computation can be much more complex than the simple example examples given above, but the basic principles remain the same.

As the soapstock represents a loss of oil it is desirable not only to keep the loss to as low a level as possible, but also to recover the fatty matter as a saleable product, even though of lower value than the original oil. In some installations a soap production unit is attached to the oil refinery, which is able to take the soapstock directly and utilize it in soap making. Many refineries do not have this facility, so it is necessary to further process the soapstock to make a product which can be handled, stored and sold. This is achieved by the addition of mineral acid to the soapstock which then separates into an oil phase and an acid water layer. In the simplest systems the soapstock is pumped from the refinery into large wooden, plastic, or lead-lined vats where it is diluted with water, and sulphuric acid added until the mixture contains free acid. It is then boiled with steam until the emulsions are broken and the soap split. At this stage the mixture is allowed to settle, the oily material rising to the top of the vat and the acid water containing some free sulphuric acid, sodium sulphate and water-soluble impurities forming the lower layer. The aqueous layer is run off, and the oil is boiled with fresh water to wash out residual sulphuric acid and other water-soluble matter, before being either sold in bulk or put into drums for sale as 'acid oil'. If the refining has been carried out skilfully this acid oil will have a very high free fatty acid content, a material which has no place in edible products but which can be used for soap production, or as a raw material for the preparation of fatty acids. Just as increased efficiency in refining can be achieved by the use of a continuous process, improvements can be made in soapstock handling and acid oil production in the same way. As the volume of soapstock is only a small proportion of the weight of crude oil being handled, continuous treatment is on such a small scale that the capital outlay may not be justified when considered in comparison with the small increase in revenue which may accrue.

1.6.3 Other refining processes

While the alkaline refining methods are by far the most widely used, other methods of removing impurities and free fatty acids have been proposed and to some extent applied industrially. Losses of neutral fat during alkali refining represent a financial loss to the refiner, hence attempts have been made to find physical methods of separating the free fatty acids from the fat.

Steam refining has been applied successfully to the de-acidification of fats which are of good colour and contain little or no other impurities. Neutral oil is not volatile and cannot be distilled but free fatty acids can be distilled in a current of steam under high vacuum, and at a temperature similar to that used in normal deodorization of fats (see section 1.6.8). To keep the time of exposure to high temperature to a minimum, stills have been designed to expose only a thin layer of fat to the distillation conditions giving the maximum distillation efficiency. In order to recover the distillate, the condensation systems for the fatty acids and steam must be separated, and an efficient high vacuum system installed to give absolute pressures of under 0.2 in (5 mm) of mercury. Fatty acids and fatty acid vapours are very corrosive, so that the plant must be made in the correct grade of stainless steel.

Refining by distillation will only remove free fatty acids although there will be destruction of carotenoid pigments which are heat labile. Other impurities such as gums, phosphatides and pigments which react with alkali and are removed in alkali refining, remain in the fat during distillation neutralization. Some of these substances are coagulated by the heat and steam to form particles of solid matter in the fat. It is necessary, therefore, to pretreat the fat before steam refining, or to give further treatment after distillation. The procedure adopted is dependent on the oil to be processed. Fats containing gums and phosphatides can be given a conventional de-gumming by hydration or phosphoric acid treatment to remove these impurities but this still leaves the pigments in the oil. It may be necessary to remove these pigments by absorption on bleaching earths using the techniques described later (section 1.6.4) before steam refining is attempted. Some of the pigments present in fats are heat labile, but others become fixed during high temperature treatment and then they cannot be removed at all. This could easily result in a fat becoming too dark-coloured for use in edible products. Another disadvantage of steam refining is that although in theory only the free fatty acid is distilled off, there is some hydrolysis of neutral oil so that the free fatty acid cannot be reduced to zero, and usually it reaches equilibrium at a slightly higher level than that achieved by alkali refining. There is also some 'carry over' of neutral oil droplets in the vapour stream going to the condensers. The efficiency of alkali refining falls as the free fatty acid content of the oil rises, but the converse applies to steam refining. In some circumstances it may be uneconomic to try to alkali-refine a fat with high free fatty acid content, but a reasonable yield of good fat could be obtained by steam refining. The latter is also a suitable technique to use with high value materials such as olive oil, where the recovered fatty acids can be re-esterified with glycerol and returned to the oil in order to increase the yield.

One commercial steam refining plant consists of three stages:

1. A purification stage in which the fat-insoluble solids are removed by filtration, followed by hydration of the gums and separation of the aqueous gum phase by centrifuging.

2. A continuous bleaching stage in which the fat is mixed with bleaching earth and then filtered to remove the solids.

3. A steam refining/deodorization stage which, it is claimed, operates at an absolute pressure of 0.04 in (1 mm) mercury. The oil flowing from this stage is reported to have a residual acidity of less than 0.1% and to be satisfactory in taste and odour.

It is claimed that this process is applicable to most oils and fats, that the cost is comparable with alkali refining, and that the quality of the finished product is better than that of a conventionally refined one. The process is, however, not widely used, in spite of its claims.

Physical refining by means of liquid–liquid extraction has been developed as a process not only for removal of free fatty acid, but also removal of phosphatides and colouring matter. The principle of the process is that when the liquid fat is mixed with a solvent only some components dissolve, the composition of the solution depending on the solvent-to-fat ratio, on the temperature of the mixture, and on the actual solvent used. A large number of solvents have been suggested as being suitable for a liquid–liquid refining process, but only two are well known. These are furfural, used in the furfural process, and hexane, used in the Solexol process. Both of these processes give a partial fractionation as well as refining the oil, although the Solexol process appears to be more effective for refining and less selective for fractionation. In this process the fat is dissolved in propane at a low temperature when the two liquids are completely miscible. As the temperature is raised the pigments, the oxidized fat, free fatty acids and gums, precipitate into a lower layer which can be drawn off leaving the fat relatively free from colour and free fatty acids. The fat can then be separated from the propane, which has a boiling point of $-49°F(-44°C)$, by allowing the solvent to distil off, after which the fat is deodorized to remove odour and residual free fatty acids. By suitable adjustment of temperature and plant design, some fractionation of the fat can be obtained. This aspect of the process will be more fully considered later (section 1.6.7.) The low boiling point of propane and the high fire risk make the cost of plant construction very high. It is doubtful that there are sufficient advantages in a liquid–liquid process to justify the high capital expenditure when compared with the more conventional alkali refining methods.

Re-esterification of the free fatty acids to re-form the triglycerides may also be undertaken as a means of refining the fat. It will be recalled that the presence of free fatty acid is the result of hydrolysis of triglycerides to liberate glycerol and free fatty acid. It would seem logical to remove the free fatty acid simply by reversing the process of hydrolysis and re-esterifying the free fatty acids with added glycerol. Unfortunately the method is not quite so simple as it may appear. While it is true that glycerol will react with fatty acids at high temperatures to form triglycerides, the reaction does not go to completion and is very slow. It is possible to overcome these difficulties by the addition of a catalyst, salts of tin and zinc being the most effective. When glycerol and fatty acids are heated under vacuum at about 200°C in the presence of about 0.2% of zinc chloride, a triglyceride containing about 2% of free fatty acid is obtained. It is necessary to provide an efficient condenser system to return the glycerol to the reaction vessel, otherwise under the conditions mentioned the glycerol will distil off into the vacuum system. After completion of the reaction, the catalyst and residual free fatty acid have to be removed, usually by alkali refining, before further processing the fat to an edible product. It is clear, therefore, that re-esterification is only a valid process when the free fatty acid content of the fat is high and when the value of the product makes it worth while. In times of fat shortage when the needs of the population for fats over-ride economic considerations, it becomes a useful process for conservation of fat resources. The re-esterified fat produced by this means will not have all the characteristics of the fat from which the fatty acids were derived. In discussing the structure of fats it was pointed out that, while some of the constants of the fats were dependent entirely on the fatty acid composition, the actual structure of the fat was controlled by the glyceride structure. When glycerides are formed by esterification of fatty acids and glycerol, the glyceride structure produced is purely random and can be predicted mathematically. It is unlikely that this structure will be the same as that of the original fat; the distinction may be unimportant for manufacturing purposes, but it could lead to difficulties if the fat was offered as genuine. Where the re-esterification is carried out in the presence of a high proportion of triglycerides, the glyceride structure of the finished fat may not differ substantially from the straight alkali refined fat.

1.6.4 Bleaching

The majority of fats contain some colouring matter either as a natural constituent or discolouration produced during the processing. Natural pigments present in vegetable oils are mainly the carotenoids, giving yellow and red colours, and the chlorophylls, which give green colours. Animal fats contain the pigments which are present in the feeding stuffs, so that they too will contain carotene and chlorophyll. Deterioration of the colour can take place during extraction of the fat, due to degradation of protein matter of the oil-bearing tissue, or extraction of pigments from these tissues by solvent, or the solvent effect of the fat itself. Oil seeds which have been damaged during harvesting or storage will yield oils that have colours that are not only darker than those from better seed, but may be more difficult to remove. The amount of colour varies very widely; a leaf lard or good quality Ceylon coconut oil can be almost colourless in the liquid state, while fresh palm oil is a deep red and crude cotton seed oil is almost black. Removal of these colours is necessary, not only because a pale-coloured fat has an appeal of 'purity', but also because the colour of the fat can influence the appearance of prepared foods, and even more important, the pigments present may affect the flavour and stability of the fats and the foods made from it. Methods of colour removal, or bleaching, vary to some extent, dependent on the type of pigment present and the fat being treated. Three groups of treatment are available: chemical treatment, heat treatment, and bleaching by adsorption.

Chemical treatment includes both alkali treatment and oxidation bleaching. It has already been stressed that one of the advantages of alkali refining using sodium hydroxide is that pigment is solubilized by the caustic alkali and passes into the soapstock, considerably reducing the colour of the fat. This effect is more important in the heavily pigmented oils, especially crude cotton seed oil.

Many of the pigments in fats are destroyed by oxidation, consequently, oxidation bleaching may appear to be a cheap and easy way to improve the colour of the fat. Unfortunately, the conditions which lead to oxidation of the pigments also favour oxidative damage to the fat itself, so that the application of these methods is not often practised in edible fat processing but is restricted to preparation of fats for soapmaking and similar non-edible outlets.

When fats are hydrogenated (see section 1.6.5), the pigments present are also reduced giving a substantial bleaching effect. Most hydrogenated fats are almost colourless and although the process is not often used directly as a means of bleaching, the substantial reduction in colour permits the processor to only lightly bleach fats which are destined for hydrogenation. Bleaching by hydrogenation using a mixed catalyst and causing very limited hardening of the fat has been patented. This method is of interest in processing very highly coloured fats, especially palm oil, although the alternative heat bleaching is probably a better economic proposition. Carotene and related pigments are sensitive to heat in the absence of air. When an oil containing a substantial amount of carotenoid pigment is heated under vacuum at temperatures of 430°F (220°C) there is a marked bleaching effect. Advantage is taken of this heat sensitivity to de-colourize palm oil either during high temperature deodorization, or during steam refining. Advantage of the heat effect can also be taken during adsorption bleaching, by operating the process under vacuum at a temperature of 300°F (150°C) instead of the more usual 210 to 220°F (100–110°C).

Bleaching, or extraction of colouring matter, was mentioned in the discussion of refining by liquid–liquid extraction. The Solexol process in particular is reported to effect a marked colour reduction during the extraction of fatty acids and other impurities. This should be regarded as a characteristic of the process rather than a bleaching method in its own right.

The most important bleaching process used in the preparation of edible fats is adsorption bleaching. Utilization of fullers' earth in the scouring of wool has been practised for centuries. The natural earths or clays used consist of various hydrated aluminium silicates. There is no evidence that the composition of a clay is related to its bleaching ability; indeed, it can be shown that samples of clays with the same chemical composition may have very different bleaching activities. In addition to the natural earths, the minerals montmorillonite or bentonite can be converted from earths with little or no bleaching activity to highly activated bleaching earths by treatment with hydrochloric or sulphuric acid. As a result of this development a whole range of natural and activated earths is now available with different properties which make them effective for the treatment of different oils. While it is true that most earths can be used to bleach any fat, the most economic and effective treatment of each fat is achieved by selection of a specific earth. In addition to the bleaching earths, adsorption bleaching can be carried out using activated carbon. The carbon is prepared from various carbon-rich substances such as coal, coconut shells, wood charcoal, peat or sawdust. After removal of volatiles by heating, the material is crushed and heated in kilns in the presence of steam or steam and air, in order to increase the adsorptive properties of the carbon, before being ground to a fine powder. The process of adsorption bleaching depends on the affinity of the adsorbent for the colouring matter, whether it is dissolved or colloidally dispersed in the fat. A great deal of effort has gone into defining the adsorption of colour in mathematical terms, the most important expression being developed by Freundlich and known as Freundlich's adsorption isotherm. The expression may be written as follows:

$$\frac{x}{m} = kc^n$$

where x = amount of pigment absorbed
c = amount of residual pigment
m = amount of adsorbent

The equation which relates the amount of substance adsorbed, the amount of residual substance, and the quantity of adsorbent, gives a linear relationship when the ratio of amount of substance adsorbed to the amount of adsorbent is plotted on a logarithmic basis against the amount of residual colour. The equation refers to constant conditions of operation, including maintenance of a constant temperature for all the measurements. It also includes two constants, k which is the adsorption activity of the adsorbent at unit concentration, and n which indicates the efficiency of the adsorption at varying concentrations of solute, i.e. the slope of the isotherm. In practical terms the equation shows that there is a limit on adsorption bleaching beyond which it is not economical to proceed. If this limit does not give a satisfactory product other bleaching earths, or carbon, or a mixture, together with suitable temperature changes, will be necessary. The best adsorbent for a particular application will have a high value for k. A study of the Freundlich equation shows that during bleaching the system comes into equilibrium while the bleaching earth still has capacity to adsorb colour and while there is residual colour in the fat. Full advantage of this observation could be taken if the oil could be made to flow counter-current to the earth. This is not strictly attainable but a multi-stage process goes a long way to achieving the benefits of a true counter-current one. In spite of the advantages of a multi-stage or a counter-current process, it is not used commercially to any great extent because of the difficulties in design of plant.

So far the principles of adsorption bleaching solely from the point of view of the removal of colouring matter have been examined, the other effects of bleaching being ignored. The adsorbent not only adsorbs the colour but also traces of soap left in the fat after refining, fat itself, and oxidized material present in the fat, anti-oxidants and trace metals. The adsorption of fat is of considerable economic importance as the value of the fat lost in this way exceeds the total cost of materials and labour for the bleaching process. For this reason, the cost of the bleaching operations requires consideration as much as the technical efficiency of the process. A compromise is usually arrived at.

Most bleaching is carried out by the batch mixing method. The usual bleaching unit is a cylindrical vessel with domed top and base. It is equipped with heating and cooling coils, an agitator, and connections to vacuum equipment. Provision is made for introducing the bleaching earth either as a powder or as a slurry. A preliminary laboratory bleach is carried out in order to establish the correct grade of earth, the concentration of earth, and the temperature, although the plant operation usually gives a better bleach than the laboratory results forecast. The fat is transferred into the bleaching vessel, and vacuum applied while it is being heated to the required temperature. During this heating period, entrained air and moisture are drawn off. It is necessary to remove all the air as under bleaching conditions of heat and agitation, any air present will cause oxidation to take place with consequent deterioration in fat quality. When the temperature and vacuum are at the correct levels, the measured amount of earth is drawn into the vessel

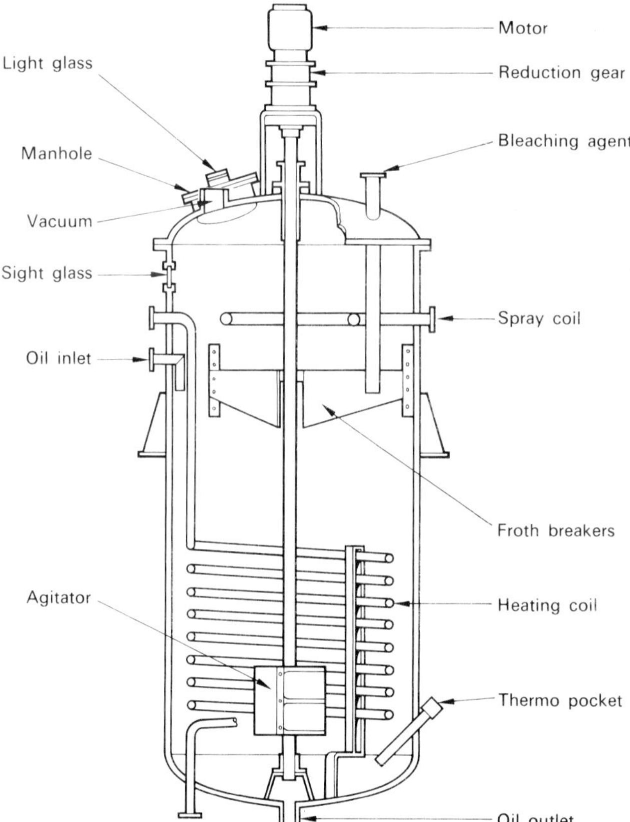

Fig. 1.30   Cross section of bleacher.

and mixed into the fat using the built-in agitator. It is essential that the agitator is designed to ensure that the earth is thoroughly mixed into the oil and kept suspended during the whole cycle. Colouring matter from the fat is adsorbed onto the earth during a period of about 15 minutes. Extension of contact time beyond this period will not give any further improvement in colour. At the end of the bleaching period the adsorbent must be separated from the fat. As it will be necessary, when using standard batch plant, to break the vacuum and allow the fat to come into contact with the air, there is a danger of oxidation of the fat at this stage. In order to minimize this effect, the fat must be cooled to a temperature of about 140°F (60°C) by passing cold water through the coils which were previously used for heating the fat. Separation of the earth and fat can now proceed with the risk of oxidation damage reduced as far as possible. This is commonly accomplished by pumping the suspension through a simple plate and frame filter press, though other types of filter are also used. As the slurry of earth and fat passes into the press, the fat is forced through the filter cloth and runs out through the channels provided. At the commencement of filtration the filtrate is still cloudy owing to fine particles of earth which have passed through the filter. This 'cloudy oil' is collected and recycled through the filter press at the end of the batch. As the amount of solids begins to build a layer on the filter cloth, the fat comes through as a bright and clear filtrate which is collected and put into storage tanks to await the next stage of processing.

The colour of the fat becomes progressively lighter as filtration proceeds. This phenomenon, which is referred to as 'the press bleach' is due to further bleaching taking place in the press itself. There is no doubt that the adsorptive capacity of the bleaching earth is far from exhausted when equilibrium has been reached in the bleacher. When the slurry starts to build up a layer of earth in the filter the ratio of earth to oil at the filter surface is substantially increased, disturbing the previously established equilibrium and allowing more colour to be adsorbed by the earth. When the batch of bleached fat has all been filtered, the press contains some residual oil and the separated bleaching earth. Unless the capacity of the press and the size of batch of bleached fat can be closely matched, the end of the batch will come while there is still room in the press for more solids. Normally, several batches of the same type of fat will be bleached in succession so that they can be filtered on the press until it is full. A second press should be available to be used for continuation of the filtration when the first press is full.

The fat which is retained in the press and adsorbed by the bleaching earth must be recovered as completely as possible in order to minimize loss. When a conventional plate and frame press is being used, the only way of recovering the fat is by blowing with air or inert gas. The pressure exerted by the gas forces the fat from the press leaving a saturated press cake which still contains about 50% of fat. If 1% of bleaching earth was added for bleaching, the filter cake will now contain a weight of fat equal to 1% of the batch. This is valuable and should be recovered if possible. Various techniques have been introduced at this stage. Blowing the press with steam, or washing the earth in the press with water or a solution of a surface active material will expel some fat although in each case the quality of the recovered oil is poor, and the water introduced will wet the filter cloths and all the surfaces and pipes. In most refineries the spent earth is removed from the press and either dumped or sent to solvent extraction plants where the fat is recovered. The greater the delay between opening the press and extraction of the fat, the poorer the quality of recovered fat. An Italian company has designed a filtration system which combines the filter press and the extraction plant in one unit. Instead of a simple plate and frame construction, the frames are lined with a fine woven wire cloth and then supported in a closed box into which the bleached mixture is pumped. As in the conventional press, the fat passes through the cloth and runs out of the press into a closed take-off channel. At the end of the process, hexane is pumped into the filter, through the layer of earth on the filter cloth and out via the fat lines to a still, where the solvent is distilled and the fat recovered. As the filter cake is not exposed to air before extraction and hexane does not remove colour and other impurities from the bleaching earth, the recovered fat is as good as the filtered material and the loss of fat during bleaching is negligible.

Bleaching processes which claim to be continuous have been introduced by several plant designers. In the Laval Auto-bleach process the bleaching vessel is divided into four compartments, each of which acts as a small batch bleacher. Refined fat is fed into the first compartment where it is mixed with bleaching earth. After a short residence time it is dropped into the second compartment and the first one refills. As the mixture of bleaching earth and fat passes from one compartment to the next the bleaching process continues, but because the small batch is restricted in volume the bleaching time is more easily controlled at the optimum value than in the conventional batch system. When the mixture reaches the last compartment it is pumped to a 'Funda' filter for removal of the bleaching earth. This type of filter consists of a tall cylindrical vessel containing a vertical shaft carrying a number of hollow horizontal plates. The fat passes through the wire cloth on the plate surface and flows through the interior of the plate and shaft to a receiving tank. Bleaching earth accumulates on the surface of the cloth until the gap between the plates has been filled. The central shaft is then rotated mechanically, spinning off the filter cake, which can be removed by a small screw conveyor in the bottom of the vessel. As the process is automatic there is a considerable saving of labour compared with the conventional batch process.

An alternative, two-stage continuous process passes the neutralized (i.e. refined) fat through a de-aerator to remove air and moisture, and then through a filter press which is full of partly used bleaching earth. Some of the colour is removed from the fat which is then heated and mixed with fresh bleaching earth. This mixture is then pumped to a clean filter press where the earth is removed, producing a fully bleached fat and filling up the press with partly used earth. When the second filter press is full, the controls are switched to allow it to become the initial one, and the number one press is then cleaned ready to re-enter the cycle as a number two. This system brings unbleached fat with all its colour into contact with part-used earth which has plenty of bleaching capacity left, fresh earth only contacting partly bleached fat. For the same degree of bleaching, this two-stage process shows savings, both in earth usage and reduced loss of fat in the spent earth.

1.6.5 Hydrogenation

Unsaturated fatty acids, as implied by their classification, have a capacity to absorb hydrogen and become saturated acids. As the addition of hydrogen to the double bonds of these acids takes place while the fatty acids are combined with glycerol in fats, the process has become one of the most important in the fat industry. Hydrogenation of fats is readily controlled, and can be stopped when the desired changes have taken place.

Natural unsaturated fatty acids and triglycerides containing them exclusively, are liquid at normal temperatures. When it is desired to prepare solid fat products from this type of raw material it is necessary either to blend a large amount of solid fat with the liquid one, or to convert the liquid glycerides into solid glycerides by adding hydrogen to the double bonds of the constituent fatty acids. Some raw materials such as fish oil and whale oil are highly unsaturated, and consequently cannot be processed and used in the natural state without oxidation causing rapid deterioration. The unsaturation can be reduced by hydrogenation to give fats which are sufficiently resistant to oxidation to be satisfactory materials for the production of edible fat products.

Hydrogenation will not take place without the presence of a catalyst, but the process is quite simple. The fat, catalyst and hydrogen are brought together under suitable conditions of temperature and pressure, and the reaction takes place. The iodine value of a fat is a measure of the degree of unsaturation, and following the changes in iodine value enables us to follow the addition of hydrogen to the fat. We can also follow the progress of hydrogenation by observing the amount of gas absorbed since we know that it takes 0.0795 parts by weight of hydrogen to reduce the iodine value of 1000 parts of fat by one unit.

The theory of catalysis is a complex subject which we will not attempt to discuss here; it is sufficient to consider the catalysts used for fat hydrogenation and the precautions which need to be taken in handling them and in preparing the fat to avoid poisoning the catalyst.

The most important catalyst used in the hydrogenation of fats is finely divided nickel. There are a number of ways of preparing the catalyst, and catalyst manufacturers have secret processes for the preparation of their brands. There are two basic methods of preparation, wet reduction or a dry process. In the wet reduction method, nickel formate is prepared from a common nickel salt and then dispersed in fat in a reaction vessel. All air is removed and a slow stream of hydrogen bubbled through the fat which is then heated to about 400°F (200°C). The suspension is cooled and some diatomaceous earth may be added in order to make subsequent handling easier. The dry process starts with the precipitation of nickel hydroxide or nickel carbonate on diatomaceous earth. The precipitate is dried and roasted in a stream of hydrogen to produce an extremely active catalyst which must be dispersed in fat without allowing it to come into contact with air.

In recent years there has been increasing interest in the use of copper chromite catalysts for special applications. It has been shown that soyabean oil can be hydrogenated to an extent that all the linolenic and most of the linoleic acids have been reduced to oleic acid, thus improving the keeping quality, without the formation of any saturated acids. The catalyst used for this process must not contain any nickel, which would continue the hydrogenation to saturation, but considerable success has been claimed for the copper catalysts. This point will be discussed more fully in a later section.

Hydrogen for the process can be prepared by electrolysis of caustic potash solution; by passing steam over iron; and by other chemical techniques. Electrolysis produces a very pure gas which makes purification unnecessary, whereas the other processes yield hydrogen containing small amounts of impurities such as carbon monoxide which would poison the catalyst unless removed.

Preparation of oils and fats for hydrogenation includes careful refining and bleaching in order to remove not only free fatty acids but also all the impurities which could interfere as catalyst poisons. Efficient removal of soap is important. It must be remembered that the bleaching process takes away colour, and also residual soap left after refining and washing. If it is considered that bleaching of a lightly coloured fat is not necessary before hydrogenation, a treatment with a natural neutral earth may be adequate to remove the offending soaps.

Batch processes for hydrogenation of fats are far more widely used than continuous processes, although developments in plant and catalysts for the latter are making it more attractive. There are two main variants of the batch process, recirculatory and dead-end methods. In all the processes there are three phases – solid, liquid and gas – which must be kept as a good dispersion of the solid and gas in the liquid phase. It is believed that the reaction takes place at the active centres on the catalyst surface and that it is only the hydrogen dissolved in the fat which reacts. The solubility of hydrogen is influenced by the pressure of the system, and the efficiency of agitation affects the area of the gas-liquid interface and hence the rate at which the gas dissolves. Interdependence of catalyst concentration, pressure, temperature, and agitation conditions present a complex series of variables, all of which can influence the rate of hydrogenation and the composition of the product. Thus, for a process to hydrogenate lard to a melting point of 50°C a catalyst representing 0.2% metallic nickel would be required. Hydrogenation would commence at a temperature of 212°F (100°C) and a pressure of 45 p.s.i. (gauge) (4 kg/cm^2 absolute). The temperature would rise to about 392°F (200°C) owing to the heat of reaction and hydrogenation would be stopped after about 120 minutes. Less catalyst would make the reaction slower, higher pressure would increase the reaction rate, lower temperature would reduce the reaction rate, and all these factors influence the characteristics of the finished fat. Hydrogenation is the addition of hydrogen to the double bonds of unsaturated fatty acids, reducing the degree of unsaturation. If we consider oleic acid, which has only one double bond, the end product of hydrogenation must be the corresponding saturated acid, stearic acid. Examination of more unsaturated acids, for example linolenic acid with three double bonds, indicates that the number of possible changes increases substantially. Not only can one, two or all three double bonds absorb hydrogen, but the addition can take place at any one of them, producing a number of different dienes and monoenes. Another complication of the process is that in addition to absorption of hydrogen, the conditions permit other structural changes, such as isomerization, to take place. In an uncontrolled reaction any, or all, of these changes could take place at random, but conditions can be controlled to make the process selective. The term 'selective hydrogenation' was applied originally to indicate the preferential conversion of linoleic acid to oleic acid, rather than oleic acid to stearic acid. It has subsequently been more loosely used to indicate a process in which the most unsaturated acids are progressively reduced in unsaturation in preference to the formation of saturated acids.

The dead-end batch process requires the simplest equipment. The reaction vessel or autoclave is a vertical cylindrical steel vessel with dished top and bottom, constructed to operate at a full vacuum and up to 200 or 300 p.s.i. (20 kg/cm^2) positive pressure. The vessel is fitted with coils for heating and cooling the fat, a good agitator, vacuum equipment, fat and hydrogen inlet pipes, an outlet pipe for the finished oil, and instruments to show the temperature and pressure inside the vessel. As hydrogen is an explosive gas when mixed with air, precautions must be taken to prevent any accumulation of gas if a leak should develop, and to exclude all possible sources of flame or spark. All electrical gear and controls must be explosion proof and are often located outside the building in which the plant operates. Auxiliary equipment includes a source of hydrogen, a catalyst mixer, and filters to remove catalyst from the fat at the end of the process.

The charge of fat is pumped into the autoclave which is then evacuated to remove all the air; the system may also be purged with nitrogen if desired. During the period under vacuum, the fat is heated to the required temperature, ensuring that all air and moisture are removed from it. The main agitator is started to assist in elimination of dissolved moisture, and then the weighed amount of catalyst is drawn in from the mixer. The vacuum is maintained until all the air has again been removed, and is then shut off. Hydrogen is admitted to the bottom of the vessel through a pressure regulator which closes off the gas supply once the operating pressure is reached. Gas accumulates in the head space above the fat and is then sucked down by the agitator and re-dispersed in the fat. As the gas is absorbed and the pressure drops, the gas inlet valve opens and admits more hydrogen, thus maintaining a constant pressure. As the reaction proceeds changes can be followed by measuring the change of refractive index or melting point of the fat. When the desired degreee of hydrogenation has been accomplished the gas supply is closed off and the vessel evacuated again to remove all the hydrogen. The charge is cooled to a temperature at which it can be safely filtered, and then the vacuum is broken and the fat pumped through the filter press to remove the catalyst. As the hydrogenation process is an exothermic reaction it may be necessary to control the temperature by passing cooling water through the coils, although some processors prefer to start hydrogenation at a fairly low temperature, allowing the heat generated to raise it and so balance an increased reaction rate due to temperature rise against the decrease in rate which occurs as the iodine value of the fat falls. During hydrogenation there is a slight rise in free fatty acid and the development of a distinctive odour and taste which are typical of a hydrogenated fat. It is sometimes claimed that these can be eliminated by good deodorization, but an alkali refining and a light bleaching will give a hydrogenated fat, which, after deodorizing, will have a better taste and be more stable than one which has only been deodorized.

The recirculatory batch process requires similar plant with the addition of a compressor and a gas purifying train. Using the same type of autoclave, the recirculatory system is charged in the same way as in the dead-end method, but instead of relying on the agitator to draw the gas into the reaction mixture, a compressor extracts the gas from the top of the autoclave and returns it through a perforated distribution pipe in the base of the vessel. A series of scrubbers between the autoclave and the intake of the compressor is designed to remove impurities from the gas stream, partly to protect the compressor from abrasive catalyst particles which may be carried over with the gas, and partly to remove odours which would otherwise be returned to the fat.

A number of continuous hydrogenation processes have been suggested but although simple in theory, they are not easily controlled. Most methods operate using a fixed bed catalyst, with the fat and hydrogen flowing through the bed. One advantage of the technique is the elimination of a filtration stage, but as the catalyst loses activity the conditions change and renewal of catalyst becomes necessary.

1.6.6 Other chemical processes

There are two other chemical processes which are of interest in the edible fat industry: the preparation of superglycerinated fats as emulsifiers and the modification of fats by interesterification.

Superglycerinated fats are widely used in the food industry as edible surface-active agents. The preparation is straightforward. As in other fat products, the fatty acid composition of the final product determines many of its characteristics. In superglyceri-nated fats, the triglycerides are largely replaced by mono and diglycerides so that to some extent they are simpler in structure than the parent fat. The selected fat is charged into a stainless steel reactor which is fitted with heating and cooling coils, an agitator, and vacuum equipment. Glycerol and an alkaline catalyst are added and after removal of air the charge is stirred and heated to a temperature of about 400°F (200°C). Under these conditions the fatty acids are redistributed between the glycerol molecules present, resulting in a mixture of monoglyceride and diglyceride with some residual triglyceride. The exact proportions in the product are dependent on the process conditions and it is reported that a continuous process will give a higher monoglyceride content than batch processes. The commercial product is sold as 'monoglyceride', or G.M.S. (glyceryl monostearate) depending on the composition of the fat charge used, G.M.S. usually being made from a fully hydrogenated fat with a high stearic acid content. Products with a higher monoglyceride content are usually prepared by distillation of the commercial monoglyceride under high vacuum, to give an isolated monoglyceride distillate.

Interesterification, or ester interchange, is a process in which the fatty acids present in a fat or mixture of fats are re-arranged to produce a different glyceride structure. In an earlier chapter we discussed the way in which natural fats are formed with varying glyceride structures which are characteristic of their origin and in which the fatty acids are not randomly distributed. Ester interchange processes can be controlled to give a product in which the fatty acids are distributed randomly in the glycerides, or in which the glycerides formed follow a directed pattern. Under the influence of a catalyst, the redistribution of fatty acids to form a random re-arrangement can be carried out at a temperature below the crystallizing point of saturated triglycerides. When this technique is adopted, the saturated glycerides separate from the system as a crystalline solid, disturbing the equilibrium so that more saturated triglyceride forms and in turn crystallizes. It is possible by this means to produce a re-arranged fat in which almost all the saturated acids are in the form of saturated triglycerides. Temperature control will influence the sequence of glycerides which crystallize in this directed interesterification.

There are a number of applications of this technique. Some fats–lard for example–crystallize as large coarse crystals which detracts from their performance as bakery products. This crystal structure is destroyed by ester rearrangement, thus improving the product. Fats which are modified by directed re-arrangement may be used to produce a plastic fat without recourse to hydrogenation or blending of hard and soft fats. This may be a valuable technique where the available fats are restricted in type and the cost of hydrogenation plant is unacceptable. Ester interchange is also being applied in conjunction with other techniques such as fractionation and hydrogenation to produce special fats with unusual characteristics.

The process requires a catalyst for the reaction. Patents have been filed to cover a wide range of materials with claimed catalytic activity. Tin and zinc preparations, sodium potassium metals, metallic alcoholates, lead and cadmium preparations all have been mentioned as having activity, but of all these, the most effective and convenient is sodium methylate. A conventional batch neutralizing vessel is suitable for carrying out the process. The fat, or fat blend, is pumped into the vessel and carefully dried. This step is vital as any moisture present will destroy the catalyst. When drying is complete, the fat is cooled to the desired temperature, about 200°F (95°C) for randomization but much lower for directed ester interchange. At the end of the reaction time it is necessary to destroy the catalyst before making any other changes. This is achieved by adding a calculated quantity of phosphoric acid which will react with the alkaline catalyst. The fat can then be given a light bleaching treatment to remove the destroyed catalyst.

1.6.7 Fractionation
Fractionation is a process in which either small quantities of undesirable glyceride components are removed, or in which a fat is separated into two or more groups of

components which have different applications from each other and from the original fat. An example of the first mentioned is the removal of small amounts of solid glycerides from salad oils so that there is no clouding of the oil at refrigerator temperatures. This application of fractionation will be considered when salad and cooking oils are discussed in section 1.8.1. Reference has already been made to the fractionation of beef fat to produce oleo stearin and oleo oil for the shortening and margarine industries. The original method of fractionation was by fractional crystallization of beef fat and this process is still used; it yields the familiar coarse, granular product which is formed when beef dripping is allowed to cool slowly. This is applied on the large scale as the basis of the batch fractionation process.

The fat is completely melted and pumped into small tanks or tubs which hold up to half a ton. The vessels are transferred to a temperature controlled room, the actual temperature used being selected so as to influence the characteristics and composition of the products. As the liquid fat cools towards the room temperature, solid glycerides start to crystallize and form large crystal agglomerates. It takes three or four days for crystallization to be completed. Then the crystallized mass is broken up, emptied into filter bags and squeezed in a hydraulic press to recover liquid and solid fractions. In more modern plants the liquid and solid phases are separated by pumping the mixture into a filter press. Precautions must be taken to prevent temperature changes in the crystallized fat during the pressing or filtration steps as this would alter the balance of phases.

Fats can be fractionated in a solvent solution, such as furfural, propane or isopropanol. In the furfural process, use is made of the fact that unsaturated glycerides are more soluble in the solvent, whereas saturated glycerides are more soluble in propane. Isopropanol is also being used to fractionate fats.

A continuous fractionation process has been developed in which the solid fat is washed out of the liquid fat phase. The liquid fat is passed through a heat exchanger where the temperature is adjusted to crystallizing conditions. Development of the crystal mass takes place in a crystallizing unit. An aqueous solution of a surface active material is adjusted in temperature to that of the crystallized fat, then the two are mixed. The solid crystals are wetted by the solution and pass into aqueous suspension; the mixture is then passed through a centrifuge to separate the liquid fat phase from the aqueous solid fat suspension. The aqueous phase is heated, then centrifuged again, thus recovering the aqueous solution for re-use and the fat fraction for further processing.

1.6.8 Deodorization

Most fats have a characteristic odour and flavour which may be pleasant like butter or objectionable like fish oils. When a fat is used unblended, its natural flavour, if pleasant, may be acceptable, but when the same fat is required for the preparation of food products its flavour may be undesirable. These flavours are removed by the process of deodorization. Most of the compounds which give rise to the flavour of fats are much more volatile than the triglycerides. They are, however, present at very low concentrations. When the fat is heated, the flavour materials start to distil and if the heating is carried out under vacuum the rate of distillation is increased and the fat is protected from oxidation. Distillation is improved further by passing steam through the fat so that deodorization may well be described as a steam distillation under high vacuum. In addition to the removal of the flavour components, which probably only account for 0.1% of the fat, there is some reduction of colour, distillation of some residual free fatty acid and breakdown of peroxides which may have formed in the fat during other processing. There is some evidence too, that substances which promote oxidation are destroyed by the conditions of deodorization. Theoretical considerations of deodorization and the laws governing steam distillation efficiency and vaporization of the flavour components have led to the design of efficient systems for the process.

The greatest cost factors in deodorization are steam consumption and loss of fat. Steam is used to heat the fat, to 'sparge' it in distillation of the volatiles, and to raise the vacuum by the ejector systems which are considered in detail later. Losses of fat occur in two ways: hydrolysis of the fat with liberation of free fatty acids which then distil, and entrainment. The latter is the more important cause of losses. As the steam bubbles pass through the fat they expand greatly and as they burst through the surface they throw fat droplets into the head space above the oil. Depending on the velocity of the vapours and the design of the plant, the droplets will be carried out of the deodorizer into the vapour main or they may fall back into the oil. The critical vapour velocity can be calculated from the steam flow rate and vacuum. From a practical point of view, the entrainment losses can be restricted by control of steam flow rate. The total amount of steam used for stripping the volatiles from the fat is also dependent on the vacuum. Not only does the volatility of the flavour materials and free fatty acids increase with higher vacua, but the amount of steam required to remove the volatiles is reduced. Many refinery processes are operated under vacuum and for many purposes the term vacuum is adequate and is well understood. It is not, however, a precise measurement as it is dependent on atmospheric pressure and varies with it. On deodorizer plant where very high vacua are used it is much less confusing to use the fundamental measurement, absolute pressure,

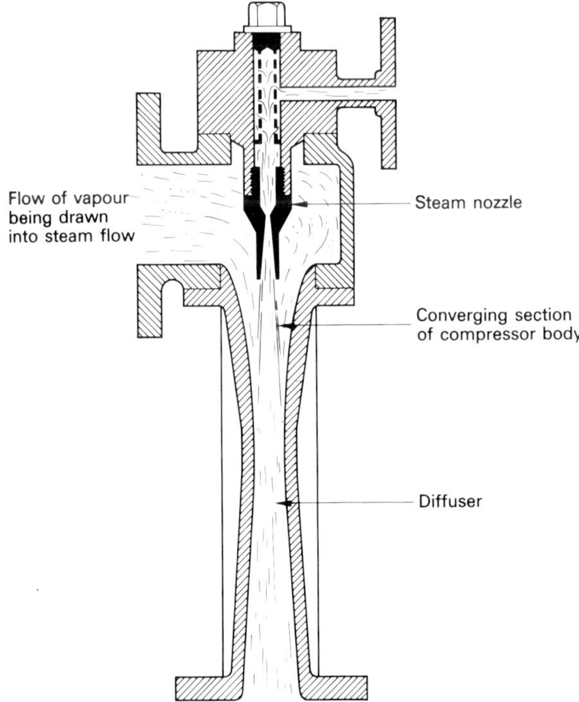

Fig. 1.31   Section of steam ejector.

which is independent of atmospheric pressure. Absolute pressure, which is measured in millimetres of mercury (mmHg), kg/cm^2 or mbar, is the converse of vacuum.

Deodorization is carried out under absolute pressures of less than 10 mm. In order to obtain this low pressure, a multistage steam ejector system is used, with spray condensers to condense the vapours. The steam ejector high vacuum system is non-mechanical and operates without any moving parts, thus simplifying maintenance. Its design is simple, but efficient. At the top of the unit there is a chamber with a vapour port at one side, and a steam jet in the top, opposite the main outlet. The body of the ejector is in the form of a double cone with the main passage way convergent at the top, the then divergent. High-pressure steam is forced through the nozzle, which has a coned outlet, giving a divergent stream at high velocity. The steam jet passes into the convergent section of the main body of the unit, entraining air from the surrounding chamber. The mixture of air and steam passes through the constriction of the ejector body and expands into the following section. If the single ejector unit is being used alone, this mixture of steam and air can be allowed to blow to atmosphere via a 'U' trap, but a single stage system is not an efficient unit for high vacuum operation, in which two or three ejectors, coupled in series, are usual. In order to decrease the potential load of gas in the successive stages of the multiple unit, the steam from each stage is condensed by a water spray, leaving only air to be passed on to the next stage. The cooling water and condensed

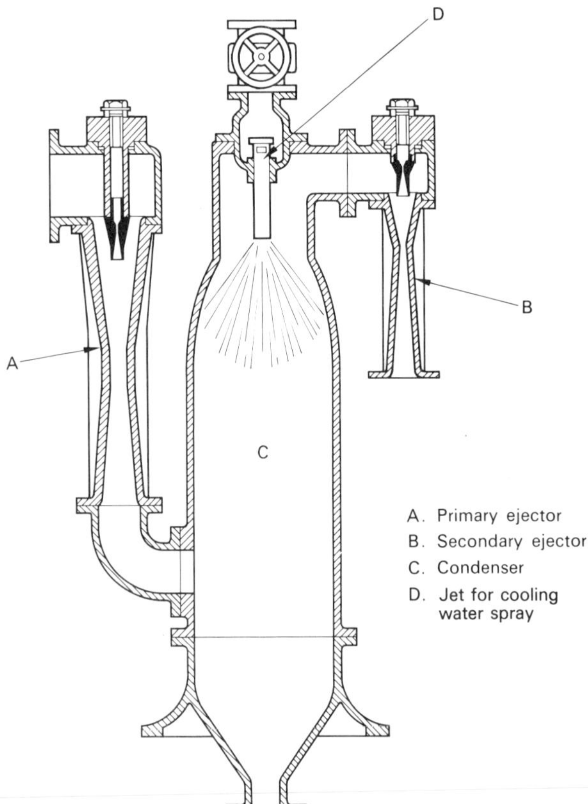

Fig. 1.32 Two stage ejector.

steam from each stage has to flow away continuously but as the condenser is operating under vacuum, the drainage pipe must be at least 34 ft (10.5 m) long. This makes it necessary for the vacuum equipment to be sited at least 36 ft (11 m) above ground level. The steam consumption of a conventional three-stage ejector system is high, and may account for as much as half of the total steam usage of a complete refinery.

As in other stages of fat processing, deodorisation plant is designed for batch or continuous operation. There are numerous designs of batch vessel, all of which attempt to perform the process as effectively as possible with the minimum loss of product. The basic unit must be vacuum tight, and is usually of all-welded construction to ensure the absence of leaks, but some types are available with flanged and bolted heads. A cylindrical vessel with dished top and base is necessary to withstand the high vacuum under which the process is operated. The unit contains coils with a large heat exchange surface to permit rapid heating and cooling of the fat. At the bottom a perforated pipe, often in the form of a spider, is used to introduce the stripping steam. Some designs incorporate a central tube, a vomit tube, which has a separate steam jet fixed centrally in its open base. A fat inlet valve is provided on the side of the vessel and a base valve for extraction of the finished charge. This base valve is a weak point in the system as any air leakage through it while the fat is being processed would seriously damage the batch, by oxidation of the fat. Careful maintenance is required and an additional safeguard is to instal two valves close together on the exit line with a connection between them, going to the vacuum system. This arrangement guarantees that air cannot leak through the base valve. Vapour and steam are extracted from the vessel through a large diameter vapour main to the high vacuum equipment. The vapour main is usually interrupted by separators which are intended to catch the carry-over and distillate, but most of the conventional separators are very inefficient. At least one vacuum equipment manufacturer offers a scrubber which, installed in the vapour main, will wash the vapours with a spray of fatty acid, condensing distillate and trapping droplet carry-over, allowing only a fat-free vapour to pass to the high vacuum set. As standard vacuum equipment becomes very badly contaminated with distillate, this eliminator unit can protect the equipment and prevent pollution of the cooling water in the spray condensers. It also provides a means of recovery of the deodorizer losses. Batch deodorizers are usually constructed of mild steel which restricts operating temperatures to about 400°F (200°C).

The process starts by filling the deodorizer to the design level with fat. Overfilling will result in excessive carryover losses, and underfilling may expose heating coils, which allows local overheating of the fat on the exposed pipe, causing charing. Variation in fill over a range of about 20% is usually available. Fat can either be pumped into the vessel, or drawn in by evacuating the deodorizer. A partial vacuum, an absolute pressure of about 25 mm Hg (33 mbar), is established in the deodorizer and steam is admitted to the heating coils. As soon as the temperature reaches 212°F (100°C) the vomit jet is opened. When steam is passed through the jet it blows the fat out of the top of the tube onto the underside of a cone so that an umbrella-shaped cascade of fat is thrown down on the surface of the fat in the main part of the vessel. This action not only agitates the charge, so assisting both heat exchange and rate of deodorization, but also helps to control carry-over of droplets of fat by trapping them in the cascade. When the temperature reaches 300°F (150°C) the full vacuum system is brought into operation by turning on the augmentor or booster. This section of the vacuum equipment will reduce the absolute pressure to a few millimetres; the actual pressure obtained depends on the plant efficiency, but it should be 6 to 8 mm Hg (8–11 mbar). The full stripping steam is now brought into operation, allowing a flow of steam of 20 to 30 pounds (10–15 kg) per hour, per ton of fat. This flow of steam is fed ot the spider and vomit, at a gauge pressure of 45 p.s.i. (3 kg/cm²). The temperature of the fat is raised to 356 to 392°F (180–200°C) and held at that temperature for about 4 or 5 hours depending on the type of oil being processed. Some deodorizers have a sampling valve on the side through which a sample of fat can be drawn and tasted to see if it has been sufficiently deodorized. When the deodorization is judged complete, the fat is cooled to about 140°F (60°C) before breaking the vacuum and discharging the batch. In order to cool the charge, cold water is passed through the coils which had previously been used for heating. When the fat temperature has fallen to 212°F (100°C) the stripping steam must be turned off in order to avoid condensation of the steam in the fat. The average cycle time is 10 hours. The finished fat should be completely odourless and tasteless. Small amounts of citric acid, 60 to 100 p.p.m., are often added to the fat either at the beginning or end of deodorization in order to sequester any heavy metal which may dissolve in the fat, reducing the stability. The vacuum equipment is kept running throughout the process, even though actual deodorizing is only about half the operation cycle. The heating and cooling stages are completed with little opportunity for heat recovery. Some plant designs include a drop tank which is also under vacuum and fitted with an agitator and cooling coils. This tank enables the process vessel to be emptied quickly before the cooling stage, so reducing the cycle time. The deodorized fat, still at a high temperature, is run from the deodorizer into the drop tank for cooling. This modification to the process has some advantages, but there is a serious risk of damage to the fat.

Continuous deodorizers, both truly continuous ones and semi-continuous variations, overcome a number of the short comings of batch processing. The designs of plant vary considerably in detail but follow a common basic pattern. Fat is pumped to the deodorizer through a heat exchanger, where it can be pre-heated by deodorized fat leaving the plant. Further heating takes place in subsequent heat exchangers using steam or other heat transfer media such as Dowtherm (an eutectic mixture of diphenyl and diphenyl oxide). When the fat has reached deodorizing temperature it passes into a series of shallow sections within a common shell. Here, the fat is under high vacuum conditions, usually at an absolute pressure of 2 to 4 mm Hg (2.5–5 mbar), and steam is blown through it as in batch deodorizers. The depth of each section is very little compared with the depth of fat in a batch deodorizer, so much more efficient vaporization of the volatiles takes place. As deodorization continues, the fat is passed automatically from section to section until at the end of one to two hours, it is fully deodorized. The product is then passed through heat exchangers to cool it before discharging the fully deodorized fat into storage tanks. During this process substantial heat economy can be achieved by passing the feedstock and finished fats counter-current through heat exchangers, so that the ingoing cold fat is heated by the hot, finished fat. Considerable economy is achieved in steam usage for vacuum equipment as the whole of the time of vacuum plant operation is for actual deodorizing. As the fat only contacts the trays within the deodorizer shell, these surfaces can be stainless steel within a mild steel shell. All the surfaces which come in contact with the fat are made in stainless steel which makes it possible to operate the process at much higher temperatures than batch processing in mild steel. Continuous and semi-continuous processes operate at temperatures of 430 to 500°F (220–260°C) and the high temperature, shallow layers, and better vacuum conditions all combine not only to give better deodorization, but also to enable a considerable reduction to be made in deodorizing time. The difference between continuous and semi-continuous plants is that in the former the fat flows in a continuous

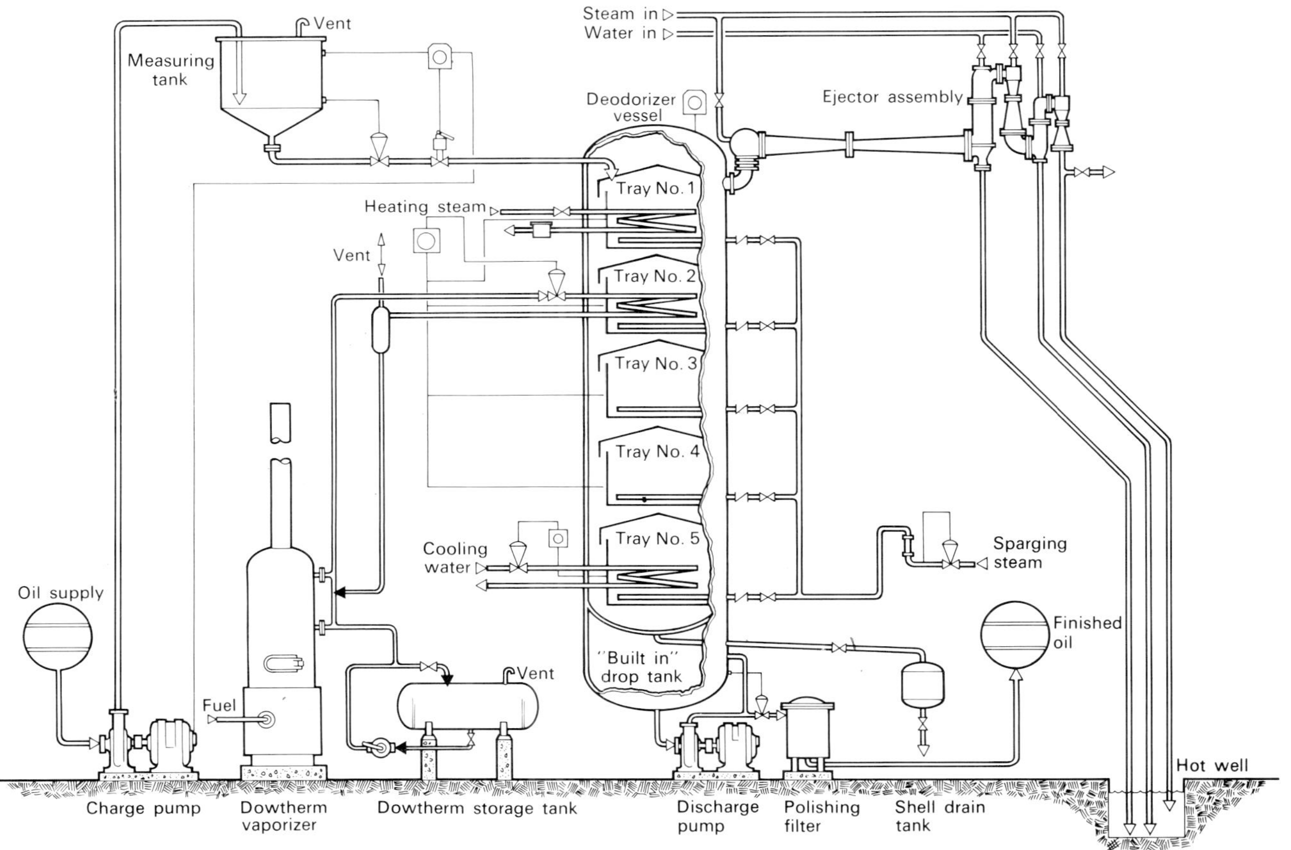

Fig. 1.33 Schematic flow diagram of semi-continuous deoderizer.

stream from section to section within the deodorizer shell, but in the semi-continuous plant fat accumulates in the top section and then is dropped as a small complete batch into the next section. A fixed residence time is scheduled for each section and the transfer is automated by a series of valves operating on a time sequence. A semi-continuous unit is more flexible than a truly continuous one, as a change of feedstock in the latter requires almost complete emptying of the plant.

All designs do not take advantage of the heat recovery possible by using counter-current heat exchangers. The Girdler semi-continuous deodorizer houses the heating and cooling sections within the shell, which avoids the use of external heat exchangers but loses the heat recovery which could be achieved.

Apart from the economy advantages of continuous or semi-continuous plant, the conditions of operation provide heat bleaching, and better flavour stability.

1.7 MARGARINE

In its original development, margarine was undoubtedly the result of a search for a butter substitute. The price of butter was high and supplies were scarce, a situation which not only made the production of a substitute necessary, but drove the French government to encourage the search. The essential concept at the outset was to try to imitate the cow in a mechanized form. It was during his experiments with the crystallization of fat that Mouriés produced a fraction which had a pearly sheen. This fraction consisted of the fatty acids isolated from the softer part of tallow which was being used for producing the experimental butter substitute, and was thought to be a C17 acid. It was considered that a new acid had been found, and because of its pearly appearance, it was named margaric acid, from the Greek word for pearl, the glyceride derived from it being margarine. The name was applied to the whole product, although not immediately. As we have already discussed, odd numbered acids are rare, and this one has been found to be a eutectic mixture of stearic and palmitic acids.

The original crude fractionation process for the fat, its treatment with enzymes and churning into a buttery fat has already been described (section 1.1.2). The manufacturing process of making a 'cream' emulsion and churning it in the conventional way of butter making was logical, and several attempts have been made, in recent years, to re-develop a churning process for margarine. Although the original process was developed in France, interest spread rapidly and soon production was taken up, first in Holland and then in many other countries. The history of the development of the market and the process, makes interesting reading: references are given in the literature section.

Since the development of the butter substitute in 1869 the product has not only changed in composition and character, but also in scope. The butter substitute was intended to be used in the same ways as butter, but the margarine market has become channeled into specialized products. The domestic market is familiar with the standard block of margarine and the soft margarine in tubs, but the baker and confectioner has a much wider range of special margarines for cakes, pastry, etc. (see section 1.7.4).

The development of modern packs for margarine has, in the United Kingdom, paralleled butter packs, starting with the large block which the grocer carved up into pats, which he then weighed and wrapped, up to the factory-produced half-pound and pound blocks in parchment and foil wraps. One manufacturer has produced margarine in 2 oz 'sticks', foil wrapped, four of these being put into a carton the size and shape of a

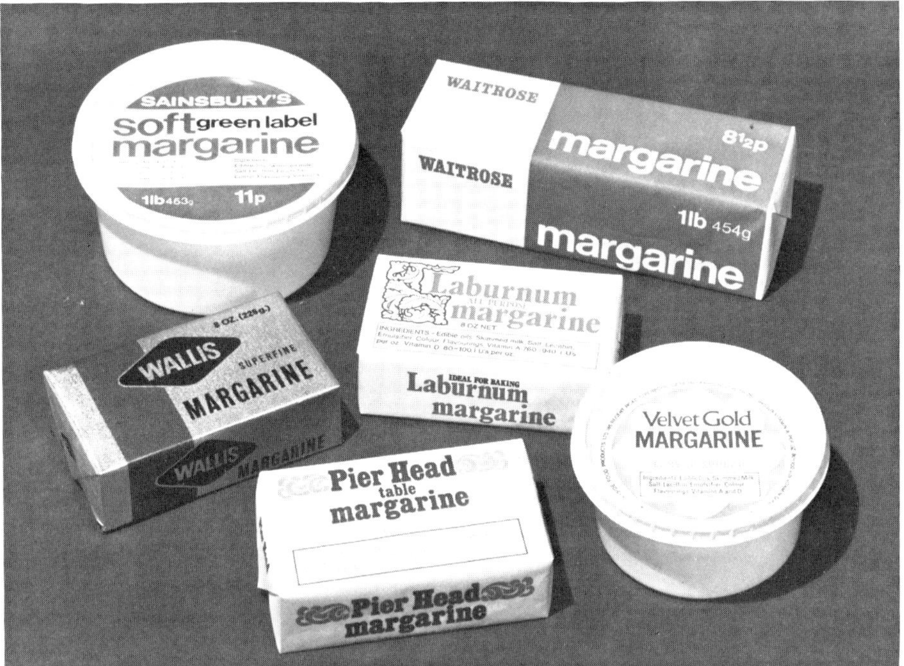

Fig. 1.34 Typical margarine packs in plastic, parchment and aluminium foil.

half-pound block. Margarine has now developed as a different product, packed in plastic tubs containing half or one pound, and the butter packers are also now starting to pack in tubs.

Ever since margarine was first produced, there has been restrictive legislation to prevent adulteration of butter with margarine or to try to make margarine look so unattractive that buyers were discouraged. In some countries there was a requirement that margarine must be uncoloured, so that it did not look like butter, that the shape of the pack should be different from butter packs (cubical for example) and in the United Kingdom it was made illegal for margarine to contain more then 10% butter fat. It was this legal requirement that restricted the butter blend margarines to the level used.

The legislation outlined above applies to the United Kingdom; other countries have similar requirements and attempts are being made to attain greater uniformity. Some of the legislation has been useful in setting standards for margarine composition. The maximum level of moisture is the same as butter – 16%. Vitamins must be added to table margarine at a stated level, which makes margarine not only a more uniform but often a richer source of vitamins A and D than butter. The amount of fat present is also controlled – it must not be below 80%. The residue, maximum 4%, is accounted for by salt and milk if present in the margarine.

1.7.1 Raw materials

A great deal has already been said about the major component of margarine; that is, the fats. We have considered the processing needed to prepare them for margarine manufacture, but now we need to look at the selection and blending of those fats to give the correct characteristics to the finished product. A number of factors must be taken into account when considering the fats to be used in the blend. Characteristics of the margarine are of course a primary consideration. A margarine to be used for pastry-making needs to be fairly tough (i.e. it does not deform and flow easily), and if for puff-pastry it must be able to withstand the high temperatures of the oven until the structure is set. Similarly, a cake margarine must cream well and a table margarine spread well. The cost of the fats also is an important factor. It has already been stated that cocoa butter cannot be considered as a fat for margarine because of its high price and limited supply. Availability of fats is also to be considered, and fat blends must change with the market, not only in order to keep production going, but to try to maintain as constant a price/cost structure as possible. The variety of processes that are now available– hydrogenation, ester interchange and fractionation–help the technologist to make the best use of available fats in order to produce a constant product.

Margarines must have plasticity in order to fulfill their function. This can only be achieved if the product has the correct structure. Plasticity is obtained by a judicious blend of solid and liquid phases, and a knowledge of the solid glyceride content of the various fats is a pre-requisite for good blending. We have already considered methods of determining the composition of fats in terms of solid fat index, either by dilatometry or by n.m.r. As the latter is a relatively recent method, most control and development of fat blends has depended on dilatometry. Determination of the distribution of solid glycerides in each available fat over a range of temperature gives a starting point for calculating the mixture which, it is expected, will give the desired overall solid glyceride structure. The blend will not precisely agree with the theoretical calculation, as the mutual solubility of the glycerides will affect the measured solids content. Measurement of solids content must be carried out on samples which have been solidified according to a rigorously standardized procedure.

Small changes in the solids content of a margarine can produce quite marked differences in the consistency of the product, partly because of the relatively low level of solids usually present. At a temperature of about 75° F (25° C), which is a reasonable room temperature, the solids content of a table margarine will be 10–20%; for other margarines a higher solids content may be needed.

Preparation of the fat blend for margarine manufacture must be well controlled. As margarine is an emulsion, any off-flavour in the fat blend may be accentuated in the margarine. Refining must be good, removing all the phosphatides and reducing the free fatty acids to below 0.1%. Bleaching must reduce the colour to a level at which it will not interfere with the colour of the final product. Deodorization should yield a bland oil with a free fatty acid around 0.05% and no peroxide value. Opinions differ concerning the best procedure for preparing the final fat blend. Two alternatives are possible, either to blend the various fats together and then deodorize the blend, or to deodorize individual fats and then blend them. Whichever course is followed the most difficult test to make on the finished blend is assessment of taste. It is rare for a deodorized fat or blend to be so perfect (i.e. free from all taste) that no one can detect a flaw in it. Experience has shown that the ability of each individual to recognize a substandard fat will change, not only from day to day, but also during the same day. It is good practice, therefore, to have a group or panel of tasters to check each batch or blend before passing it into margarine production. The best results are obtained when the freshly deodorized fat can be turned into margarine immediately it becomes available, although storage of up to 24 hours should be acceptable if the conditions are good. When a margarine manufacturer does not have his own refinery, care must be exercized in the transport of the fats to the factory.

When margarine was first made the fat was emulsified with milk before churning. Although the method of manufacture has changed and the basis of margarine formation is much better understood, it has not been found possible to exclude milk from the margarine without some loss of palatability. There is no doubt that milk in the margarine contributes to the overall flavour, and it also influences the texture and stability of the product. Milk of different types may be obtained from various sources. One of the greatest dangers is the introduction into the margarine of off flavours and high bacterial contamination with the milk. The nutritive qualities of milk are well known, but these very qualities make difficult the protection of milk from rapid bacterial growth. Factories which use liquid whole milk or skim milk must make provision to store the milk in refrigerated tanks until it is required. Whole milk will introduce a small amount of butter fat, less than 1%, and it is doubtful if this has a beneficial effect. Skim milk is a much cheaper raw material and is considered to be just as good. Both types of liquid milk must be checked on receipt in order to ensure that they are suitable for incorporation in margarine. The flavour must be satisfactory, free from taints and foreign flavours. Acidity should be low (high acidity indicates growth of souring bacteria) and the bacteriological quality, as shown by the methylene blue test, should be checked.

It is not always possible to have supplies of liquid milk in adequate quantities and of uniform quality just whenever they are needed and in these circumstances spray-dried skim milk powder is often used, and reconstituted just when required. Whichever type of milk is selected, the bacterial population must be reduced or removed by pasteurization or sterilization before use. In modern plants, plate heat exchangers with heating and cooling sections are used for either pasteurizing or sterilizing the milk. This type of plant gives flash heating and cooling of the milk.

Although some manufacturers use sweet milk in margarine manufacture, the more usual practice is to culture, or ripen it in order to improve the flavour. A great deal has been written about the ripening of milk and the development of flavour and aroma. Its origins are in the dairy industry and stem from investigations into the improved flavour of butter made from soured cream. A study of the system revealed that certain bacteria acting together converted lactose in the cream to lactic acid and then produced various flavour materials, some of which were not identified and others, such as diacetyl and dimethyl carbinol, were known. For many years, flavour manufacturers have sought to identify the components of the flavour of butter, in order to make a suitable additional flavour which would make ripening unnecessary. The process which was developed to control souring of cream for butter making has been adapted to preparing ripened milk for margarine production. Cultures of the ripening organisms, such as *Streptococcus cremoris*, *Str. diacetylactis*, and *Leuconostoc cremoris* are available commercially as a freeze-dried, balanced, mixed culture. A starter culture is obtained by adding the freeze-dried culture to a pint of sterilised milk and incubating it at 73 to 77°F (23–25°C) for 18 to 24 hours. A good growth of organisms will take place and the starter can be used to inoculate a series of second generation starters. The number of propagation stages depends on the total amount of cultured milk required.

After the milk has been sterilized it is pumped into culture vats. Some manufacturers use vertical cylinders for ripening milk, but the alternative horizontal semi-cylindrical type are probably more widely used. The vat is built in stainless steel and jacketed. Inside is a sweep agitator which often doubles as a cooling coil. The temperature of the milk is adjusted to the suitable one for the culture, about 73 to 77°F (22–25°C), and one of the starter cultures is added and distributed by the agitator. Incubation is continued for 12 to 18 hours, and then the temperature is lowered to stop bacterial activity. Some manufacturers use a two-stage ripening process in which the ripening is started at about 77°F (25°C) to encourage rapid acid development during the first 6 to 8 hours. The temperature is then lowered to 68°F (20°C) to encourage the aroma development over another 10 hours. At the end of the ripening period the milk is cooled to about 41°F (5°C) and held at this temperature until used.

Another milk product which is sometimes used in margarine is whey, the product left after removal of curd in cheese making. This material can contribute to a margarine flavour as an alternative to cultured milk. It can be used sweet, or cultured in the same way as skim milk. There is a risk of cheese flavours appearing in the margarine and the use of whey has not become wide-spread.

Salt is used in margarine as in butter for its influence on the taste. The level of salt used is dependent on market demand; it may range from less than 1% to 3 or 4%. Appreciation of the salt content is not simply dependent on the amount present. The salt is in the aqueous phase which is dispersed in droplets throughout the fat phase of margarine so that the response of the palate is dependent on the size of droplets and the rate of melting of the fat. A coarse dispersion will give a more intense reaction than a fine one, and a fat phase which has a low melting point and melts rapidly will also taste more salty than one heavier on the palate. In the United Kingdom the most popular salt content seems to be about 2%. There is a second function of salt in the margarine. It has a preservative action by suppressing the growth of bacteria. As the moisture content of margarine is only 16%, the actual concentration of salt in the aqueous phase is much higher than the 1 or 2% as the product suggests. If there is 2% salt in the margarine, the aqueous phase contains 12.5% salt, a concentration that inhibits the growth of many spoilage organisms. This factor is important in milk margarines in which the milk solids could provide a good nutrient to assist bacterial growth. Most producers of unsalted margarine do not use milk in their products because of this hazard.

The quality of the salt used should be controlled strictly. Traces of metals such as copper and iron will make the fat liable to rancidity, and other impurities can cause instability. Salt for use in butter-making has been specified in a British Standard and in the UK this is the grade to be used for margarine. Usually, a pure vacuum-dried salt from underground deposits is used. This grade is available in bulk delivery by tank waggon, from which it can be blown directly into either a heated silo, or a saturator. Batch processes for margarine manufacture using liquid milk may add dry salt directly to the churn, but continuous processes use saturated brine, especially if re-constituted milk is being used. This ingredient can be reconstituted to a higher milk solids content than liquid milk contains so that salt solution can be used without reduction of the amount of milk used.

The quality of water is also important. Both reconstituted milk and brine need water for their preparation, while unsalted margarines have all water for the aqueous phase. A check must be made regularly on the quality of the water available and if there is any indication that it contains suspended soilids (peaty water) or poor colour, or is of low bacteriological quality, steps must be taken to either obtain better supplies or to purify the available one. Suspended solids must be removed by filtration; bacteriological quality can be improved by sterilising filters by heat, or by chlorination.

Flavours for the margarine are important ingredients. Efforts have been made in the processing of the fats to produce a blend which is as free as possible from any flavour. This fat made into margarine, without added flavour, would be unacceptable as margarine. Some flavour is introduced by the addition of cultured milk but for many palates the total flavour concentration from this source is inadequate. Studies of the flavour components of ripened milk, ripened cream, and butters, have slowly revealed the identity of the complex mixture of compounds which, in small amounts, but blended together, combine to give the flavour which is associated with butter. Strong butters owe part of their flavour and aroma to small amounts of butyric acid, and other organic acids have been detected. More recent work using highly developed equipment has shown that lactones also arè present in butter. These compounds probably result from the attack of bacteria on the fatty acids during ripening. Special flavour blends have now been made available for the margarine manufacturer to add, in small quantities, to his product in order to reinforce the flavour produced from the ripened milk. Control of such complex mixtures is difficult and a great deal of confidence is placed in the flavour producers who undertake strict quality control over the flavour. Regular tasting of the finished margarines is probably the best safeguard against changes in the flavours.

Emulsification of the aqueous phase within the fat phase is assisted by the presence of milk, but the standard additive to produce a suitably stable emulsion is monoglyceride (G.M.S.). This surface-active fat derivative has been mentioned previously. It should contain not less than 45% monoglyceride and the taste should be fairly good. It is difficult and not very pleasant to try to taste G.M.S., which has a melting point well above that of the human body. It must be free from any suspicion of rancidity, even though it is used in table margarines only at about 0.5%.

Lecithin, usually from soyabeans, is added to help emulsification and also to control the spattering during water evaporation when the margarine is used for frying. Soyabean lecithin usually contains 60–65% of the soya phosphatides, most of the balance being oil. Odour, taste, and the amount insoluble in acetone are the only check tests carried out on lecithin.

Colours used in margarine have been the subject of investigation and legislation in many countries. Up to 20 years ago, it was widely believed that the only way to produce a stable colour in margarine was by the use of synthetic dyes. Then the use of coal tar dyes in foods generally came under close scrutiny, and the publication of lists of acceptable synthetic dyes followed. With a very restricted choice of dyes available and different lists for each country it became almost impossible to produce a good butter colour in margarine using synthetic dyes. Natural colouring matters were available, but they were not all suitable. One natural colour which was always available was in red palm oil. Normally the colour of palm oil is removed by bleaching and heat treatment, but it was shown that careful refining, just sufficient bleaching to ensure absence of soap, and a low temperature deodorization produced a palm oil with sufficient colour to provide a natural colour additive for margarine. The seeds of a tropical shrub, *Bixa orellana*, contain a carotenoid pigment which has long been known as annatto. It has been used as a food colour in cheese, butter and margarine. There are a number of disadvantages in the use of annatto; it is sensitive to acid and alkali, leading to variations in colour if the acidity of the milk in the margarine changed. Under some conditions it can stain the wrappers used for the margarine; it is light sensitive; the colour produced is usually rather pinker than is expected in margarine. A number of synthetic carotenoid pigments were produced, including β-carotene. Synthetic β-carotene was attractive not only by virtue of the good colour it gave and its acceptability as a synthetic copy of a natural pigment, but also because, unlike all the other colours except palm oil, it contributed to the nutritional value of the margarine. This will be discussed further when we consider vitaminization. Most margarines today use β-carotene as the colouring matter.

Preservatives and antioxidants are minor ingredients which are not widely used in margarine. Various chemicals have been used as preservatives, but most of them are no longer permitted. No preservative is permitted in the United Kingdom. Some countries allow the addition of boric acid, benzoic acid, or sorbic acid, the last named being the least objectionable. As preservatives are added to prevent growth of bacteria in the aqueous phase, there is no need for them if about 2% of salt is used and production is carried out hygienically with good plant cleaning, and special attention to all milk-handling stages. Antioxidants are a different problem. The oxidative deterioration of fats is controlled by antioxidants and it may be considered that the addition of these substances would be advantageous. Usually margarine made with well-refined fats and stored properly will become unpalatable owing to various changes, including loss of flavour components, long before it is noticeably rancid. In view of this and the desire to present the public with a fresh product, little use is made of antioxidants even though many countries permit them.

A few other ingredients are used in margarine. Starch is an indicator additive which some authorities call for, in order to make it easy to check for the presence of margarine in butter. It is an unsatisfactory ingredient to add to margarine as it is rarely sterile, but is difficult to sterilize. The only form of treatment is to disperse it in oil and subject it to a heat treatment immediately before use. Another marker ingredient is sesame oil which is easily detected by the Baudouin test. In some countries sesame oil forms an obligatory part of the fat blend but it must be handled carefully if it is to be used as a marker. Sesamol, which is present in the oil, is the reactant giving the distinctive red colour. It is adsorbed on bleaching earth and distilled during normal deodorization. For these reasons sesame oil as an indicator must be lightly bleached and deodorized at a low temperature, 120 to 130°C; it must therefore be separately deodorized and added to the rest of the blend at the margarine manufacturing stage.

1.7.2 Vitaminization

Although a vitamin concentrate added to margarine can be dismissed as just another ingredient, it is such an important one nutritionally that it deserves more attention. It was considered that butter provided an important contribution to the daily requirements of vitamins A and D, but when margarine was first marketed no vitamins were added. This omission enabled nutritionists to draw attention to the inferiority of the product compared with butter. Manufacturers of margarine overcame adverse publicity by incorporating vitamins to a level which was above that of butter. The addition of vitamins A and D to table margarine was made compulsory in the United Kingdom in 1940 when the levels were set at 450 to 550 i.u. per ounce (15.9–19.4 i.u. per gram) for vitamin A and 90 i.u. per ounce (3.2 i.u. per gram) for vitamin D. Later, these levels were increased and in 1954 legislation not only set the standards, but specified the method of calculating the concentrations. The new levels were set at between 760 to 940 i.u. per ounce (27–33 i.u. per gram) for vitamin A, and 80 to 100 i.u. per ounce (2.8–3.5 i.u. per gram) for vitamin D. Other countries have made their own regulations, and at different levels. The first difficulty to be overcome in vitaminizing margarine was to find a suitable source of vitamins. Fish liver oils, especially cod liver oil and halibut liver oil, were the best-known sources, but it was not practicable to add sufficient fish liver oil to give the margarine a reasonable vitamin content without making it taste very fishy. The fat-soluble vitamins form part of the unsaponifiable matter of the fish oils, so a process was devised to separate the unsaponifiable from the oil and re-disperse it in deodorized vegetable oils. It was later found that whale liver oil and shark liver oils contained so much more vitamins than the fish oils that they could be diluted with vegetable oil and given a mild refining and deodorization. This change reduced the fishiness but still did not eliminate it. A new process was developed by Unilever Ltd. A solution of the liver oil was passed through an adsorption chromatographic column, the vitamins and most of the oil percolated through, leaving the fishy taste adsorbed on the alumina. Liver oils processed in this way had a much better flavour than those processed by any other method. Methods of high vacuum distillation were also successful in separating a vitamin concentrate.

Synthesis of the vitamins followed intensive research efforts. Vitamin D was obtained by irradiating ergosterol (a sterol from yeast) and crystallizing the synthetic vitamin from the irradiated mass. The product was named calciferol. The process for synthetic vitamin A (retinol) followed much later, but now that both of these fat-soluble vitamins are available commercially in a pure form, they are the ingredient of choice for vitaminizing margarine. The quantities required can be dissolved in oil and diluted to a known standard which can be added, as a concentrate, to the margarine blend. Palm oil was known to contain a quantity of carotenoid pigments which includes β-carotene. It was established that in the animal and human body the β-carotene was converted into vitamin A. Use of unbleached palm oil as a colouring matter provided some β-carotene, but palm oil also contains other, non-active pigments. The synthesis of β-carotene solved this problem. Pure β-carotene could be used to colour the margarine and all the colouring matter would also be pro-vitamin A, which enables the high cost of β-carotene to be partly offset by its contribution to the total vitaminization costs.

Prepared concentrates of vitamin A, vitamin D, and β-carotene, dissolved in refined oil, are available from reliable suppliers. Checks of the potency of vitamin A and carotene can be carried out using a spectrophotometer to measure the light absorption at appropriate wavelengths. Checks on vitamin D are not so simple and are usually taken as correct if the other two components are. For this reason the manufacture of vitamin concentrates is controlled by licence, the licencee being required to add the correct proportion of vitamin D to the vitamin A concentrate. It is usual for the finished concentrate to be dispensed into bottles, each of which holds just enough for a batch of margarine.

1.7.3 Production processes

In the preceding pages we have considered the preparation of various ingredients for the manufacture of margarine. Now we must discuss the techniques of combining these materials to make the product.

The ingredients are naturally divided into two phases – the fat phase and the aqueous one. These two phases do not normally mix, and when they are dispersed in one another, they separate quickly. In order to make a stable mixture, emulsifiers are added to hold them together in an emulsion. When fat and water are emulsified, one of them forms a continuous phase, the other the dispersed phase; if the water is the dispersed phase, then the type of emulsion is referred to as water-in-oil. This is the one required for margarine and it is also, of course, the type of emulsion which we know as butter. When making the emulsion the conditions must be selected so that the droplet size of the dispersed aqueous phase is just correct. If the droplets are too large the emulsion will tend to break, or 'weep'. Water loss will take place and the risk of bacterial growth will increase. At the other extreme, too small a droplet size will lead to complete loss of flavour due to the inability of the droplets to reach the palate. The margarine would then give a sensation of being pure fat. The intermediate condition will give just enough emulsion stability for the margarine to go through the process without loss of moisture and still yield its flavour in the mouth. During the processing the fat blend must be converted from its liquid state into a smooth, plastic solid, with the aqueous phase dispersed in the structure.

We have discussed the effect of the ratio of solid and liquid glycerides on the margarine and referred to the conversion of the emulsion into a smooth, plastic, solid. Before considering in detail how this is achieved we must examine, in a little more detail, the mechanism of the process.

Slow cooling of a fat, or fat blend, permits the higher melting glycerides to form crystals which grow as the fat cools, so that when cold there is a mass of large crystals and a quantity of uncrystallized glycerides. Not only would these conditions produce a very poor and unstable dispersion of the aqueous phase, but the product would be semi-fluid, granular, and unacceptable. The cooling of the emulsion must therefore be rapid in order to produce a large number of small crystals, which, if they grow, will not become large and coarse as in slow cooling conditions. An important aspect of margarine manufacture is the operation of the equipment so as to cool the emulsion as fast as possible. The most important developments in the industry have been aimed at this objective.

Methods of manufacture have progressed through a number of different stages from the original method of Mouriés to the automated continuous processes being installed in modern factories. Some of the intermediate methods are still in use in small factories in various parts of the world, or in larger factories for special products; consequently, there is no sharp division between batch processing and fully continuous methods. In many plants some stages have become continuous and work in conjunction with batch units, a combination which can, with good planning of operations, give an over-all continuous output.

Before the crystallization process can start, the ingredients must be combined, and the emulsion made. In batch plants the vessel used for making the emulsion is called either a churn, or a blender. The older plants used, and still use, a churn, a term which reflects the original connection between butter and margarine making. It is necessary for the vessel to be jacketed so that the temperature of the contents can be adjusted. Older churns were made of iron, tinned on the inside, and oval in shape, but more modern units are made in stainless steel, and are usually cylindrical. The oval type of churn is fitted with twin paddle stirrers, the blades of the paddles being perforated in order to produce a shearing action in the emulsion. This action helps to improve emulsification and the two agitators ensure good mixing and keep the emulsion in a liquid state. If the fats have been deodorized individually they are now blended together, either in a special blending weigh tank, or in the churn or blender into which the rest of the ingredients will also be weighed. At the time of blending, the temperature of the fat should be adjusted to between 100° and 120°F (40–50°C), although this temperature must be adjusted according to the melting point of the blend. The emulsifier is added to the fat at this stage and then dispersed in it before adding the aqueous ingredients. It will be recalled that the ripened milk is at a low temperature and the water will be at ambient temperature, so that when these two are mixed to give the aqueous phase, they assume a mean temperature of 45 to 50°F

(7–10°C). With the agitators running in the churn, the milk is added, followed by the water required, either as sterile water or as crushed ice. The form of the water is dependent on the requirements of the process of crystallization which follows the churning. The older methods of crystallization, especially the 'wet' method and the brine chilling drums, require pre-crystallization of the emulsion. The 'wet' method provides poor chilling of the final emulsion, so that unless the crystal formation is taken as far as possible in the churn, the texture of the product will be poor. Cooling by cold water in the jacket of the churn would not give a sufficiently rapid temperature drop, and the cooling would soon be slowed down by the build-up of solids on the side of the churn, even with efficient agitation. The cooling is achieved by adding crushed ice as part of the aqueous phase. When available, flaked ice is even more efficient as a chilling agent as the large surface area improves the rate of heat transfer. Salt is added to the blend in the dry state rather than as a brine, in order to allow the maximum amount of flexibility in ice addition. The amount of ice which can be added is limited by the permitted volume of aqueous phase, which sets a maximum of 16% water. Therefore temperature of the blend at the time of mixing can only be three or four degrees above its melting point if crystallization is to be promoted in the churn. As the glycerides begin to solidify the viscosity of the emulsion increases and when it has reached the desirable consistency it is passed from the churn to the next stage.

'Wet chilling' of the emulsion was the only method in the early days of margarine manufacture, and although it has been replaced in most places by more efficient techniques it is still used in some small factories. As the thick emulsion runs from the churn, it is sprayed with ice cold water which further chills it and removes the heat of crystallization. The emulsion solidifies into a granular mass floating in the water. The excess water must be separated and this is accomplished in open trucks equipped with weirs and sieves to allow the surplus water to drain away, while retaining the margarine granules in the truck. Crystallization continues to take place while the margarine is standing, and it is often necessary to leave it for several hours for crystallization to become complete.

The cooling which takes place in the churn converts some of the high melting glycerides into a large number of small crystals, which grow during the solidification and resting periods. Time must be allowed for this to take place, and it is common for the margarine to rest overnight before attempting to pass it through the working processes. Margarine granules made in this way occlude water in addition to that added as the aqueous phase. This excess water is removed by mechanically squeezing it away while the margarine is kneaded. As the glycerides crystallize during the rest period they tend to form a rigid interlocking pattern which must be broken down into a plastic mass. A series of plasticizing units have been developed to perform this operation. Lumps in the margarine are broken down by passing the mass through a series of rollers. Improved plasticity is produced by kneading, often under vacuum, and using machines similar to butter working equipment. After the working and kneading is completed, the margarine is moulded into blocks by a packing machine.

The process uses large quantities of cold water; the product is held in open containers for many hours and then worked in machines which are difficult to sterilize. Under these conditions it is impossible to prevent infection with spoilage organisms which start a deterioration of the product before it is even packed. Use of refrigerated water which runs to waste is inefficient. Investigations of methods to replace the cumbersome wet process with a more convenient and efficient one led to the development of the twin cooling drum system patented by Schou. The drums, each about 5 or 6 ft ($1\frac{1}{2}$–2 m) in diameter, were mounted side by side in parallel with a narrow gap of 10–20 thousandths of an inch (0.2–0.5 mm). They were constructed in mild steel with a carefully machined surface in order to maintain a uniform gap when they were rotated. Provision was made to chill the inside of the drums with refrigerated brine, which was circulated rapidly through the drums so that only a small temperature rise took place in the brine. A large temperature difference would allow uneven cooling of the drum surface and this would affect the product. While the brine circulated through the drums, the drums were rotated, one clockwise, the other anti-clockwise, and the margarine emulsion was pumped into the gap between them. The emulsion rapidly crystallized in the gap, and then as the surfaces of the drums reached the minimum space, the solidified margarine was spread in a uniform film on both surfaces. As the film was slowly carried round a half circle, crystallization continued and the temperature of the margarine was reduced. At a point diametrically opposite the feed position a scraper blade removed the margarine from the drum and dropped it into trucks where it could complete crystallization before working and kneading as in the wet process. Since there had been no contact between the margarine and the cooling medium there was no surplus water to be removed during the working stages. Only half of the surface of each drum was utilized in this plant, with a consequent failure to reach full efficiency. This was remedied by increasing the diameter of one roll and replacing the other with a simple transfer roll. The emulsion from the churn was pumped into a trough near the bottom of a large-diameter horizontal drum. A small roll running in the trough transferred emulsion to the large drum surface. The operation was the same as for the double drum, the scraper knife being located just below the feed trough, so that almost all of the drum was utilized for chilling. Again the film of margarine was collected in trucks and rested before being mechanically worked.

Both drum methods are cleaner and more efficient than the wet method. They also give much more scope for adjustment of operational conditions. At the churning stage, the degree of crystallization cannot be as high as the older method as the emulsion must be able to flow from the churn to the feed mechanism and remain fluid until it is picked up on the surface of the drum. This can be ensured by having a water jacket on the feeding trough to keep the emulsion at a set temperature. As the melting point of the glycerides in the fat blend covers a wide range, the trough can be warm enough to keep the emulsion fluid without melting the higher melting point crystals, which have already been produced in the churn. Adjustment of the emulsion temperature can then be easily carried out, as can the speed of drum rotation, and the thickness of film on the drum. The temperature of the drum surface is dependent on the brine temperature and this too can be changed, within limits, to give the optimum chilling conditions to the emulsion. All these factors also control the throughput of the drum. Good maintenance of the drum surface and the edge of the scraper blade are necessary if the performance is to be maintained. A mild steel surface, especially one that has been in contact with emulsion containing salt, will rust and corrode if left at the end of a production run. The surface of the drum should be swabbed to remove residual emulsion and then coated with a liquid oil to prevent condensation on the cold metal surface. A cold drum must never be washed with hot water as there will be serious risk of distortion of the surface. This type of drum process is still in operation.

Improvement of chilling came by eliminating the use of brine as a heat transfer medium and bringing the refrigerant (ammonia) to the drum itself. Direct use of liquid ammonia in the drum increased the efficiency of refrigeration and enabled lower drum surface temperatures to be reached. Gerstenberg and Agger used this cooling method in the development of a completely improved process for margarine manufacture. Margarine produced on the drum systems already described had to be rested for crystallization to become complete and then transferred to working units. By taking advantage of the increased chilling available with a drum cooled by direct expansion ammonia, rest periods could be reduced considerably and working processes brought closer to the crystallizing stage.

Preparation of the emulsion for this system takes place in a cylindrical vacuum blender. The fats are weighed into the blender, either as individual fats or as a blend, and the fat soluble additives, flavours and vitamins are added. A vacuum system is generally used to pull the ingredients into the blender in place of a pump. Agitation of the mixed fat phase is started, using a high-speed stirrer. The milk, brine and additional water (if needed) are then drawn into the blender and the emulsion is formed by agitation under vacuum. Mixing under vacuum prevents the incorporation of air into the emulsion. This is important as air bubbles in the emulsion reduce the efficiency of heat exchange. The Gerstenberg drum rotates faster than the older designs and a good heat exchange through a thin emulsion film is required for optimum performance. The blender is fitted with a water jacket so that the temperature of the emulsion can be controlled, and adjusted to the requirements of the type of blend. Some formation of crystal nuclei from the higher melting components will take place, but not to the same extent as in ice-chilled emulsions.

The drum is built of steel, usually with a hard-chromium-plated surface, and precision ground. It is mounted with its long axis horizontal. Liquid ammonia is pumped into the drum, where it is directed to the cooling surface. As it evaporates it abstracts heat from the surface, producing a very low surface temperature. The vapours are drawn back to the ammonia compressor for liquefying and re-cycling. It is essential, as with all direct expansion ammonia cooling systems, that the design of refrigeration system ensures that no lubricating oil can be carried over onto the cooling surface. If the compressor allows any oil to pass into the ammonia, it will be deposited on the inside of the drum surface, seriously reducing heat exchange efficiency. A feed trough is mounted on one side of the drum, on a level with its longitudinal axis. The surface of the the the drum forms one side of the trough which is jacketed on the other side and bottom. Warm water, thermostatically temperature controlled, circulates through the jacket to maintain the emulsion temperature. A float valve controls the level of emulsion in the trough. Emulsion is transferred from the blender via an emulsifying pump into the trough. The pump action helps to stabilize the emulsion so that separation of the phases does not take place in the pipe lines or the trough. As the chilled drum rotates, it picks up a thin film of emulsion from the trough. The thickness of the film is controlled by a spreader bar mounted above the trough. This unit, which is adjustable, skims the surface of the film on the drum, returning excess emulsion to the trough. During its travel of one rotation of the drum, the emulsion solidifies, crystallizes, and leaves the drum surface as a very thin continuous sheet of margarine, which is removed by a knife mounted underneath the trough. This film of product falls into a hopper which is mounted below the drum, and is designed to hold the margarine produced during a run of about half-an-hour. When the hopper is full, margarine flakes are drawn from the bottom, at the same rate as they are falling in at the top, so that once production is established the margarine flakes flow continuously through the hopper, having a resting period to complete crystallization and attain a suitable temperature for mechanical treatment. As the film of margarine falls into the hopper it breaks up so that by the time it reaches the bottom it is in the form of flakes or powder.

The design of off-take from the bottom of the hopper is varied to cope with different margarines which may require a longer crystallizing time. If the blend is ready for immediate working, the hopper is mounted directly onto the feed worms of the 'Complector'. The latter comprise two parallel worms rotating in opposite directions, and so transporting the flakes from the hopper into the Complector barrel. The main unit contains a shaft which is fitted with a single worm at the in-put end. At the end of the worm the barrel widens into a number of working units each of which contains a fixed perforated plate and a rotating knife, the knife being mounted on the same shaft as the worm. The barrel finishes with a coned section. A vacuum pump which is part of the plant, extracts air from the main barrel of the Complector so that the margarine does not become aerated during the working. As the product is fed into the worm conveyor, the air is removed and the flakes start to be compacted. The product is then forced through the perforations and is chopped by the rotating knives until it reaches the coned section, where it is compressed and extruded as a ribbon. It may be found that the margarine is being overworked in the Complector, but this can be corrected by removing some of the working units and replacing them with blanks. It is usually necessary to allow the margarine to rest again before packing, so that extruded ribbon is picked up on a belt, the speed of which is adjusted to exactly match the extrusion speed of the Complector. The length of belt, which was formerly made of neoprene reinforced with canvas, but now may be stainless steel, can be selected to give an adequate resting time before the margarine reaches the packing units.

In the alternative system, which is needed for pastry margarine in particular, the worms at the bottom of the hopper are replaced by a pair of rollers which rotate intermittently in opposite directions, so that the margarine flakes are drawn out and dropped onto a conveyor. The flakes are lifted by the conveyor and dropped into the top of a second hopper which is mounted on the Complector unit as already described. When this unit is used it is usually desirable to feed the output of the Complector directly into a packing machine. Another facility which is built into the Complector is a warm oil circulating system which makes it possible to heat the worms in order to hasten crystallization. The system is made continuous by having two or more blenders available so that a series of blends can be prepared in order to keep a constant flow of emulsion to the drum. Provision is made for temperature control or variation to be introduced at the blender and the feed trough. Chilling can also be altered by control of the flow of ammonia in the refrigerating drum. The rest time in the hopper can easily be shortened, but not lengthened, and the working temperature can be controlled by the warm oil circulation system. The production rate can also be changed by altering the thickness of the film on the drum and adjusting the speed of the feed worms on the Complector. This flexibility has enabled better use to be made of all the fats that become available than was possible with other methods. As all the surfaces in contact with the product are made of stainless steel, cleaning of the plant is simplified and contamination can be reduced. Plant designs similar to the Gerstenberg have been offered by other suppliers such as Schröder.

In spite of the reduction in the operations during which the product is exposed to atmosphere, there are still sections where contamination of the product by air-borne organisms can take place, and even better production units, with less time between blending and packing, have been sought. The latest advance in margarine manufacturing techniques is the use of a tubular chiller, in which margarine emulsion passes through the inside of a tube, with the refrigerating medium on the outside. In itself a good development, in that the margarine is not exposed during the chilling period, it also allows the whole process to be telescoped into one compact plant. A number of plant manufacturers have developed their own versions of the tubular chiller including the Votator, Perfector, Kombinator, and Astra. These various plants, though based on the same principle, vary in detail, and the process is still in the course of modification and development. The older plants using drum and Complector, drum and separate working units, and even the wet process, are still being used but most of the larger manufacturers have already changed over to tubular chiller units, except for the production of some special products such as margarine for puff pastry, which some producers still believe can only be produced on a drum system unless certain characteristics are to be lost. Tubular chillers utilize a number of principles which could not be used in the previous types of plant. High efficiency of heat transfer is maintained by using a scraped film heat exchange technique (see below). Rapid crystallization is ensured by keeping the crystallizing mass in constant agitation and the amount of mechanical working can be considerably reduced if this can be applied during crystal formation, rather than breaking down the crystal structures which grow in a long quiescent crystallization period.

The main unit of the tubular chiller consists of a stainless-steel tube with a large diameter 'mutator' shaft running down its axis. The shaft carries scraper blades down its length. Between the shaft and the tube surface is a narrow annular space which is the working area of the unit. There is a wider tube surrounding the central one providing a uniform annulus for the flow of refrigerant. The whole unit is insulated and covered with a protective casing. During operation, the margarine emulsion is pumped into the central tube at high pressure. The wall of the tube is chilled by the direct expansion ammonia

Fig. 1.35 Two Perfectors (left) direct coupled to fully automatic moulding and wrapping machines (centre). On the right two case filling machines.

system, so that as the emulsion comes into contact with the cold surface it crystallizes and solidifies. The mutator shaft rotates at a high speed forcing the scraper blades against the wall of the tube and scraping the margarine off, leaving a clean surface. This process of solidification and scraping continues and the pressure of the pump forces the now cooled and solidifying mass down the tube. Meanwhile, the margarine is also being subjected to vigorous agitation which keeps the crystals from forming a firm interlocking structure. Provision is also made for mechanical action to further work the margarine.

The Votator was developed by the Girdler Corporation in the USA and has become one of the most widely used tubular chilling plants. It is built in two stages, the 'A' unit which is designed on the basis already described and the 'B' unit in which the crystallization is allowed to come to completion before passing on to the packing machine. One of the design characteristics is the floating action of the scraper blades which allows the centrifugal force of the rotating mutator shaft to throw the blades into close contact with the wall of the tube so scraping off the solidified margarine. Margarine blend, at a temperature of about 100°F (38°C), is pumped into the narrow annulus of the unit at a pressure of 300 p.s.i. (21 kg/cm²). As it is chilled the temperature drops to about 50°F (10°C), and then it passes into the 'B' unit where crystallization continues while the product is gently agitated to keep it fluid. In order to keep the unit compact, the heat exchange tube is divided into three short lengths with inter-connecting pipe-work. The

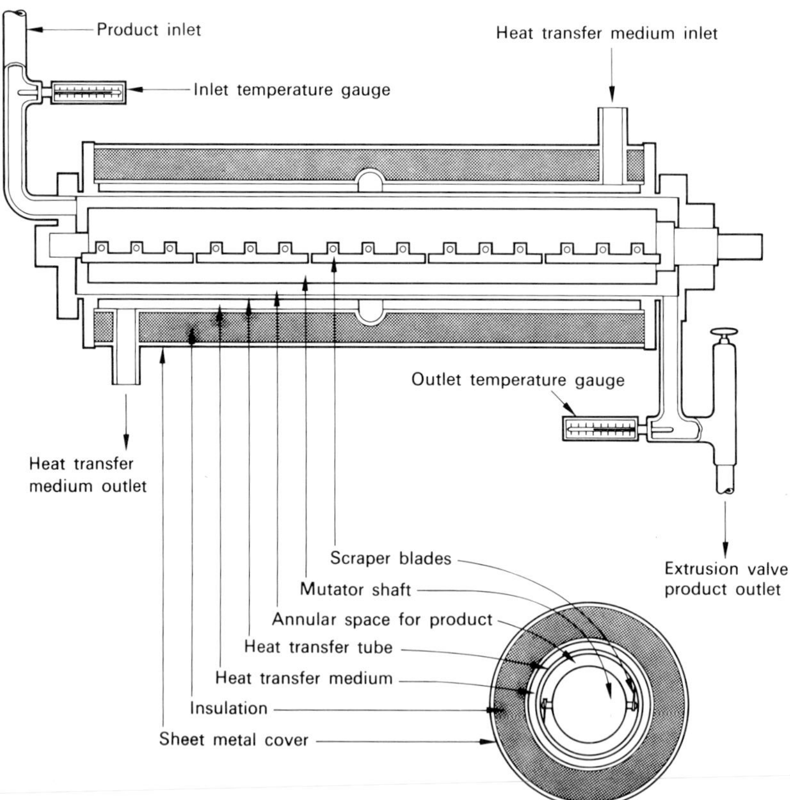

Fig. 1.36 Cross-section of Votator tubular chiller.

outlet of the last tube passes through an extrusion valve into the 'B' unit from which the product flows in a soft, semi-solid state. This material is fed into the packing machine where it is formed into the familiar blocks of margarine.

Although the process as described follows the guide lines laid down by the manufacturers, users of the plant have patented a number of variations of the process, including schemes for producing some crystallization in the blend before feeding through the Votator. Schemes for after-treatment have also been proposed. The Gerstenberg 'Perfector' is also based on the system described, but the makers claim advangtages over the others. One of the patented features is the design of the mutator shaft and scraper blade assembly. The scrapers are mounted so that spring action presses the blade against the tube wall and in operation the centrifugal force presses it more firmly against the surface, giving a very efficient cleaning of the heat exchange area. The plant is available in various sizes, the number of tubes being increased in ratio with the desired through-put at the rate of about 12 cwt (600 kg) per hour, per tube. As with the Votator, the tubes are interconnected by pipework, and finally the margarine passes into a resting tube where crystallization is completed. Between the last chilling tube and the resting tube there is a compensator which controls the flow of the margarine through the resting section and into the packing machine. Excess margarine is either fed back to be re-processed (i.e. melted and passed into the emulsion line) or to a second packing unit. The motive force to drive the emulsion through the plant is a pump which can work at a pressure of up to about 600 p.s.i. (40 kg/cm^2). This pump can also be modified to act as a metering pump feeding the oil and aqueous phase into the unit without any pre-blending.

The development of the tubular chiller process has not only made obsolete the kneading and working machines, but has also made it possible to dispense with conventional blenders. They can still be used as units for weighing and mixing all the ingredients, and feeding the prepared emulsion to the tubular chiller, but other devices can be used with advantage. The simplest way of feeding the chiller unit is to prepare the fat soluble materials in one tank and the milk, brine and aqueous materials in a second. A proportioning pump can then be set to transfer measured quantities of both phases, in a reliable ratio, to the chiller, where mixing takes place on a small scale immediately before passing into the chilling tube. In this way the uniformity of the margarine emulsion is more reliably controlled than when it has to flow through pipelines where there is a risk of separation of the phases. This metering system can be carried even further by the use of a multiple head metering pump. In order to ensure uniformity of flow in the metering pump systems, all the pump heads are mounted on a common drive shaft connected to one motor. Each ingredient is fed by its own variable flow pump-head so that there is no need to separately measure any of them. The total flow can be combined in a mixer and fed directly into the tubular chilling unit. With these advances in plant design it is now possible for all the ingredients to be blended, fed into the crystallizing unit, passed through the working stages, and be packed, without any exposure to the air, and the whole operation completed in a period of under half an hour instead of several days.

In a tubular chiller plant there is always some imbalance between the chilling and crystallizing units and the packing machine, resulting in the need to by-pass some of the processed emulsion back to a 'rework' tank. If any problems develop in the packing machine, or if a stoppage is necessary to fit a new reel of wrappers, the tubular chiller output must either be fed back to a rework tank, or the whole machine stopped. Rework tanks are not very satisfactory units: the product in them must be kept mixed to prevent separation, it has to be heated to melt the returning margarine, and if the flow is irregular it is not simple to maintain the rework tank at a constant temperature if cold margarine is being added at intervals. The rework material has to be fed into the main stream of margarine emulsion passing to the chilling unit. If this supply comes as emulsion from a blender, the addition of a proportion of rework material is not difficult, but when the emulsion is being made by the proportioning pump system it is more difficult to make provision for handling the rework material. A special unit has been developed to solve this problem. When the margarine is by-passed from the tubular chiller, instead of feeding it into a rework tank it is fed into a re-melting and tempering unit. The connecting pipework is enclosed in a hot water jacket so that the margarine is melted on its way. Melting is completed in a plate heat exchanger, which is heated by hot water circulation. The molten emulsion then passes to a tempering unit which cools it down again to about two degrees above its melting point, before passing it back to the tubular chiller. Fresh emulsion is also fed through the same tempering unit so that the two flows merge. As the remelt flow increases, the system controls the production of new emulsion to maintain a correct throughput.

We have referred on several occasions to packing the product, but have not yet discussed the units used. In the UK the simplest plant is used for bulk margarine in 28 lb ($12\frac{1}{2}$ kg) blocks. A hopper to receive the margarine is built over a worm which conveys the product into a cone-shaped section which terminates in a nozzle. As margarine is forced into the cone by the action of the screw it is compressed and extruded as a bar, which is chopped into blocks by a cutting wire, wrapped in parchment paper, and put into boxes. This type of machine can be used with almost any production unit. Margarine for catering establishments and bakery use is usually packed in this form. Some margarines can be packed in a fluid state straight from the tubular chiller, in which case the pack is usually a polythene-lined box or a metal can. The margarine sets solid within a few minutes of filling. Most of the domestic products are packed in half-pound (227 g) or one-pound (454 g) blocks or into plastic tubs. As the soft tub margarine is produced under slightly different conditions from packet margarine, we will consider the packet material first.

Half-pound and one-pound blocks of margarine are wrapped in vegetable parchment or foil wrappers which are supplied in reel form. The wrappers are drawn continuously into the machine until a hole or marker on one side of the wrapper is detected by a controlling device in the machine. This device actuates a guillotine which cuts the wrapper to the right size and feeds it into the actual wrapping unit. The packing machine can be mounted directly onto the resting tube of a tubular chiller, or fed from a hopper when a drum system is being used. In the bottom of the hopper is a pair of feed worms which force the margarine into a compression chamber terminating in a nozzle, which is again connected to the block-forming and wrapping unit. The central part of the packing machine is a rotating drum in which are a number of rectangular pockets. A piston head forms the bottom of the pocket. The drum rotates until one of the pockets comes opposite the end of the filling nozzle and margarine is forced into it. A sliding shoe ensures a good joint between the filling nozzle and the drum. When the pocket is full the drum rotates, presenting another pocket for filling. Meanwhile, as the drum rotates, the machine feeds a cut wrapper over the open end of the filled pocket and the piston pushes the block of margarine out of the pocket. Various arms pick up the block of margarine, complete the folding of the wrapper around it and then fold in the ends. The amount of margarine in the pack is controlled by the depth of the pocket. This can be varied, and on the current models the adjustment can be made while the machine is running. An operator takes the wrapped blocks and packs them into a carton. Even this stage can now be automated by the use of a case-packing machine which collects the blocks of margarine as they come from the packing machine and slides them into boxes. As the individual packs come out of the packing machine, samples must be taken at random and weighed. The packing machine should be adjusted to give the minimum overweight, but no underweight is permitted.

It will be noted that the dispensing of the margarine for formation of the blocks is entirely volumetric and depends on the precise filling of each pocket. The blocks of product are sold by weight, so that variations in the density of the margarine or failure to fill the pockets completely will lead to low weights. For these reasons the performance of the equipment must be studied and a quality control chart drawn up. This ensures that no underweight packets are produced, that the minimum overweight is packed, and that the machines are only adjusted when the tolerance limits are exceeded. The importance of restricting overweight to a minimum can be best appreciated by considering the actual 'loss' of product which appears to take place. A modern packing machine can have an output of about 120 blocks per minute, so that an overweight of one-sixteenth of an ounce (1.8 g) on each half-pound (227 g) block amounts to 0.5 tons of product on each packing machine, in a 40-hour working week.

The most recent development in margarine manufacture and packing is the production of soft margarine, which is packed in plastic tubs. There are few changes in the production procedure, although the technique and composition differ. This product can only be made on the tubular chiller type of plant. Only a limited amount of crystallization must take place after packing so that more working and chilling is needed before the actual packing operation. As a result of these operations the margarine is extruded from the resting tube as a thick cream. The tub-packing machine operates with a

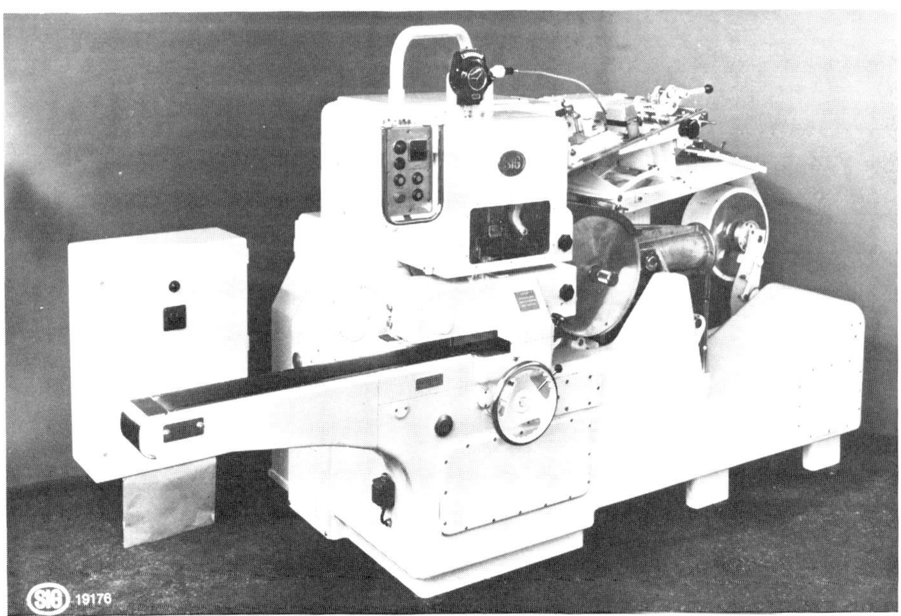

Fig. 1.37 Automatic SIG packing machine which delivers packs of margarine to the conveyor on the left.

flat turntable which has a number of sockets round the outer edge. Empty tubs, fed from an overhead magazine, are dropped into the sockets one by one. As the turntable rotates, a sensor checks that a tub is in place and then a metered amount of margarine is extruded into the tub. At another station round the turntable the lids, again fed from an overhead dispenser, are dropped into position and then pressed on. As the turntable continues its rotation the filled and lidded tubs are lifted out of their sockets and pushed onto a conveyor belt which takes them away to be packed into cartons. The plant can fill up to 180 tubs per minute.

Before leaving the subject of packaging, some consideration should be given to the materials used for packs. In the UK vegetable parchment is the established material for packet margarine. Bulk blocks are also wrapped in this material, which should conform to the British Standard. A cheaper wrapping material which is sometimes used is high wet-strength grease-proof paper. Care must be taken with this material as it is treated with resins during manufacture, and these sometimes liberate formaldehyde, an undersirable substance to be in contact with foods. The wrappers are decorated with various designs and lettering which must be printed with carefully selected inks, so that there is no risk of the margarine being tainted with a printing ink odour. The inks must not be soluble in margarine or pinholes in the wrapper could lead to marking of the product and the ink would smear in use. Laminated foil wrappers must have a suitable adhesive between the foil and the parchment; a mineral wax such as microcrystalline is frequently used.

At present, most tub margarines are packed in a special grade of P.V.C., but other plastics such as A.B.S. or polystyrene are being used. At least one manufacturer is using a two-part tub with a waxed cardboard inner liner and a P.V.C. outer. The lids are of similar plastics to the tubs.

Margarine is a food which is consumed without cooking. It is therefore of vital importance that a very high standard of hygiene is practised in the margarine factory. The dairy where the milk is prepared and ripened and the pipelines along which the milk is pumped are areas which require special attention and careful cleaning. The materials used in building the plant and any auxiliary equipment are also important. Stainless steel is the material of choice wherever possible.

1.7.4 Various types and their properties

Domestic margarine is now available in two types, a hard margarine which is still regarded as a butter replacement, and a soft margarine. The harder product is used for all domestic purposes and must, therefore, be formulated to be versatile. It must spread fairly well, at least as well as butter, and must have a good butter-like flavour. The same product is expected to make good cakes, so that it must possess adequate creaming properties, a function which is considered in more detail for cake margarines. In some countries where margarine is used for frying foods an anti-spattering agent is incorporated. These requirements are met by selection of a fat blend with a good plastic range and a solids content of about 15 to 20% at 68°F (20°C), and an upper limit of the melting range at 95°F (35°C), so that it melts rapidly and completely in the mouth.

Soft margarine is formulated so that it can be stored in a refrigerator until required and then can be spread straight away without any heating; this demands a low solids content at 41°F (5°C). At the same time the margarine must retain sufficient firmness at normal room temperature so that it does not melt on the bread. This type of formulation requires a large amount of liquid components together with some well crystallized glycerides to give sufficient body to the product. Although primarily designed for spreading, a good soft margarine will also cream well and produce good cakes.

Two other types of domestic margarine products are promoted as health foods. One of these contains a high proportion of poly-unsaturated acids and is considered especially suited for reducing the risk of heart disease. When we were considering the biochemical and nutritional aspects of fats, the current hypothesis of the importance of poly-unsaturated fatty acids in the control of heart disease was mentioned. Some soft margarines made entirely from vegetable fats have been formulated to have a very high poly-unsaturated fatty acid content. The method of manufacture is the same as for other soft margarines, but the formulation is restricted to a few oils. In order to get a product with a high poly-unsaturated acid content and retain flavour and stability, oils with a high proportion of linoleic acid, but low linolenic acid content, are needed. Sunflower seed oil with about 60% linoleic acid and safflower seed oil, with 70%, are the obvious best choices to provide a high linoleic acid content. Soyabean oil also has a high linoleic acid content (54%), but the presence of about 5% linolenic acid makes the oil less stable and at high concentrations there is some risk of reversion flavours developing.

Products which have a 'low calorie content' have been put on the market. They cannot be sold as margarine because they do not fulfil the requirements of legislation to contain not less than 80% fat and not more than 16% water. This type of spread, which is unsuitable for cooking, achieves a 'low calorie' level by replacing part of the fatty matter with water. It is an emulsion with about half of the normal margarine fat content, which when spread on bread in the customary quantity, only adds to the meal half as much fat as a conventional margarine would. This product is packed in tubs similar to ordinary soft margarine.

Turning from domestic products to the requirements of the bakery trade, we find that there are a number of special products required to perform a specific role. One that has been mentioned several times is pastry margarine. Puff pastry is required to have a structure consisting of distinct layers of crisp material well separated by air spaces. Both the technique of baking and the margarine used play important parts in achieving the correct end product. The baker makes his dough with a little soft fat, or even no fat at all, and rolls it out. Lumps of pastry margarine are deposited on the slab of dough which is then folded, turned through 90° and rolled again. The rest of the margarine is placed on the rolled dough which is again folded, turned, and rolled. This turning and rolling process repeated until it has been done eight or nine times. A cross-section of the finished dough shows a system of alternate continuous layers of dough and margarine. During cooking, the margarine acts as a seal and a barrier, permitting the moisture and raising agents to lift the layers apart, and preventing them from sticking together. The margarine layer maintains this open structure until the dough has set into its rigid form. In order to fulfill all these requirements the margarine must be tough and have a high melting point. The toughness is helped by reducing the moisture content to about 10% instead of the conventional 16%. It is also usually made without milk. One of the fats which was found most useful for pastry margarine was oleo-stearin, which has a melting point of about 130°F (54°C), but this can be replaced with hardened vegetable or fish oils of similar melting point. With this type of product longer crystallizing and resting times are needed. This leads manufacturers to use the older types of process which gives much more time for resting than the continuous process. Although designers of tubular chillers claim that pastry margarines can be made on their equipment, most margarine producers find that drum coolers give a more satisfactory product. Even then it is found that the margarine improves considerably if it is tempered after packing. Although pastry margarine fulfils its purpose very well, it has one disadvantage. If the puff pastry is kept too long before eating, the high melting glycerides recrystallize. In their recrystallized form the glycerides do not completely melt in the mouth, and therefore they tend to give a waxy sensation on the palate. The term 'palate cling' is used to define this effect. Freshly prepared pastries should not have this defect.

In the production of cakes, the function of fat is very different from its function in puff pastry making. Cake margarine contributes to two aspects of cake structure: its creaming properties and its effect on crumb structure. When flour and water are mixed to form dough, the gluten in the flour is formed into strands which constitute an elastic network enclosing the starch granules. The inclusion of fat in the system shortens the gluten strands giving a more crumbly product. Variation in structure and type of end product is partially dependent upon the fat and the method of incorporation. Obviously this is very much a simplification of conditions and there are a number of other factors which play their part. The more critical function of the cake margarine is its creaming performance, which contributes a great deal to the raising behaviour of the cake. The ability of the margarine to be whipped up and hold air is a result of its crystal structure. If the product contains a proportion of well-formed fairly high melting point crystals, while the rest of the fat is in a liquid phase, it will be plastic and a product into which air can be incorporated. Creaming is a process in which margarine and sugar are beaten together in a high-speed mixer. The presence of the sugar crystals helps to increase the incorporation of air. When the aqueous ingredients of the cake, eggs and milk, are added to the creamed fat and sugar, the mix should be able to hold the air. Addition of small quantities of certain emulsifiers to the margarine will not only increase the incorporation of air, but also will help it to emulsify the liquids without loss of the air. This is important because when the flour is added and the cake batter is baked, the air bubbles entrained in the fat emulsion expand with the heat of the oven, raising the cake, opening the texture, and improving the eating quality. This last point is difficult to define, but easy to observe. The crumb becomes more tender and flavour is improved, the flavour of the margarine blending with the flavours of the other ingredients to give a fuller, richer, taste. The cake size, lightness, and texture, are all dependent on the margarine, and these results can to a great extent be judged by a test of its creaming properties. The manufacture of cake margarine is not different from the techniques already described, but the composition of the fat blend is most important. A good distribution of melting range with a small amount of high melting components and plenty of liquid fat provide a good start. Various emulsifiers may be used as additives to enhance the creaming properties and air retention.

A cream replacement for cakes is made from a special margarine and sugar, with or without the further addition of milk powder. The margarine for this cream-filling product is a special uncoloured margarine which is formulated to give a very high over-run; that is, it is very good creaming margarine. Care in formulation of this product is necessary so that there is good resistance to oxidation. This is achieved by the use of hydrogenated oils.

1.8 OTHER OIL AND FAT PRODUCTS

The use of locally available fats in the preparation of food has established a specific pattern of traditional fat usage in different areas of the world. In this section we will examine the preparation of oils for cooking and salads, and shortenings and lard products.

1.8.1 Salad and cooking oils

There is no real difference between oils to be used for the two purposes; the distinction is based only on the appearance of the oil when stored at refrigerator temperatures. A salad oil should remain bright and clear, whereas a cooking oil may be very cloudy or even solidify under these conditions. The same oils, with one exception, may be used for both purposes by introducing a process of 'winterization' for oils intended for use as salad oils. The exception is groundnut oil which cannot be winterized, because at low temperatures the oil tends to set to a gelatinous solid.

The demand for salad oil to be clear at low temperatures probably originates with the introduction of cotton seed oil as a replacement of olive oil. As olive oil remained clear at low temperatures, any oil intended to replace was expected to behave similarly. Hence all salad oils must be clear at low temperatures. There is also a reluctance among the buying public to accept as a salad or even a cooking oil, a product which is non-uniform and which appears to have a deposit in it. As a consequence, there has developed a convention that bottled oils should remain clear, so that de-waxing of sunflower seed and corn oils is considered necessary even for cooking oils, especially when sold in bottles. Olive oil is one of the few oils which in the untreated state can be used successfully as a salad oil. Its flavour is stable and in the unrefined state is regarded not only as acceptable, but desirable, and it remains clear in the refrigerator. Soyabean oil also remains clear at low temperatures but it is not so stable to oxidation, and flavour reversion sets in giving it a pronounced 'beany' tang. The flavour can be stabilized by hydrogenation, but as the conventional hardening process forms solid glycerides as well as reducing the unsaturation, soyabean oil treated in this way must be winterised. There has been some development in this field using alternative catalysts to the normal nickel one. It has been shown that if a copper chromite catalyst is used, the poly-unsaturated acids are converted to less unsaturated ones without the formation of any solid acids. If this type of process is used, all the tri-unsaturated acids can be reduced, giving a much more stable oil, suitable for use as a salad oil, without any subsequent winterisation. Although this system has been shown to be possible, there are a number of practical obstacles to the widespread use of the technique. It is necessary to have a completely separate hydrogenation plant for the process as there must not be any risk of the copper catalyst becoming contaminated with nickel. Also, and perhaps more difficult, all traces of the catalyst must be removed before the oil comes into contact with the air.

Sunflower seed oil and corn oil are also suitable for preparing salad oils but both contain a small amount of wax which will crystallize out at low temperatures giving a cloudy appearance. Cotton seed oil produces a great deal of solids when chilled, so it too must be winterized to be acceptable.

Winterization is the removal of highly saturated triglycerides from liquid oils. The process is normally carried out on the oil after refining, but a process has been developed in which the winterization is carried out in a solvent.

For conventional winterization oil is pumped into tanks equipped with cooling coils through which refrigated brine is passed. The oil is cooled to approximately 55°F (12.5°C) over a period of 6 to 12 hours. At this temperature crystallization commences and the rate of cooling is reduced so that over the next 12 to 18 hours the temperature falls to 45°F (7.5°C). As crystallization proceeds there may be a delay in temperature drop or even a slight rise in temperature due to heat of crystallization, but cooling is continued to about 42°F (5°C). It is then held for a further 24 hours at this low temperature before being filtered at a low pump pressure through a chilled filter press. The yield of winterized oil may only be about 80 to 85%, but the recovered solids can be used as a solid component of a shortening or margarine.

Other techniques for winterisation use plate heat exchangers or tubular chillers to reduce the temperature of the oil before putting it into insulated tanks for the crystallization to proceed. Variations on type of filter are also introduced to give easier filtration. One type of horizontal filter uses wire mesh plates which can be heated as well as cooled. When the filter area is blocked with accumulated solids the filtration is stopped and the plate warmed to melt the accumulated solids which are pumped away. The filter is then re-cooled and filtration continued. One of the difficulties of winterization is the high viscosity of the cold oil which makes filtration very slow. Though not widely used, miscella winterization is an answer to some of these problems.

Winterization in solvent has a number of advantages:

1   Equilibrium between solid and liquid phases is established more rapidly.
2   The crystals formed are large and more easily separated.
3   The viscosity of the liquid is reduced, making filtration easier and quicker.
4   Winterization can be carried out on solvent-extracted oils following miscella refining and bleaching before recovery of the oil.
5   It is possible to winterize groundnut oil using this process.

The oil, dissolved in a mixture of 85% acetone, 15% hexane, is chilled to $-10°F$ ($-12°C$) and held for about 2 hours before filtration. The yield of winterized oil is higher than with the first method as no oil is entrained in the crystal mass.

While de-waxing can be carried out in the same way as winterization, alternative methods have been proposed in which the waxes are crystallized and washed out with a detergent solution. The process starts by pumping the oil through a heat exchanger in which the temperature is reduced to between 40° and 60°F (5–15°C), causing the waxes to precipitate as fine crystals. The cold oil is mixed with a solution of detergent which wets the surface of the waxes so that they become suspended in the detergent phase. When the mixture of oil and detergent are centrifuged, the treated oil is obtained free from waxes. The detergent suspension can now be heated to melt the wax, and centrifuged to separate the detergent for re-use and the wax for disposal.

After winterization or de-waxing, the oil is deodorized in the usual way. It is cooled, any additives put in, and filled into drums, cans, or bottles. Large quantities of oils are used for cooking purposes and are packed into 45 gallon (200 litre), 5 gallon (22.5 litre), and 1 gallon (4.5 litre) containers, but most oil for domestic use is filled into quart and pint bottles or litre and half-litre bottles. Glass bottles are normally used, although various clear plastic bottles have been offered. The main advantage of plastic bottles is their lightness and the lower transmission of ultra-violet light. Against this is their greater permeability to oxygen and other gases, which can lead to deterioration and taints developing. Although it is intended that bottled oils should be used while fresh, if there is a delay the plastic bottles tend to soften and distort, owing to slow attack on the plastic by the oil. The bottling machines used for salad or cooking oil fill (by means of a volumetric filler) cap, and label the bottles in one operation.

Oils that are being used for cooking and deep frying tend to foam when food is fried. Oxidation of the hot oil first of all forms peroxides which rapidly break down to produce surface active substances. These form persistent bubbles when the water vapour is trying to escape from the food. In deep frying units, foaming of the oil not only presents a hazard, but interferes with the volume of food which can be processed at any one time. The addition to the oil of a small quantity of dimethylpolysiloxane (about 2 p.p.m.) considerably reduces the foaming and also extends the useful life of the oil by reducing the rate of oxidation. It is important to note that any plant which is used for handling oils containing anti-foaming agents must not be used for handling bakery fats. Very small amounts of the anti-foaming agent will reduce the creaming properties of shortenings and cake margarines.

It has been suggested that the useful life of frying fats can be increased by the addition of antioxidants but as the antioxidants are steam volatile the concentration in the oil is very rapidly reduced below the effective level once the food is put into the fryer.

1.8.2 Shortenings

Bakery operations require other fats in addition to margarines. This need was traditionally fulfilled by lard, which has been supplemented and replaced by shortenings. In some cases the new shortenings make it possible to produce baked goods which could not be made from lard.

We have considered the techniques of obtaining lard from animal tissues, but not methods of lard packing. Lard produced from different animals varies greatly in melting point and it may be necessary to blend lards from different sources in order to get a consistent product. An alternative way of stiffening a soft material is to add a small quantity of hydrogenated lard. If 5 to 10% of lard, hardened almost to saturation, is added to plain melted lard, a considerable stiffening effect will be obtained. One difficulty of this procedure is that legislation in some countries does not permit a blend of lard and hydrogenated lard to be sold as pure lard. Under these circumstances the only satisfactory way of stiffening it is to blend stiffer lard with the soft material. As the fat is frequently packed and used without any processing, a stringent check on specification is needed. Peroxide value and free fatty acid must be low, and the keeping quality checked by means of an accelerated test such as the Sylvester test or the Schaal test. Colour and odour are also a guide to quality. Animal fats do not naturally contain any antioxidants and therefore they do not have as good a keeping quality as the vegetable oils and fats. This disadvantage can to some extent be offset by adding a quantity of one of the permitted antioxidants such as butylated hydroxytoluene, butylated hydroxyanisole, or a gallate. These materials, added at a level of 100 p.p.m. can make a substantial difference to the keeping quality of lards.

The propyl, butyl, and dodecyl gallates all give very good antioxidant protection to lard, but do not 'carry through' to baked goods; i.e. they do not give effective protection to fat after it has been incorporated in the food and the product has been cooked. An antioxidant with good carry-through properties will extend the storage life of the finished product and also increase the stability of the fat before use. A second disadvantage of gallates is that in the presence of iron they can form blue-black iron compounds. Even a wire staple in a carton can be responsible for producing a discoloured area in lard containing gallates. Butylated hydroxyanisole (B.H.A.) not only gives fairly good stability to the lard but it carries through well. Butylated hydroxytoluene (B.H.T.) confers stability but, like the gallates, does not give good carry through. Citric and phosphoric acids are also added as metal sequestering agents to inhibit catalytic activity by traces of iron and copper. A great deal of information has been published on the tests carried out by various organizations, and a number of blends of antioxidants have been recommended for use.

The liquid lard is blended to give the correct melting point, the antioxidants are mixed in, and the product is ready to solidify. Several continuous processes are available for solidifying the lard, the simplest being a water cooled system which operates in conjunction with the Titan rendering plant. The system starts at a feed tank from which the liquid lard is pumped through a series of water cooled chillers of a similar design to the tubular chiller with rotating scrapers in a jacketed tube. The lard emerges from the chiller like a thin cream, partly crystallised but still fluid, and passes on to the filling points. As the product is still fluid the filling unit is fairly simple, either an on/off valve from which cartons can be filled manually, or a volumetric filler for filling into small containers. The crystallized lard sets very quickly in the pack and it can then be further chilled, and stored in a refrigerator. In order to prevent setting in the cooling tubes it is necessary for the feed line to pass the filling units and return to the feed tanks, making a continuous circuit. The last section of the return line is jacketed with hot water in order to re-melt the lard before it returns to the feed tank. This system is fairly cheap to run and easy to operate, but the pack, whether large of small, must comprise a pre-formed carton, either liquid tight or with a bag liner. This design of plant is particularly suited to the small producer, especially if the filling unit is integrated with a rendering plant. The large-scale packer needs a lard drum or a tubular chiller.

The construction of the lard drum is very similar to that of the single margarine drum, although for a given capacity it is probably smaller in diameter. Cooling of the drum may be by refrigerated brine or direct-expansion ammonia. Liquid lard is fed into a trough, one side of which is the surface of the drum. Lard is picked up on the surface where it immediately solidifies and is carried round the drum to a scraper blade mounted just below the trough. As the lard is removed in the form of a thin sheet it falls into a receiving trough where it is carried, by a screw conveyor, the length of the trough, into a working unit in which the product is homogenized. The homogenized product is packed by a machine operating in the same way as for margarine packing.

Tubular chillers have been produced to crystallize lard and shortenings in a similar manner to margarine. Provision is made on a lard and shortening chiller, for air or inert gas (nitrogen) to be incorporated in the product during chilling. The incorporation of gas in the product helps to soften the texture and make subsequent manipulation easier. The amount of working and resting in the second stage of the tubular chiller unit can be varied to produce a lard which can be filled into cartons while it will still flow, or made up into packets on the same type of packing machine as is used for margarine.

When judged by its creaming properties, lard is not a very good bakery fat. This is a result of its tendency to crystallize in large coarse crystals, a characteristic which still persists in the hydrogenated product, and when it is chilled rapidly on a drum or in a tubular chiller. The first stage of creaming, in which the fat and sugar are beaten together, may result in a reasonable incorporation of air, but the aerated mixture collapses as soon as egg and moisture are added. This problem has been overcome by interesterification of the lard, so modifying the glyceride composition and with it the crystal form.

The replacement of lard by shortening started with mixtures of refined cotton seed oil and oleo stearin. This type of product made by blending, or compounding, liquid or soft vegetable oils and fats with hard animal fats or high melting fractions, found a ready market as lard compound, or compound cooking fats. With the development of hydrogenation techniques the 'all-hydrogenated shortening' arrived on the American market. Based on cotton seed oil and later on soyabean oil, the product was made by hydrogenating one oil to a suitable melting point. A better product was obtained when two hydrogenated fats were blended. From these beginnings developed the numerous shortenings which are now available for many bakery purposes. Not only are the shortenings more consistent than lard, but they are made from deodorized fat blends giving a completely flavourless fat. This can be a great advantage in the production of sweet and delicately flavoured cakes in which lard may be regarded as having an unacceptable 'foreign' flavour. The formulation of shortenings and their production does not now differ substantially from the production of margarine. Fats are selected and blended in order to give the right plastic range for the baker, and are crystallized over a drum or through a tubular chiller in a similar way to the margarines.

Shortening which is being produced for the baking industry must have good plasticity. A plastic fat is one which is rigid enough to hold its form and appear solid under small stress, such as its own weight, but which will flow without fracture when a stress exceeding the yield value is applied, yet ceases to flow as soon as the stress is removed. The act of spreading butter or margarine is a demonstration of flow under stress. This structure is dependent on several factors, of which the ratio of solid to liquid glycerides is a determining one. If the solids content is too high a brittle fat will result, one in which the block will crack and break into pieces when a stress is applied, instead of flowing. Setting the correct level for the solids ratio does not, on its own, ensure a plastic material; the size and number of fat crystals also have a significant effect. Large crystals do not restrict the movement of the liquid portion of the fat and the product can quite easily appear as a suspension of solids in a liquid, with all the characteristics of a liquid suspension. A large number of small crystals give the right structure for a plastic fat. Under these conditions the crystals interlock to some extent, holding the fat rigid, but the application of a stress to the fat breaks the bond between the crystals, allowing flow to take place. The magnitude of the stress required to reach the yield value is expressed as the firmness of the fat.

Plasticity is of importance when using the fat to produce bakery products. A plastic fat when mixed into a cake batter will become distributed as thin films coating the starch and flour particles. It also functions to retain air bubbles within the batter. Observations have been made confirming that air bubbles are not dispersed in the aqueous phase of the cake batter, but are dispersed in the fat where they are retained during the early stages of baking.

It is the objective of producers of plastic fats to obtain as wide a plastic range as possible.

As shortenings are chilled the fats crystallize in an unstable state, and the plasticity and performance of the product can be much improved by storing the chilled and packed product at a relatively high temperature. This storage period is usually up to 72 hours duration at a temperature only a few degrees below its melting point. Under these conditions the crystal structure changes and stabilizes, so that on cooling to room temperature the product has a better performance and texture then it would have if merely stored after chilling. This process, known as tempering, is practised in most production units.

No aqueous phase is required, the product being 100% fat, nor is it usual to add any flavouring to shortenings. Although the absence of an aqueous phase makes shortening less vulnerable to deterioration by bacterial attack, it is more likely to be subjected to poor storage conditions, and therefore its resistance to oxidation should be good.

Shortenings are frequently used as an alternative to margarines in many bakery operations and for these purposes very similar fat blends are used. The major characteristics are the same, but the shortenings cannot supply the flavour given by the margarine. One type of shortening which has increased in usage over recent years is the 'high ratio' type. The bakery products for which this shortening is required have a high sugar and moisture content and a shortening with good emulsifying properties is required. This is achieved by adding relatively high concentrations of G.M.S. or superglycerinated fats.

There are a number of fat products which play a minor role in bakery processes and in cooking. Bread fat is added in small quantities to doughs which use yeast as a raising agent. There is no necessity to add fat to this type of product, but the addition to the bread of 3 or 4% bread fat will not only increase the loaf volume but give a better eating quality to the crumb. Incorporation of G.M.S. in the bread fat yields a softer loaf. As the fat acts as a lubricant and shortening agent, its composition is not important, so that it becomes the cheapest product in the range of cooking fats and shortenings.

Tin greasing emulsions are minor, but interesting, products. Their function is to prevent the adhesion of baked products to the baking tins. They usually consist of about 50% fat and 50% water in the form of a water-in-oil emulsion, part of the fat being emulsifier. The emulsion is sprayed onto the baking tins before the cake mix is put in. As the fat is in the continuous phase, it forms a barrier between the cake and the tin, and as the cake mix has a continuous aqueous phase the barrier does not break. Ideally, as the baking proceeds the crust of the cake forms and as the temperature of the tin rises, the emulsion breaks down, liberating the water as steam which, as it escapes, helps to break any adhesion which may have taken place.

Our consideration of margarine and cooking fats has been concerned with the products which replace or supplement the traditional butter and lard. In the Indian sub-continent the fat of buffalo milk is converted into a water-free fat product known as Ghee. The fat is prepared by churning the milk, separating the butter, and then heating until the water is driven off. The dried fat is allowed to cool, forming a coarse, granular product. Vegetable oil products made as a supplement to, or replacement of ghee, are known as vanaspati. The fats used are defined, hydrogenated and deodorized as already described, and blended to give a product of melting point between 86°F (31°C) and 98°F (37°C). The blended fats are then cooled and filled into tins while still liquid. The tins are cooled slowly to allow the growth of the desired granular structure.

1.8.3 Mayonnaise

The derivation of the name of this product is obscure, and a number of suggestions have been put forward with, it seems, little supporting justification. It is however, a fat-based product of importance in some areas; a product which may be described as a semi-solid emulsion. Its high viscosity is due to the large amount of fat phase which is dispersed in minute droplets throughout a rather small aqueous phase. Emulsification is assisted and the product stabilised by the phosphatides of egg yolk. As with other food products there are different regulations specifying the composition of mayonnaise, but for our purpose it is sufficient to note the proposed standard set by the European Economic Community. These require that the product should contain not less than 6% of egg yolk and not less than 80% total fat, including the fat present in the egg.

In order to make a good stable product the fat should be a liquid vegetable oil that is free from stearin, or which has been winterised. Crystallization of solid glycerides can be responsible for emulsion instability. Winterized cotton seed oil, maize oil and sunflower seed oil are satisfactory for the purpose.

The source of egg yolk can be either fresh egg or frozen yolks. It is reported that dried yolks can be used, but that they do not give such good results.

The type of vinegar used will influence the flavour of the finished product. Unless the malt or cider vinegar or tarragon flavours are wanted, the use of spirit vinegar will give the most satisfactory results. Other ingredients are sugar and spices. As in the production of all fat products, a check must be made on the ingredients to ensure the absence of trace metals, especially copper and iron.

Mayonnaise can be produced using a planetary stirrer in a large stainless-steel mixer. The egg yolks, vinegar, sugar and spices are blended together in the mixer, then the oil is added slowly and dispersed as fast as it is added. Increased stability of the emulsion is ensured by homogenization. In the homogenizer very high shearing forces are produced, which break down the droplets, reducing their tendency to coalesce.

1.9 STATISTICS

The general trend of world markets shows a continued rise in the production of fats. About two-thirds of the total is suitable for conversion into edible fat products, in addition to the butter fat produced. As butter contains 16% moisture and milk solids it is necessary, for the tables, to convert the butter output into terms of butter fat for the sake of uniformity, and to simplify comparisons on a common basis. In table 1.24 the data have been arranged to show the distribution of fats in accordance with origin and use. The group of animal, fish and marine fats contains both edible and inedible grades.

Changes in production are due to a number of factors. As with all agricultural products, vegetable oil seed production is dependent on the effects of weather and disease. Acceptability of fats for edible purposes is influenced by the observations of nutritionists, and the availability of fats from 'hunted' animals, whale oil and fish oil, is dependent on reproduction rate of these marine creatures. Because of over-fishing of whales the production of whale oil has fallen markedly, and the theories propounded concerning the links between animal fats and arterial disease has reduced the acceptability of such fats in manufactured fat products.

About two-thirds of the world population live in the poorer and undeveloped countries, and prospects in the world markets of oils and fats are influenced by progress there. It has frequently been pointed out that as the standard of living rises in a country, the demand for fats also grows. The widespread increase in demand for animal protein (meat) results in an increased availability of animal fats. It has been proposed that

Table 1.24 World supplies of oils and fats (percentage of total production)

	1968	1969	1970
Vegetable fats suitable for edible use	56.9	57.4	58.4
Butter fat	13.0	12.7	11.9
Other animal fats, including fish and whale oils	25.7	25.1	25.1
Non-edible vegetable oils	4.3	4.7	4.6

Source: Shortening and Margarine Manufacturers Association.

processes for converting animal fats into more acceptable fractions should be investigated more thoroughly.

The pattern of oil milling is being changed by the increasing growth of the oil-milling industry in countries where the oil seeds are grown. This change has been steadily proceeding for a number of years so that the oil-milling capacity of the United Kingdom has declined. One notable exception to this trend has been the building of a large extraction plant to produce soyabean oil. The trends can be seen in table 1.25.

Forecasting of future developments in the production and utilization of edible fats is almost impossible. The increasing emphasis on health aspects of foods is likely to produce a greater demand for the unsaturated liquid vegetable oils, especially those with a high linoleic acid content, and the continued research into the moulding of plant genetics to the needs of the population will undoubtedly bring new oils with new characteristics into the edible fats industry.

Table 1.25 Oil seeds imported into the UK for processing (tons)

	1968	1969	1970	1971
Cotton seed	20847	21176	2573	–
Groundnuts	119721	73306	62296	50925
Soyabeans	240662	324350	365985	306561
Copra (coconut)	48208	45773	31366	34211
Palm kernels	51702	43985	38021	48664
Rape seed	80607	77871	49916	65018
Others	–	–	–	179
Total weight of oil seeds	561747	586661	550457	505538
Oil equivalent	187958	175422	152108	148340

Source: Shortening and Margarine Manufacturers Association.

Table 1.26 World production of edible oils and fats ('000 metric tons)

	1968	1969	1970
Coconut oil	2225	2065	2120
Palm kernel oil	380	375	375
Babassu oil	65	85	80
Palm oil	1290	1390	1495
Olive oil	1475	1580	1355
Cotton seed oil	2330	2455	2360
Groundnut oil	3195	2775	2985
Soyabean oil	5650	6090	7400
Sunflower seed oil	3275	3400	3360
Sesame oil	585	550	585
Maize oil	295	300	305
Safflower oil	220	205	195
Rape seed oil	1785	1660	1810
Lard	3995	3910	4080
Tallow	4980	5055	5290
Whale oil	215	210	210
Fish oil	1095	895	990
	33055	33000	34995
Butter fat	5205	5065	4975
	38260	38065	39970

Source: Shortening and Margarine Manufacturers Association.

Table 1.27 Edible fats imported into the United Kingdom (metric tons)

	1968	1969	1970	1971
Cotton seed oil	11733	11982	41105	34633
Groundnut oil	126712	85236	95779	67771
Soyabean oil	14833	25194	61437	90700
Sunflower seed oil	66314	99504	34229	23906
Rape seed oil	10828	12041	14714	5895
Coconut oil	47633	43172	48632	42696
Palm oil	108739	139400	162426	222678
Palm kernel oil	24285	31702	34698	38667
Olive oil	3110	2826	2999	3181
	415187	451057	496019	530137
Lard	193409	178093	189735	203715
Edible tallow	2723	2741	3144	2825
Premier jus	6074	6498	6456	5504
Whale oil	10415	2257	1534	830
Herring oil	–	–	–	85573
Hydrogenated fish oil	42218	49940	47534	42497
Other fish and marine fats	183487	180082	158179	98982
	438326	419611	406672	439926
Total imported vegetable oil	415187	451057	496019	530137
Vegetable oil from oil seeds	187958	175427	152108	148340
Total vegetable oil	603145	626484	648127	678477
Total animal and marine fats	438326	419611	406672	439926
Grand total	1041471	1046095	1054799	1118403

Source: Shortening and Margarine Manufacturers Association.

Table 1.28 Production of margarine ('000 tons)

	1967	1968	1969	1970
United Kingdom	304.1	297.6	312.6	309.2
Australia	55.7	60.9	63.9	65.6
Canada	84.7	85.9	91.0	89.1
United States	943.8	955.8	974.1	995.7
Austria	39.6	38.1	38.9	38.7
Belgium	127.7	128.8	132.6	133.6
Denmark	85.8	87.4	86.6	86.4
Finland	21.8	25.5	27.1	33.4
France	146.6	151.5	155.6	157.7
West Germany	557.4	563.1	550.6	539.4
Greece	–	–	–	–
Irish Republic	11.2	12.5	13.2	–
Italy	35.0	35.0	37.0	41.0
Netherlands	252.7	254.8	237.8	230.1
Norway	88.7	82.7	82.8	82.2
Portugal	18.5	20.4	22.9	25.7
Spain	24.8	24.1	29.3	31.0
Sweden	118.8	122.8	126.1	133.0
Yugoslavia	23.9	22.3	27.0	33.2
Czechoslovakia	44.3	48.4	48.1	50.7

Table 1.28 (contd.)

	1967	1968	1969	1970
East Germany	187.8	185.3	183.3	184.7
Hungary	7.6	7.5	–	–
Poland	134.0	136.0	143.0	163.0
Soviet Union	614.0	637.0	680.0	745.0
Israel	19.9	22.2	23.1	24.9
Japan	77.7	87.5	101.4	115.0
South Africa	12.6	14.5	15.0	–
Venezuela	9.9	12.4	12.8	–
India (vanaspati)	386.0	467.0	477.0	514.0
Pakistan (vanaspati)	93.7	88.8	117.8	129.0

Source: Shortening and Margarine Manufacturers Association.

1.10 LITERATURE

1.10.1 General

K. A. WILLIAMS. *Oils, Fats and Fatty Foods*. Churchill, 1966.

D. SWERN. *Bailey's Industrial Oil and Fat Products*. New York, Interscience, 1964.

1.10.2 Analysis

Methods of Analysis of Oils. British Standard 684. London, British Standards Institution.

Official and Tentative Methods of the American Oil Chemists' Society, American Oil Chemists' Society.

V. C. MEHLENBACHER. *Analysis of Fats and Oils*. Garrard Press, 1960.

H. R. BOEKENOOGEN. *Analysis and Characterisation of Oils, Fats and Fat Products*. New York, Interscience, 1964.

1.10.3 Botany

J. W. PURSEGLOVE. *Tropical Crops – Monocotyledons*. London, Longmans 1972.

J. W. PURSEGLOVE. *Tropical Crops – Dicotyledons*. London, Longmans, 1968.

1.10.4 Composition

T. P. HILDITCH and P. N. WILLIAMS. *The Chemical Constitution of Natural Fats*. London, Chapman and Hall, 1964.

KUKSIS. 'Newer developments in determination of structure of glycerides and phospholipids', in *Progress in the Chemistry of Fats and other Lipids*. Vol. 12, London, Pergamon, 1972.

1.10.5 Margarine manufacture

A. J. C. ANDERSEN and P. N. WILLIAMS. *Margarine*. London, Pergamon, 1954.

M. K. SWITZER. *Margarine and other Food Fats*. London, Leonard Hill, 1956.

1.10.6 Refining

A. J. C. ANDERSEN. *Refining of Oils and Fats*. London, Pergamon, 1953.

M. K. SWITZER. *Continuous Processing of Fats*. London, Leonard Hill, 1951.

CHAPTER 2

Protein-Containing Foods in General

2.1 DEFINITION

The term 'protein' includes a group of complex nitrogenous organic substances which form an important part of living tissue.

2.2 THE GENERAL STRUCTURE OF PROTEIN

All simple proteins yield mixtures of amino acids and ammonia when hydrolysed by acids, alkalies or by enzymes. Conjugated proteins also exist. These are proteins which contain other organic compounds such as carbohydrates, lipid or fatty substances, purines and porphyrins. Proteins are composed of amino acids which are linked together by peptide linkages (—CO—NH) or, in the case of the amino acids proline and hydroxyproline, by the linkage —CO—N <. Because the amino acids, of which the protein is made, contain both acid carboxyl (—COOH) groups and weakly basic amino (—NH) groups, the composite protein will reflect the effect of the sum of these two groups in its so-called iso-electric point. This is the pH at which the molecule in aqueous solution just fails to migrate to either the positive or negative pole on passage of an electric current. The protein at its iso-electric point carries no unbalanced positive or negative charge. At this pH the solubility and water-holding capacity are lowest. The last-mentioned factor is of importance in its affect on the cooking yield of meats and tenderness of meat.

A composite protein molecule contains peptides which are in turn, combinations of amino acids. The physical and chemical nature of the protein depends on the nature and quantity of the constituent amino acids (see section 2.7.) Early investigations into the composition of proteins indicated that most contained about 16% of nitrogen. Analysts, therefore, use the factor of 100/16, i.e. 6.25, to convert the nitrogen determined by analysis to 'crude protein'. However, the nitrogen content of the proteins of some foods has been found to differ from 16% and consequently different conversion factors are used to calculate the 'crude protein' content of such foods. Table 2.1 gives the factors. The protein value computed by this method is only accurate when the foodstuff does not contain appreciable quantities of other nitrogenous substances.

Table 2.1   Factors for calculating protein from nitrogen content of food

Food	Factor	Food	Factor
Animal origin		*Plant origin*	
Eggs	6·25	*Legumes*	
Meat	6.25	Beans (Castor)	5.30
Milk	6.38	Beans (Jack)	6.25
Gelatin	5.55	Beans (Lima)	6.25
		Beans (Soya)	5.71
Plant origin		*Nuts*	
Cereal grains		Groundnuts	5.46
Barley	5.83	Almonds	5.18
Maize	6.25	Brazil nuts	5.46
Oats	5.83	Chestnuts	5.30
Rye	5.83	Coconuts	5.30
Rice	5.95	Hazelnuts	5.30
Wheat (whole kernel)	5.83	Walnuts	5.30
Wheat (bran)	6.31	*Seeds*	
Wheat (embryo)	5.80	Cottonseed	5.30
Wheat (endosperm)	5.70	Linseed	5.30
		Sesame	5.30
		Sunflower	5.30

Data from *Composition of Foods*, Agriculture Handbook No. 8, United States Department of Agriculture.

The official handbook of the American Association of Agricultural Chemists lists the following factors: wheat grains 5.70, nuts and nut products 6.25.

2.3   SOURCES OF PROTEIN

Plants and yeasts synthesize proteins from simple inorganic nitrogen and nitrogenous salts, whereas animals, birds, fishes and fungi require an adequate supply of ready-formed amino acids for growth and replenishment of tissue. Some proteins of plant origin are deficient in one or more special amino acids which are essential for healthy growth of animal tissue (see section 2.5).

Table 2.2   Survey of sources of protein

Material	Variety	Condition	Crude protein: g/100g of edible protein ($N \times 6.25$ or other factor where applicable)
Animal origin			
Beef	Chuck	raw (82% lean, 18% fat)	18.7
		cooked (81% lean, 19% fat)	26.0
	Porterhouse steak	raw (64% lean, 36% fat)	15.3
		grilled (58% lean, 42% fat)	20.5
	Lean from rump	raw	21.2
		roasted	29.1
Veal	Chuck	raw (86% lean, 14% fat)	19·4
		cooked (85% lean, 15% fat)	27.9

Table 2.2 (contd.)

Material	Variety	Condition	Crude protein: g/100g of edible protein ($N \times 6.25$ or other factor where applicable)
Pork	Loin	raw (80% lean, 20% fat)	17.1
		cooked (80% lean, 20% fat)	24.5
Lamb	Shoulder	raw (74% lean, 26% fat)	15.3
		roasted (74% lean, 26% fat)	21.7
Chicken	Roaster	raw	18.2
		roasted	25.2
	Breast	raw	20.8
		fried	32.5
Fish	Cod	raw	17.6
		grilled	28.5
	Herring	raw	17.3
Eggs	Chicken	raw, whole	12.9
		fried	13.8
		poached	12.7
		omelette	11.2
	Duck	raw, whole	13.3
Milk	Cows	wholemilk	3.5
	Goats	–	3.2
	Human	–	1.1
Plant origin			
Wheat	Flour	80% extraction	12.0
	Breakfast cereal	shredded	9.9
		puffed	15.0
Maize	Cornflour	–	7.8
	Breakfast cornflakes	–	7.9
Rice	White	raw	6.7
		cooked	2.0
	Breakfast	puffed	6.0
Oats	Oatmeal or rolled oats	dry	14.2
		cooked	2.0
Groundnuts	With skins	raw	26.0
	flour	defatted	47.9
Soyabeans	Mature	raw	34.1
	flour	full fat	36.7
		defatted	47.0
	Protein	concentrate	66.2–69.6
		isolate	90.0
Peas	Immature	raw	6.3
		cooked	5.4
	Frozen	cooked	5.1
Beans	Snap, green	raw	1.9
		cooked	1.6
	Broad, immature	raw	8.4
	Runner	raw	1.1
		boiled	0.8

2.4 CLASSIFICATION OF PROTEINS

2.4.1 Animal proteins

Next to water, proteins are the most abundant substance in animal tissues. They are involved in functions such as muscle contraction, respiration, genetic characteristics, enzymic reactions, as well as constituting the major component of skin, hair, bones, muscle, blood, glands and other internal organs.

Proteins have been classified into *globulins* and *albumins* according to their solubility in water and in salt solutions – albumins are water soluble and globulins are salt soluble.

The most abundant protein in muscle is a globulin, *actomyosin*. This consists of two proteins, *actin* and *myosin*, filaments of which, under the influence of a nucleoprotein (see below), adenosine triphosphate (ATP) slide one into the other, so causing contraction of the muscle.

The contractile (muscle) proteins in meat are separated from each other and joined to the bone by connective tissue proteins, *collagen, reticulin* and *elastin*. The hair, horns, hooves, nails, claws and the outer layer of the skin contain the protein *keratin*. The respiratory pigments include the protein complexes *haemoglobin* in the blood and *myoglobin* in muscle.

The living cells contain *nucleoproteins* which are important in the biosynthesis of proteins and also enzymes which also are primarily protein in nature and which are found wherever biochemical reactions are taking place in the living cell and even after death.

a. Actin is present in muscle protein to the extent of about 13%. Its iso-electric point is pH 4.7, which is lower than that of myosin at pH 5.4. Actin can exist in two forms, G-actin, a monomeric globular protein, and F-actin, a fibrous polymer. G-actin polymerizes in the presence of certain salts and adenosine triphosphate (ATP) or adenosine diphosphate (ADP) to form F-actin:

$$\text{G–ATP-actin} \quad \xrightarrow{\text{salt}} \quad \text{F–ADP-actin} + \text{phosphate.}$$

The polymerization is accelerated by Mg^{++} ions at neutral or alkaline pH, but at a neutral pH polymerization is inhibited by Ca^{++} ions.

b. Myosin is present to the extent of about 38% of the muscle protein. It has a strong affinity for the divalent cations Ca^{++} and Mg^{++}. Myosin hydrolyses ATP to ADP and phosphate. Ca^{++} ions activate the hydrolysis and Mg^{++} ions inhibit the reaction at low concentrations. *Muscle contraction* brings about the formation of a complex between actin and myosin to form actomyosin. The sliding rod theory of Hanson and Huxley (1953) is still held as the most acceptable explanation of muscle contraction. According to this theory, the filaments of actin and myosin slide over one another as contraction occurs. Bendall (1963) suggests that in the relaxed condition the cross bridges of the myosin filaments are kept from engaging with the actin filaments by the Mg–ATP complex. During contraction, however, the Mg–ATP complex is broken and the cross bridges engage with the actin filaments. The result is a shortening as the actin filaments are pulled towards the myosin.

c. Nucleoproteins. In 1868 a substance having a basic protein and acidic component containing a high level of phosphorus was discovered. Many years later, investigations

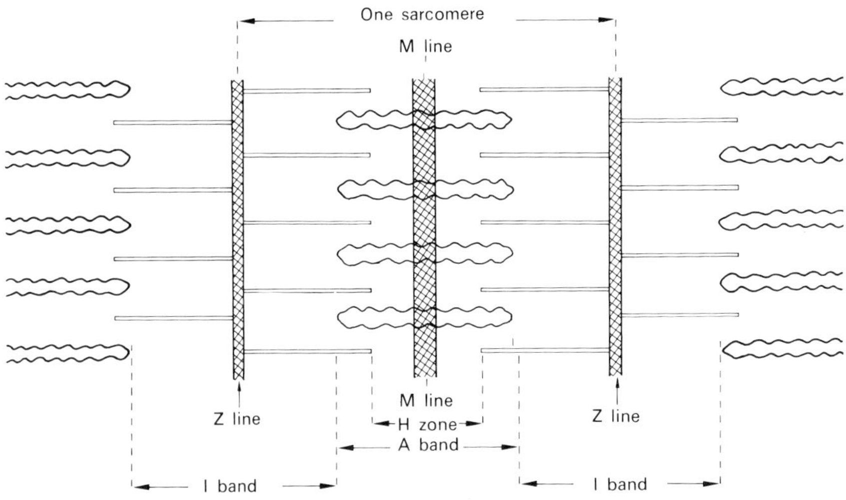

I Band of light colour - thin filaments of actin
A Band - dense colour where actin and myosin filaments overlap
H Zone - lighter part of A band due to myosin filaments only
M line - caused by a bulge in the centre of each myosin filament
Z line - dense boundary line of the sarcomere

Fig. 2.1   Schematic diagram of a muscle sarcomere in the resting condition.

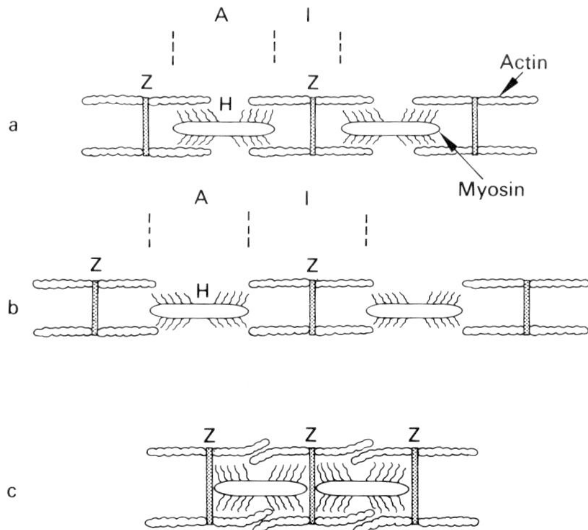

Fig. 2.2   Fine structure of a sarcomere in longitudinal section, showing myosin filaments with their cross-bridges and globular actin. The length of the sarcomere has been reduced tenfold for ease of drawing. Other features are approximately to scale: (a) Muscle at rest length = $l_0 = 2.4\mu$ per sarcomere (b) Muscle stretched to $l = 1.3\,l_0$, (c) Muscle contracted to $l = 0.6\,l_0$.

showed the presence of purine and pyrimidine compounds associated with either D-ribose (ribonucleic acid or RNA), or D-2-deoxyribose (deoxyribonucleic acid or DNA). The purines in mammalian muscle nucleic acids are mainly guanine (2-amino-6-oxypurine) and adenine (6-aminopurine). The pyrimidines commonly ocurring are cytosine, uracil and thymine. The sugars are bound to the bases in an N-glycosidic linkage to the nitrogen of position 9 in the purines and to the nitrogen of position 3 in the pyrimidines. According to whether the sugar unit is ribose or deoxyribose, RNA or DNA is formed. These nucleosides are joined to one another by phosphate-sugar (pentose) linkages to form nucleotides. DNA is involved in cell reproduction in the genetic pattern of an animal. Flavine adenine dinucleotide (FAD) is a co-enzyme involved in electron transfer. Adenosine diphosphate (ADP) and adenosine triphosphate (ATP) are energy acceptors and donors. Di- and tri-phosphopyridine nucleotide (DPN and TPN) are coenzymes involved in oxidation-reduction systems.

d. Collagens occur in the connective tissues of the animal body. They are of great importance in gut wall, bladder wall, tendons, skin, dentine of teeth and connective tissue sheathes of muscle. Collagen is unique amongst the proteins in having a high hydroxyproline content. It lacks the essential amino acid tryptophan (see section 2.5) and also the amino acid cystine.

e. Reticulin is another connective tissue protein which is associated with collagen. It can be identified histologically by its different ability to stain with acid fuchsin and the branching pattern of its fibres.

f. Elastin is a fibrous yellow protein mainly found in ligaments. It stains selectively with Weigerts Stain, becoming blue-black. Whereas collagen yields gelatins when subjected to severe conditions of moist heat (as in the braising of meat or pressure cooking) elastin does not.

g. Keratin. Hair, wool, horn, hoof, feather and the horny layer of skin are composed almost entirely of keratin. Although the chemical composition of the individual keratins varies, all are characterized by possessing a high cystine content. Likewise all keratins are resistant to digestion by the enzymes pepsin and trypsin.

h. Haemoglobin and myoglobin. (see fig. 2.3). Both these protein complexes act in oxygen transfer, the haemoglobin as an oxygen carrier in the blood and myoglobin as an oxygen store in the muscle-cells. In addition to the protein portion of the molecule known as the globin, another portion consists of two parts; an iron atom and a large planar ring known as the porphyrin. The porphyrin is in turn composed from four pyrrole groups linked together by methene bridges.

i. Egg proteins. Since the function of egg substance is to nourish the embryo it is to be expected that it will contain several proteins. The composition of egg yolk, which is primarily a source of food, is markedly different from that of the white which serves firstly as a protective barrier and finally as a food.

The yolk contains in addition to the proteins, free fats or lipids and also fat-protein complexes (lipoproteins). Complexes of phosphorus and proteins (phosphoproteins) are also components of egg yolk. The high fat proteins in egg yolk have been called lipovitellenin. The major phosphoprotein is phosvitin and the low fat proteins α, β and γ livetins.

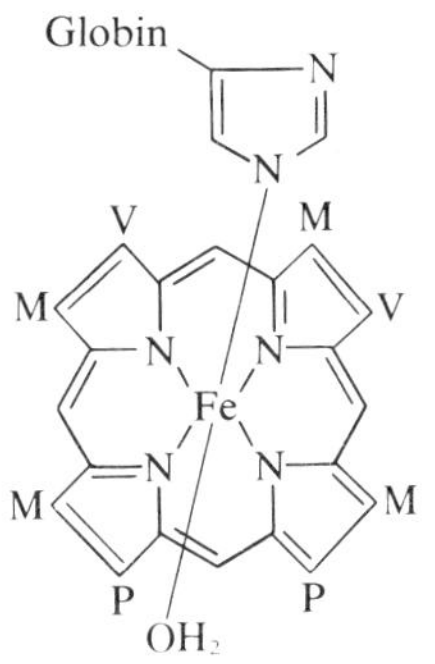

Globin

M = methyl group
V = vinyl group
P = propyl group

Fig. 2.3   Schematic representation of the haem complex of myoglobin.

The egg white proteins are more interesting biochemically, being important as embryological material for culturing.

Ovalbumin is present to the extent of half the total proteins in egg white. Its molecular weight (M.W.) is approximately 45 000. It denatures easily and contains several sulphydryl (—SH) groups. The other egg white proteins are as follows:

Conalbumin (13% approx.) Approximate M.W. 77 000. It is an antimicrobial protein. It possesses the property of binding metals such as iron, copper and zinc. It has a homologue in blood serum known as transferrin.

Ovomucoid (11% approx.) Approximate M.W. 28 000. It inhibits the action of the enzyme trypsin. It contains about 20% carbohydrate and lacks tryptophan in its amino acid makeup. The trypsin inhibition is very specific: chicken ovomucoid, for example, will not inhibit human trypsin.

Lysozyme (3.5% approx.) Approximate M.W. 15 000. It is an enzyme for polysaccharides and has antimicrobial properties.

Ovomucin (1.5% aprox.) is a mucoprotein (protein-carbohydrate complex) which has not been obtained in a pure state. It is viscous in alkaline solutions and its molecular weight has not been determined. It is responsible for the viscosity of egg white.

Flavoprotein (0.8% approx.) Poorly characterized to date. It binds riboflavin (vitamin B1).

Ovoinhibitor inhibits proteinase enzymes of bacterial and fungal origin.

Avidin is a trace protein in egg white. It binds the B vitamin, biotin.

j. Milk proteins. Cows' milk contains approximately 3.5% proteins composed of 3% casein and 0.5% whey proteins. The casein system is very complex and the following forms of casein have been separated: α_s, β, γ, κ and λ. Each of the three main components (α_s, β and γ) can polymerize and under certain conditions complexes with one another are formed. Calcium caseinate is used dietetically as a rich protein food whilst sodium caseinate is sometimes used in comminuted meat products as an emulsifying or as a binding agent. Besides its food use, casein, at one time, was much used in the manufacture of artificial ivory for knife handles and buttons. It still enters into the preparation of some adhesives and paper or textile sizes.

Of the whey proteins, β-lactoglobulin is the most abundant. Again it is a group of proteins. The molecular weight at the iso-electric point (pH 5.1–5.3) is of the order of 36 000 but when solutions between pH 3.7 and 5.1 are cooled below room temperature a tetramer having a molecular weight of 144 000 is formed.

α-Lactalbumin, also a whey protein, has a molecular weight of 16200 at its iso-electric point (pH 5.1). On the acid side of the iso-electric point, polymerization resulting from inter- and intra-molecular interactions occurs at rates which are time and temperature dependent. The polymerization can be reversed by addition of alkali. On the alkaline side of the iso-electric point weaker intermolecular polymerization takes place.

Minor proteins in whey comprise a number of enzymes and also the proteins transferrin and lactollin.

2.4.2 Fish proteins

Although fish muscle contains similar sarcoplasmic and myofibrillar proteins to mammalian muscle, the post-rigor changes in fish differ significantly from those in warm blooded animals. In addition there are many varietal differences amongst fishes with varying muscle biochemistry. The fall in muscle pH post-mortem in most species is less than that of mammalian muscle slaughtered under commercial conditions. This is probably due to lower glycogen reserves, but it is likely that during trawling or even line fishing, the glycogen reserves are to some extent depleted ante-mortem.

A muscle fibre is composed of many myofibrils embedded in sarcoplasm. The *sarcoplasmic* proteins constitute 20–30% of the total protein and in general properties resemble meat sarcoplasm. They are soluble in water or dilute buffers and contain many of the enzymes which are involved in muscle action.

The *myofibrillar* proteins (65–75% of the whole muscle protein) give the muscle its texture, characteristic of the species. *Myosin* constitutes 40% of the total protein. Detailed investigation of fish myosin has been hindered by the fact that extraction procedures which give pure myosin from mammalian muscle give mixtures with actin or actomyosin from fish. A method which has proved useful is first to extract actomyosin and then disassociate the latter with ATP or pyrophosphate, finally separating the components by ultra-centrifugation.

Comparatively little work has been done on fish *actin*. It appeared to be similar to rabbit actin, but it is now known that early rabbit actin extracts were contaminated with tropomyosin.

Little is known of the *connective tissue proteins* in fish except that they amount to about 3% in gadoid species and up to 10% in elasmobranch species. The shrinkage and gel temperatures of fish skin and swim bladder collagen are lower than those of mammalian collagen.

2.4.3 Seed proteins

The major seed proteins are classified in table 2.3.

The albumin and globulin proteins are valuable sources of enzymes which become very important during germination.

2.5 THE ROLE OF PROTEINS IN NUTRITION

Dietary proteins are split during digestion, firstly into peptides and then amino acids. About twenty amino acids are commonly found in food proteins. Man can convert many amino acids into other kinds that he needs, but there are eight which are necessary

Table 2.3 Classification of seed proteins

Type	Solubility	Examples
Albumins	Water-soluble	Enzymes
Globulins	Insoluble in water	Edestin (hempseed)
	Soluble in dilute neutral	Arachin and
	salt solutions	conarachin (peanuts)
		Glycinin (soybeans)
		Legumin (peas)
Prolamins	Ethanol-soluble	Gliadin (wheat)
		Hordein (barley)
		Zein (maize)
Glutelins	Insoluble in neutral salt or	Glutenin (wheat)
	in alcohol	
	Soluble in dilute alkali or	
	acid	Hordenin (rye)
		Avenin (oats)

for health, but which he cannot synthesize. These must, therefore, be supplied by the diet. The eight amino acids necessary for adults are:

Iso-leucine	Phenylalanine
Leucine	Threonine
Lysine	Tryptophan
Methionine	Valine

Two additional ones necessary for growing children are:

Arginine   and   Histidine.

Animal proteins usually contain adequate proportions of all the essential amino acids, but gelatin, which is derived from gristle, bone marrow and certain parts of some fish, is completely lacking in tryptophan. Vegetable proteins may be lacking in some essential amino acids. Cereal proteins tend to be deficient in lysine and methionine and maize protein is deficient in tryptophan. The protein of leguminous pulses such as peas and beans is richer in lysine than cereals but deficient in methionine. Yeast protein in deficient in tryptophan. Ruminant animals can synthesize all the essential amino acids from non-essential amino acids and other nitrogenous compounds.

In addition to providing amino acids for growth and repair of the body, proteins supply energy. Their energy value is in fact higher than that of carbohydrate.

1 g of carbohydrate (as glucose) produces 3.75 kcal. (16 kJ)
1 g of protein produces 4 kcal. (17 kJ)
1 g of fat produces 9 kcal. (38 kJ)

The efficiency of protein utilization is closely connected with the energy value of the diet. If the proportion of protein is too high it increasingly becomes used as an energy source. If the energy intake is restricted due to too low a quantity of carbohydrates, fats or alcohol, then protein becomes used primarily as an energy source. The levels of vitamins and minerals in the diet also affect protein utilization.

2.6 BIOCHEMISTRY AND METABOLISM

In order to determine the amino acid composition of proteins, it is necessary to split up the protein into its constituent amino acids. This can be done by hydrolysing the protein with either acid or alkali, or by treatment with an appropriate enzyme. The salt formed by neutralization with an alkali or acid after acid or alkaline hydrolysis respectively, can be removed by dialysis through a membrane.

An amino acid cannot be estimated by direct titration with standard alkali because the amino acid contains basic —NH_2 groups and acidic —COOH group(s). Sorensen devised a method of blocking the amino group by treatment with neutral formaldehyde, when the

Table 2.4   Tests for specific amino acid groups

Amino acid	Reactive group	Tests and reagents	Solution required*
Arginine	Guanidino	Alkaline NaOCl and α-naphthol → red colour	PH.
Cysteine and cystine	Sulfhydryl	Reduce with Zn, then β-naphthoquinone-sodium sulphonate → red colour; or sodium nitroprusside + NH_3 → red colour	H PH
Glycine		Remove NH_3 and tryptophan, then o-phthaldialdehyde → violet colour.	H
Histidine	Imidazolyl	Diazotized sulphanilic acid + $HgCl_2$ → red colour (tyrosine interferes)	H
Hydroxyproline + proline	Pyrrole (after oxidation)	Oxidize with PbO_2, treat with isatin → red colour, or p-di-methylaminobenzaldehyde → red colour.	H
Methionine	Methylthioethyl	Sodium nitroprusside + 14 N NaOH, then acidify → red colour, or α-naphthylamine diazonium HCl → orange colour.	H H
Serine	Hydroxymethyl	Periodate oxidation → HCHO; distill into chromotropic acid solution → violet pink colour.	H
Threonine	Hydroxyethyl	Periodate oxidation → CH_3CHO; aerate into p-hydroxy-biphenyl solution → red-purple colour.	H
Tryptophan	Indolyl	p-dimethylaminobenzaldehyde → red purple colour.	P.H.
Tyrosine	p-Hydroxyphenyl	Mercuric sulphate + nitrous acid → red colour. α-Nitroso-β-naphthol → red colour.	P.H. P.H.
Tyrosine + tryptophan		Phosphomolybdotungstic acid → blue colour.	H

* P = protein solution;   H = protein hydrolysate

basic amino group is replaced by a neutral CH_2: N group enabling the —COOH group to be directly titrated with standard alkali. The following reaction occurs with glycine (amino-acetic acid).

$$HCHO + H2NCH_2COOH = CH_2:NCH_2COOH + H_2O$$
formaldehyde   glycine

$$NH_2CH_2COOH + HONO = HO·CH_2COOH + N_2 + H_2O$$
glycine   hydroxyacetic acid

Another method of following protein hydrolysis is due to Van Styke and consists of measuring the nitrogen evolved when an amino acid is treated with nitrous acid

$$NH_2CH_2COOH + HONO = HO·CH_2COOH + N_2 + H_2O$$

The tests listed in Table 2.4 are the classical tests for certain reactive groups, but they are rather old fashioned now. The modern technique is to separate the amino acids of an hydrolysate in a chromatograph, fixing them with a colour developer such as ninhydrin. Alternatively, fluorescence of the chromatogram can be used to identify the presence of certain amino acids. The majority of amino acids fluoresce and the presence of non-fluorescent bands together with their position on the chromatogram, serve to show the presence of, and identify, those which do not fluoresce.

Owing to a general world shortage of palatable protein and the increasing relative costs of animal proteins, increasing research effort is being applied to producing edible protein, for animal and human needs, from the following: groundnut, soyabean, broad bean (*Vicia*

Fig. 2.4   Chunks of 'Kesp', a meat-like protein based on soya bean.

Process from n-alkane feedstock

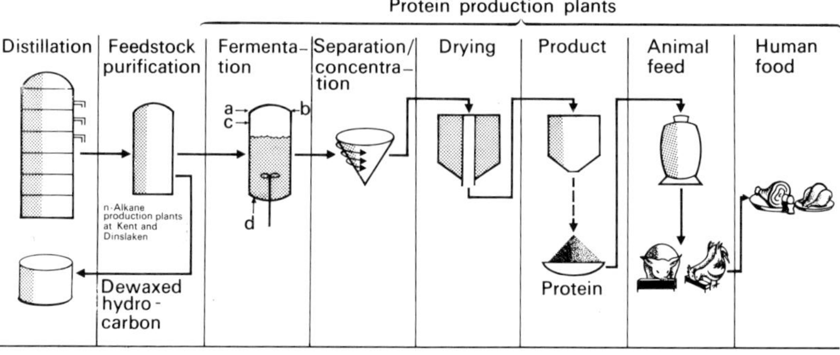

Fig. 2.5 Schematic layout of a plant at Grangemouth, Scotland, producing animal feedstuffs (Toprina) from paraffinic hydrocarbons.

faba), rape seed, coconut, algae, leaves and micro-organisms grown on substrates such as wheatstarch, coal, oil and natural gas.

The British Ministry of Agriculture, Fisheries and Food set úp in 1972 within its Food Science Division, a novel Protein Intelligence Unit to collect, review and publish information on various aspects of novel proteins.

Amstein (1965) elucidated many of the biochemical reactions which bring about the synthesis of peptides from the 20 amino acids in food proteins. The amino acids are first converted into energy-rich intermediates by reaction with adenosine triphosphate and are then attached to 'transfer' ribose-nucleic acids, Transfer RNAs are polynucleotides containing 70 to 80 residues per chain, and each RNA will accept only an appropriate amino acid. The use of isotope-labelled proteins has facilitated the *in-vivo* study of protein metabolism.

During 1973 it has been discovered that under certain conditions, proteolytic enzymes which normally hydrolyse proteins to peptides and amino acids, can form proteins known as 'plasteins'. A plastein formed from fish waste protein hydrolysate contained a higher proportion of essential amino acids than did the hydrolysate of the waste.

2.7 AMINO ACIDS

The following list enumerates the amino acids which are hydrolysis products of proteins.

1. Monoamino dicarboxylic acids.

Aspartic acid	$HO_2C-CH_2-CH(NH_2)-CO_2H$
Glutamic acid	$HO_2C-CH_2-CH_2-CH(NH_2)-CO_2H$
Hydroxyglutamic acid	$HO_2C-CH_2-CHOH-CH(NH_2)-CO_2H$

Fig. 2.6 Textured strands of Mycoprotein, a fungal-based protein.

2.   Basic amino acids.

Arginine $NH_2C(=NH)-NH-CH_2-CH_2-CH_2-CH(NH_2)-CO_2H$

Lysine $NH_2-CH_2-CH_2-CH_2-CH_2-CH(NH_2)-CO_2H$

Histidine

$$\begin{array}{c} \text{CH} \\ N\!\!\diagup\!\!\diagdown\!\!NH \\ HC\!=\!C-CH_2-CH(NH_2)-CO_2H \end{array}$$

3.   Monoamino monocarboxylic acids.

   (i) Extractable by wet butyl alcohol; readily soluble in water; insoluble in anhydrous alcohols.

Glycine $CH_2(NH_2)-CO_2H$

Alanine $CH_3-CH(NH_2)-CO_2H$

Valine $(CH_3)_2CH-CH(NH_2)-CO_2H$

Leucine	$(CH_3)_2CH—CH_2—CH(NH_2)—CO_2H$
Norleucine	$CH_3—CH_2—CH_2—CH_2—CH(NH_2)—CO_2H$
Isoleucine	$CH_3—CH_2—CH(CH_3)—CH(NH_2)—CO_2H$
Phenylalanine	$C_6H_5—CH_2—CH(NH_2)—CO_2H$
Serine	$CH_2OH—CH(NH_2)—CO_2H$
Threonine	$CH_3—CHOH—CH(NH_2)—CO_2H$
Cysteine	$CH_2(SH)—CH(NH_2)—CO_2H$
Methionine	$CH_2(SCH_3)—CH_2—CH(NH_2)CO_2H$

Tryptophan

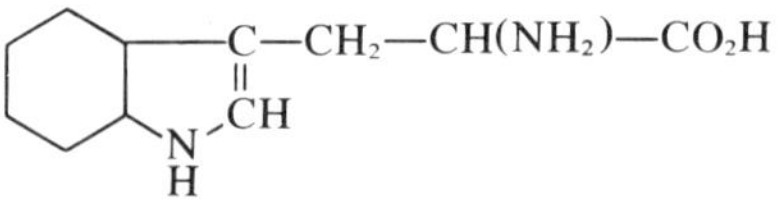

(ii) Sparingly soluble in water

Tyrosine

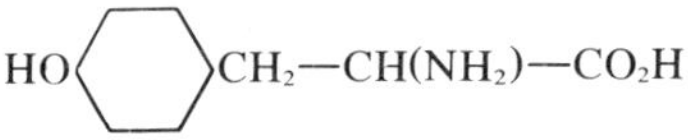

Di-iodotyrosine

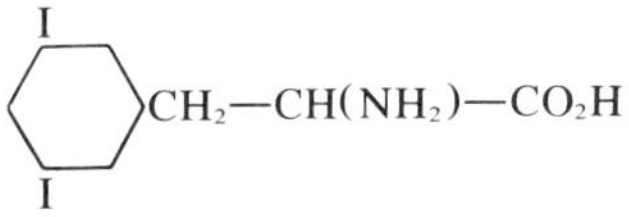

Thyroxine

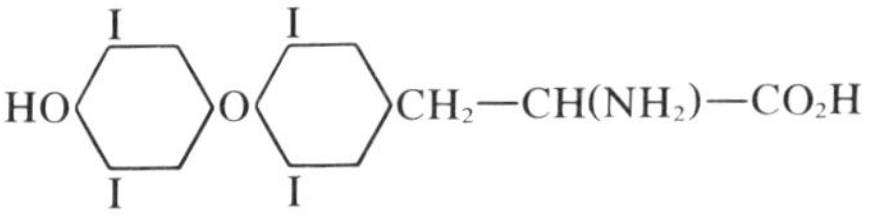

Cystine $[—S—CH_2—CH(CH(NH_2)—CO_2H]_2$

(iii) Soluble in alcohols

Proline

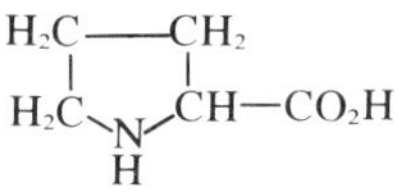

Hydroxyproline

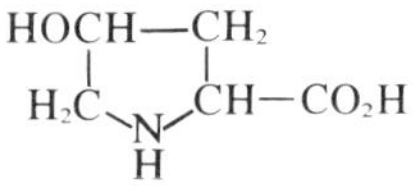

2.7.1 Glutamic acid

This is produced by enzymic digestion or chemical hydrolysis of vegetable proteins and the monosodium salt is widely used as a seasoning agent in foods. In 1970 its use in baby foods in the USA was banned because it was found to be the cause of flushing due to constriction of the blood capillaries of the neck. Over-liberal use of soya sauce (which contains glutamate) can also bring on the symptoms in certain individuals. It stimulates strongly the salivary glands and intensifies the natural flavour of certain foods. When added to food the usual level is about 0.02%. Glutamic acid is a non-essential amino acid. It is involved in transamination (movement of—NH_2 groups) in living tissue, this reaction being brought about by an enzyme, transaminase, e.g.

Glutamic acid $\xrightarrow[\text{transaminase}]{\text{glutamic-alanine}}$ $\xrightarrow{\text{pyruvic acid}}$ alanine + keto-glutaric acid

2.7.2 Lysine

With the exception of glycine (amino acetic acid) which contains no centre of asymmetry, all amino acids of protein origin occur in optically active forms. L(−) lysine is an essential amino acid but the D(−) isomer is nutritionally inactive. The DL form of lysine can be synthesized and from the mixture of isomers a product containing 95% of the L(−) form can be extracted. The latter is available commercially for admixture to cereal proteins which are deficient in lysine. Because lysine is strongly basic the synthetic acid is usually produced as a hydrochloride.

Table 2.5 demonstrates the effect of adding synthetic lysine on the nutritive value of cereals, as measured by the protein efficiency ratio (P.E.R.). The P.E.R. = (Increase in weight of rats in 28 days)/(weight of protein) using milk casein as a standard (casein P.E.R. = 2.5).

2.7.3 Methionine

Although other sulphur-bearing amino acids such as cysteine and cystine occur generally in food proteins, methionine is a limiting essential amino acid, in certain foods such as cereals, legumes, fish, egg and milk (McLaughlin, 1963); i.e. it limits the nutritional value of the whole protein. The conversion of methionine to cystine has been proved by feeding rats with methionine containing radioactive sulphur. This has been recovered in the tissues as radioactive cystine. Cystine itself has a sparing effect on methionine; i.e. it reduces the necessary daily requirement of methionine. The daily requirement of methionine for a man has been put at 1000 mg/day but this figure drops to 200 if ample cystine is present. Synthetic methionine made commercially from acrolein and methylmercaptan is used in animal feeds. Both the laevo and the dextro forms are nutritionally active. Supplementation of methionine, especially in the case of humans, is a complicated matter. Bressani found that adding methionine to cereal protein for infants produced a detrimental effect. One cannot always correct inadequate proteins with synthesized supplements. The level of protein intake in the diet and the limiting amounts of other essential amino acids are critical factors.

2.7.4 Threonine

This can be synthesized but the biologically inactive DL-allothreonine is more readily produced than the active DL threonine.

Table 2.5 Effect of added synthetic lysine on nutritional value of cereals, measured by Protein Efficiency Ratio (P.E.R.)

	P.E.R. of original grain	L-lysine-HCL added, %	P.E.R. of fortified grain
Millet	0.7	0.30	2.1
Sorghum	0.7	0.20	2.2
Wheat	1.3	0.10	1.7
Maize	1.5	0.10	2.0
Rice	1.5	0.05	2.3
Barley	1.6	0.05	1.9

* P.E.R.s obtained after feeding the respective grains to baby rats.
Source: E. E. Howe, Report of the Presidential Scientific Advisory Committee No. 2, p. 319.

2.7.5 Tryptophan

This can be synthesized from indole which is an expensive starting material. An alternative route, via phenylhydrazine and acrolein presents certain difficulties. The protein gelatin is completely lacking in tryptophan. In intestinal putrefaction, tryptophan breaks down into indole and skatole which partly diffuse into the blood stream. Indole is oxidized to indoxyl which is eliminated from the body by the kidneys as indoxyl suphuric acid (indican). In animals, much of the dietary tryptophan is used for protein synthesis. A large number of derivatives of tryptophan can replace the effect of tryptophan itself on growth. Another function of tryptophan is to bring about the excretion of kynurenic acid in the urine. Again, many tryptophan derivatives can cause this, but not necessarily the same ones which are growth stimulators.

2.8 PROTEIN ALLERGIES

The body may react to ingestion of or even contact with certain proteins or polysaccharides. These so-called antigens induce antibodies which in turn, react with subsequent antigens or even homologous forms of the antigen and remove them. This reaction is used medically to combat infections and poisons such as snake venom. The antibodies so formed are usually in the γ-globulin fraction of the blood serum proteins.

2.9 LITERATURE

MC CANCE and WIDDOWSON. *Chemical Composition of Foods.* London HMSO.

Composition of Foods. Agriculture Handbook No. 8, United States. Department of Agriculture Washington D.C.

Manual of Nutrition. London HMSO, 1970.

Amino Acids Content of Foods and Biological Data on Protein. Food Consumption and Planning Branch Nutrition Division, Food and Agriculture Organisation Rome, 1968.

J. R. BENDALL. *Proceedings Meat Tenderness Symposium.* Campbell Soup Company, New Jersey, 1963.

Proteins and their Reactions. 3rd Symposium on Foods, Oregon State University, Westport, Connecticut, AVI Publishing Co. 1963.

J. J. CONNELL, *Recent Advances in Food Science. Fish Muscle Proteins,* Vol. 1. London, Butterworths, 1962.

N. R. JONES. *Recent Advances in Food Science. Fish Muscle Enzymes and their Technological Significance,* Vol. 1. London, Butterworths, 1962.

W. J. DYER and J. R. DINGLE. *Fish Proteins with specific reference to freezing, Fish as Food.* New York, 1961.

A. M. ALTSCHUL. *New Protein Foods Technology,* Vol 1a, Academic Press, 1974.

ONONE and RIDDLE. 'Use of plastein reaction in recovering protein from fish waste', *J. Fish Res. Board Can.,* 1973, 30(11), 1745–7.

A. KOIICHI, Y. MICHOKO, A. SOICHI, and F. MASAO. 'General properties of a plastein compound synthesized from a soya bean protein hydrolysate', *Agr. Biol. Chem.,* 1973, 37(11), 2505–9.

Meat, Fish and Similar Products

3.1 BEEF

Domestication of wild cattle probably first took place in India, the Near East and Egypt around 6000 BC. They are known to have existed in Babylon as early as 5000 BC. In Europe the earliest domesticated type (*Bos taurus brachyceros* or *Bos longifrons*) was a small, slightly built, black or brown animal with short horns. It is assumed that it was brought to southern Europe from the Near East by tribes moving westwards during the earliest neolithic (polished stone) period. In Britain this primitive type, known as the Celtic Shorthorn, remained basically unchanged until Roman times whereas on the European mainland it was displaced by a large, powerful, strong-boned and long-horned animal (*Bos taurus primigenius*) probably brown in colour and popularly known as the European wild ox. The last known specimen was killed in Poland in 1627.

In Britain the Roman invasion and subsequent occupations were responsible for introducing cattle from a number of different sources. Isolation and regional variations in soil and in methods of agriculture and husbandry contributed towards making various types of cattle distinct from one another. However, it was not until the eighteenth and nineteenth century that the development and recording of a number of variously productive breeds took place, probably encouraged by the continuous in-breeding method devized by Robert Bakewell (1725–1795). For the preservation of the established breeds an artificial barrier had to be erected to prevent cross-breeding. This took the form of herdbook regulations. An official herdbook was created for each recognized breed and contained the characteristics peculiar to the breed and the lines of descent. Registered animals were regarded as 'pure-bred'. In Britain registration is still carried out by various cattle breed societies, mostly affiliated to the National Cattle Breeders' Association. British pedigree cattle have been exported for the last century to other parts of the world where further changes have taken place according to soil, feeding and environment.

3.1.1 Breeds and rearing

Cattle breeds can be classified according to their producing abilities: beef, beef and milk (dual purpose) and milk.

In the UK important beef-producing breeds are: Aberdeen Angus, Devon, Hereford, Lincoln Red, Galloway, Belted Galloway, Highland, Sussex, Beef Shorthorn, Charolais, Luing and Limousin.

The main dual purpose breeds are: Dairy Shorthorn, Welsh, Black, South Devon, Red Poll, Dexter, Meuse-Rhine-Ijssel and Simmental. The beef and some of the dual purpose breeds of cattle are described below.

As meat producers cattle can be divided into three main groups: baby beef, prime bullocks, and old bullocks and cows. Baby beef are killed at 12–14 months when, because of intensive feeding on concentrates and grass from birth, they have usually attained a live weight of 400 to 500 kg (880 to 1 100 lb). Prime bullocks are killed at $2\frac{1}{2}$ to $3\frac{1}{2}$ years when their live weight is about 550 to 850 kg (1 200–1 880 lb). They usually suckle for the first 6 months on the pasture and are then yarded or put to grass according to the food supply available. Old bullocks, which with cows provide about 50% of British beef, are kept with relatively little attention until 4 to 5 years of age before slaughter, and cows are kept until they are past their prime milk-producing age.

Intensified livestock production has resulted in a greater use of concentrates to supplement the available grazing. They should be given regularly and change of feed should be gradual. For beef cattle on grass the appetite should be controlled either by grazing for a limited time or by electric fencing. Some exercise is good for cattle but too much toughens the meat. Male beef cattle should be castrated for ease of handling and, where necessary, animals should be dehorned.

a. Aberdeen Angus. The breed, officially recognized by the Highland Society in 1835, originated from the counties of Aberdeen, Banff, Kincardine and Forfar.

The coat is black, the skin black-pigmented and the hair short to medium in length. The head is small and the cylindrical body is long and straight. The average live weight of mature bulls is 800 kg (1760 lb) and of cows 500 to 550 kg (1 100–1 200 lb). They graze well and are docile.

Aberdeen Angus cattle can adapt to various altitudes and soil types as has been shown by their being successfully reared in Canada, Argentina, Australia and New Zealand.

Calving takes place in the spring when the cows are about $2\frac{1}{2}$ years of age. The calves are suckled during the summer and weaned in the autumn. In the winter they are often yarded and fed hay, straw, grass silage, roots and bruised oats. At birth the live weight of calves is low (male: 28 kg (62 lb), female: 26 kg (57 lb)) but they gain quickly. Cows produce enough milk to start the calves off well. Thereafter the level of nutrition should be kept high. Animals are sold for beef at $2\frac{1}{2}$ years. The breed is an excellent beef producer – the muscles are marbled and the joints have a fine bone structure. Bulls are often used for cross-breeding because they label their offspring so well with good beef characteristics. They are also frequently used for producing beef calves from dairy cows.

b. Devon. The breed was developed in Devon, Cornwall, Somerset and Dorset and was first recorded in 1850.

The coat is dark red, sometimes dappled, and the hair is fairly long and often curly. The skin has an orange-yellow pigmentation which is visible round the eyes and muzzle. The head is broad and the horns are cream with black tips, medium in length, and move upward and outward from the head. The average live weight of steers at $1\frac{1}{4}$ years is 550 kg (1 100 lb).

The breed can be reared successfully independent of soil type and has therefore become popular as a ranching animal in many countries. Calving can take place out of doors all the year round, and because cows feed their young well calves are suitable for early slaughter. Steers are often sold for fattening on pasture land before being slaughtered for baby beef. It is also possible to rear cows for dual purpose, killing them for beef after a period of milk-producing. The beef produced by the breed is well marbled and the joints are small. It has a fine grain and a good flavour.

   c. Hereford. The Hereford Herdbook Society was formed in 1876, 30 years after the establishment of the first official herdbook.

   The skin of the animal is thick and the hair is soft, silky, curly and of medium length. Red and white markings are characteristic of the breed. The head is short with cream horns. The average live weight of a mature bull is in the region of 830 kg (1 830 lb) and of a mature cow 540 kg (1 190 lb). The Hereford is probably the hardiest of English breeds. It thrives on rich pasture but can also be reared successfully on extensive ranch grazing or under drought conditions. It resists disease well. The breed has been reared outside England more extensively than any other.

   Cows calve first at about $2\frac{1}{2}$ years and continue to do so until 15 years or even older. Although cows suckle their young well, the animals mature more slowly than Beef Shorthorn. The meat produced by the Hereford is well marbled, slightly coarse grained, and carries more fat than that from the Beef Shorthorn or Aberdeen Angus. Like the latter breed, bulls label their progeny well and are often used for producing crossbred animals for beef.

   d. Lincoln Red. The breed was developed from the large cattle native to Lincolnshire and various shorthorn stock introduced to the area in the nineteenth century. It became officially distinct from Lincoln Red Shorthorn only in 1960.

   The coat is deep red with a few streaks of white on the underside. The skin is lightly pigmented, the muzzle pale, and the hair fairly thick and long. The head is short and wide with horns which grow outward, forward and downward. The average weight of a well fed two-year-old is 525 kg (1 157 lb).

   The breed is best reared on rich pasture land but can withstand wind and cold, needs little shelter during the winter and can be fed economically on barley straw, hay, turnips and fodder beet. The cow has a good supply of milk and calves can be fattened early, economically and rapidly. Because of the rapid development of calves the breed has a good record as a producer of baby beef. The Lincoln Red was originally a dual-purpose breed but most breeders now prefer the meat-producing type of animal. After the Aberdeen Angus and Hereford, Lincoln Red bulls are the next most widely used for cross-breeding.

   e. Galloway. This polled breed originates from the south-west of Scotland but is also found in the north and east of England. The Galloway Cattle Society was established in 1877.

   The coat is usually black but may be dun-coloured or have brown tinges. The skin is moderately thick and is dark-pigmented and the hair is soft and long. The average weight for mature bulls is 550 to 600 kg (1 200–1 323 lb) and for cows 450 kg (992 lb).

   The breed feeds and fattens well on poor hill grazings and can winter outdoors with a little shelter and supply of hay and oat straw. In this respect it is hardier than the Aberdeen Angus. The animals are fairly slow to mature, but the quality of their beef is excellent. Galloway cows and White Shorthorn bulls can be crossed to give a blue-roan animal known as the Blue-Grey.

Fig. 3.1   Aberdeen Angus Bull.

Fig. 3.2   Devon Cow.

Fig.   3.3   Champion Hereford Bull. 'Rowington Vernon'.

Fig. 3.4   Lincoln Red Bull.

Fig. 3.5   2 year old Sussex Bull.

Fig. 3.6.   Galloway Cow.

f. Belted Galloway. The breed originates from the same areas as the Galloway but it is claimed that the cows are better milkers than Galloways.

g. Highland. The West Highland or Kyloe is the native breed of the Western Highlands and the Scottish Isles.

The recognized colours are red, yellow, dun, creamy brindle and black. The coat is shaggy and slightly wavy. The body is moderately sized and the head has large spreading horns.

The breed is very hardy and can subsist on poor grazing. Calves grow slowly, the cow yielding little more milk than is necessary for the rearing of offspring. Steers are rarely marketed before $3\frac{1}{2}$ years. The beef of well fattened animals is of high quality.

Fig. 3.7 West Highland cow and calf.

h. Sussex. The breed is probably derived from the cattle native to Britain at the time of the Roman invasion. It was originally reared in the south-east of England as a draft animal, but later its beef characteristics became prominent. The Sussex Herdbook was established in 1874. The breed is deep red in colour with a white brush on the tail. The skin is yellowish-white and the hair is soft and of medium length and thickness. The horns are small, curving outward and forward.

Animals graze well on any pasture and can withstand exposure to winter winds with only natural shelter. Little or no supplementary concentrate is needed as the breed thrives on home-grown winter feed. Calves can be born in the open and mature reasonably early, producing baby beef at 1 to $1\frac{1}{2}$ years and beef at 3. The meat is distributed well on the carcass, is tender and well marbled. In many countries the bulls are used for upgrading scrub stock and for improving the beef producing ability of calves from dairy cows.

i. Beef Shorthorn. The breed is derived from early cattle stocks in the northeast of England–Northumberland, Durham and Yorkshire. The Herdbook was established in 1822.

The coat can be red, white or roan, the latter being the most common. The skin is a non-pigmented creamy colour with flesh coloured muzzle, lips and eyelids. The hair is medium to long and of medium thickness. The horns are wax coloured, blunt at the end and curve forward from the crown. Some polled strains have also been developed. The average live weight of bulls is 700 to 900 kg (1 544–1 984 lb) and of cows 500 to 600 kg (1 100–1 322 lb).

The breed can be reared in most environments. It grows best in temperate regions, however, and has been successfully imported to North and South America, Australia, New Zealand and South Africa. Beef Shorthorn have a rapid growth rate and mature early producing excellent baby beef. Older animals, if kept actively growing from birth to maturity, produce a lean carcass with small joints and well marbled, tender muscles. Though bulls do not colour-mark their offspring they can enhance their ability to mature early and produce a considerable amount of beef.

j. Charolais. The Charolais is the second most numerous breed in France. It has also been exported throughout Europe and the USA. The breed provides excellent beef animals with exceptionally well-developed loin and thigh muscles. The animals are efficient grazers with a quiet disposition and rapid growth rate. The bulls colour-mark and are popular for cross-breeding. In the United Kingdom trials for cross-breeding purposes were approved in 1961 and the Charolais is now the third most important beef breed.

k. Luing. The breed was developed from Beef Shorthorn and Highland cattle. Its name is taken from the island of its origin. The breed society and Herdbook were recognized by the Secretary of State for Scotland in 1965.

l. Limousin. This is a beef breed characterized by high live weight gains and high meat yield. It is claimed that the breed is comparatively free of calving difficulties. The import of Limousin from France was approved in 1970.

m. Meuse-Rhine-Ijssel. In the Netherlands, as no specialized beef breeds are kept, emphasis has been placed on combining good musculature and beef quality with high milk production. The predominating breed is the *Dutch Friesian*. The Dutch Friesian cattle are fairly uniform in colour (usually black and white coat with the colours separated into well defined areas), well muscled and have a smooth, rounded contour. Although in the past meat has been of secondary importance to milk, recently selection had been directed towards an improved beef carcass. In the province of Groningen, in the northeast of the Netherlands, the *Groningen White-headed cattle* breed has developed from stocks whose origins were similar to those of the Dutch Friesian. The animals are black on the body (in some cases red) with white on the head, brisket, underline and fetlocks. They show a better muscular development than the Dutch Friesian. In the east and southeastern regions of the Netherlands, in the districts bordering the great rivers in the provinces of Overijssel, Gelderland, North Brabant and Limburg, the *Meuse-Rhine-Ijssel* breed has evolved from the same stock as the Friesian. It is a red and white, dual-purpose breed, very similar to the Red and White cattle found in the contiguous areas of Belgium and Germany. It has a superior carcass quality to that of the Friesian and there is no tendency for excessive subcutaneous fat to be deposited. Dutch breeders are organised into two associations, the Friesian Cattle Herdbook Society in the province of Friesland and the Netherlands Cattle Herdbook Society in the remaining ten provinces of the Netherlands.

n. Simmental. Simmental is a dual-purpose breed from Swiss cattle and is found throughout Europe and the USA. In Germany the breed has been developed for beef production.

3.1.2 Diseases and combating

The main diseases of cattle which are scheduled as notifiable in the United Kingdom are: *foot-and-mouth disease, anthrax, rinderpest* and *bovine pleuro-pneumonia*. If the presence of a so-called notifiable disease is suspected or detected the proper authorities should be informed without any delay.

Though usually not fatal, *foot-and-mouth disease* causes serious economic loss owing to fever and disruption of feeding. It is carried by a virus, is highly infectious and occurs in at least six different types. After an attack of one type an animal does not become resistant to other types. In infected animals the highest concentration of the virus can be found in the soft tissues about the hoofs or the mucous covering of the tongue. The virus is resistant to common disinfectants such as carbolic acid and can survive for months in the bone marrow or other parts of frozen or chilled carcasses. Formalin or an alkali such as caustic soda may destroy the virus. Vaccines against the disease are available but their efficacy in combating all the types of the disease has not yet been proven. They are expensive, do not become operative until 10 days after injection and give full protection for only 4 months. At present the official policy in the United Kingdom is to try to prevent the occurrence of the disease by strict quarantine regulations and, when an outbreak occurs, to take immediate action by killing all affected stock. The mortality rate is about 3.5%. Animals can recover from between 10 days and a year after an attack.

The acute form of *anthrax*, causing sudden death, occurs mainly in adult cattle. The anthrax bacilli present in the blood at the time of death disintegrate within a few days but if blood is spilled they form resistant spores which can survive for several years. Conditions in Britain are on the whole not favourable for sporulation. The usual source of the disease is from spores introduced on imported materials. The British anthrax order

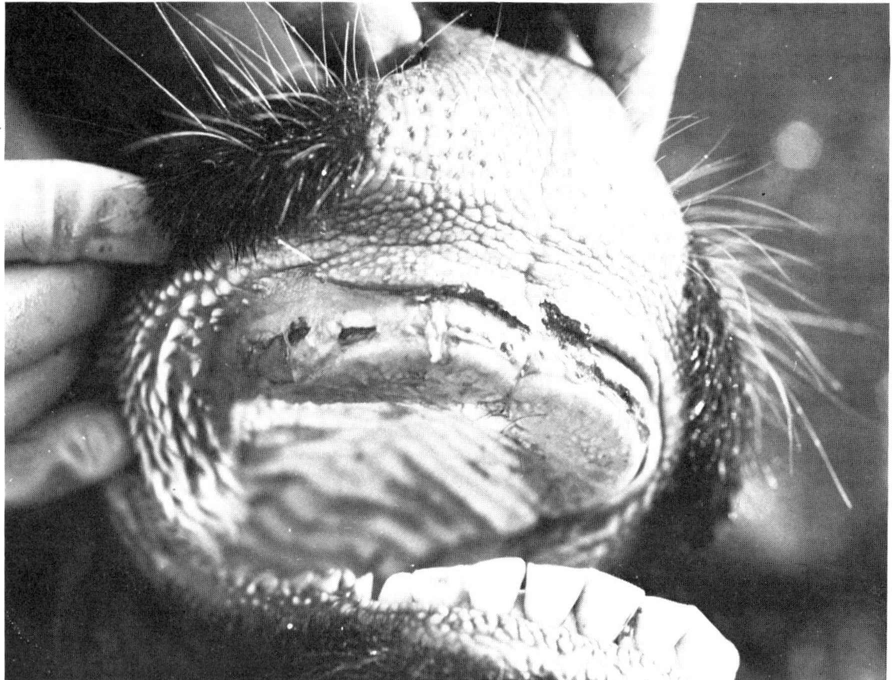

Fig. 3.8 Upper lip and jaw showing blisters of animal with foot and mouth disease.

of 1928 stipulated that any death suspected to have been caused by anthrax should be reported to the police and that if anthrax is present the carcass should be burned and buried and the place disinfected. The carcass should be opened only for diagnosis. An efficient vaccine, which can prevent outbreaks of the disease in a previously affected area, has been produced in South Africa. The disease can spread to man.

Rinderpest, or cattle plague, and *bovine pleuropneumonia* were eliminated from Europe and the British Isles towards the end of last century and now occur only in Africa and Asia.

Of the non-notifiable diseases *bovine tuberculosis* is the most common and can be fatal to cattle whereas the human type causes only localized lesions which can heal spontaneously. The bovine type may cause pulmonary, alimentary or udder tuberculosis. There are few obvious clinical signs of the disease at first but later breathing becomes constricted and an intermittent cough occurs. A frequent source of the disease is the expelled mucous which contains tuberculosis bacilli. The most effective control is the tuberculin test which shows whether tuberculosis is present in an active or latent form. Purified Protein Derivative (PPD) in doses of 0.1 ml is injected intradermally in the neck. After 72 hours the test is read and any swelling which causes a change in the skin thickness greater than 4 mm is considered a positive reaction. In Britain a Tuberculosis Eradication Scheme was introduced in 1950. To eradicate the disease, infected cattle should be destroyed and all possible places of contamination disinfected.

Certain diseases of cattle can be classified according to the stage of development at which animals are particularly susceptible to them–from birth to 6 months (see 3.2.2 *Veal diseases*); young cattle and adult cattle. After 6 months cattle can be affected by *blackquarter* caused by a gas-producing micro-organism which produces swelling in subcutaneous tissues and muscles, *necrotic and suppurative conditions* in the liver, muscles, kidneys, lungs and udder, *salmonellosis* which causes fever, diarrhoea, and often death and can cause food-poisoning in man, infestation with *external parasites* such as lice, mange, mites and ticks, and various *nutritional disorders.*

Diseases which affect mainly adult cattle are *Johne's disease* which causes chronic enteritis; *actinomycosis* and *actinobacillosis* which cause lesions of the mouth and tongue, respectively; *red-water disease*, prevalent in districts where ticks exist and with symptoms such as dark colouration of the urine, jaundice and diarrhoea; *cow pox*; *warble-fly infestation* where the larvae of gadflies bore through the skin of the animal and pus forms in the cavity so made; *cerebrocortical necrosis* from unknown causes but with effects of wandering, paralysis and blindness; various *metabolic disorders* such as *milk fever* and *grass tetany*, and *disorders of the reproductive system* such as *mastitis.*

3.1.3 Slaughtering, butchery

The economic significance of disease in cattle is shown by the high number of totally or partly condemned carcasses. (Partly condemned here means that certain portions of the carcass and the offals are unfit for human consumption.)

Ante-mortem inspection of cattle takes place as the animals leave the transport truck to enter the lairage pen at the abattoir. Experience shows that a large number of the animals suspected of disease during the ante-mortem inspection are subsequently condemned during post-mortem examination. An ante-mortem inspector will look for tuberculosis, actinobacillosis, emaciation, blackquarter, ringworm, mange, mastitis, transit fever and tumours.

A minimum period of 12 to 24 hours rest is essential before slaughter. If the animals are not allowed to rest the keeping quality of the meat is reduced because the level of acidity in the muscles does not have a chance to rise and the putrefactive micro-organisms from the intestinal tract can invade the carcass. There should be a good supply of drinking water available to the animal so that the bacteria in the intestine are reduced and so that the increased water content of the carcass allows the hide to come off more easily when the carcass is dressed. Withholding food also decreases the risk of infection from micro-organisms in the intestine. Since all animals are susceptible to cross-infection, especially with *salmonellae* from other animals, they should not be kept in the lairage pens longer than necessary.

In England and Wales the Slaughterhouse (Hygiene) Regulations of 1958 lay down certain requirements for layout, construction, equipment and practices for abattoirs. The following sections are regarded as necessary for an abattoir to run efficiently: lairage pens, an isolation block for infectious or injured animals, a slaughter hall, a cooling hall, a hide store, a gutter and a trippery.

In the UK the prevention of cruelty to animals before and during slaughter is ensured by The Slaughter of Animals Regulations of 1958. Legislation in Britain and in most European countries requires that all cattle slaughtered in an abattoir, except those slaughtered by the Jewish or Mohammedan ritual, should first be rendered unconscious.

In a well-organized slaughterhouse the animals are led from the lairage to a holding pen and are then driven along a passage to the stunning pens. The usual method of stunning cattle in Britain is by captive-bolt pistols which destroy the brain immediately, rendering the animals unconscious by intracranial pressure and a sudden jerk which produces 'acceleration concussion'.

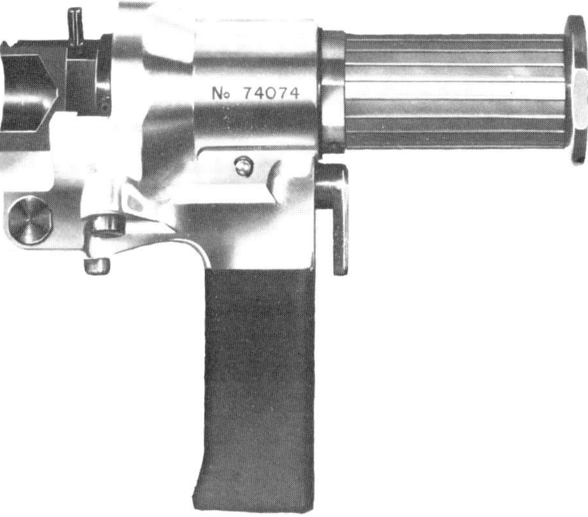

Fig. 3.9   Cash captive bolt pistol.

A method of electrical stunning has also been devised – the Elther method. Energy of 285 J (1J = 1watt-second) is passed during one second through apparatus resembling a pair of earphones.

The animals are led in individually and stunned in a standing position. To prevent reflex muscular action cattle are sometimes pithed before the next stage, bleeding.

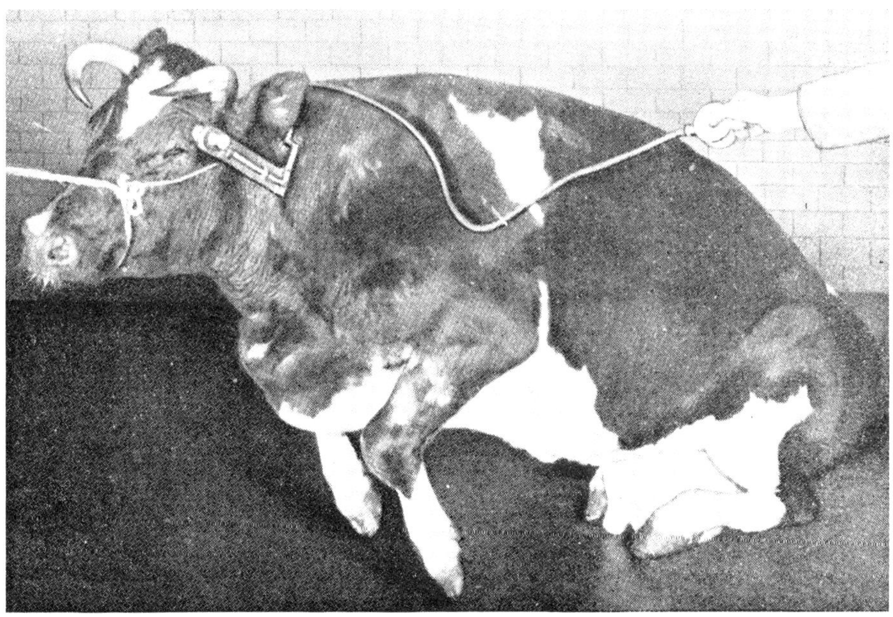

Fig. 3.10   Stunning of cow by the Elther electrical stunning method.

Pithing is the insertion of a rod into the hole made by the bolt of the pistol to destroy the medulla oblongata, the part of the brain which controls motor action.

After stunning the animals are hoisted by one or both hind legs to an overhead rail leading to the sticking point where the throat is cut, usually by severing all the neck vessels with one transverse cut. In some abattoirs bleeding is carried out with the animals on the floor. The advantage of a rail system, however, is that it allows centralized collection of the blood. The animals should be bled for 6 minutes before the head is removed from the body and the horns sawn off.

In most slaughterhouses the animals are then washed either in cold water or under a spray at 21 kg/cm^2 (300 p.s.i.) pressure heated to 35°C (95°F) for the reduction of surface bacterial contamination and better carcass appearance. Next the carcass is lowered, by means of a mechanical dropper on to a dressing bed where the feet are removed and the hide is partly removed. The carcass is then raised by a hoist to a dressing conveyor and the hide is completely removed. The hind legs are separated to a distance of about a metre (about a yard) to make the removal of the internal organs easier. The viscera drop on to a table covered by washable material. With the hind legs further extended the carcass is sawn into two halves and then four quarters. The internal organs (heart, lungs, kidneys, spleen and liver) and the intestines are put in a tray and the four quarters are hung up on hooks ready for inspection. All parts of the animal should be properly labelled so that they can be identified without any doubt.

Finally the carcasses may be wrapped in sterile linen sheets soaked in a 10% salt solution for reducing cooler shrink and improving the external appearance.

A stunning pen with bleeding rail and carcass droppers can deliver 15 cattle an hour, and, after dressing, a single mechanical saw can split about 25 carcasses an hour. The process is accelerated by 'line dressing' in which the carcass is propelled on an overhead rail and the dressing is divided into various single stages with an operator for each stage. Some large factory-abattoirs can process 5000 cattle an hour.

Routine post-mortem examination should take place as soon as possible because the carcass tends to set rapidly, especially in cold weather, making examination of the lymph nodes more difficult. In the slaughterhouse adequate space and facilities should be provided for the meat inspector. The parts of the carcass should be examined in the following order: the head and its lymph nodes (retropharyngeal, submaxillary and parotid) for foot-and-mouth disease, inflammation of the mucous membrane of the mouth, actinomycosis (lumpy jaw), actinobacillosis (ulcers on the tongue), cysticerci or beef measle (the cystic stage of the tapeworm) and tuberculosis; the lungs and their lymph nodes (bronchial and mediastinal) for inflammation of the pleura, tuberculosis, the early stages of liver fluke, and hydatic cysts (cystic form of the tapeworm of the dog); the heart (including its membranous sac) for haemorrhages indicating the presence of septicaemia and an acute infective disease, cysticerci, hydatic cysts and tuberculosis; the liver for fatty acid degeneration indicating the presence of an acute febrile and toxaemic condition or chemical poisoning, actinobacillosis, abscesses indicating intensive feeding ('barley beef') or bacterial infection ('plum pudding liver'), and parasitic infestations such as hydatic cysts, cysticerci, liver fluke and linguatulae (the larval stage of *Oesophago-stomum radiatum*); the stomachs and intestines (including the mesenterium and its lymph nodes); for tuberculosis, actinobacillosis and linguatala nodules; the spleen for tuberculosis, anthrax, blood tumors and coagulation necrosis (or infarct). In female animals the uterus should be examined for septic conditions and the udder, including the supramammary lymph nodes, for abscesses usually caused by *Corynebacterium pyogenes* infection, septic inflammation and tuberculosis. The whole carcass should be examined for bruises (resulting from maltreatment), inflammation (resulting from injury), abscesses and tuberculosis. The cut surface of the bones (in the spinal column) should also be examined, the kidneys loosened and visually inspected for abnormal conditions and the lymph nodes incized for tubercles, haemorrhages, etc.

In Britain the main causes of condemnation apart from bovine tuberculosis are the presence of tumours, inflammation of the lungs and pleura, emaciation, pericarditis and septicaemia.

3.1.4 By-products

Those parts of the animal that cannot be sold directly as food are known as by-products. Of these, some can be made into food for human consumption after various kinds of processing–tripe, rennet and fats–while others can be made into a wide range of non-consumable products. The reduction of waste increases the economic return from each animal.

By-products can be obtained from the stomach and intestines, the hide, the hair, the fat, the blood, the bones, the horns, the hoofs and condemned carcasses.

The *stomach* can be made into tripe and the *intestines* into sausage containers.

The *hide* is the most economically important by-product. Hides are sent from the abattoir either fresh or salted and dried, and are processed into leather over a period of about three months. During the process the hides are soaked in water and placed in pits filled with milk of lime from one to four weeks. At this stage the hair is loosened and the fibres are opened, facilitating the scraping-off process. The lime is removed by washing in weak acid. The hides can be tanned either by a vegetable process using tannin or a chemical process called chrome tanning where the reaction with the chromium salts gives a stable hide fibre resistant to bacterial attack and high temperature. In the case of chrome tanning, however, which is faster than vegetable tanning and relatively cheap, the leather does not have many of the qualities required for useful articles without further processing. Chrome tanning is the most common method of tanning light leather and shoe upper leathers. Gelatine can be made from trimmings from the hide.

Hair. After separation from the skin by treatment with lime or other chemicals, the hair can be used in a crude state by plasterers as a binding agent. Cattle hair has excellent insulating properties and can be used for the manufacture of underfelts.

Fat is the second most important by-product after the hide. It can be trimmed from the intestines and other internal organs. The trimmed fat can be rendered down to yield pure fat which is marketed as premier jus or tallow. The premier jus can be used as lard by pastry makers, in the manufacture of breakfast foods, in compound lard and in margarine making. The fat can also be processed into oil or stearin. The oil is used in lubricating oils and stearin can be used in the manufacture of compound lards and margarines prepared from hydrogenated vegetable oils. The non-edible fat can be used for the dressing of leather (fat liquoring), or further rendered down and processed to give commercial glycerol. The meat fibres which are left after the extraction of fat, known as cracklings or greaves, can be ground and used as poultry feed. The so-called neat's foot oil is prepared from cattle feet. The feet are washed and scalded at 77°C (170°F) to loosen the hair, boiled for 15 minutes to allow loosening and removal of the nails and then split. Prolonged cooking and subsequent settling produces a supernatant oil which is skimmed off and filtered. This oil, which is odourless, is used mainly in the manufacture of leather and as a lubricant for delicate machinery.

Bones. Fresh bones can be used for edible fat extraction, but the bones of diseased or condemned animals cannot. After the extraction of fat and gelatine the bones are in a soft chalky state and can easily be ground into bone meal. If the bone meal is dried rapidly in a kiln equipped with sheet iron trays and through which hot air is blown it yields a product called feeding bone flour; on the other hand if the bone meal is not dried rapidly enough the final product will contain about 10% of moisture and is called steamed bone flour. In contrast to feeding bone flour, the steamed bone flour may contain anthrax spores and can only be used as fertilizer (the main constituent of bone meal is calcium phosphate). *Blood* can be used either in edible products such as black pudding or processed into animal feed or fertilizer. If the blood is to be processed the collection should be made at central points in the slaughterhouse to ensure hygienic conditions. Fresh blood must be processed at the earliest possible moment to prevent decomposition and loss of nitrogen content. On average, 5 tonnes (5 tons) of raw blood yield 1 tonne (1 ton) of dry product. Fresh blood can also be converted into dry fibrin, blood powder or albumen. In order to extract fibrin the collected blood must be agitated immediately while the plasma can still be easily separated from the fibrin. The fibrin is then dried in a heated vacuum chamber. The final product can be stored for a long period without undergoing any change. The *horns* are sawn from the skull and used for producing buttons, combs and hair ornaments. Low grade horns can be used for fertilizer or horn meal. Horn meal can also be used in the preparation of foam-type fire extinguisher fluid.

The *hoofs* are removed from the feet by immersion in hot water. White hoofs are especially favoured by manufacturers of articles such as buttons and combs. The less valuable black and striped hoofs can be dried by steam and ground into horn meal fertilizer which is used particularly for grape vines.

Condemned material. It is essential that condemned material is collected and processed in such a way that infection of edible material is avoided, particularly since some of the end products, such as meat meal and bone meal, may still contain active micro-organisms like *salmonellae.* The best and most economical method of disposal of condemned carcasses and offal is by heat treatment in an enclosed vessel, a process which ensures sterilization and allows a good return of marketable products. In general by-product plants should be divided into two parts, the unclean where the infected materials are introduced and the clean where the sterilized materials are treated and stored. Some countries, Denmark for example, require that imported meat, bone and blood meal, or similar products, must be re-sterilized before being offered for sale.

3.2 VEAL

3.2.1 Rearing

In Europe top quality veal is produced from calves fed on milk and killed at 2 to 4 months old. Calves reared for veal should be well developed and relatively fat at birth. Male calves intended for veal are not castrated since this would interfere with body development.

In Britain, however, veal is almost entirely supplied by so-called 'bobby calves'. They are mainly bull calves of dairy breed killed at 1 to 7 days because with a daily milk requirement of 4.5 l (1 gallon) it would be uneconomical to keep them. Because of early slaughter bobby calves show little muscular development and their flesh is used only in manufactured products.

3.2.2 Diseases

Diseases which most often affect calves are tuberculosis, septicaemia (navel ill), calf diphtheria, haemorrhagic enteritis (calf dysentery and white scour), peptic ulcers, pneumonia, leptospirosis (giving rise to haemorrhagic jaundice), ringworm, coccidiosis (red dysentery) and worm infestation of the alimentary canal, respiratory tract and liver.

Tuberculosis is characterized by the predominance of infection of the lymph nodes and lesions in the abdominal cavity. If the tuberculosis was acquired *in utero* the whole carcass should be condemned because the infection is generalized.

Septicaemia affects mostly 1 to 4 day old calves and is usually umbilical in origin. The infection may spread from the navel area to the liver and other internal organs, making the whole carcass unfit for human consumption.

Calf diphtheria is mostly caused by the consumption of infected food or drinking water. The grey necrotic areas which can be scraped or peeled off only with difficulty during meat inspection occur together with ulcerated areas mainly on the tongue, pharynx or gums. Usually the head only is condemned but if the disease has spread to the stomach, intestines or lungs the whole carcass should be condemned.

Haemorrhagic enteritis is caused either by *salmonellae* (calf dysentery) which affect the mucous membrane of the alimentary tract resulting in emaciation and anaemia, or by *Escherichia coli* resulting in a form of scour (white scour). In the latter case usually calves born in the first 4 months of the year are affected. The carcass should be condemned. *Peptic ulcers* are found in the fourth stomach (abomasum) of 8 to 10 week old calves.

3.2.3 Slaughtering, butchery and by-products

During the ante-mortem inspection the following conditions should be looked for: immaturity, calf diphtheria, ringworm, cowpox and white scour.

Calf pens are usually provided as an integral part of sheep lairage as it is common practice for calves to be slaughtered in the same abattoir section and by the same slaughtermen. According to the requirements of the Slaughter of Animals (Prevention of Cruelty) Regulations young animals unable to take swill or solid food must be slaughtered as soon as practicable.

In the slaughterhouse the calves are hung up by their hind legs, stunned (usually with a wooden hammer) and their throat cut. Electrical stunning can also be used.

Calves yield about 2.8 kg (6 lb) of blood. In the past it has been the practice to make an incision in the jugular vein at the side of the neck and to bleed the animals slowly in order to obtain whiter meat. This practice has been abandoned, mainly for humanitarian reasons.

The former practice of inflation of the carcass by mouth or by mechanical means for making dressing easier is forbidden in many countries (USA, Germany, etc.) since it may misrepresent the condition of the animal and can make the examination of the joints difficult.

During post-mortem examination the inspector should give special attention to the abomasum for evidence of peptic ulcers, to the small intestines for evidence of white scour or dysentery and to the umbilical region and joints for evidence of septic inflammation.

Veal calves have an average carcass weight of 68 kg (150 lb). The proportion of bone is about 25% of dressed carcass weight. In relation to live weight the weight of the calf's stomach and intestines is less (11%) than in adult cattle (16%) because of the fluid nature of the digestive contents. The carcass yield is, therefore, correspondingly greater in calves (63% of live weight) than in adult animals. In bobby calves the proportion of bone to dressed carcass weight is usually 50%.

The by-products are similar to those obtained from cattle. Additional by-products are rennet and sweetbread. Rennet is extracted from the fourth stomach of calves fed exclusively on milk. The stomach is washed and macerated in salt brine. The rennet thus isolated in a dilute, impure form is concentrated to a powder by evaporation. Rennet is used in cheese-making to coagulate milk. It can also be used for making junket. The true sweetbread is the thymus gland. In the calf the thymus is at its greatest development at 5 to 6 weeks when it weighs about 0.45 to 0.75 kg ($1–1\frac{1}{2}$ lb). In older animals a gradual atrophy takes place and by the onset of sexual maturity only a small part remains. The sweetbread consists of two portions, the 'heart bread' (thoracic portion) which is rich in fat and roughly the shape of the palm of the hand and the 'neck bread' (cervical portion) which is poor in fat.

3.3 PORK

3.3.1 Breeds and rearing

It is probable that the domesticated pig is descended from two closely related species, *Sus scrofa* indigenous to Europe and North Africa and *Sus vittatus* indigenous to eastern and south-eastern Asia. The earliest known domesticated pig, *Sus scrofa palustris*, can be traced back to neolithic times. In Europe the domesticated pig of early historic times appears to have been closely related to the local wild type. In Britain during Saxon times large herds were maintained in the extensive oak forests.

The tendency in pig breeding over recent years, particularly in northern and western Europe, has been for the number of lard breeds to be reduced and the number of meat breeds, either for pork or bacon, to be increased. Artificial selection, therefore, has favoured those animals whose progeny carried as little fat and as much meat as possible at a given live or dressed carcass weight. The proportion of fat to lean meat in the carcass depends not only on hereditary factors but also on the live weight of the pig at the time of slaughter. At slaughter a meat-type pig may produce much fat if it has a high live weight and a fat-type pig may produce a relatively high proportion of meat if it is killed at a low weight. The ratio of fat to lean meat can also be controlled by the intensity of feeding. Pork breeds and bacon breeds are reared, fattened and marketed differently. The various breeds are described in sections 3.3.1 *a* to 3.3.1 *j* below.

Pork breeds, such as the Berkshire and Middle White are slaughtered at a low live weight (50–70 kg, 110–154 lb) to prevent the development of too much fat. The preferred body conformation is short, broad and deep. They should be able to mature early and produce sufficiently large joints.

Bacon breeds such as the Large White, the British Landrace, the Tamworth and the Welsh are usually slaughtered at a live weight of 90 to 100 kg (198–220 lb) producing a dressed carcass weight of 60 to 70 kg (132–154 lb). The typical body characteristics are long, light shoulder and heavy, well-rounded hams. They mature later than pork-type pigs and at slaughter the thickness of fat on the back should not exceed 50 mm (2 in); at the middle of the back and at the loin it should not exceed 30 mm ($1\frac{1}{4}$ in). The back fat is usually removed from fatter carcasses for use in sausage production or in the canning industry.

For the production of both pork and bacon several breeds are reared in the United Kingdom. These dual purpose breeds are: British Saddleback, the Gloucester Old Spots, the Large Black and the Long White Lop-eared. Since the requirements for pork and bacon are different the quality of the product is often lower than in single purpose breeds.

The organization of pig breeding differs from country to country according to its importance to the economy. In the United Kingdom and in many other countries the breeding programs are based on officially recognized pedigree herds. The ten breeds now recognized by the National Pig Breeder's Association of Britain fall into three main groups: the white, black and black and white, and red breeds. In the first group are the Large White, the Middle White, the Long White Lop-eared, the Welsh and the British Landrace. In the second group are the Large Black, the British Saddleback, the Gloucester Old Spots and the Berkshire. In the third group is the Tamworth.

a. The Large White. The Large White (or Large Yorkshire) has pedigree records dating back to 1884 and was developed near Leeds towards the middle of last century. It is now common in the northern half of the United Kingdom and has been exported to most pig breeding countries. The breed is the largest among British breeds and its body structure is typical of a bacon breed. The coat is reasonably abundant and consists of fine straight hair. The average number of offspring per litter is high, usually averaging eight. It is hardy and active but not very docile. It crosses well with the Middle White and the Berkshire.

b. The Middle White. The Middle White is derived from the crossing of the Large White and the Small White (or Small Yorkshire) which is now extinct. The breed has been bred 'pure' now for over 50 years. It is one of the smallest of British breeds, being slightly smaller than the Berkshire, and has a short broad and deep body with short legs. The hams are full and the shoulders are prominent. The coat is more abundant and the hair is longer than that of the Large White. The breed matures early and is reasonably prolific. It can be fattened and marketed at almost any age and produces excellent pork, either pure or crossed with the Berkshire. On the other hand it produces good bacon if it is crossed with the Large White or the Large Black.

c. The Long White Lop-eared. This breed is similar to the Large Black in all but in colour. It originates from Devon and Cornwall and is bred for both pork and bacon.

d. The Welsh. The Welsh breed is somewhat similar to the Middle White but tends rather more towards the bacon conformation. The body is longer and slightly less rounded. It produces better bacon than pork.

e. The British Landrace. The Landrace was introduced to Britain from Sweden in 1949. It is a hardy animal descended from the Scandinavian and eastern European white pig. In Britain and Denmark (the Danish Landrace) the breed was developed for bacon production by selection based on progeny testing.

f. The Large Black. The Large Black was officially established as a breed in 1899 and was derived from foundation stock in Devon, Cornwall, Suffolk and Essex. It is slightly shorter than the Large White with a deep side, light smooth shoulder and long thick ham. The face is of medium size, and the ears are long and floppy with points closing

Fig. 3.11   Large White Boar.

Fig. 3.12   Middle White Sow.

Fig. 3.13   Long White Lop-eared Sow, 4 years old.

Fig. 3.14 Large Black Sow.

Fig. 3.15 Welsh gilt or young sow.

Fig. 3.16 Saddleback Sow.

in at the level of the tip of the nose. It is black in colour and hardy and docile in temperament. It grazes freely outdoors but when reared indoors it may suffer leg weakness. It is a prolific breed and crosses well with the Large White for bacon and the Middle White for pork. It is found in most parts of the United Kingdom and is also widely exported.

g. The British Saddleback. The British Saddleback is derived from the Essex and Wessex Saddleback breeds which were amalgamated in 1968, 50 years after their establishment. It is claimed that both breeds are almost wholly the descendants of the Old English Pig with little or no admixture of east-Asian blood.

h. The Gloucester Old Spots. The Gloucester Old Spots is native to the Gloucester region. It is a large breed reaching weights comparable to those of the Large White. The head is of medium size and similar in shape to that of the Large Black. However, the ears, though floppy, are shorter and wider set than those of the Large Black. The second part of the name of the breed is derived from its colour which may be white spots on black or vice versa. The skin colour corresponds to that of the hair. The coat is thick, the hair long, straight and curly. The breed is hardy and accustomed to being reared outdoors. It grows well with grass as a large proportion of its food. It is prolific and can be fattened and marketed at any age.

i. The Berkshire. The Berkshire was officially established as a breed is 1884. The face is short, the forehead dished and the snout not turned up. The ears are short and pointed. The body is medium in length, wide and deep, and with short legs. The loin is thick and the ham full. The bone structure is fine. The animals are black with white feet, a white tip on the tail and a small but varying amount of white on the face. The thick coat is medium in length and the hair is fine. The breed combines earliness of maturity with quality of carcass (like the Aberdeen Angus in cattle and the South Down in sheep). However, the sows are less prolific than in most other breeds and the piglets do not grow quickly at first. The boars cross well with Middle White sows for young pork.

j. The Tamworth. The Tamworth, like the Saddleback, has close association with the native English pig. The breed can be traced back to 1884 or even earlier. The body is long with a fairly deep side and width and long legs. The snout is very long and straight. The ears are long and point forward. The coat colour is reddish or chestnut and the skin is flesh coloured. The breed is hardy and fairly prolific though slow to mature. However, at slaughter the proportion of lean meat on the carcass is high and it produces excellent bacon.

3.2.2 Diseases

In the United Kingdom infectious atrophic rhinitis, foot-and-mouth disease, swine fever, swine vesicular disease, and anthrax are notifiable.

Infectious atrophic rhinitis (or snuffles) mainly affects the head of the animal but may also cause systemic changes. The main symptoms are that the nasal bone tends to turn upwards from the lower jaw, the snout becomes dished and the skin over the affected part becomes wrinkled. In porcine *foot-and-mouth* disease mainly the feet are affected. 'Casting of the claw' often takes place. The so-called 'thimbling' is caused by the formation of new horn-like tissue while the old claw is still there.

Fig. 3.17 Gloucester Old Spots Sow, 5 years old.

Fig. 3.18 Berkshire Sow.

Fig. 3.19 Tamworth Boar.

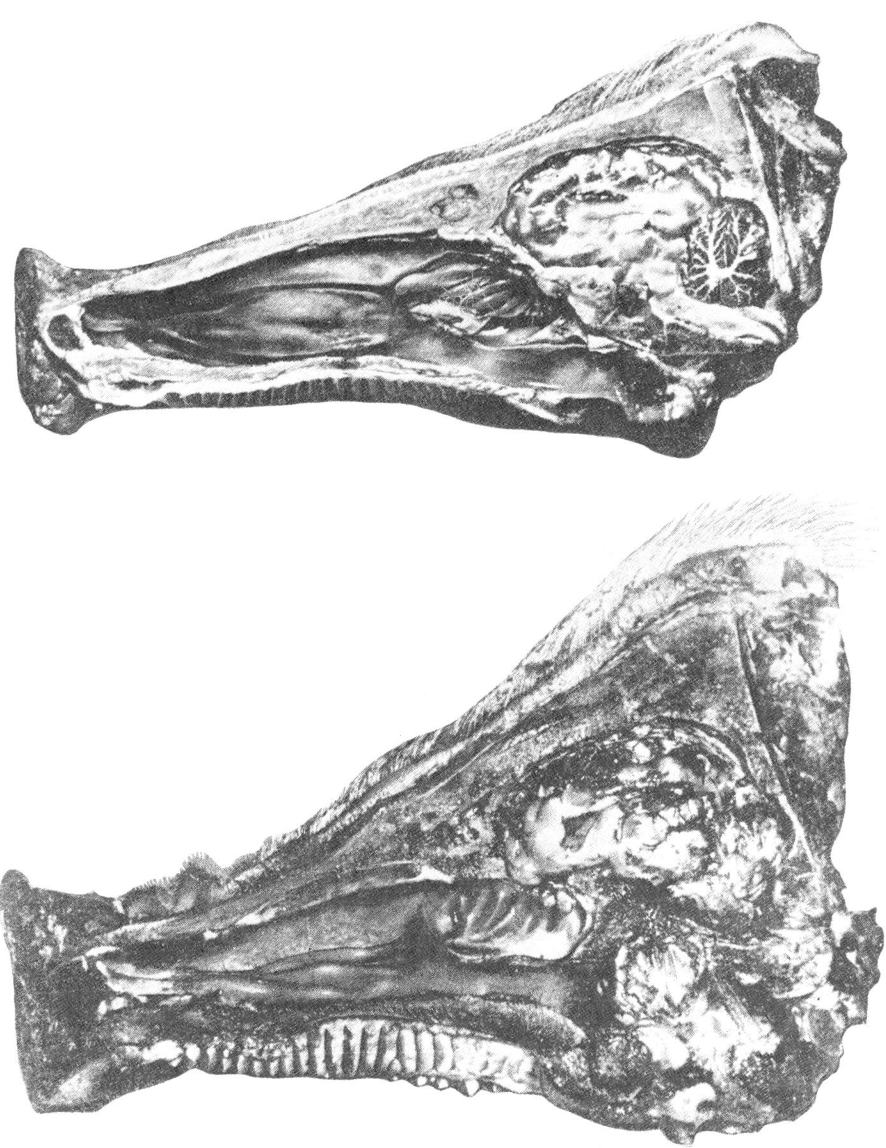

Fig. 3.20 Atrophic rhinitis of pig head. (*Top* = healthy head, *Bottom* = diseased head).

Vesicular disease of pigs. In 1968 a new syndrome resembling foot-and-mouth disease was observed in Lombardy, Italy. The condition was of low morbidity and short duration. In 1970 a similar condition was observed in the New Territories, Hong Kong. Recently the disease has also been recognised in the British Isles. In the field it is difficult to differentiate between the new syndrome and foot-and-mouth disease because the clinical evidence is consistent with a mild foot-and-mouth disease virus infection.

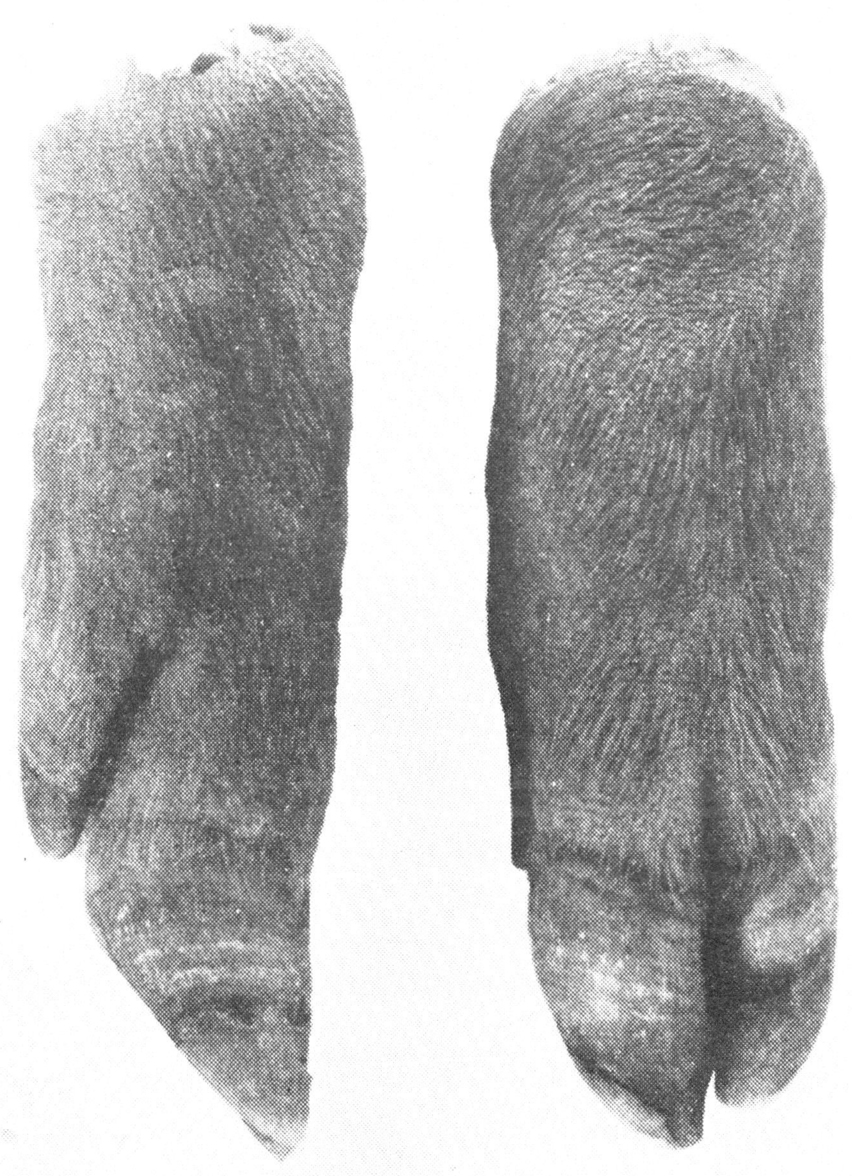

Fig. 3.21 Pig's feet showing thimbling.

Laboratory investigations have shown that the virus responsible for the disease has different physical and chemical properties from those of the foot-and-mouth disease virus. There are indications that a porcine enterovirus may be involved. Therefore in future outbreaks of the vesicular disease in pigs it would be advisable to consider the porcine enteroviruses as causative agents in the differential diagnosis in addition to the viruses of foot-and-mouth disease, vesicular exanthema and vesicular stomatitis.

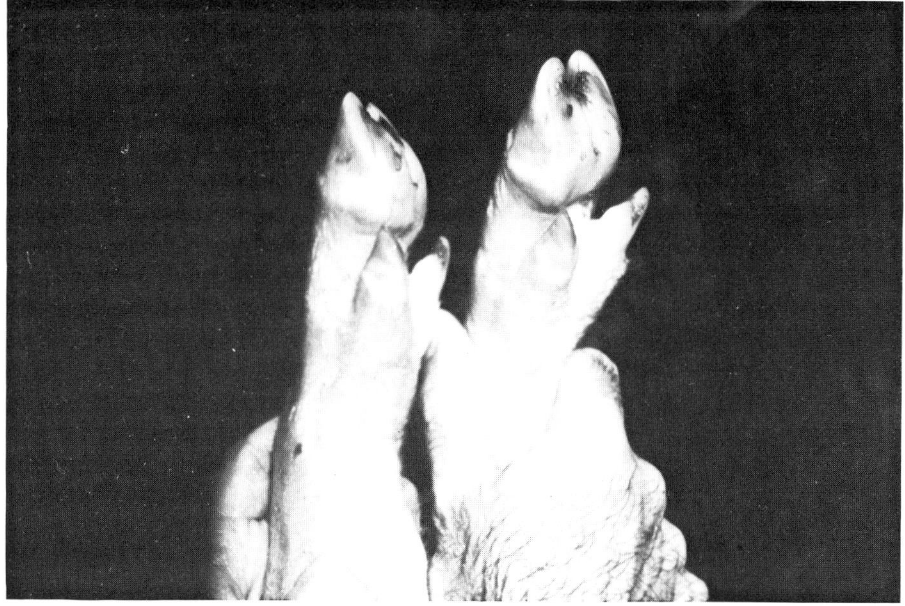

Fig. 3.22   Feet of a pig showing unruptured lesions of Swine Vesicular Disease (12–24 hours).

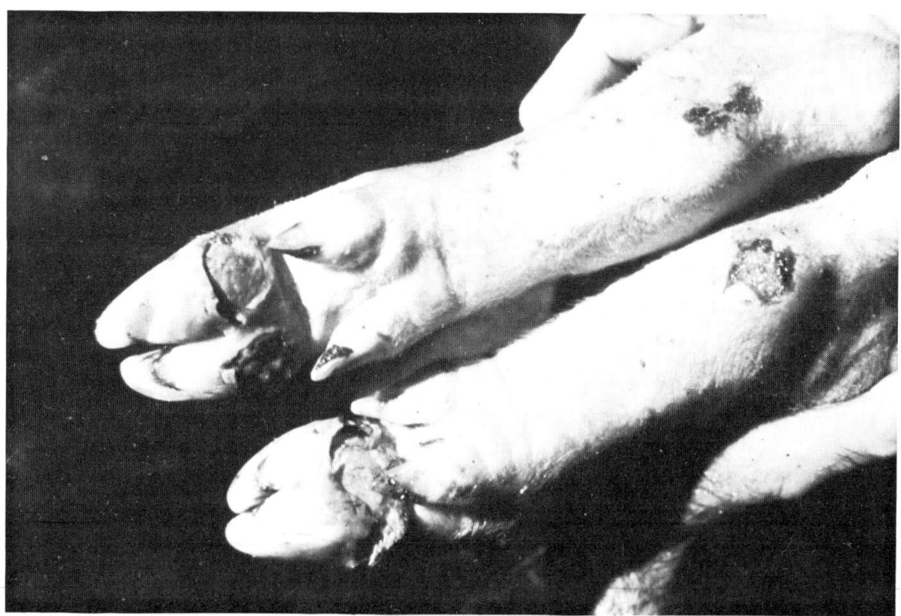

Fig. 3.23   Feet of a pig showing recently ruptured lesions of Swine Vesicular Disease. (48 hours).

Swine fever (or hog cholera) is characterized by the symptoms and lesions of septicaemia (raised body temperature, congestion of the mucous membranes, multiple haemorrhages in various parts of the body, lymph node enlargement and enteritis). It affects mainly the alimentary tract but can involve other parts of the body (giving pneumonia, for example). The disease is likely to be more acute in younger than in older pigs though it can occur in animals of all ages. In acute cases the carcass should be condemned because the virus can remain viable for long periods of time even in chilled, frozen or salted pork. In certain cases the carcass may not be passed for food because of emaciation. Other animals and man are not affected by the disease.

Anthrax affects pigs less often than it does other farm animals. If the disease occurs in Britain it is usually in the acute form and the affected animals may be found already dead. Therefore sudden deaths in pigs should be reported immediatley to the health authorities. The disease can be transmitted to man mainly through an abrasion in the skin. The chronic form of the disease is characterized by swellings in the throat which may cause death by suffocation. Malignant carbuncles on a swelling of the skin may also occur. Sometimes the alimentary tract can be affected.

Other diseases which can affect pigs are *gut oedema, actinomycosis of the udder* and *swine erysipelas.*

Gut oedema is sometimes fatal to young pigs and is probably caused by certain types of *Escherichia coli.* The first diagnosable symptom is swelling of the eyelids and the concomitant lesion is swelling of the stomach wall. A carcass showing an acute state of the disease should be condemned.

Actinomycosis usually affects the udder tissues of sows, causing the udder to be converted into a hard tumour. The resulting fibrous tissue contains abscesses which can be as large as a walnut. About 75% of chronic mastitis cases are caused by actinomycosis. Affected udders should be condemned.

Swine erysipelas often affects pigs when the keeping conditions are poor (bad housing, lack of hygiene, and fatigue). The slender rod-like micro-organism (*Erysipelothrix rhusiopathiae*) which causes the infection is often present in apparently healthy pigs. The disease is spread by the ingestion of urine and faeces from affected or recovered animals. It can also be transmitted to man by inoculation through skin lesions. The infection is most common among slaughterhouse workers. In pig the most characteristic lesions of the disease are pinpoint haemorrhages in the lungs, kidneys, and the serous covering of the stomach and intestines. On the skin sometimes swollen red rhomboid or diamond-shaped areas with a pale centre may appear. The chronic form can cause cardiac disturbances. Carcasses affected by the septicaemic form of the disease should be condemned. A pig which has suffered from swine erysipelas may be susceptible to septic or non-septic arthritis. Even in non-septic cases the affected joint should be removed from the carcass.

Further diseases which can affect pigs are *tuberculosis, virus pneumonia, pig paratyphoid* (or necrotic enteritis) with symptoms resembling those of swine fever and causing emaciation, *icterus* (or jaundice) derived from cirrhosis of the liver or leptospirae ingested from the faeces or urine of infected rats and *swine pox* which is rare in Britain. Various forms of *nutritional deficiencies* can occur as can various *parasitic diseases.*

Parasites can affect the lungs (especially *Metastrongylus elongatus*), the intestines (especially *Ascaris lumbricoides* or round worm), and the stomach (especially *Hyostrongylus* or threadworm) as well as the intestines and muscles (*Trichinella spiralis*;

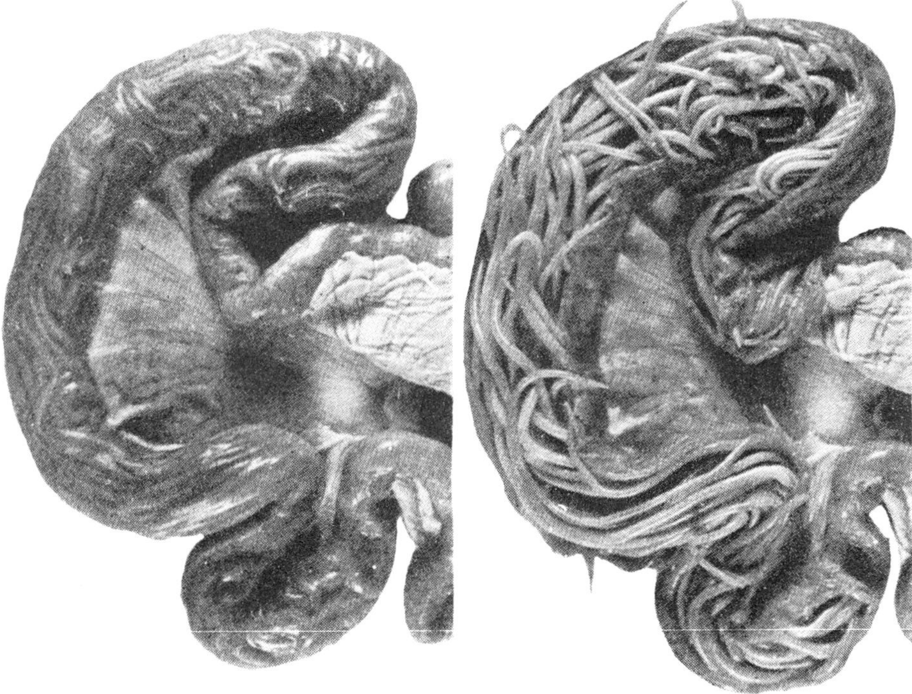

Fig. 3.24 Intensive infestation of the small intestine of pig with *Ascaris lumbricoides.*

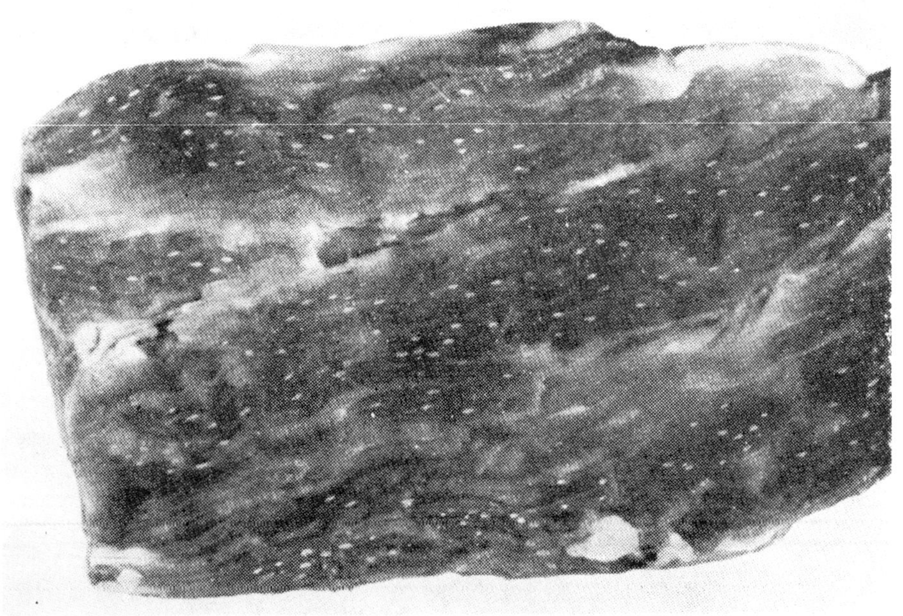

Fig. 3.25 Trichinellosis of muscle of pig.

adult form in intestines, larval form in muscles). The latter form of infestation, known as *trichinellosis*, can affect all carnivorous animals and man. In man a massive infestation could lead to severe muscle pain or death. Unfortunately it is not always possible to detect *Trichinella* cysts in pork during meat inspection even with laboratory methods such as the trichinelloscope or digestion. However the cysts can be destroyed by ordinary cooking. Pork imported to Britain must be inspected for the presence of *trichinellae* except where the exporting country can establish that no cases of trichinellosis have occurred during the preceding three years.

3.3.3 Slaughtering

Pigs should be examined ante mortem especially for atrophic rhinitis, gut oedema, foot-and-mouth disease, rabies, actinomycosis of the udder, herniae, tumours, abscess formation and injuries.

In Britain the legislation of 1959 (see section 3.1.3) requires that pigs should be mechanically stunned prior to slaughter. In European countries, however, it is a common practice to slaughter pigs without first making them unconscious. The methods generally used for stunning are: hammer, carbon dioxide anaesthetization or electrical stunning. The most commonly used apparatus for carbon dioxide anaesthetization is the oval tunnel, designed to stun 240 bacon pigs an hour. As the method is ineffective with large boars and sows they must be stunned by the captive-bolt pistol used for cattle (see section 3.1.3).

The electrical method is widely used in bacon factories. Given that certain conditions are observed this method is effective and humane as it gives rise to cerebral stimulation which causes inco-ordination of the cerebral nerve cells. The animals should be bled immediately after unconsciousness has been effected, otherwise, though still paralysed, the animals may regain consciousness.

After being led into the slaughter hall in a queue, stunned and shackled by the hind legs to an automatic rail, the animals are bled. An incision is made in the middle line of the neck at the depression in front of the sternum and the knife is then pushed forward to sever the anterior vena cava at the entrance to the chest. Occasionally, when the incision is inexpertly made, the carotid artery may also be pierced.

The blood is caught in a tunnel beneath the rail. The pigs move forward on the rail at such a pace that when most of the blood is lost (after about 6 minutes) the carcass is dropped mechanically into a scalding tank (water temperature 60°C, 140°F) where it remains immersed for 1 minute. The effect of scalding is to loosen the hair. Care must be

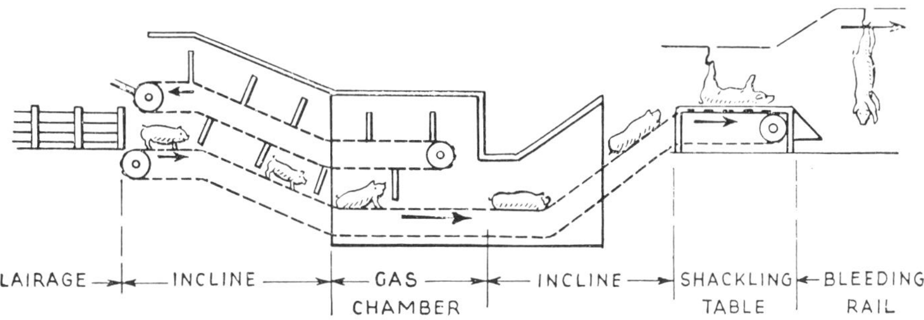

Fig. 3.26 Tunnel for anaesthetization of pigs by carbon dioxide.

taken that the carcass does not emerge from the tank 'cooked'–a state in which the bristles and scurf become set, making scraping and cleaning impossible. To prevent bacterial contamination the scalding tank should be drained and filled with fresh water daily.

Next the pigs are moved mechanically to a dehairing machine, a cabinet with fast revolving brushes which partially scrape the carcass. A large dehairing machine is capable of processing 300 pigs an hour. The carcasses are then transferred to a singeing plant which consists of two vertical half-cylinders lined with fireproof bricks forming a cylindrical chamber. The carcasses are singed at a temperature of 1371–1538°C (2500–2800°F). The singeing turns the colour of the skin brown, removes any remaining hair, and has a preservative effect on the carcass. About 140 carcasses an hour can be processed in this machine.

The carcasses are then sprayed with cold water and scraped clean by hand or by a back-scraping machine before being dressed. The head and vertebral column are removed and the carcass is split in two. The fat in the abdominal cavity is loosened, the internal organs and intestines are removed and the kidneys are freed for inspection.

During post-mortem inspection special attention should be given to skin lesions which may indicate swine erysipelas, swine fever and urticaria. The skin should also be examined for 'spotty eruption' (caused by faulty development of hair follicles) and transit erythema (caused by exposure of live animals to cold air). The internal organs should be examined particularly for pneumonia, pleurisy, pericarditis and peritonitis. The kidneys especially should be examined for blood spots indicating septicaemia. The lymph nodes (submaxillary, bronchial and mesenteric) should be examined for tuberculosis and abscesses.

a. Bacon preparation. The weight of two sides of pork prepared for curing is about 60% of the live weight of the animal. When the sides of pork are cured, either by dry-salting or by pickling in a tank, the process can be accelerated by preliminary injection of the pickle into the blood vessels.

Bacon is imported into Britain largely as pickled sides, undried and unsmoked. Drying and smoking are subsequently carried out by wholesale merchants. Bacon is smoked to preserve the colour and to give a special flavour to the meat. The bacon is first dusted with pea-meal and then smoked. The flavour derives partly from the pickling and partly from the wood shavings used in smoking. The combined application of heat and smoke usually significantly reduces the surface bacterial population. In smoke generated from wood the main bacteriostatic and bactericidal substance is formaldehyde. The process of smoking takes up to three days at a temperature of 30°C (85°F).

3.3.4 By-products

About 25% of the live weight of an average pig can be converted into economically viable by-products. The parts of the animal which yield the most important by-products are the blood, the hair, the fat and the internal organs.

Blood. By-products from blood are produced economically only where the daily killing in a slaughterhouse is large since the necessary drying equipment is expensive to install. The primary edible by-products from blood are black sausage and various types of black pudding. Black pudding is usually a dark-coloured sausage made from defibrinated blood, suet, flour and boiled barley. The mixture is poured into a casing (ox runners or middles–the gut of the animal) which is tied with string every 228 mm (9 in). The sausages are then cooked for half an hour at 88°C (190°F).

Fresh blood can also be made into dried blood by various methods such as steam coagulation, direct drying, spray drying or pan drying. Soluble dried blood can be obtained by drying defibrinated blood. Blood albumen is the product obtained by removing the blood cells and fibrin from fresh blood and drying the resulting serum. The yield from 100 pigs of average size would be about 35 kg (78 lb) of fibrin and blood corpuscles and 14 kg (30 lb) of albumen. In the preparation of soluble dried blood and blood albumen, the fibrin should be removed from the blood by manual or mechanical agitation before coagulation sets in. The dried blood can be used as fertilizer whereas the blood albumen can be used in various manufacturing operations (production of air-craft, plywood, adhesives, etc.), as a clarifying agent in wines, in the preparation of certain sugars and photographic papers, and as a colour fixative in cotton goods like gingham.

Hair. An average pig carcass yields about 1 kg (2 lb) of bristles. The long bristles from the back and tail are used for brush-making (shaving brushes, for example). The shorter hair can be used, after chemical purification, for stuffing mattresses and padding chairs and cushions in upholstery manufacture.

Fat. Up to 15% of the live weight of an average pig is surplus fat which can be rendered into various qualities of lard. The highest quality fat is leaf fat obtained from the peritoneal lining and the next highest quality fat is obtained from the mesentery and omentum (or peritoneal fold). The fresh, clean, raw fat is cut up into small chunks, heated and stirred in steam-jacketed containers. The containers are jacketed to prevent the discolouration and overheating (with subsequent lowering of value) that can occur when the fat is heated over an open fire. The liquid fat is run off into a jacketed tank where it is cooled and settled. If the cooling process is rapid, a smooth, white, uniform appearance is ensured for the final product. However, if the cooling is slow, hard white stearin crystals are formed.

Stomach. Pepsin can be produced from pig stomach by removing the mucous membrane from the outer wall and washing it gently in cold water to remove impurities. The cleaned mucous is kept under refrigeration until the beginning of the manufacturing process. The linings of the stomachs of 4 to 5 pigs yield about 0.5 kg (1 lb) of pepsin. In veterinary practice, pepsin is used to increase the digestive power of the gastric juice where there is a deficiency of pepsin in the animal.

Intestines. The intestines are cleaned and scraped before being used as containers for sausage meat.

Spleen. The spleen is used as food for cats and dogs.

Tongue. The tongue is used for food either fresh or salted.

Lungs. The lungs weigh about 0.5 kg (1 lb). They are generally used as an ingredient in white pudding and for cat and dog meat.

Heart. The heart, which weighs about 171–200 g (6–7 oz), is sold fresh.

Liver. The liver, which weighs about 1 kg (2 lb) in pork pigs and about 3 kg (7 lb) in sows, is usually sold fresh. A liver extract can also be prepared from pig liver for the treatment of pernicious anaemia. It contains folic acid and vitamin B_{12} as active ingredients. The *bile* is used as a cleanser and by the leather industry for polishing.

Kidneys. The kidneys, which weigh about 171–342 g (6–12 oz), are sold fresh.

3.4 LAMB

3.4.1 Breeds and rearing

The sheep population of the world is about 1 000 million and in Britain alone about 11 million lambs are born every year. About half this number of animals are slaughtered as lamb, about a quarter as mutton, and about a quarter go into breeding flocks and eventually reach the market as 'ewe mutton' when 5 to 7 years old.

In both lamb and mutton carcasses of any breed the desirable characteristics are short, stocky legs, a thick full loin, a broad full back, thick, fleshy ribs with a wide breast and shoulder, a deep chest cavity and a short plump neck. Lamb is more tender and less fat than mutton and is therefore in greater demand at present. Fat sheep in wool should have a carcass yield of at least 50%. The Suffolk and Scottish halfbred can have a yield as high as 54%, while old thin ewes in full fleece may have a carcass yield of only 35%. As a rule, longwooled sheep tend to produce fat carcasses whereas the best quality mutton is produced by mountain sheep.

The chief breeds of sheep are described in sections 3.4.1*a* to 3.4.1*i*.

The aim in sheep breeding is to produce a type of lamb which matures early for summer slaughter or alternatively can be kept and fattened economically during the winter. For lamb and mutton production various sheep breeds are crossed. The quality of the crossbred flocks from which both lamb and mutton are obtained depends mainly on the pure breeds from which they are derived. In the British Isles there are various *longwool breeds*, such as the Border Leicester, the Improved Dartmoor, the Whiteface Dartmoor, the Devon Longwooled, the Kent or Romney Marsh, the Leicester, the Lincoln Longwool, the South Devon, the Wensleydale and the Teeswater; various *short-wool breeds* or Down breeds such as the Clun Forest, the Devon Closewool, the Dorset Down, the Dorset Horn, the Hampshire Down, the Kerry Hill, the Oxford Down, the Ryeland, the Shropshire, the Southdown, the suffolk and the Wiltshire Horn; and various *mountain breeds* such as the Black Welsh Mountain, the Cheviot, the Derbyshire, the Gritstone, the Exmoor Horn, the Herdwick, the Lonk, the North Country Cheviot, the Rough Fell, the Scottish Blackface, the Shetland, the Swaledale, the Dales-bred, the Welsh Mountain, the South Welsh Mountain and the Radnor.

Breeding flocks may be classified according to the terrain in which they are reared. *Mountain flocks* are usually made up of small, hardy, pure breeds like the Scottish Blackface, the Cheviot, the Swaledale, the Welsh and the Exmoor Horn. They graze on mountain and moorland.

Upland flocks may consist of purebred or crossbred sheep. For the most part the flocks are made up of purebred mountain ewes and carry first class lambs. In Scotland these ewes are crossed with Border Leicester rams, in the north of England with Wensleydale and Teeswater rams and in Wales with Kerry Hill rams. Purebred flocks are intermediate in size between mountain and lowland breeds and resemble, in many respects, the first crosses obtained from mountain ewes.

Lowland flocks, if they are purebred, consist mainly of Down and Longwool breeds. They provide rams–Suffolk, Hampshire, Oxford, Shropshire and Leicester–for crossing on lowland farms to produce fat lambs.

In Britain the most important utility sheep is a cross between a Border Leicester ram and a Cheviot ewe. Since the Border Leicester carries an excessive amount of fat and is generally lacking in lean meat and the Cheviot has a firm and relatively lean carcass, their

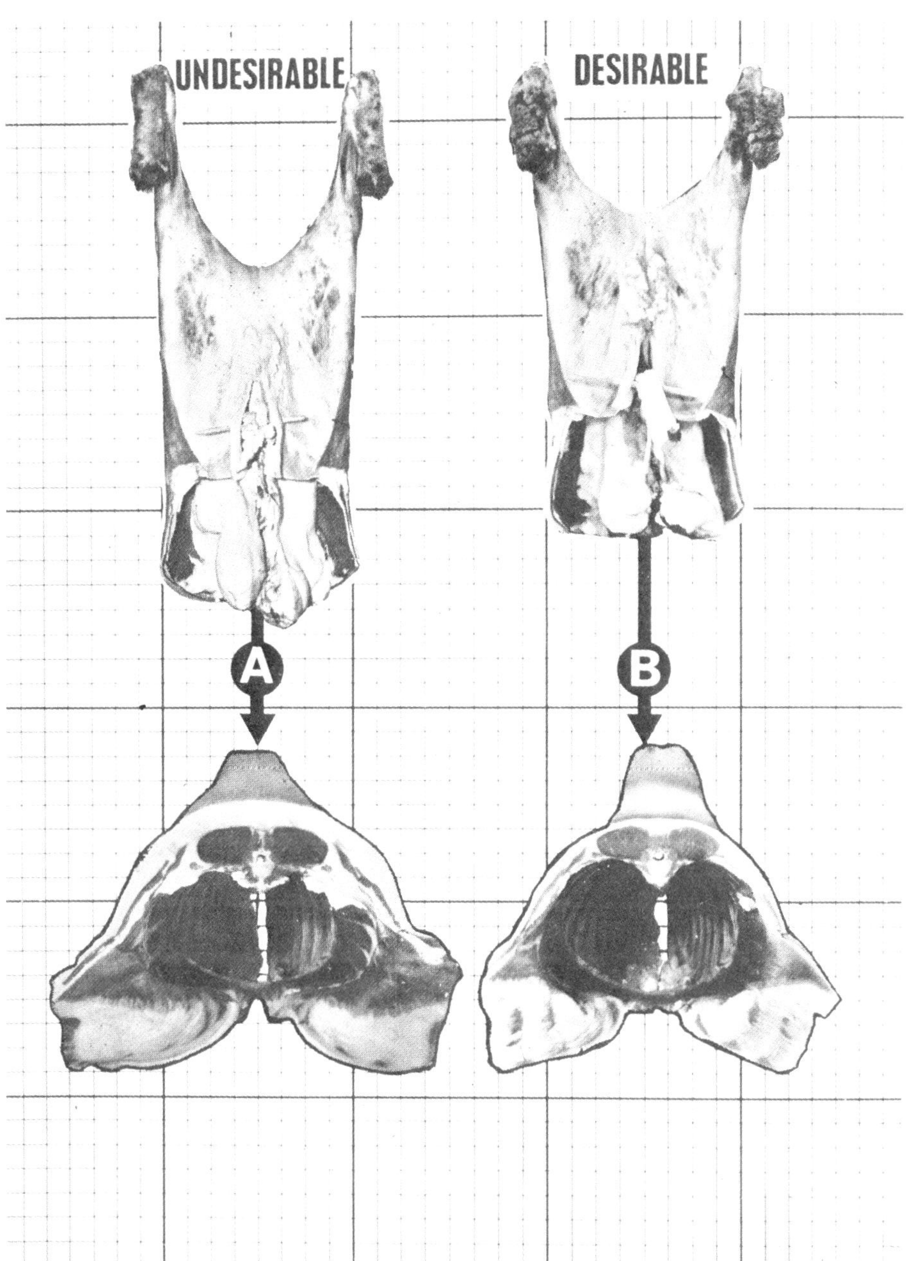

A. Leg narrow, bones long; loin too fat

B. Leg well filled out, bones short; loin not too fat, eye muscle deep

Fig. 3.27 Carcases of mutton showing desirable and undesirable types of leg and loin.

cross produces tender and succulent meat. When a halfbred ewe is further crossed with a Suffolk, Oxford or Hampshire ram, lambs are suitable for both summer slaughter or winter keep. Occasionally Ryeland and Southdown rams are also mated with halfbred ewes. The crossing of Scottish Blackface or Hill ewes with Border Leicester rams, produces the so-called 'mule lamb' which is in great demand for summer slaughter because of its tender flesh.

The most important lamb and mutton producing breeds in Britain are: the Border Leicester, the Cheviot, the Suffolk, the Oxford Down, the Hampshire Down, the Ryeland, the Southdown and the Scottish Blackface.

a. Border Leicester. The Border Leicester is a longwool breed reared mainly in the north of England and the lowland area of Scotland. The most valuable characteristic of the breed is the use of the ram as a crossing sire. Being comparatively fine boned the ram may be used safely on small sized ewes. The most popular crosses are with the Cheviot ewe (producing the Scottish Halfbred) and with the Blackface ewe.

The head is medium in size, hornless, and covered with short white hairs. The nose is aquiline and the ears, which are large and prominent, are white in colour with occasional black spots. The neck is long and the head is carried in a characteristically jaunty manner. The back is exceptionally wide and level. The ewes are good milkers and under favourable conditions a flock can produce 170% of lambs.

b. Cheviot. The Cheviot is a very old mountain breed reared mainly in the Cheviot hills between England and Scotland. The face is white with an arched nose and large, dark, bright eyes. The ears are pointed. The body is deep chested and jauntily carried. Originally Cheviot rams were horned, but the majority are now hornless with the exception of the Lockerbie type popular in Wales in which a small horn is preferred. In the Brecon Beacon hills in Wales Cheviots and Cheviot-Welsh crosses are well established. Cheviot ewes crossed with the Border Leicester produce the popular Scottish Halfbred ewe.

c. Suffolk. The Suffolk breed, established in 1886, is derived from crossing Southdown rams with blackfaced Norfolk horned ewes. It is the only breed to be characterized by black face and ears. The breed is outstanding for crossing since the animals have a broad and level back with strong loins, and are well covered with lean meat. Suffolk rams are often crossed with hill or lowland breeds. A popular cross is with the Scottish Halfbred.

d. Oxford Down. The Oxford Down is derived from the crossing of Hampshire and Southdown ewes with Cotswold rams. It is the largest and heaviest of the Down breeds. It was established in England in 1830 and introduced to America in 1853. It can now be found in almost every sheep breeding country in the world. The Oxford Down is a big animal with a bold appearance. The head is covered with wool which forms a topknot. The body is broad and symmetrical and is carried on legs set well apart. It is one of the most prolific of the Down breeds. In the north, rams are used to a great extent for crossing with Halfbred or Greyface ewes.

e. Hampshire Down. The Hampshire Down has been exported to most parts of the world. It is an old established breed, well adapted to the mixed down and arable holdings of the chalk uplands in Hampshire and Wiltshire. The face and ears are a rich brownish or blackish-brown colour, and the head and forehead are well covered with wool. The animals are rapid growers and their lambs reach high weights. When reared under favourable conditions they can increase their live weight at the rate of 0.5 kg (1 lb) a day for the two months after birth. The shoulders are wider than in other breeds and the hindquarters are particularly full. The rams cannot therefore be crossed with small ewes. The ram marks its progeny with size and early maturity.

f. Ryeland. The Ryeland is one of the oldest breeds in Britain, originating from Hertforshire. It resembles the Southdown in appearance but has longer legs and a bigger frame with a wide back and substantial girth. The face is whiter than that of the Southdown and the ears are large. The animals are good foragers and docile in nature. The breed matures quickly but it is necessary to control feeding to prevent lambs from getting too fat.

g. Southdown. The Southdown was the earliest improved breed of the shortwool type. The breed, reared for centuries on the chalk hills of Sussex, has been developed into one of the most famous breeds of British livestock. The face is rather short, woolly and mouse-coloured. The upper portion of the ears is covered with close-growing wool. The body is compact, the chest wide and deep and the legs short and woolly. The breed produces a top quality lamb carcass, with small bones and finely grained meat. Apart from the Welsh Mountain breed it produces the smallest of all joints. A Southdown sheep of 51 kg (112 lb) can produce as much prime meat as a 63 kg (140 lb) sheep from any other breed. The lambs mature so early that a lamb born in January can be sold for meat in April. This characteristic makes the breed excellent for crossbreeding. Southdown rams label their progeny well and are used more often than any other breed for producing small, high quality lamb for export. The New Zealand lamb trade became successful mainly through its Southdown stock and the suitability for public demand in Britain of the weight and grade of meat produced from the breed. In the South Island of New Zealand the Southdown is crossed with the Romney Marsh breed.

h. Scottish Blackface. The Scottish Blackface probably originated in the border counties between England and Scotland and then moved up to the rough heather (calluna) covered hills of Scotland. It is also found in Devonshire and in the Dartmoor area. The ewe is a good milker and it is possible for a lamb to gain 227 g (8 oz) a day for the first 4 months.

The name of the breed is derived from its distinctive black and white markings. Other distinguishing characteristics are the Roman nose, the large horns sweeping downwards and outwards, the slightly crested neck and the general carriage. Crossed with the Border Leicester it produces the Greyface.

i. Kent or Romney Marsh. The Kent or Romney Marsh breed is very adaptable and has adapted well to most sheep breeding countries in the world. The animals forage well and, in marshland, are highly resistant to parasitic infestations. The wide face is white, medium in length and broad with a black nose. The head is well covered with wool. The neck and shoulders are thick and strong, the body deep and the legs strongly boned. The breed is not very prolific, producing under normal conditions about 120% of lambs from a flock. The lambs produced from ewes crossed with Southdown rams fatten readily and yield excellent meat.

Fig. 3.28 Border Leicester Ram.

Fig. 3.29 Suffolk Ram.

Fig. 3.30 Pen of 3 Hampshire Down ram lambs.

Fig. 3.31 Ryeland Ram.

Fig. 3.32 Southdown Ram.

Fig. 3.33 Greyface ewes.

Fig. 3.34   Typical Blackface ewes with lambs grazing on high ground.

3.4.2 Diseases

In *Diseases of Sheep*, a booklet issued by the British Veterinary Association in 1955, 65 diseases of sheep and lambs are reported. Of these 22 are caused by bacteria, 8 by viruses and rickettsia, 6 by external parasites, 6 by parasitic worms, 7 are associated with metabolic disorders and 4 are of unknown origin. Diseases caused by bacteria include tetanus, caseous lymphadenitis, blackquarter, braxy, black disease, pulpy kidney disease, tuberculosis, pyaemia and enterotoxaemia. Diseases caused by viruses include foot-and-mouth disease and sheep pox. The most important internal parasites are liver fluke, hydatic cysts, roundworms, lung worms and tapeworms. The most important external parasites are mites, ticks, maggot fly and nostril fly.

Sheep scab, foot-and-mouth disease and sheep pox are notifiable diseases.

Sheep scab is a highly contagious disease caused by mites, the commonest variety in Britain being the *Psoroptes communis* var. *ovis*. The female mite can lay up to 40 eggs. After 1 or 2 months the sheep begins to rub and bite the affected spots causing inflammation followed by vesicles and the formation of scabs. The wool becomes ragged and patchy.

Foot-and-mouth disease in sheep usually occurs subsequent to an outbreak of the disease in adjacent cattle or pigs (see sections 3.1.2 and 3.3.2). At the start of infection the animals are dull and do not feed. Later, vesicles develop on the foot and lameness becomes evident. Salivation may or may not occur. Although most affected sheep can survive, in Britain the Diseases of Animals Acts require that all affected sheep be killed, and that affected carcasses should be burned or buried.

Sheep pox is caused by a virus. After an incubation period of 6 to 8 days, papules (or skin eruptions) and vesicles develop on the skin where there is little wool (eyes, cheeks, lips, inner thighs, posterior of belly and under the tail) and on the mucous membranes. In

Britain the Sheep Pox Order of 1938 stipulates that carcasses intended for human food may be removed from an infected place only if they are certified free from the virus and if the skin is removed. As with foot-and-mouth disease, affected carcasses should be burned or buried.

Tetanus is caused by *Clostridium tetani* and affects mainly lambs up to about three months of age. The toxin ultimately affects the spinal column of the animal causing rigidity of the muscles, especially those of the jaw (hence 'lock-jaw') and tetanic spasms. Tetanus causes losses among sheep but it is extremely unlikely to affect a human who has eaten the flesh of an affected animal. The whole carcass should be condemned. Inoculation with anti-tetanic serum is a successful countermeasure if it is given in the early stages of the infection before the toxin has affected the spinal column.

Caseous lymphadenitis, caused by *Corynebacterium pseudotuberculosis*, is characterized by the presence of caseous material in the affected lymph nodes and may be detected by the observation and palpation of accessible lymph glands. When the condition is generalized, emaciation occurs. Where it is local, the affected part should be condemned, but in cases of emaciation the whole animal should be declared unfit for human consumption.

Blackquarter outbreaks often follow shearing, docking or castration since wounds can be affected by the causative micro-organism, *Clostridium chauvaei*, which is present in the soil in spore form. Rapidly spreading crepitant swellings appear in the subcutaneous tissues and sometimes in the jaw, head and tongue. The liver and kidneys become enlarged and congested. The carcass should be totally condemned.

Braxy is caused by *Clostridium septicum*, a micro-organism that commonly occurs in the gut. The disease occurs when the vitality of the abomasal mucous membrane is lowered and is characterized by sudden death and rapid decomposition of the carcass. An affected carcass should be condemned. *Black disease* is a related disease caused by *Clostridium oedematiens* and occurs mainly in Australia.

Pulpy kidney disease, caused by *Clostridium welchii*, occurs in lambs from $1\frac{1}{2}$ to 4 months old and causes the kidneys to become soft, pulpy and haemorrhagic. The same micro-organism causes *lamb dysentery*. Death is caused by septicaemia and toxaemia with or without ulceration of the intestines.

Tuberculosis has a low incidence in sheep. In Britain it is caused mainly by the bovine type of mycobacterium but in the USA most cases have been found to be caused by the avian type. The disease is spread to sheep either by letting them graze on pastures after cattle or poultry with tuberculosis or by the consumption of affected cows' milk.

Pyaemia, sometimes associated with thick infestation, is caused by *Staphylococcus aureus*. It is estimated that about 5% of all lambs are affected by the disease and on some farms the death rate can reach 20%. Another form of pyaemia, *joint ill*, can occur after birth, causing the infected joints to become swollen.

Enterotoxaemia, or turnip sickness, is common among store or fattening sheep. It is caused by the absorption of the toxin of *Clostridium welchii* into the bloodstream. The effect is paralysis of the vital centres of the central nervous system and, usually, death.

Roundworms, such as *Haemonchus contortus*, *Trichostrongylus axei* and *Ostertagia circumcincta*, found in the abomasum and the small intestines cause anaemia, oedema and finally emaciation.

Pimply gut, caused by *Oesophagostomum*, affects the intestines and makes them unfit for use as sausage casings.

Lung worms of sheep, *Dictyocaulus filaria*, *Protostrongylus rufescens* and *Muellerius capillaris*, cause nodules and patches on the lungs. Extensively infested lungs should be destroyed.

The commonest *tapeworm* of sheep in Britain and Australia is *Moniezia expansa* which has little importance in meat inspection. On the other hand it is necessary to examine carcasses for the cystic form of the tapeworm of the dog. In sheep the following cystic diseases can be found: *Cysticercus tenuicollis* (the larval form of *Taenia marginata*), *Cysticercus ovis* (the larval form of *Taenia ovis*), *Coenuris cerebralis* (the larval form of *Multiceps multiceps*) and *Echinococcus* (the larval form of *Echinococcus granulosus*). The cysts may occur in the liver, heart and diaphragm except in the case of *Multiceps* which invades the brain and of *Echinococcus* which invades the lungs and liver only. *Echinococcus* plays an important role in the occurrence of the hydatid disease in humans. In every case only the affected organs should be condemned with the restriction that in the case of *Echinococcus* the organs must be condemned even when only a single superficial cyst is present.

The *liver fluke, Fasciola hepatica*, is very common in sheep. The young flukes invade the liver causing acute swelling and congestion and leading to acute parenchymous hepatitis. Adult flukes settle in the bile ducts, leading to chronic cirrhosis and the formation of connective tissue in the walls of the bile ducts and surrounding liver tissue. The effect on the carcass is oedema and emaciation.

Among *Arthropoda, nostril fly* (or *Oestrus ovis*) larvae affects sheep causing a catarrhal inflammation of the nasal mucous membranes and adjacent cavities. The head of an affected animal should be condemned.

Maggot fly (*Lucilia sericata*) deposits its larvae on sheep. Badly infested sheep do not thrive and death may occur in a few days if the animals are overlooked or neglected.

3.4.3 Slaughtering

Ante-mortem inspection should identify those animals with emaciation, sheep scab, tetanus, foot-and-mouth disease, caseous lymphadenitis and pneumonia.

The sheep are driven, or sometimes led by a bell-wether, to a passageway adjoining the slaughter hall and are carried by hand to the slaughter hall where they are put on a wooden or metal crate before stunning and dressing. Metal crates are preferable because they are easier to clean. According to the Slaughterhouse (Hygiene) Regulations the equipment and fittings in the slaughterhouse should be of an easily cleanable, non-corrosive material–wood should be used only for broom handles, and chopping and cutting blocks. The rate at which carcasses can be handled when line slaughter with 10 to 12 men is used is about 60 to 70 sheep an hour.

The animals are stunned either by a wooden hammer or by electrical stunning at 198 J for 1 second. In sheep proper unconsciousness may be assumed if there is immediate flexion of all four legs followed by the closing of the eyes and the extension of the hind legs a few seconds later.

A metal-covered bleeding bench, big enough for 10 sheep and with a gutter along one side to catch the blood, should be provided for bleeding. However, sometimes a bleeding rail similar to that used for cattle (see section 3.1.3) is used. An incision is made in the jugular furrow close to the head and both carotid arteries are severed. The head is jerked back to rupture the spinal cord. Bleeding of the carcass lasts on average about 5 minutes and yields for an average sheep 2 to 2.5 kg (4 to 5 lb) of blood. The carcass is opened and the internal organs are prepared for inspection.

During post-mortem inspection the carcass should be examined for satisfactory bleeding and setting, the lungs for parasitic infestations, especially hydatic cysts or lung worms, and liver for fascioliasis. In Australasia the carcass is also palpated for evidence of arthritis, caseous lymphadenitis, inoculation abscesses and lesions due to grass seed awns.

3.4.4 By-products

The parts of the sheep from which edible by-products can be obtained are the blood, the tongue, the stomach, the intestines, the heart, the lungs, the kidneys, the head and the fat. Non-edible by-products can be obtained from the bones, the spleen, the ovaries and the skin.

The *sheep tongue* is used for food either fresh, salted or tinned.

The *head* can be used fresh for broth or for potted meat (minced and salted or seasoned and placed in a pot or other vessel for preservation).

Sheep blood is one of the ingredients of black pudding (see section 3.1.4). After drying it can also yield blood meal for animal feed.

The *spleen* is used as food for cats and dogs.

The *lungs* are used for white pudding and, like the spleen, for cat and dog meat.

The *heart, liver* and *kidneys* are usually sold fresh. Sheep's liver can be used in the preparation of faggots which are popular in the north of England and in Wales. Liver can also yield a medical extract for the treatment of anaemia, and *bile* can be used as a general cleanser for polishing leather.

Fat, processed in the same way as beef fat, yields lard. The use of sheep's fat is restricted because of its strong flavour. Blended with other fat it can be used for dripping and, because of its hardness which results from a high stearin content, it is often used as the preservative layer on the top of glass jars of meat paste.

Of the sheep stomachs, the first, second and fourth are cleaned and used for tripe. The first stomach (rumen) is used as a container for haggis in Scotland. The ingredients of haggis are chopped sheep liver, heart and lungs, oatmeal, fat and spices. They are packed into the stomach which is then sewn and cooked for 5 hours.

The *sheep intestines* (the runners, or small intestines, and the caecum and colon, or large gut) are cleaned, scraped, sterilized, dried and polished until smooth so that they can be used as containers for sausage meat. The runners, prepared in 90-metre (100-yard) bundles, are used for ordinary beef or pork sausages, and those of lambs or small sheep, filled with seasoned prime pork, are used for chippolata sausages.

The *bones* yield fat which can be used for soup-making, and gelatine.

A pharmaceutical product, such as follicular hormone, can be extracted from the follicular liquor of the *ovaries*.

Gloves and chamois leather are the main by-products of *sheep skin*. Kid gloves are made from the skin of lambs while ordinary leather gloves are made from sheep skin. For chamois leather the skin is impregnated with fish oil to make it soft and pliable, a process known as shammoying.

3.5 POULTRY

3.5.1 Origins

It is generally accepted that the modern domestic hen has over thousands of years been developed from the wild Jungle Fowl, which in the course of a season laid perhaps two or even three small clutches of eggs and if she was lucky, hatched out the baby chicks and reared them herself. However, owing to the continual presence of predators it is doubtful whether as many as half the broods ever survived to reach maturity. The important point to note is the enormous contrast between the performance of the origin of the species and its present-day descendant, which is being bred to lay 250 to 300 eggs in its first year of production; i.e., ten or twelve times as many as its forbears, which are said to have come originally from the jungles of Burma, India and southern parts of China.

Another type, the Malay Fowl, is also believed to have been used in the make-up of some of today's popular breeds, as it is known to have been bred in India over 3000 years ago, probably for culinary purposes. The fact that poultry was recognized in Biblical times as being of value in terms of both eggs and meat is clear from references in the Gospels and also from the list of provisions supplied daily for King Solomon's table more than 3000 years ago, which included 'fatted fowl'. Later on, Shakespeare in *As You Like It* refers to the justice's 'fair round belly with good capon lined', thus showing that the art of caponizing – or castrating – as well as the delicacy obtained thereby was known and appreciated even then.

Artificial incubation of eggs and hatching of chicks was also evolved in Ancient Egypt, where the required constant temperature was maintained by the setting of the eggs on racks or shelves in underground heated caverns and by turning the eggs daily. The results obtained, however, appear to compare unfavourably with today's norm for modern incubation, which requires at least 90 healthy chicks for every 100 fertile eggs set. It is in fact only during the present century that the keeping of poultry has really progressed from the farmyard and backyard flock to become a vast and growing industry of national importance, second only to beef and milk in the value of its products. This rapid development has been made possible by the increase in scientific knowledge which has taken place during this century, in the field of genetics, nutrition, control of disease and husbandry methods.

3.5.2 Breeds

a. Domestic fowl. A particular breed of poultry as in any other class of livestock is one possessing certain characteristics which distinguish it from the others. Thus when mated to birds of the same breed these same main characteristics are reproduced in the progeny, and the extent to which they are, or are not reproduced determines the degree of the breeds purity. There are, of course, varieties within the breed, as for example White and Black Leghorns, and these can be sub-divided further into 'strains'. A 'strain' is in fact a 'family' which has been selected over a number of generations for some specific purpose, such as high egg production, or perhaps to produce extra-heavy birds in a shorter growing period for the poultry meat trade. However, it is rarely possible to have the best of both worlds. Thus, when one is selecting for higher and ever higher egg yield, the value and suitability of the carcass as poultry meat is diminished accordingly, and vice-versa.

For the foregoing reasons, therefore, the various breeds of poultry may be divided broadly into three groups: laying breeds, table breeds, and general purpose breeds. The laying breeds are those in which selection has been primarily towards more and larger eggs per bird, with some deterioration in carcass weight and quality as a result. These breeds of prolific egg production – 280 to 300 eggs of good size per bird during its first laying year – are mostly small, weighing only 1.6 to 1.8 kg ($3\frac{1}{2}$–4 lb) at maturity, and are therefore called 'light' breeds. On the other hand table breeds, as the name implies, have been selected with the aim of producing as much meat of good quality as possible, preferably in as short a time as possible. These are known as 'heavy' breeds. Breeders of this type of bird have, however, always been beset by two problems; firstly the tendency towards lower egg yields for hatching purposes, and secondly, the natural effect of large, meaty frames on the ability of the male bird to mate, resulting in a high proportion of infertile hatching eggs. Suffice it to say that breeders of the more popular breeds of this type have produced a hen weighing 2.7 to 3.2 kg (6–7 lb) at maturity with a yield of about 140 eggs suitable for incubation in the laying period, of which an average of an least 80% will hatch into chicks capable of growing to an average for both sexes of over 1.8 kg (4 lb) within 56 days. Of equal, if not greater importance in achieving this result is the type of sire used in mating, and here, amongst other things, the weight requirement at maturity is generally between 3.6 and 4.5 kg (8–10 lb).

The third group consists of so-called general purpose breeds. These are intermediate types with good laying capabilities, while not quite in the same class as the light breeds, yet superior to the latter in table qualities. They have come very much to the fore in recent years owing to the consumer preference for brown and tinted eggs, since these are characteristic of most general purpose breeds and heavy breeds. Since the end of the Second World War a vast amount of experimental work has been undertaken on both sides of the Atlantic, in the first instance by cross-mating two or more of the standard pure breeds with the aim of improving egg production, table qualities or vigour, or all three. The use of the computer has enabled some of the guesswork to be eliminated, with the result that it has become possible for the geneticist to obtain the result he is seeking with far more certainty and in less time. In consequence, several of the more popular types now in use for egg production and also to some extent for meat production bear only a small resemblance to the pure breeds from which they were developed. There is some indication, however, that the danger of extinction which threatens some of the more valuable pure breeds has now been recognized by geneticists, some of whom are producing stock with many of the familar characteristics of the original standard breeds along with improved performance.

The group known as light breeds are often referred to as Mediterranean breeds since most of them originated from that area. These include various coloured varieties of such breeds as the Leghorn, Minorca and Ancona. They are mainly prolific layers of large, white-shelled eggs, but owing to their naturally nervous disposition they tend to panic all too easily, which can on occasions seriously affect egg production. In their homelands bordering the Mediterranean, egg production during the winter months falls only when weather conditions are abnormally cold. In colder climates winter egg production of these breeds tends to be poor unless the birds are protected from the effects of adverse weather conditions, as in fact they may be if kept in intensive houses incorporating some means of controlling the interior environment. A non-Mediterranean light breed worthy of mention is the Welsummer, which depends for its popularity on its production of attractive brown eggs. However, by far the most economically important light breed is the White Leghorn which has been used more than any other in the development of modern laying strains. Present-day trends would seem to indicate that it will remain so.

Between the two World Wars when the poultry industry was getting into its stride, the distinction between heavy breeds and general purpose breeds was fairly clear cut. The former were big-framed and slow growing, but at maturity were heavily fleshed and were therefore used almost exclusively for the table, since egg production was not one of their strong points. Unlike the light breeds, which are practically non-broody, the heavy

Fig. 3.35   Rhode Island Red pullet, typical of 'general purpose' breeds.

Fig. 3.36   White Leghorn pullet, typical of 'light breeds'.

breeds have broody tendencies and are therefore naturally good mothers. The best known and most popular heavy breeds of the period include different coloured varieties of Dorking, Sussex, Orpington, Old English, Cornish and Indian Game; also, in France, Faverolles and Bresse. Although most of these breeds are no longer to be seen in large numbers, they have been used extensively in the make-up of modern poultry meat strains, providing parent stock to produce the enormous chick requirement for the vast and growing broiler industry.

Much the same can be said of the so-called general-purpose breeds, which include various coloured varieties of Plymouth Rock, some strains of Light Sussex and the Rhode Island Red. Here again the geneticist has made extensive use of these breeds in the make-up of modern egg-laying and meat-producing strains. For example, the female lines of most of the world's broiler breeding strains contain the blood of the White Plymouth Rock in very high proportions. Another 'general purpose' breed which has contributed much to broiler female lines is the New Hampshire Red. On the other hand, the male lines in the more popular types of broiler breeder matings owe much of their origin to the Game breeds, notably Cornish Game, the aim being to bring together in the resulant broiler progeny quick growth, rapid feathering–preferable white–a large, compact frame carrying a high proportion of breast meat, together with the hardiness and docility which are characteristics of both the heavy and general purpose breeds.

b. Broilers. The name 'broiler' originally referred to a method of cooking a young chicken weighing about 1.6 kg ($3\frac{1}{2}$ lb), but this name has since been applied to the production of all young chickens for table of approximately equivalent weight. Originating in New Jersey, USA about the turn of the century, production was small and relatively unimportant until after the First World War. About 1920, production began in earnest, but did not expand rapidly owing to limited knowledge in the field of vitamin requirements and nutrition. However, it was not long before scientific knowledge in this

Fig. 3.37 Typical broiler breeding trio.

respect made expansion possible and from about 1930 onwards broiler production in the USA has progressively increased, to the point where consumption per capita is currently around 23 kg (50 lb) per annum.

The growth of the broiler industry in the United Kingdom and also in some European countries, whilst equally impressive, does not conform to the American pattern. In the UK, severe rationing of foodstuffs throughout the war years and up to the summer of 1953 virtually prevented the production of poultry for the table, let alone expansion. However, from then onwards expansion began in earnest and accelerated, so that within 20 years the industry has grown from next to nothing to one producing currently over 300 million birds per year and increasing by some 10% annually. Per capita consumption of poultry meat in the UK is now over 11 kg (25 lb), 80% of which is broiler meat.

c. Turkeys. Any account of the poultry industry as a whole would be incomplete without reference to turkeys and also – though these are of lesser importance numerically and economically – to ducks and geese. In the case of turkeys, here again the pace was set by the United States industry, which with abundant supplies of home-grown feeding stuffs was in no way inhibited in its expansion. In the USA for decades past turkey has been the traditional dish on Thanksgiving Day and to a lesser extent at Christmas or New Year. However, sales promotion campaigns inducing consumers to eat turkey meat more frequently have been successful to the point where it is now becoming an all-the-year-round dish. For many years the predominant breed was the American Mammoth Bronze, but public demand for a smaller alternative made way for the Beltsville White. During the last 20 years or so these breeds have given place to Broad Breasted Bronze and Broad Brested Whites, covering the complete range of weight requirements, and as their name implies, carrying a very high proportion of breast meat – a factor that was lacking previously.

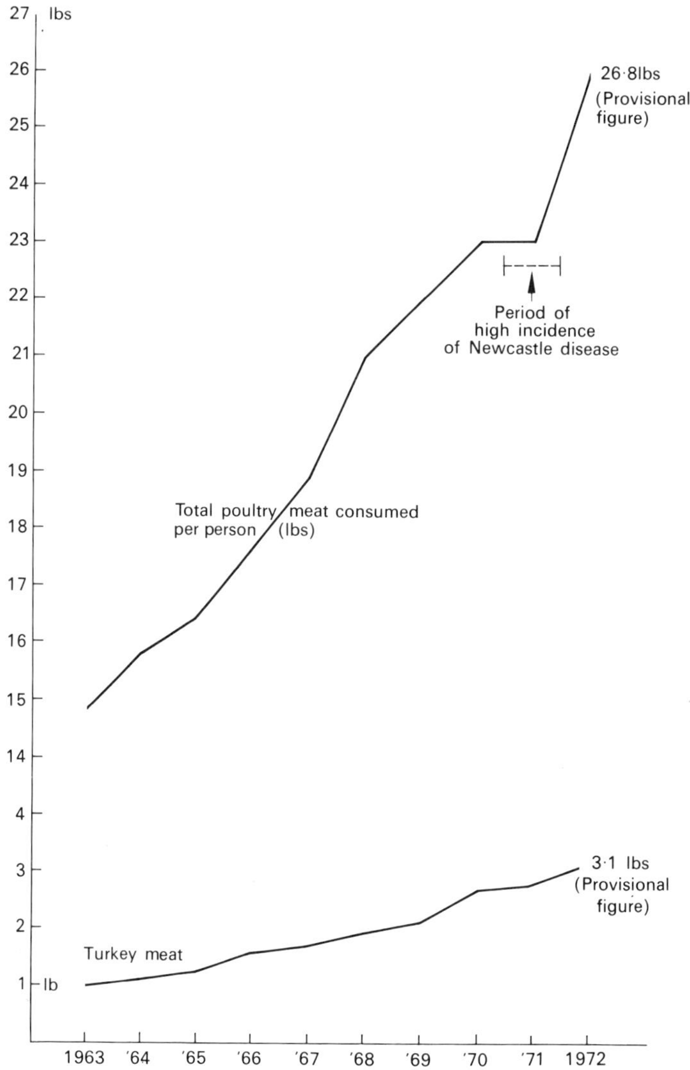

Fig. 3.38   UK poultry meat consumption per capita 1963–1972.

The growth and development of the turkey industry in the UK, like the broiler industry, dates from the de-rationing of feeding stuffs in 1953. Up to the outbreak of war in 1939, the most widely used breeds included the Bronze, Norfolk Blacks and British Whites. However, production was on a relatively small scale and almost exclusively aimed at the traditional Christmas market. Since 1953, however, by cross-mating with stock of American origin, the industry has made rapid progress, not only in stimulating public demand to the extent that over 18 million turkeys are reared and sold in the UK yearly, but also because British breeders have now established for themselves a reputation and a demand for stock such as the Dimple Bronze and Dimple White which is world-wide. One must, however, recognize the fact that as with the broiler industry, expansion of production has been greatly assisted by the introduction of deep-freezing and cold-storage techniques, thus permitting continuous turkey production all the year round, rather than merely for the Christmas trade.

   d. Ducks. Not many years ago duck eggs were very popular and regarded as quite a delicacy. The most widely-used breed was the Khaki Campbell, and individual production records of 300 eggs were by no means exceptional. Certain cases of food poisoning were, however, traced to duck eggs, which since then have fallen into disfavour despite the fact that adequate cooking dispenses with any risk. For table purposes the White Aylesbury is the most popular breed in the UK, followed by the White Pennine. Over the last ten years or so the duck industry has become organized and efficient and is now putting on the market oven-ready duckling in attractive packs at prices about 25% higher than for oven-ready chicken. The main centres of production have become established in Norfolk and Lincolnshire.

   e. Geese. The home of the goose is said to be Italy, where at one time it was held to be sacred, as it also was in Egypt several thousand years ago. The keeping of geese in Britain has continued for centuries but at no time has it become an industry in its own right. By nature excellent grazers, they find a high proportion of their own food and in so doing help to keep grass under control. Like ducks they are very gregarious and tend to keep close together as a flock. They are also first-class 'watch dogs' and give strident warning at the approach of a stranger. The table qualities of the goose need small comment other than the fact that goose meat properly cooked is as succulent and tasty as any other type of poultry meat, if not more so. Artificial incubation of goose eggs has never been a commercial success, which implies that breeding and hatching is best carried out by natural methods. For this reason flocks of geese have rarely been large and have mostly been regarded as a useful sideline on farms and smallholdings. The smaller breeds include the Chinese, Roman and Brecon Buff–the latter being of British origin and the better layers, producing 60 or 70 eggs per bird each season. The egg production of the heavier table breeds such as the Embden and the Toulouse is roughly half as many, but their table qualities are superior. A common practice is to cross-mate these two breeds for the best overall results.

3.5.3 Rearing

Methods of rearing poultry artificially at first sight appear to be poles apart, from the hot-water bottle method used by the back-yard poultry keeper to the pressurized, intensive brooder house designed to rear chicks in tens of thousands in a controlled environment. They all have, however, one aim in view, namely, to provide the baby chick with similar conditions of warmth, fresh air and protection as a mother hen would, or should provide for her brood under natural conditions. The same holds true for artificial hatching, since the fertile egg must have heat and a certain amount of moisture. It requires to be turned regularly and needs an adequate supply of fresh air in order to hatch–all of which functions are performed naturally and automatically by the mother hen throughout the three-week period (four weeks in the case of turkeys, ducks and geese) of sitting and hatching.

   a. Incubation. The storage of hatching eggs for incubation is important, since it is often found necessary to hold eggs up to seven days or even longer in order to fill the incubator. Ideally, hatching eggs should be stored at 13 to 16°C (55–60°F), but just before being put into the incubator they should be brought up to room temperature. Between the wars, the 'Mammoth' forced-draught incubator became popular with hatcherymen, each machine holding thousands of eggs in a compact cabinet. Up to that time the only available alternative for hatching large numbers of chicks was to link together rows of small, table-type incubators, each with its tray of a hundred or so eggs, all served by a common hot-water pipe, thermostatically controlled to provide and maintain a constant centre-of-egg temperature of 38°C (100°F). Water trays were provided since the egg must not be allowed to dry out too quickly. (Ideally, the relative humidity should be about 55%). In the huge forced-draught incubators of today, all these functions are performed automatically, including turning the eggs. In the case of any slight fault or deviation from normal, alarm bells ring until the fault is rectified.

b. Chick rearing. A popular idea has it that all chickens are stupid, senseless creatures, yet what other class of livestock is capable of fending for itself at birth? Given heat, light, food and water the baby chick will attend to its own needs even without parental or human assistance. The first item of paramount importance is the source of heat without which the baby chick will die within hours, hence on the one hand the back-yarder's hot-water bottle or electric light bulb, and on the other, the thermostatically-controlled brooding system or 'hover', providing heat for thousands of chicks. The ideal starting temperature is 31 to 32°C (88–90°F) and this should be reduced at the rate of about one degree per day down to about 21°C (70°F) by which time the chicks will be three weeks old. At this point heat requirements differ according to the type of chick being reared. If broilers, then the room temperature should be maintained as evenly as possible between 18 and 21°C (65–70°F) in order to encourage maximum growth. The local heat supplied under the brooders or hovers is normally dispensed with during the fourth week.

If on the other hand the chicks are for egg production the room temperature should be reduced to 15.5°C (60°F) during the fourth or fifth week in the process of hardening them off. If the chicks are being reared as replacement pullets for egg production they generally remain in the house until being moved to their permanent laying quarters at about 16 to 18 weeks of age. If, however, they are required for breeding purposes it is common practice to put them out on grassland and house them in range shelters, so as to increase their stamina and resistance to disease.

Adequate space for feeding and drinking is important. Whether the birds are broilers or replacement pullets, linear space for feeding should increase from 25 mm (1 in) per bird at day-old to 75 mm (3 in) by the time the birds are 8 weeks old. The corresponding drinker space should be 6 mm ($\frac{1}{4}$ in) to 19 mm ($\frac{3}{4}$ in) respectively. If tube feeders and circular drinkers are in use, normal requirements should be regarded as 25 tube feeders and 8 circular drinkers per 1 000 birds. Normal practice is to use a proprietory brand of feed, generally in pellet form designed specifically for the job, and to follow the feeding programme recommended by the feed compounder.

Intensive rearing houses of modern design, whether for broilers or replacement pullets, not only require means of controlling both brooding (or foster mother) and house temperatures, but also of controlling ventilation, as only in this way is it possible to ensure the environmental conditions to enable the growing birds to live and thrive. A typical brooder temperature is 32°C (90°F) at hatching, reducing to 21°C (70°F) at three weeks old. A practical forced ventilation rate which has proved itself over many years is to supply 0.03 m^3 (1 ft^3) of fresh air per minute for every 0.45 kg (1 lb) of live weight chicken in the house. Thus, a house containing 5000 birds averaging 0.9 kg (2 lb) live weight would require 300 m³ (10000 ft³) of fresh air per minute. During very cold weather this rate should be reduced slightly and in hot weather it should be increased. Whether the birds are being reared under whole house heating systems, in pressurized houses (the air in which is maintained at a pressure slightly above atmospheric, to reduce draughts), or even in battery cages the same basic principles apply. In the case of turkeys both feeding and watering space is increased in accordance with their size and weight. Ducks also require extra drinking space. Floor space allowances per bird for birds reared in houses are generally as follows: for broilers, day-old to 8 weeks, 0.065 m² (0.75 ft²); for growing pullets, 0.09 to 0.13 m² (1–2 ft²) according to age; for turkeys, 0.14 to 0.55 m² ($1\frac{1}{2}$–6 ft²) according to age and size. The most satisfactory floor litter is soft-wood shavings, starting off at a depth of 75 to 100 mm (3–4 in). In the case of birds grown for slaughter, the old litter is usually removed, the house and appliances cleaned and disinfected at the end of each crop, and fresh litter put down for the next batch. In batteries, the floor space allowed is about half that which is provided for floor-reared birds.

3.5.4 Diseases of poultry

A somewhat disturbing feature of the poultry industry at home and abroad is the readiness of poultry and egg producers and breeders to look to the veterinarian to get them out of trouble rather than the geneticist. This is perhaps excusable in view of the small profit margins available, but when the poultry keeper is faced with a serious epidemic, the tendency is to clutch at any straw the veterinary profession has to offer, be it vaccine or antibiotic or other current treatment without always counting the cost. This is a natural reaction in an emergency, but is all too infrequently followed up by invoking the aid of the geneticist in seeking to build resistance to disease into the poultry stock, since the need for profitability demands the undivided attention of the poultry man towards increasing output or growth rates. It is not possible here to discuss all the diseases and ailments to which poultry are prone. The most important ones are described below.

a. Marek's Disease. A good example is Marek's disease, a disease of the avian leucosis complex; the latter also includes lymphomatosis (characterized by an enlarged liver and tumours affecting lungs, kidney etc. and leukaemia). The importance of this disease, which causes lameness, paralysis, wing-drooping, tumours in various parts of the body and consequently high mortality, cannot be over-estimated since it also seems to predispose the survivors to various other diseases. That natural resistance to Marek's disease can be introduced into a breed by selection has been proved, but when it was demonstrated recently that the causal agent was cell-associated, and that if an appropriate vaccine were administered, preferably at day-old, the bird stood a fair chance of complete immunity, the breeders accepted the remedy eagerly. Since then the results have been striking, but there have also been a number of inexplicable breakdowns of vaccine. An interesting point is the similarity of Marek's disease to a pre-war condition at the time known as fowl paralysis, which was widespread and caused very high mortality in birds from about three months of age and onwards, right through the laying period. During the war, fowl paralysis virtually disappeared, probably because of the shortage of feeding stuffs and the necessity to eliminate weaker specimens, also greater

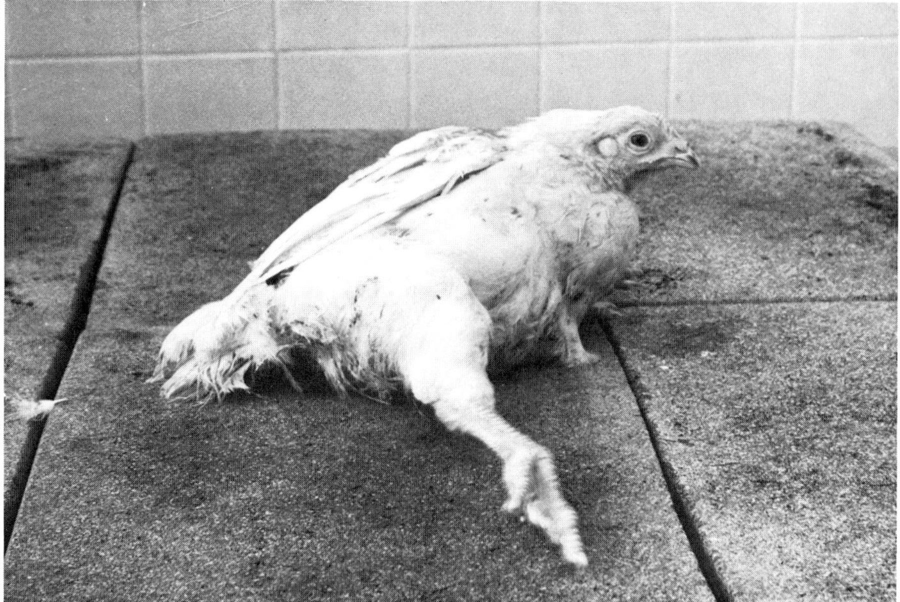

Fig. 3.39 A nine-week old pullet completely paralyzed; a suspected case of Marek's disease.

knowledge of the incidence of the disease. However, it reappeared from about 1960 onwards in the form of Marek's disease.

Marek's disease can probably be regarded as the chief scourge of poultry, but many others cause severe losses. Those mentioned below have created the greatest havoc and have proved most costly to the industry. They are as follows, though not necessarily in order of importance: Newcastle disease (fowl pest), infectious bronchitis, mycoplasmosis (C.R.D.) *E. coli* infections, coccidiosis, salmonellosis, and blackhead (mainly affecting turkeys).

b. Newcastle disease. Otherwise known as fowl pest or fowl plague, this disease spread throughout Europe and was first recognized in northern England in 1927. It has all the attributes of a plague and in its acute form 100% mortality occurs. Caused by a highly infectious and contagious virus, there is no known treatment. Prior to 1963 in the hope of eradicating the disease the British Ministry of Agriculture operated a slaughter-with-compensation policy. A notifiable disease, it became an embarrassment owing to the number of outbreaks which occurred and the amount of public money paid out in compensation. In March of that year the slaughter policy was dropped in England and Wales and replaced with a voluntary vaccination policy using inactivated vaccine. This was reasonably successful in controlling the strain of virus which was then prevalent, but was found to be quite ineffective with young stock in the late summer of 1970, when a much more virulent strain swept the country, causing the death of some 200 million birds of all ages. Since that time the use of 'live' vaccines has been permitted and is largely successful when used intelligently. Turkeys are also affected, but not ducks or geese.

c. Infectious bronchitis. As its name implies, this is a respiratory disease caused by a virus which is also contagious but does not affect turkeys, ducks or geese. The disease itself causes wheezing and coughing but does not normally last more than a few days and may or may not be fatal. In laying birds it can cause a fall in egg production as well as affecting egg quality and shell texture. However, an undesirable characteristic of this disease is that it frequently leads to other diseases, such as mycoplasmosis and *E. coli* septicaemia. There is no treatment for the disease and therefore control measures, as for fowl pest, include the use of both live and inactivated vaccines, of which the former is the more successful.

d. Mycoplasmosis (C.R.D.). Often referred to as chronic respiratory disease, this condition is caused by an organism called *Mycoplasma gallisepticum* in chickens and *Mycoplasma meleagridis* in turkeys, causing coughing and sneezing in both species, with a nasal discharge in the case of chickens and swollen sinuses in turkeys. Not infrequently this disease follows an attack of infectious bronchitis in young chickens and broilers. Since it is prevalent during winter and when ventilation rates are inadequate, extra warmth and ventilation are obviously indicated. Medication is effective in controlling the disease.

e. E. coli infections. The organism *Escherichia coli* is present in the intestines of all farm animals and poultry, also in man. Most strains are harmless, some even beneficial. Others are pathogenic. However, even these do not normally affect the host unless it has been subjected to some stress factor, in the case of young chickens and broilers for example, infectious bronchitis or C.R.D. This gives the organism the opportunity to multiply, which it then does and passes from bird to bird via the droppings. As it is also airborne it not only spreads within the house but also to adjacent houses, often causing high mortality, loss of condition and in the case of broilers a high proportion of rejects at the processing plant. Medication is effective provided it is applied during the early stages of the disease.

f. Coccidiosis. Commonly affecting chickens and turkeys, though rarely ducks and geese, coccidiosis is caused by one or more of the *Eimeria* group of microscopic parasites. So far as chickens are concerned there are two main groups, *E. tenella*, which affects the caecal tubes, and the intestinal group, of which there are six or seven types. In pre-war days there was no effective treatment for this disease which wrought havoc in young stock, other than attempting to interfere with the organism's life cycle by clearing out the litter and renewing it every second day. Since then control has been established with sulpha drugs; more recently, feed is medicated at preventive level with drugs known as coccidiostats. It is, however, an ever-present menace unless these precautions are taken.

g. Salmonellosis. Two members of this group of bacteria, *Salmonella gallinarum* (fowl typhoid) and *S. Pullorum* (B.W.D.) are pathogenic only to chickens and turkeys. Medication is helpful, but it is essential to eliminate survivors of an outbreak from the breeding flock as these diseases are egg-transmitted and cause high mortality in baby chicks. Both types can be detected by the rapid blood agglutination test, in which a drop of blood is mixed with a suitable antigen. Coagulation indicates the presence of salmonella.

There are, however, numerous other types of bacteria of the *Salmonella* group, many of them harmless, but there are some like *S. typhimurium* which not only affect all domestic farm animals and poultry but have caused many cases of food poisoning in

man, some of which have been traced back to poultry. Ducklings are particularly susceptible to *S. typhimurium*. Much work is currently being done to control the spread of this
disease to human beings by improving farm hygiene, and particularly by monitoring
imported feedingstuffs, some consignments of which have been implicated.

h. Blackhead. At one time this disease was the scourge of the turkey industry. It is
caused by a microscopically small parasite, *Histomonas meleagridis*, which works
through the agency of bacteria of *E. coli* type, mostly affecting turkeys but also
occasionally young pullets and broilers causing a characteristic darkening of the
head – hence the name – and sometimes high mortality. Medication is now effective in two
ways, prevention by a drug in the feed at low level, and curative dosing via the drinking
water.

i. Deficiency diseases. In addition to the foregoing and other diseases of lesser
significance, there are a number of diseases caused by a deficiency of some vitamin or
other ingredient in the diets of chicks and adult stock. Under entirely natural conditions
these problems rarely occur because all that is necessary for growth and health is present
in the natural habitat. This may not be so under artificial conditions. For example, just
after the First World War when the poultry industry began to develop quickly and
intensive rearing was gaining popularity, poultry keepers were unable to understand why
young chicks being reared indoors when about three weeks old, developed lameness and
were unable to stand through having contracted rickets, until it was realized that lack of
direct sunlight was responsible, or alternatively, that vitamin D was lacking in the feed.
When cod liver oil was added at the rate of 1 to 2% of the food ration, rickets was no
longer a problem.

3.5.5 The slaughter and dressing of poultry

At one time the accepted method of killing all poultry for table was by dislocation
of the neck just below the head without breaking the skin. The bird was then hung up by
the feet to ensure that all the blood drained into the neck cavity where it then congealed.
The next step was then to pluck off the feathers, generally by hand. It was necessary to do
this immediately after killing while the bird was still warm to prevent tearing the skin,
which would otherwise be likely to happen, particularly in the case of young birds and
broilers. When plucking was completed, including the removal of pin feathers or stubs in
young birds not fully matured, they were then hung again by the feet and allowed to cool
before being sent off to the retailer. Prior to killing it was necessary to starve the
birds for about 24 hours to ensure that the gut was completely empty, otherwise
discoloration or 'greening' would begin to appear within a matter of hours, particularly in
warm weather, rendering the bird unfit for human consumption. The butcher would
normally store the carcasses in his cold-room or hang them in his shop window for sale.
He would then eviscerate the bird as and when sold to a customer and include the feet,
heart, liver, gizzard and neck in the sale – unless of course the customer preferred to do
the job. This method is still practised on a very limited scale by back-yarders or small
holders but is definitely dying out.

On the other hand, it is estimated that fully 25% of all poultry in the UK, including
turkeys, are being sold at wholesale level uneviscerated, or 'New York Dressed' – N.Y.D.
In this case killing, after stunning, is by bleeding, that is, by cutting the jugular vein in the
neck just below the head. By clearing all the blood from the carcass, keeping qualities are

improved. This has become a factory operation, in which the birds are hung by the feet on a conveyor line, stunned, bled, then passed through a dip tank with water at 53°C (128°F) swirling around them. The conveyor line then passes the birds through one or more automatic plucking machines with long rubber flails revolving at high speed which effectively remove the feathers and the stubs. Rapid cooling of the carcasses is then necessary for adequate shelf life and this is done either by cold air or by immersing the birds in ice-slush. Once cooled, the birds are hung once again on a conveyor line for separating into weight grades and finally for packing.

The majority of birds, however, are sold eviscerated; that is, oven-ready. A small but significant proportion are being sold fresh oven-ready, but most are deep-frozen, since they can if necessary be stored over long periods in this state without deterioration. Killing and removal of feathers is as previously described, followed in all the larger plants by a highly mechanized factory operation, much of which is carried out automatically. The birds are conveyed through machinery which cuts off the feet, slits the neck and removes the head. Removal of eviscera and giblets then follows, mostly performed manually, after which the carcasses are automatically washed and passed on to a spin-chiller, in which they are slowly spun through ice-water to bring down the body temperature quickly. The carcasses are then hung on a 'drip-line', long enough for surplus water to drain off. A pre-frozen giblet pack (heart, liver, gizzard and neck) is then inserted into each carcass which is then packed in a sealed plastic bag, sorted into its

Fig. 3.40 Part of the production line in a poultry processing factory.

weight grade, labelled, and passed into the the blast-freezer. After freezing the birds are packed in boxes or cartons for despatch and sale or cold storage. Outputs of 4 000 birds per hour are quite commonplace.

A European Common Market directive has decreed that eventually, all birds for human consumption must have had their internal organs inspected during processing and passed as fit for human food, which by implication would rule out the sale of N.Y.D. poultry. This is causing great concern in some sections of the industries on both sides of the English Channel.

3.5.6 By-products

Apart from the obvious product of the poultry meat industry, i.e., packs of cut-up and jointed chicken and turkeys–which strictly speaking are the result of further processing and therefore not truly by-product–there are two of significance and growing importance in view of the world shortage of proteins and fertilizers. These are offal meal, prepared from the otherwise waste products of a poultry processing plant, and processed waste litter from broiler and intensive houses.

a. Offal meal. There are four main types of waste products in a poultry processing plant, namely, blood, feathers, feet and eviscera. Many modern factories have installed equipment to render these materials into a valuable feeding meal with a crude protein content of over 60% and an oil content normally of 20%. Briefly, the process is to pressure-cook these waste products at a high temperature and for long enough to ensure destruction of any disease organisms that may have come in with the birds being processed. A standard practice is to pressurize the container at 2.8 kg/cm^2 (40 p.s.i.) and hold at 121°C (250°F) for about 15 minutes, then over the next hour to lower the pressure gradually down to atmospheric, by which time the moisture content has fallen to about 10% and the material is ready when cool to be bagged. It is often included in animal feeds at up to 10% of the total ration.

b. Processed broiler litter. The use of broiler litter in its raw state as a valuable, balanced fertilizer is by no means new and it has been applied on farmland ever since the broiler industry got under way. However, if left for any length of time its properties decline steadily. Over the past few years interest in drying and grinding the litter has been growing, as in this form it is far more stable. Machinery is available for dealing with the material. Recently the British Ministry of Agriculture initiated experimental work on the use of broiler house litter as a protein feed additive for ruminants, i.e., for cattle and sheep. These were so successful that in its dried and ground state many thousands of tons have been processed and used as a protein supplement in the winter rations of cattle and sheep. The machinery in use for processing the material incorporates a burner unit which subjects the material to hot gases at high enough temperatures and for long enough to ensure adequate sterility, which is so necessary for a product used for feeding purposes. The product is sometimes referred to as 're-cycled nitrogen' and has a protein content generally between 25 and 30%, which makes it an attractive proposition in view of its low cost and the continuing steep rise in prices of conventional proteins, owing to world shortage. Experimentally, the material has been used successfully at 50% of the total ration, but the usual rate of inclusion is around 15%. The material is also sold to gardeners as a fertilizer.

3.6 EGG PRODUCTION
3.6.1 Shell eggs

Eggs in shell have been regarded as a vitally important and valuable article of human food from time immemorial. Rightly or wrongly, it has been claimed that the egg-white or albumen is the nearest natural product to human blood, and that when a baby's stomach is too upset to digest milk it can safely be fed on white of egg.

The rearing of pullets for egg production and some of the diseases to which they are prone have already been dealt with briefly in the section 3.5.3 – rearing. The present section is devoted to the main methods currently in use for the production of eggs in shell for the UK market. It is important to recognize the change in the pattern of egg production which has taken place within the last 50 years. In the 1920s when the poultry industry was beginning to grow rapidly, intensive methods were gaining popularity not only because of the saving in labour, but also, because birds could be confined to the house where they were protected from the effects of bad weather; this helped to maintain production at more satisfactory levels. However the pattern changed completely throughout the war and for eight years afterwards owing to the strict rationing of feedingstuffs, which had the effect of discouraging the specialist egg producer and encouraging the general farmer to take his place, since it was only the latter who was likely to have sufficient second quality grain to spare for poultry feeding. But at this time the general farm flock was rarely allowed to roam the farmyard as in days of old. The value of poultry manure was being realized and therefore the flock was either folded over the pasture land in fold-units moved daily, or in colony houses moved five or ten yards twice a week. When poultry units were used in this way the quality of grassland was greatly improved. After the de-rationing of feed in 1953, the specialist once more began to take over and from that time onwards has produced the bulk of the national output of eggs. Production rose so rapidly that within a few years the seller's became a buyer's market and in April 1957 the British Egg Marketing Board was established to take over the marketing of eggs and the administration of the subsidy on eggs payable to producers, both of which functions had previously been performed by the Ministry of Agriculture. The Board ceased to exist on March 31st, 1971, from which time there has been a free market for eggs; at the same time the UK Egg Authority was established under an Act of Parliament. The main function of the Authority has been to operate a market intelligence service to assist the industry. Of the total eggs produced nationally, the bulk continues to be collected, graded, quality tested, packed and sold through the egg-packing stations.

a. Poultry in battery cages. In the early 1930s the trend towards intensive poultry keeping led to the ultimate step of housing laying hens in single-bird battery cages. On the premise that a hen will only lay if it is warm, dry and well fed, and therefore satisfied with its lot, this method, despite the set-back of the years of feed rationing, has steadily gained ground so that today it is estimated that over 90% of eggs passing through the packing stations are from hens in batteries, mostly in cages of three or more birds apiece. In the interests of animal welfare, minimum standards of space per bird are laid down by Government regulations. Originally hand-fed cage by cage, manual labour is now reduced to a minimum with automatic feeding, watering, removal of droppings and collection of eggs. Egg production per bird has been proved many times over to be higher under this system than any other, and over 90% production (that is, more than 90 eggs per 100 hens per day) can be maintained for weeks on end. Battery cages are mostly arranged in rows down the length of the house, up to four tiers high. In order to maintain even

Fig. 3.41　Egg collecting time in a hen battery.

house temperatures, not unduly influenced by external weather conditions, the insulation properties of the house are important, as is also the ventilation system. Fan extraction at the apex is not ideal since the lower tiers of cages are often appreciably colder then the higher tiers. Pressurized ventilation would seem to be the more satisfactory method. Control of lighting is also important, since too much light is liable to cause 'cannibalism', and too little on the lower tiers discourages feeding.

b. The deep litter system. A logical step in intensive keeping is the deep litter, or built-up litter house. As its name suggests, the litter is not removed during the time the birds are in the house, but fresh litter is added from time to time, i.e., built up as and when required. The litter used is generally woodshavings with sawdust, starting with a depth of not less than 200 mm (8 in), but chopped straw and peat moss are good alternatives. Ideally, the birds should be introduced into the house during the summer months, or alternatively, when the house has been warmed up. Under conditions of warmth the litter together with the birds' excreta will begin to 'work', or break down as in a compost heap. This has not been found to be detrimental to the health and well-being of the birds, rather the reverse. A common practice is to instal a droppings pit running down the length of the house, generally in the centre, over which are fixed slatted platforms 600 to 900 mm (2–3 ft) from floor level. The natural tendency to roost at night encourages the birds on to the slatted area above the pit and so helps to preserve the floor litter in good condition for a much longer period than would otherwise be the case. Although some commercial egg production is still carried out in deep litter houses, the superior results obtained in hen batteries has tempted most producers to adopt that system. However, almost all breeder flocks are at the present time being run on deep litter in windowless houses or with the windows blacked out, so that the necessary light control can be maintained, without which egg production would become too seasonal. The desirability of light control applies to battery production as well as to deep litter systems. Sufficient space for most, if not all the birds to feed or drink at the same time time is important, the more so if the birds are breeders. Failure to supply enough space often leads to cannibalism. For trough space 100–150 mm (4–6 in) and drinking space 25 mm (1in) per bird is normal. One nest box to five birds, or the equivalent if using large, communal nests, are standard requirements. Floor space (including slatted area) requirement per bird is about 0.3 to 0.4 m^2 (3–4 ft^2) according to the size of the birds.

In the UK the yearly consumption of eggs per capita has varied only slightly above or below 270 during the past nine years. This figure is likely to increase in the future owing to the rising world prices of meats and other protein foods.

3.6.2 Liquid eggs

Before the Second World War the use of whole eggs in liquid form became popular with bakeries and caterers because they were convenient and labour saving. Another advantage was the growing practice of freezing the liquid egg in large metal cans, so that its keeping qualities could be prolonged. Home production of frozen egg was small, and a substantial proportion of the imported product originated from China, where at that time effective enforcement of safety regulations to control hygiene was lacking, so that, rightly or wrongly, outbreaks of food poisoning were frequently blamed onto frozen egg. In consequence, the popularity of the product suffered.

It was not until the advent of the British Egg Marketing Board that liquid egg really came into its own, this time as a means of dealing with market surpluses of home-produced first and second quality eggs. 'Breaking out', and the production of liquid egg, was put into the hands of a limited number of licensed processors working to strict standards of hygiene. Sales were developed under the name of 'Lion Brand' and as such the product was readily accepted by the bakery and catering trades. During the final year

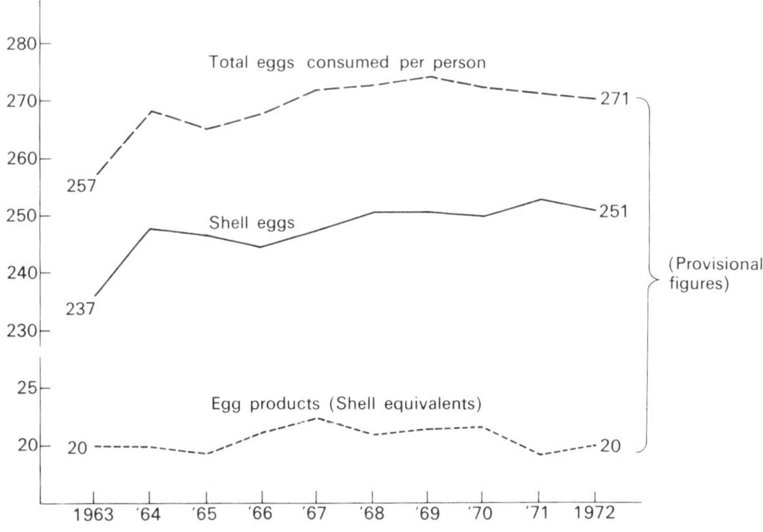

Fig. 3.42 UK egg consumption per capita 1963–1972.

Fig. 3.43 Egg-breaking machinery in a liquid egg plant.

of the Board's operation nearly $3\frac{1}{2}$ million boxes of eggs (1 box = 360 eggs) were broken out for the liquid egg trade, in about equal quantities of first and second quality eggs.

Since the Board's demise and the advent of a free market in eggs, the situation has changed considerably, and at present only about 2 million boxes are being broken out annually for the liquid and dried egg market, most of which are second quality eggs. This represents approximately 5% of the total annual UK output of eggs. It is estimated that about 60% of this business is in the hands of five large organizations supplying the bakery and catering trade, to a large and growing extent in bulk. Thus, when supplies of liquid egg are in excess of normal bulk requirements, the usual procedure is to freeze and store the surplus against anticipated periods of shortage.

Regulations governing the manufacture and handling of liquid egg hygienically require the inspection of all eggs being broken out and the rejection of any substandard specimens. Pasteurization then takes place in which the liquid is held for not less than $2\frac{1}{2}$ minutes at 64°C (148°F), following which the liquid is cooled to 2°C (36°F) and either fed into large tankers for bulk delivery or frozen for storage and future use.

3.6.3 Dried eggs

At one time the use of dried egg was extensive in the bakery, catering and food processing industries, particularly in wartime when available shipping space and transport was strictly limited. The process of drying was usually the 'spray' method – as used in the preparation of spray-dried milk (see Volume 8, Chapter 4). The liquid is sprayed into a stream of hot air which evaporates the moisture and produces solid particles which are collected, forming a dry powder. However, in course of time the extra cost of drying the liquid product has become uneconomic so that today it is estimated that only about 1% of liquid egg is spray-dried.

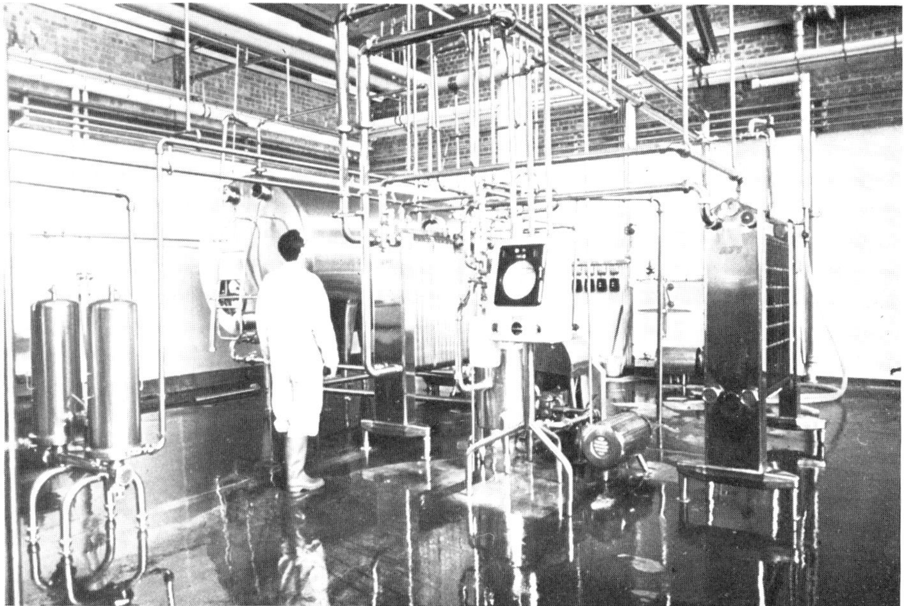

Fig. 3.44   Pasteurizing plant in a liquid egg factory.

This is also true of the process known as 'accelerated freeze drying', which up to 1972 was claiming about 6% of the liquid egg market. Since that time egg prices have increased, bringing the price of liquid egg into direct relationship with that of shell eggs for the first time. As a result this method of drying liquid egg has also become uneconomic within the prevailing price structure and may well remain so unless export or other outlets are found.

3.7   SAUSAGES AND THEIR PRODUCTION

This section covers the production of British-type fresh sausages, British-style cooked sausages and frankfurters. It is not possible in a short outline of the subject to include the European (continental) sausages.

3.7.1 Legal requirements

The minimum meat content of sausages in the UK was specified in 1967 in Statutory Instrument No. 862, *The Sausage and other Meat Products Regulations 1967.* These regulations were modified slightly in 1968 by Statutory Instrument No. 2047, *The Sausage and other Meat Product (Amendment) Regulations 1968.* Space does not permit more than a brief mention of some of the points raised in those regulations and the reader is advised to read the original text, and also the following regulations:
1.   *Labelling of Foods Regulations,* S. I. No. 400, 1970,
2.   *The Preservatives in Foods Order 1962,* S. I. No. 1532, 1962, which permits the addition of 450 parts per million of sulphur dioxide as a preservative in fresh (i.e. uncooked) sausage.

3. *The Colouring Matter in Food Regulations*, S. I. No. 1203, 1966, and S. I. No. 1102, 1970.

4. *The Offals in Meat Products Order 1953.* S. I. No. 246, 1953 *Statutory minimum meat contents* (as laid down in S. I. No. 862).

(i) Any sausage or sausage meat or polony or hog pudding shall have a meat content of not less than 50%.

(ii) Any sausage or sausage meat, described directly or by implication as a pork sausage or as pork sausage meat, shall have a meat content of not less than 65%.

(iii) Any sausage (other than a canned sausage) which is described directly or by implication as a 'frankfurter', 'Vienna sausage' or 'salami' shall have a meat content of not less than 75%.

(iv) Any sausage canned in brine or other liquid medium which is described directly or by implication as a 'frankfurter', 'Vienna sausage' or 'salami' shall have a meat content of not less than 70% calculated on the drained weight of the sausages.

Meat is defined in the regulations as being the flesh, including fat, and the skin, rind, gristle, and sinew in amounts naturally associated with the flesh of any animal or bird which is normally used for human consumption, and includes cured meat and permitted offal, but does not include fish.

Any egg which is present in any meat product, not exceeding one fifth by weight of the meat content of such meat product, may be reckoned as part of the meat content if the meat product bears as part of or in immediate proximity to its name, a description indicating the presence of egg. As mentioned above, the fat in the sausage or meat product is part of the meat content but in a sausage or sausage meat the lean meat content must not be less than 50% of the minimum total meat content, and in the case of other meat products not less than 60% of the minimum total meat content.

3.7.2 Sausage ingredients

a. Meat. Whilst meat for sausage manufacture can be brought into a sausage factory, far more control on the trimming standards, freshness and hygienic preparation of the meat is possible if the sausage manufacturer also slaughters and dresses meat. This is particularly so of pork meats. A consistent quality end product requires reasonably consistent raw materials and apart from merely maintaining a constant ratio of lean to fat, the nature of the fat is important. Pork head meat fat for example, is softer and oilier than back fat. Fresh meats give a sausage with better binding and water retention during cooking than frozen meats, even though the meats may be fully thawed before use. The best quality pork sausages are made with high proportions of shoulder meat and pork back fat. Medium class sausages contain a higher proportion of pork belly meat, and the poorer class sausage a high proportion of pork head meat. The use of polyphosphates (which are briefly discussed later) enables manufacturers to produce sausages with minimum frying losses even when the quality of the meat varies in its natural water-holding capacity.

b. Binder or filler. British fresh sausages contain appreciable levels of cereal binders which give the sausage its characteristic juiciness and texture. Soaked bread, which was once the usual binder, has been replaced by granular yeastless rusk. Boiled rice was also used but this too has more or less been completely replaced by rusk. Farina or potato flour is sometimes used either directly or as a component of a proprietory binder mixture. Sodium caseinate has found favour with some sausage manufacturers but owing to the world rise in the cost of milk products in 1972, it became an expensive ingredient. Similarly, skimmed milk powder which in some cases has given improved colour stability to sausages when displayed under fluorescent lights, is too expensive when world prices for milk products are high. Soya flour can be used in small quantities (2 to 3%) and owing to its natural tocophorol content, confers stability against oxidation and possible development of white spots therefrom. Soya protein is, however, frequently introduced in the shape of proprietary additives and may be present in the form of concentrates of soya flour having a protein content of the order of 70%, or isolated soya protein at 90 to 95% protein. Some proprietory products are mixtures of soya proteins with cereal starches. Wheat gluten is another protein which is used to a limited extent as a protein binder in pork sausages rather than beef. Blood plasma has a beneficial effect on frying losses and texture. The bacteriological problems associated with handling liquid blood plasma limit its use, but there are fewer problems with spray-dried plasma.

c. Polyphosphates. These are commonly used in commercial sausage production. They have a much stronger effect, weight for weight, than the salt in the sausage in extracting myosin from the lean meat. It is believed that this extracted myosin, which is distributed over the meat, fat and even the moistened rusk, retains water when the sausage is cooked.

d. Seasonings. Salt constitutes about 90% of the usual pork or beef sausage seasoning. The final level of salt in the sausage should be between 1% and 2.0%. The salt should be free of any nitrates or nitrites, since the former can be reduced to the latter and nitrite can destroy the 'fresh' colour of meat by oxidizing the red meat pigment oxymyoglobin to the grey-brown metmyoglobin. Further reaction, particularly under the reducing action of the sulphur dioxide present in the sausage, can produce the red-coloured nitric oxide myoglobin (sometimes called nitrosylmoglobin) which when heated, gives pink-red coloured nitric oxide myochromogen. Pepper usually constitutes the second major seasoning in sausages. Because it creates black specks in the sausage, black pepper is only used in certain coarsely cut sausages. Usually the more expensive white pepper is used, but nowadays, for reasons of consistency of flavour and bacteriological considerations, oleoresin and essential oil extractives mixed into a salt, rusk or glucose base are frequently used. When extracted pepper is used the extract is made from black pepper, but does not cause discolouration of the sausage meat. Other supplementary spices used in various sausages are mace, nutmeg, cayenne pepper, ginger, coriander, pimento, cardamon, cinnamon, clove garlic, sage, thyme and rosemary.

For convenience, the salt, seasonings, polyphosphate, metabisulphite preservative (see below) and sometimes dry colour, are blended together and added to each chopping as a premixed and preweighed unit.

e. Preservative and colouring. As mentioned above, 450 parts per million of sulphur dioxide may be added to sausage meat to extend its shelf life and delay colour loss. It is usually introduced in the form of sodium metabisulphite, $Na_2S_2O_3$, containing 64% of available sulphur dioxide, as a mixture with the seasonings and polyphosphates.

f. Casings. Sheep casings are used for small fresh sausages linked 16 to the pound, and for frankfurters. Hog casings are used for larger fresh sausages linked 8 to the pound.

Hog fat ends (bungs) are still used for good quality liver sausage, but artificial casings are now popular. Beef runners are used for black puddings and liver sausage, and beef weasands (oesophagus) for luncheon-type sausage, although again, the last two types of casing are now largely replaced by synthetic casings:

3.7.3 Equipment

The minimum equipment for a sausage production line is:

1. Chillroom for meat and for finished goods. The finished goods should preferably be stored in a separate chiller and cooked meats should not be stored in the same chiller as raw materials.
2. Boning and meat preparation tables.
3. A meat grinder (optional, depending on the type of sausages, i.e. coarse or fine textured, and on the raw materials).
4. A bowl chopper. These vary from very small, 7 kg (15 lb) batch size, to 200 kg (400 lb) size.

Fig. 3.45 Continuous Filler feeding two linkers.

5. Filling machine.
6. Linking tables (linking machine optional).
7. Finished sausage cooler. This may be of the tunnel type, with the product hung on trolleys, or it may be a continuous cooler in which the sausages are hung on conveyor hooks. In either case the air flow, humidity and temperature should be such as to reduce the internal temperature of the sausages to 1.7°C (35°F) within about 20 minutes without excessive drying of the sausages.
8. Scales and packing tables, including metal detector, and wrapping aid or machine.

3.7.4 Methods of production and recipes

The method of manufacture varies according to the type of sausage being produced. The coarser cut type favoured in the north of England can be made with a mincer and mixer. The latter machine may be either a planetary-beater bowl mixer or a horizontal twin-arm trough mixer. For the smooth textured sausage favoured in the midlands and south, a bowl chopper is essential. For frankfurters, the fine cutting of a bowl chopper is often supplemented by a high speed vertical cutter/mixer mill.

The basic principle is to obtain the maximum water uptake of the lean meat and at the same time to extract the maximum amount of myosin to the surface of the meat pieces through the action of the salt, the polyphosphate and the physical kneading and/or cutting which the meat receives.

Where the meat and fat are pre-minced and part of the fat addition consists of 100% pork back fat, a suitable order of addition is: lean meats, water, seasoning (containing polyphosphate and metabisulphite), rusk and other optional binders, and finally, diced or minced fat.

Where the meat and fat cannot be separated in the raw materials and where the end product is a smooth textured sausage, the order becomes: meats, water, seasoning and rusk. In this situation it is not necessary to pre-mince the meat and fat.

The following recipes are given as examples. For further details the reader should refer to the literature.

Product	%
Cambridge-style pork sausage (Meat content 66%)	
Lean pork	21.8
50/50 fat/lean pork	14.2
Back fat	19.4
Head meat	6.6
Cooked rind emulsion	5.6
Ice water	18.6
Rusk	10.3
Seasoning (including 500 p.p.m sulphur dioxide, 0.3% polyphosphate and colouring)	2.5
Protein additive	1.0
Beef sausage (standard quality)	
Lean trimmed beef flank 70:30 (lean/fat)	20
Processed pork rinds	5
Brisket (point end)	5
Mutton breast	5
Pork fat	15

Product	%
Rusk	14.5
Water/ice	30
Skimmed milk powder	1.5
Seasoning including 500 p.p.m. sulphur dioxide	2.25
Promine D. (soya isolate)	1.5
Hyphos (polyphosphate mixture)	0.25

Frankfurters (75% meat content required)
These sausages can be made with all beef meats, all pork meats, or a mixture. A typical mixed meats recipe is as follows:

Beef trimmings	25
Veal trimmings	28
Lean pork trimmings	21.3
Pork heart	2.3
Ice water	15.4
Soya flour	2.5
Skimmed milk powder	2.3
Salt	2.3
Sugar	0.4
Seasoning: ginger	
coriander	
mace	0.25
white pepper	
garlic	
Polyphosphate	0.3
Sodium nitrite	0.02

Source: Oppenheimer Casing Co. (UK) Ltd.

A suitable method of production is to pre-mince the meats separately through a 3 mm ($\frac{1}{8}$ in) plate, then chop with seasoning salt, phosphate, half of the water, nitrite, milk powder and soya flour, sugar and the rest of the water until a finely chopped emulsion is formed. Fill into sheep casings or synthetic casings. Smoke for $1\frac{1}{2}$ h at 43 to 54°C (110–130°F). Cook by raising the temperature of the smokehouse or by immersion in hot water 82°C, (180°F) until centre temperature is 63°C (145°F). Cold shower or transfer to a cold water tank. When synthetic casings have been used, strip before packing the frankfurters.

3.7.5 Analysis of sausage products

Details of the analytical techniques for the following estimations will be found elsewhere in this volume or in literature references. In order to determine the apparent meat content and the ratio of fat to lean meat, the following analyses are required:

Analysis	Method
Moisture	Drying to constant weight after mixing with a known weight of dry sand.
Protein	Usually by the Kjeldahl method, although rapid methods using selective dyestuffs can be used.
Ash	Incineration at 500°C
Fat	Soxhlet or Werner Schmidt extraction.
Carbohydrate	By difference, i.e. by subtracting the sum of the above percentage components from 100.

In addition to the above determinations, the salt content should be determined. This can conveniently be carried out on the ash, titrating with silver nitrate using potassium chromate as indicator. Sulphur dioxide should also be determined to test for compliance with legal limits. The usual method is that of steam distillation, titrating with N/20 iodine in the receiving flask as the distillation proceeds, using starch as indicator.

In the case of cured products such as frankfurters, the nitrate and nitrite expressed as the sodium salts should be determined. (The legal maxima are 500 and 200 p.p.m. respectively).

3.8   MEAT PRESERVATION

3.8.1 Canned meat

Certain canned cured meat products, i.e. those containing salt and sodium or potassium nitrite, are given a heat treatment which only pasteurizes them. Such products must be kept in refrigerated storage, since at higher temperatures heat resistant spores which survive the pasteurizing process may grow. An example of this class of meat products is large canned hams. These are processed by heating in an open-kettle water bath to an internal temperature of about 71°C (160°F). Another class of canned cured meats, such as small canned hams, luncheon meat and chopped pork, are commercially sterile after a thermal process which gives a sterilizing value of F_0 0.5 to 1·0. (The F value of a process is defined as the period in minutes of exposure to 121°C (250°F)–assuming instantaneous heating and cooling at the beginning and end of the process–which will have a sterilizing effect equivalent to that of the process. When the sterilizing value of a process is calculated using a basis of $z = 18$, the symbol F_0 is used. Here z is defined as the slope of the thermal death time curve and is also the number of degrees Fahrenheit required to produce a ten-fold increase or decrease in thermal death time.) Sterilizing at F_0 0.5 to 1.0 is effective; although small numbers of spores of mesophilic bacilli and of clostridia survive in canned cured meats, their germination is inhibited by the nitrite or more correctly by the undissociated nitrous acid. Perigo *et al* discovered in 1967 that when nitrite was heated in a bacteriological medium, an inhibitory effect on certain bacterial spores greater than that due to the nitrite alone could be demonstrated. Also, whereas the effect of nitrite alone depends on the pH, the 'Perigo effect' was largely independent of pH. (See literature references at end of chapter). Uncured meat products such as stewed steak in gravy, steak and kidney pies, meat and vegetable stew, require a process equivalent to a minimum of F_0 8 to sterilize heat resistant spore-forming bacteria to a degree which is commercially acceptable. Nearly all canned meat products are solid packs; i.e., there is no heating by convection currents in a liquid phase as with canned vegetables in brine (an exception is canned sausages in brine). Heat transfer is therefore by conduction and the process time at a given steam temperature is dependent upon the size and shape of the can as well as the thermal diffusivity of the contents. The installation and control gear on retorts have been specified in a memorandum published by the National Canners Association of America (See Literature references).

Space does not permit a detailed description of the technique for canning each product but table 3.1 gives brief details of typical methods.

The minimum meat content of canned meats in the UK is specified in *The Canned Meats Products Regulations* 1967, Statutory Instrument 1967, No 861.

Table 3.1 Production methods for canned meat

Product	Ingredients	Method of preparation	Can	Typical process[1]
Large hams	Ham, gelatine, salt, phosphate, nitrite	The old method of arterial injection of legs of pork still containing the rind fat and bones has largely been replaced by multi-needle injection of boneless defatted meat followed by 'tumbling' or kneading the meat.	Pear-shaped or long, usually fitted with aluminium sacrificial anode	Open kettle to centre temperature 71°C
0.45 kg (1-lb) hams	Ham, gelatine, salt, phosphate, nitrite	The old method of arterial injection of legs of pork still containing the rind fat and bones has largely been replaced by multi-needle injection of boneless defatted meat followed by 'tumbling' or kneading the meat.	Pear-shaped.	75 min at 110°C
Chopped pork and ham	Cured pig-meat, seasonings, polysulphate	Meats, seasonings, water, etc. chopped together in a bowl-cutter, then vacuum mixed.	340 g (12 oz) rectangular. Meat lacquered or meat-release lacquered.	75 min at 110°C
Canned frankfurters Sausages in brine		Sausage emulsion heat-set in synthetic casing, which is peeled off before sausages are filled into can.	300×310 (3 inches in diameter, $3\frac{10}{16}$ in. in height)	25 min at 121°C
Corned beef	Beef, salt, nitrite and sometimes sugar.	Cubed meat precooked, mixed hot with curing salts, filled hot into can.	340 g (12-oz) taper can, unlacquered interior	2 h at 115.5°C
			2.7 kg (6 lb) taper or parallel sided, plain tinplate interior.	4 h at 114°C
Ox tongues.	Tongue, curing salts, agar.	see section 3.8.2.	Round 2.7 kg (6 lb): Plain tinplate.	5 h at 108°C
			Round 0.45 kg (1 lb): Plain or meat release lacquered.	90 min at 114°C

| Meat in gravy. | | Partially thawed or fresh cubed lean beef filled with hot gravy, vacuum seamed. | 0.45 kg (1-lb) 300 × 310 lacquered internally with a meat lacquer. | 95 min at 118°C |
| Steak and kidney pies. | Meat, gravy, and pastry | For technical and quality reasons the use of bottom pastry was generally discontinued in about 1960. Partially thawed or fresh cubed lean beef placed into can. Gravy added and a disc of flaky (puff) pastry laid on top. Lidded and vacuum seamed. | Plain or meat lacquered tapered flat can. | 2 h at 115.5°C |

[1] The typical process times listed above are used in commercial production, but much depends on the salt in water phase, on the presence or absence of nitrite, on the texture and on the normal pre-canning bacterial population of the product. They should therefore be only used as guides and reference should be made to the laboratory of the can suppliers or to a reputable food technology institute or laboratory, before embarking on the canning of any item.

3.8.2 *Cured meat*

In former times, bacterial reduction of nitrate was necessary to produce the nitrite which is essential for the colour-forming reaction, but nowadays it is usual to control the nitrite level in the brine by directly adding sodium or potassium nitrite. Too high a nitrite level in the brine causes a green discolouration, and the amount of nitrite in the product becomes too high from a food safety point of view. For many years the poisonous nature of nitrite has been recognized and latterly the formation of carcinogenic nitrosamines from the reaction of nitric oxide with secondary amines, tertiary amines and secondary amides, has been established. Although no correlation between residual nitrite in the food and the level of nitrosamines has been found, regulating authorities in many countries are decreasing the amounts of nitrate and nitrite in raw and cooked cured meats. In the UK *The Preservatives in Foods (Amendment) Regulations*, S. I. 1971, No. 882 limited the nitrate level to 500 p.p.m. and the nitrite to 200 p.p.m. There are indications that these limits may be further reduced in the future.

The chemistry of the formation of the cured meat colour, nitric oxide myoglobin, has been the subject of considerable study. In 1901, Haldane suggested that nitric oxide was involved in forming the pink-red colour of cured meat. A great deal of research effort was put into this area during the 1960 decade, with the result that much more is now understood. For details of these researches, the reader should consult the papers by Tarladgis, Taylor and Walters, and by Fox and Ackerman. The following simplified diagram will, however, serve to illustrate the mechanism of the reaction:

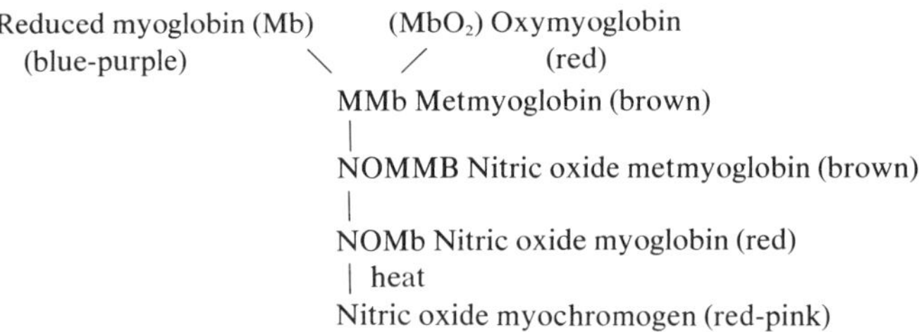

The bulk of the bacon eaten in the UK is cured by the Wiltshire method, although some is cured by multi-needle injection after deboning and derinding. Also, a small amount is made by a patented process of slice-curing rashers of pork through a brine bath or under a brine spray. In the Wiltshire process the whole side is cured without any derinding or defatting. This requires that the pigs be lean and the usual 'dead weight' of the carcass is 7 to $7\frac{1}{2}$ score (about 65 kg, 140–150 lb).

The day after slaughter, the chilled sides are trimmed and the shoulder blade is withdrawn. The cavity is usually packed with salt or a salt/sodium nitrate mixture and brine is then introduced with the aid of a brine needle containing a series of lateral holes. The operator pumps in about 8 to 10% of chilled brine, making approximately 30 injections. The strength of the brine, which contains about 0.5% sodium nitrate and 500 to 1000 p.p.m. sodium nitrite, is usually in the region 25 to 27% w/v and the hydraulic pressure 5.5 to 6 kg/cm² (80–85 p.s.i.). The sides are then layered rind side undermost into vats, each layer being sprinkled with salt. Battens are then fixed over the top and 'cover' brine containing 24 to 26% w/v salt, 0.5% sodium nitrate, and 800 to 1000 p.p.m. sodium nitrite is run in. The temperature is maintained at 5°C ± 1°C. After four to five days the brine is run off and the sides removed for draining and maturing, rind side

Fig. 3.46 Bacon sides being cured in fibreglass vats and matured in stacks.

Fig. 3.47 Multi-needle brine injector injecting pickle into pork legs for ham manufacture.

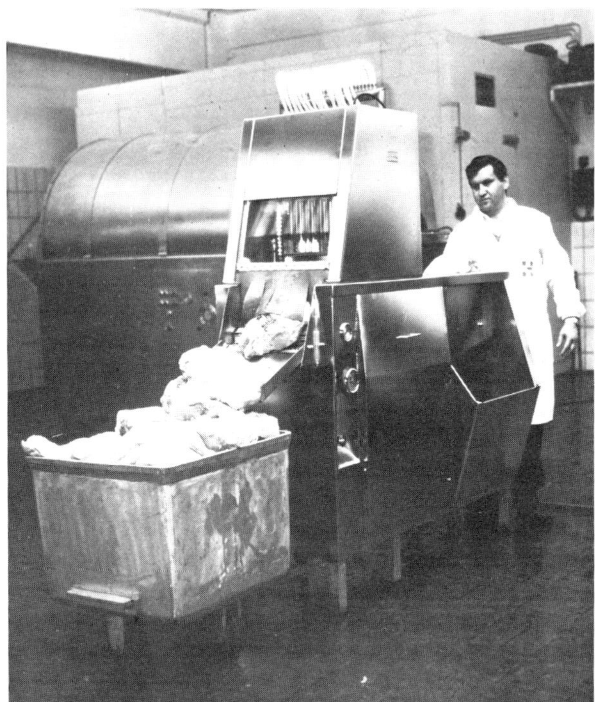

uppermost, in stacks about eight sides high, for 4 to 7 days, at $5°C \pm 1°C$. Multi-needle machines are avillable from several manufacturers which enable bone-in whole sides to be injected, the needles being spring loaded but retractable when they strike bone.

Hams and cured shoulders are nowadays usually cured by a combination of multi-needle injection of boneless joints followed by tumbling in a special machine. Some tumbling machines work on the principle of impacting the meat pieces together by continuously lifting the meat in a revolving drum type or end-over-end type mixer. For smaller pieces of meat where the kinetic energy of the pieces after falling a short distance is low, paddle type machines which slowly knead the meat pieces are used. In addition to increasing the penetration and distribution of the brine, tumbling causes myosin to migrate to the surface of the cuts and this acts to cement the pieces together when cooked, thus giving a sliceable product from several small pieces. Polyphosphates are usually added to the brine to improve the water-holding capacity of the meat when it is subsequently cooked. Hams and some shoulders are packed into aluminium moulds which may be pear-shaped, D-shaped, or 'leg-of-mutton'-shaped. Alternatively, they may be packed by means of a combination vacuum/compression press into long stainless steel moulds fitted with a spring-loaded end plate to give compression whilst the meat is cooked. They are water cooked, or hot moist air cooked, to an internal temperature of $65°C$, and finally cooled either by immersion in cold water or by cold water sprays. Care must be exercized to prevent contamination of the cooked meats with bacteria, both from a food safety standpoint and also because certain lactobacilli and streptococci cause greening discolouration. Some cured shoulder meat is packed into synthetic casings about 125 mm (5 in) in diameter, and after a mild heat treatment to firm up the meat, cut into boiling joints of about 0.5 kg (1 lb). These may then be vacuum-packed.

Ox tongues and briskets are usually cured by injection and tanking for several days and are pre-cooked before being packed into cans. Tongues can be skinned easily if they are scalded and skinned within an hour or so post-mortem, but if this is not done the skinning must take place after curing and cooking, just prior to being canned.

The pH of the meat is important in cured products. In long-keeping products such as York ham, Suffolk ham and Westphalian ham, a low pH slows up bacterial spoilage and also allows the curing salts to penetrate and disperse evenly through the meat. In canned hams, a low pH accentuates jelly formation in the product. This is to some extent buffered by the polyphosphates used in cooked ham and canned ham production. Also the method of cure (i.e. multi-needle injection plus tumbling) overcomes resistance to brine diffusion caused by a high pH and consequently a close textured meat. A defective condition known as P.S.E. (Pale Soft Exudative) muscle occurs in a varying percentage of pigs soon after slaughter. The condition is associated with a rapid fall in pH to 6.0 within an hour post-mortem. This rapid fall precipitates some of the sarcoplasmic proteins, causing a decrease in their water-holding capacity. The condition is accentuated by anxiety and stress in the pig ante-mortem and by certain breed characteristics. Rapid cooling to slow up the glycolysis reactions, so that the pH does not fall too rapidly, decreases the incidence of the defect.

3.8.3 Smoked meat

The smoking of meats probably originated with cave dwellers who found that meats put to dry near the roof of the cave and above the fire, resulted in increased keeping quality of the meat. The flavour of smoked meat was probably an acquired taste. Smoke confers an improvement in keeping quality greater than that due to simple reduction in the water activity of the surfaces owing to the drying effect. It also contains antioxidants which help to inhibit the oxidative rancidity which cured meats undergo, especially as a result of the presence of curing salts. The colour, even of deeper parts of the meat not in contact with the smoke, tends to be enriched because the last stages of the curing reactions are accelerated by the warm temperature. Above a temperature of around 53°C ('hot smoking') the meat proteins begin to denature and assume a cooked appearance.

For smoking whole sides of bacon or hams, the usual temperatures employed are 32° to 43°C, but frankfurters are usually smoked and cooked in dual-purpose smoker/cooker cabinets. The meat Inspection Division of the US Department of Agriculture requires that the internal temperature of frankfurter sausages shall reach 58.3°C. The meat industry prefers to cook to between 64.4 and 68.3°C to reduce bacterial spoilage. Since the process requires several hours at controlled temperatures, the introduction of the desired degree of smoke flavour is not a problem with frankfurters in natural casings, but synthetic casings of cellulose or other films are not as permeable to smoke as natural ones. Various attempts have been made to overcome this. A patent was granted to J. A. Ziegler (U.S.P. 2944909) in 1960, whereby the sausage meat is subjected to smoke treatment whilst it is being comminuted in a chopping device. A patent covering neutralized smoke flavouring was granted to H. E. Wistreich (U.S.P. 3467527) in 1969. The acidic components of smoke extracts cause instability in sausage emulsion and the patent covers the neutralization of the acids with food-grade alkalis.

In the direct smoking method, the smoke is generated outside of, but adjacent to, the smokehouse in special burners using hardwood sawdust of oak, beech or hickory. Resinous woods such as pine must be excluded. The use of liquid smoke extracts, although not general, has the following advantages:
1. Easier to control than direct smoke production and deposition.
2. Eliminates costs of smoke generating equipment.
3. In certain cases permits of less weight loss in the smoking-cooking process.
4. Undesirable carcinogenic or tarry hydrocarbons can be kept in an insoluble form.

A patent was granted to W. M. Allen (U.S.P. 3503760) in 1970 covering the use of liquid smoke regenerated into a cloud or mist for application to sausage products.

3.8.4 Dried meat

The preservation of meat by drying has been practiced since very early times in remote communities but has largely given way to other forms of preservation. A considerable amount of research was put into the drying of cooked meat during the Second World War and in the 1950 decade, resulting in vacuum contact dehydration (VCD) and accelerated freeze drying (AFD). Table 3.2 (Sharp, 1953) shows the saving in space by dehydrating and compressing meat, and hence its value in emergency and military feeding systems:

Table 3.2 Reciprocal density of dehydrated and compressed meat

Commodity	ft^3/ton	$m^3/tonne$	ft^3/ton meat solids	$m^3/tonne$ meat solids
Frozen quarters	95	2.7	264	7.5
Frozen quarters (boneless)	78	2.2	173	5.0
Canned corned beef	54	1.5	157	4.5
Dehydrated beef compressed to a specific gravity of 1.000	54	1.5	75	2.1

At the present time, accelerated freeze-dried cooked meat is used to a limited extent in certain high quality dry soup mixes and in convenience foods. Chicken pieces are especially suitable for these products and large pieces can be dried and still reconstituted within a reasonable time. Well prepared hot-air dried cooked minced beef has an acceptability which is comparable with that of freeze dried, as illustrated in table 3.3. (Ballantyne *et al*, 1958).

Table 3.3 Acceptability of freshly prepared dehydrated cooked ground beef

Method of drying	Rehydration time (min)[1]	Acceptability rating	Tenderness[2]	Flavour[2]	Juiciness[2]	Rehydration
Freeze	5	6.2	5.5	5.5	6.0	120
Hot air	5	6.9	3.8	6.4	5.9	110

[1] Rehydration was by simmering in 1% sodium chloride at 93 to 99°C.
[2] Ratings were on a scale from 1 (extremely bad) to 9 (excellent).

Both in the case of hot air drying and freeze drying, the higher water holding capacity of fresh meat with a high ultimate pH is still apparent in the dried meat.

In general, low grade meat from lean carcasses is preferred for air drying. After boning, the meat is pressure cooked and trimmed. The cooking liquor is partially evaporated separately and added to the cooked meat. After a brief mixing period, during which partial drying may be achieved by heating the jacket of the mixer, the mixture is minced through a 4.8 to 6.3 mm (3/16 to $\frac{1}{4}$ in) plate. The moisture content at this stage may be 45 to 50%.

Tray loading densities of about 0.1 kg/m^2 (2 lb/ft^2) are used in horizontal tray driers and an air velocity in the region of 200 m/min (600 ft/min). Temperatures as high as 80°C can be used in the early stage but as the meat becomes drier a lower air temperature is desirable to prevent a gritty texture and burnt flavour.

Freeze drying permits the production of dry steaks or chops which, when certain techniques are applied, are comparable after rehydration and cooling, with similar cuts cooked from fresh meat. The limitations of their appeal as commercial products are the expense of the process and packaging and the low acceptability rating, due to the unusual appearance in the dry form.

During normal air drying, evaporation of water from the fluid phase is associated with a movement of salts together with the remaining water towards the surface of the pieces. This results in contraction of each piece. In the freeze drying process, sublimation of water vapour directly from the ice phase yields a product which has undergone little or no shrinkage and which is honeycombed with pores. The size of these pores is controlled, *inter alia*, by the rate of freezing–slow freezing gives larger ice crystals, hence larger pores and quicker rehydration of the product. It is possible to freeze the product in the drier by evaporative cooling but it can freeze too quickly, forming too small intra-myofibrillar ice crystals. Also, evaporative cooling from the water phase at the surface before freezing commences can form a skin of concentrated salts which hinders rehydration. Heat must be applied to the product to provide energy for the sublimation but the amount must not be sufficient to cause thawing. The various methods adopted to apply the heat include water heated plates, spiked heater plates and microwave energy.

Both air dried and freeze dried bulk packed meat must be stored under an inert gas such as nitrogen, to inhibit oxidative rancidity, and in moisture impermeable containers.

3.8.5 Chilled and frozen meat

Man soon discovered that the body heat of a slaughtered animal must be allowed to dissipate quickly or the carcass will soon putrefy. There is little doubt that reduction of the water activity in the surface tissues during the early stage of cooling accounts for much of the bacteriostatic effect of cooling, especially under ambient conditions. The factors involved in the technology of meat chilling are indeed complicated. Too rapid a cooling rate immediately post-mortem in lamb, for example, causes marked contraction of the muscles ('cold shortening') whereas rapid cooling of pork *inhibits* the development of Pale Soft Exudative muscle (P.S.E.) in certain pigs where the condition may be latent. Too slow a cooling rate, especially of beef carcasses, can bring about the condition known as 'bone taint'. This is manifested by a putrid-sewage like odour in the region of the hip joint, and the head of the femur may be greenish brown. Unlike muscle, the pH of bone marrow does not fall post-mortem and hence remains a more suitable medium for bacterial growth. A full discussion of this subject is reported in *Meat Chilling – Why and How*, Meat Research Institute, Symposium No 2, Meat Research Institute, Langford, Bristol, 1972.

A major outbreak of foot and mouth disease in the UK in 1968 led to the prohibition of bone-in chilled beef from South America. A swing then took place to shipping chilled boneless primal cut joints in vacuum packed units. Two major systems are available for vacuum packaging meat. One utilizes bags of heat shrinkable polyvinylchloride/ polyvinylidene chloride copolymer which are evacuated and sealed with an aluminium clip. The package is then contacted momentarily with hot air or hot water to shrink the film tightly onto the meat. The other system uses laminated bags of nylon/poly- ethylene which are vacuum sealed in special machines where both the inner and outer sides of the bag are under vacuum when the seal is made. In addition to shipping of boneless meat at $-1.4°C$, vacuum packaging is being used for ageing and distribution of meat through the chain from abattoir to shop. For this trade, temperatures in the region of $+2°C$ are used.

The storage of frozen meat has long been practised. The disadvantages are: exudation of fluid on thawing, oxidative rancidity if not packaged and stored correctly, especially in the joint form, and a poor visual acceptability. Nowadays more imported frozen meat is cut into steaks whilst still partially frozen. Also a considerable amount of meat is now cooked from the frozen state in the form of beefburgers, steaks, reformed steaks (steak-shaped portions moulded from meats which have been cut into small flakes) and even joints. This overcomes to a large extent one of the major disadvantages of frozen meat, namely drip. Lamb carcasses are still frozen whole in air blast tunnels whereas beef is usually boned and cartoned before freezing. Large quantities of boneless frozen beef are used for manufacturing other meat products such as beefburgers, canned steak in gravy, etc.

If muscle is frozen before the adenosine triphosphate (ATP) content has fallen to the rigor-mortis level, a very fast activity of the enzyme ATP-ase which converts ATP to ADP ensues on thawing to cause marked shortening of the muscle and drip loss (thaw-rigor). These effects are greater with boneless meat. On whole carcasses, the muscles are to a large extent held in a stretched condition thus lessening the effects of thaw-rigor.

Fig. 3.48 Large meat factory in Melbourne, Australia, in which meat is prepared and frozen prior to shipment overseas.

Even when the freezing rate has been sufficiently fast to produce small *extracellular* ice crystals, if the meat is subsequently stored at too high a temperature the small ice crystals will sublime and recrystallize on the larger ones, thus causing damage to the tissue and increased drip on thawing. The wrapping material for frozen meat should be strong enough to withstand puncture or tearing under the conditions of use, moisture-vapour impermeable and preferably light proof. It must also be acceptable from a contaminant or toxicity standpoint for contact with food stuffs. Oxidation and discolouration can be caused by light of wavelength 5 600 to 6 300 Å. Fat rancidity and lean discolouration take place sooner in frozen pork than in frozen beef.

3.9 FISH

3.9.1 Classification and distribution

Fishing is a multi-million dollar world-wide industry and one of mankind's essential sources of animal protein, providing some 3% of the world's total protein supply. In addition, fish are also an important source of the raw materials needed for animal feedstuffs, oils, medicines and other products essential to industrial processes.

The sea itself covers some 71% of the earth's surface and fish of one sort or another inhabit every part of it from the littoral shallows to the abyssal depths (10 000 metres and more), and from the tropics to the poles, where fish which have adapted themselves to an extremely cold and lightless environment have been found beneath the antarctic ice. In fact over 10 000 quite distinct species of fish inhabit the world's oceans, seas, rivers and lakes, without including crustacea, mollusca and other marine life.

Of the total number of fish species, some 1 500 are of commercial importance, i.e. caught in sufficiently large quantities to have an appreciable economic value.

Faced with such a large variety of fish species it is obviously necessary to employ some scheme of classification so as to differentiate between them. The scientific or taxonomic classification sub-divides fish into classes, sub-classes, families, orders and species of related fish according to their biological and morphological characteristics. In this classification, most commercially important fish fall into two major groups: those with bony skeletons (osteichthyes); and those with cartilagenous skeletons (elasmobranchae). Most commercially valuable fish belong to the higher orders of bony fish (Teleostei), e.g. cod, haddock, hake, plaice, tuna, sardine and herring. There are, however, commercially valuable fish in the second group, like sharks, skates, rays and dogfish (sub-class Selachii) and some too with bony cartilagenous skeletons like the sturgeon (sub-class Chondrostei).

The visual differences between the two main groups of fish are easy to see. The bony fish have the mouth located at the tip of the snout, the gill openings are shielded by a single gill cover on each side, the skin is covered with scales and the rays of the caudal (tail) fin are symmetrical. The cartilagenous fish, however, have the mouth situated on the under-side of the head at some point midway between the snout and the back of the head, the gill openings are not covered but appear as a series of slits on either side of the head, the skin is covered not by scales but by scutes (small, sharp bony plates), and finally the upper ray (a continuation of the backbone) of the caudal fin is longer than the lower.

Fish can also be classified in a general fashion according to their habitat into three basic groups:

Marine fish which live and spawn only in the sea. This particular group can be further subdivided into the pelagic species which spend the greater part of their lives in the upper reaches of the oceans, like tuna, herring, anchovy, and sardine, and the demersal species which spend the greater part of their lives on or near the bottom, like cod, hake, haddock and plaice.

Freshwater fish which live and spawn only in fresh water (rivers, lakes, etc.). The principal commercial species in this group are the carp, perch, catfish and trout.

Anadromous and catadromous fish which live in the sea but migrate to the rivers to spawn or vice-versa. Anadromous species migrate from salt water to fresh water to spawn; the catadromous species migrate from fresh water to salt water. The anadromous species include the salmon and trout (Salmonidae), the sturgeon (Acipenseridae); the catadromous species include eels.

Again, we can look at the physical appearance of fish and divide them again into two main groups, the round and the flat. Round fish are those with symmetrical torpedo and snake-like bodies. Flat fish, however, have compressed bodies and are of two kinds: those which spend their lives on one side and which live on the bottom (plaice, sole, witch, etc.) and those which swim vertically (i.e. normally) and which are usually pelagic.

The world fish catch in 1971 was 69 400 000 tonnes; of this 8 850 000 tonnes were freshwater fish (see table 3.4), and 2 999 000 tonnes were catadromous fish. Of the marine fish catch about 15 500 000 tonnes were demersal fish and the remainder were pelagic. Freshwater and catadromous fish therefore account for only a relatively small part of the total world fish catch. The freshwater species caught in the greatest numbers were the

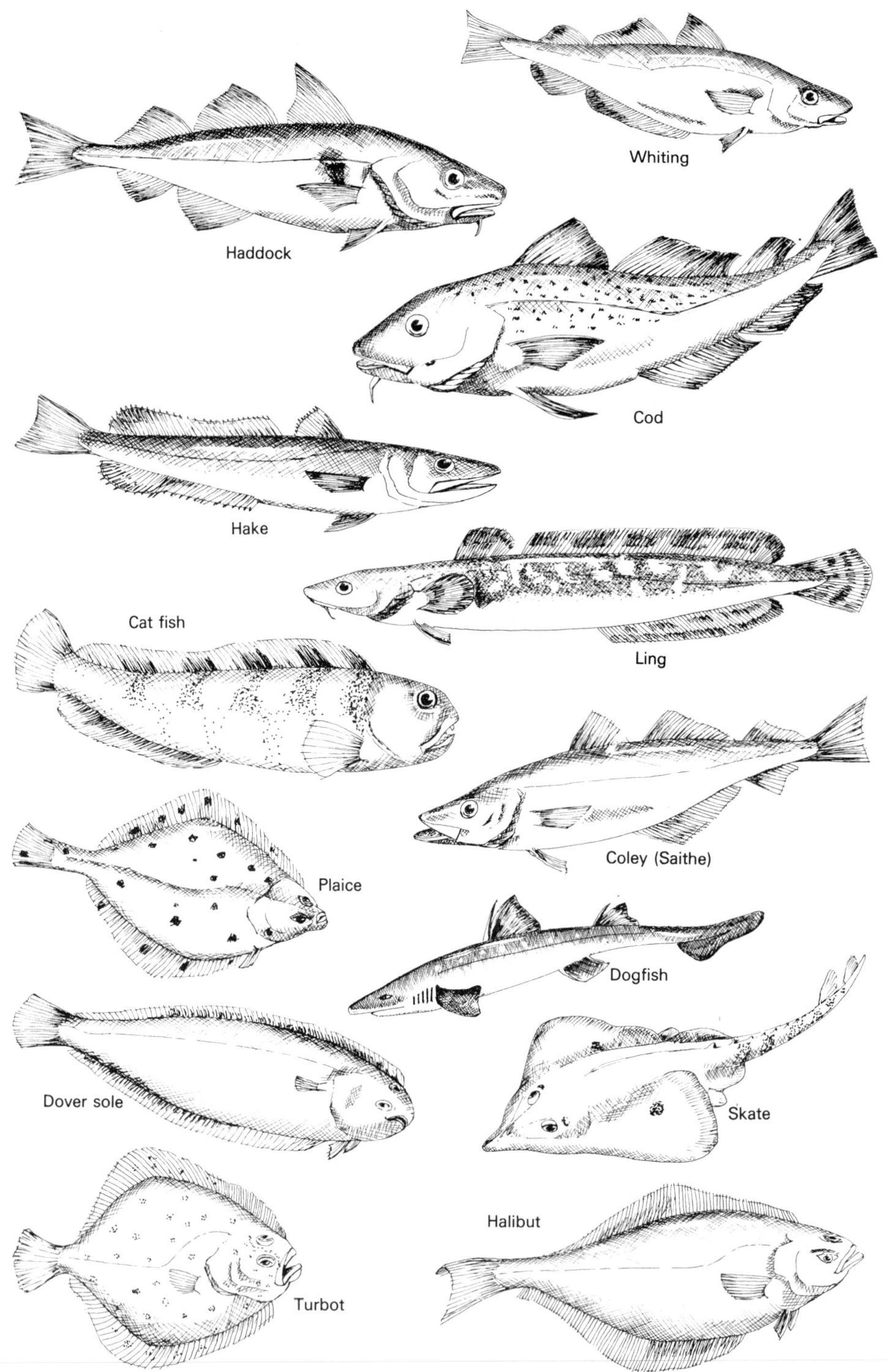

Fig. 3.49 Some species of marine fish.

Table 3.4 World catch by species (in thousand metric tonnes)

Species	1965	1966	1967	1968	1969	1970	1971
Freshwater fish	7030	7370	7240	7410	7690	83210	8850
Diadromous fish	1530	1870	1770	1890	2100	2940	2990
Marine fish	39600	42950	45930	48670	47190	52360	51620
Crustaceans	1200	1290	1370	1480	1510	1630	1630
Molluscs	2940	3000	3160	3480	3240	3320	3210
Miscellaneous	79	830	980	990	880	1000	1050

Source: *Yearbook of Fishery Statistics*, Vol. 32, Rome, Food and Agriculture Organization.

carps, barbels and other cyprinids (*Cyprinidae*) at 320000 tonnes; salmon, trout, smelts etc. provided the largest catch–2160000 tonnes–in the catadromous group. In the marine fishes, the species contributing the greatest part of the demersal catch were the cods, hakes, haddocks etc. at 10730000 tonnes, while the herrings, sardines and anchovies contributed the greater share of the pelagic catch at 18960000 tonnes.

Of the world's total catch of fish, shellfish, marine mammals etc., inland waters contributed a total of 9530000 tonnes and the oceans provided 59910000 tonnes. As table 3.5 shows, the largest catch was taken in the Pacific Ocean (33530000 tonnes) with 23320000 tonnes being taken from the Atlantic and 3060000 tonnes from the Indian Ocean.

3.9.3 Fishing

a. The gear. Fishing is one of man's oldest food-gathering activities and in one essential it has barely changed from the time man first stabbed at a fish with a spear; for the fisherman is still a hunter and fishing today is the only food industry which exploits a completely wild stock.

At the present time 90% of all fishing is conducted in coastal waters, the continental shelf area, which accounts for only some 7% of the total area of the oceans and where the water depth is 200 metres (656 ft) or less. There are two basic causes for this; the limitations of the gear and the limitations of the equipment used to find the fish.

Until very recently demersal fishing gears were only capable of fishing to a maximum of around 500 metres (1640 ft) which naturally limited such fishing to the continental shelf areas of the oceans. However, in the last year or so, demersal fishing has been extended to depths of 2000 metres (6561 ft) and more. This extension of practicable fishing depth is so recent as to have had little appreciable effect on the world's production of fish from the sea, although the USSR has established a successful fishery for the

Table 3.5 World catch by fishing area (in thousand metric tonnes)

Area	1965	1966	1967	1968	1969	1970	1971
Inland waters	7630	7990	7830	8060	8330	9020	9530
Atlantic ocean	19930	20850	22180	23110	22640	23560	23320
Indian ocean	2000	2190	2290	2460	2610	2770	3060
Pacific ocean	23680	26260	28140	30320	29040	34280	33530

Source: *Yearbook of Fishery Statistics*, Vol. 32, Rome, Food and Agriculture Organization.

'Grenedier' (*Macrurous rupestris*) at this depth off the coast of Labrador and Newfoundland.

In the case of pelagic fish it is the shortcomings of fish location which has limited the expansion of the fisheries into the deep oceans. Here again, however, advanced technology has begun to solve the problem of finding and identifying fish schools by the development of advanced ship-borne sonar systems and air-borne (and even satellite) remote sensors which can 'see' and film the bioluminescence of fish schools, thus pinpointing their exact position in the vast expanse of the ocean.

Even today primitive fishing tools like spears, arrows, rakes and even birds like the cormorant are still used to catch fish. However, the use of such devices contributes only a very small part of the world's catch, the overwhelming majority of which is caught by net and a smaller proportion by hook and line.

In a brief survey it is impossible to give anything more than a very cursory description of the principal gears employed. They can be divided into one of the two following basic groups:

Hooks and lines which comprise either long or short lines carrying a number of baited or unbaited hooks which are laid in the path of the fish, or a hand (or mechanically-) held pole to which one or more baited or unbaited lines are attached.

Nets. There is a large variety of nets which can be subdivided as follows: stationary or bottom-set nets (generally used in inland fisheries); drift nets which are of two basic kinds – the gill net and the trammel net – and are laid suspended by floats, vertically across the path of the fish; drag nets which include the trawls and seines; and finally encircling nets like the ring net and the purse-seine.

By far the most important of the two groups of gears listed above are the nets. Nets have been woven by machine since the early nineteenth century although some nets are still woven by hand even today. Traditionally, they were constructed of natural fibres such as manilla, cotton and hemp. In recent years however, the man-made fibres like nylon and polypropylene have almost completely replaced the traditional materials.

Most netting has a four-sided mesh with each side of equal length. The knots are known as points and the length of twine (mesh side) between knots is known as a bar. The actual size of mesh employed depends on which part of the net the section of netting is to be fitted, the species and size of the fish being hunted and, of course, on any legal requirements for minimum mesh sizes which may be stipulated by the regulating authority responsible for the particular fishing area in question.

Netting is cut to the particular size and shape required by straight edge cuts (either 'all-point' or 'all-bar') and by tapers (different combinations of points and bar edges). The edge of a piece of netting is termed the selvage and is usually reinforced with heavier or double twine thus creating a double selvage. A framing is attached to the netting to give it the required shape. The attachment of this framing and the attachment to the framing of any necessary lines, rings, chains, floats, etc. is termed 'hanging the netting'.

For our purposes we shall divide fishing gears into two groups: those for demersal fishing and those for pelagic fishing.

Demersal fishing gears

The 'otter' trawl. The most important gear in this group is the bottom or 'otter' trawl.

The trawl net is a roughly triangular shaped bag of netting which is dragged along the bed of the sea by the fishing vessel. The net itself is made up of different sized panels which make up the wings (net-sides which extend forward of the net-mouth), the sides,

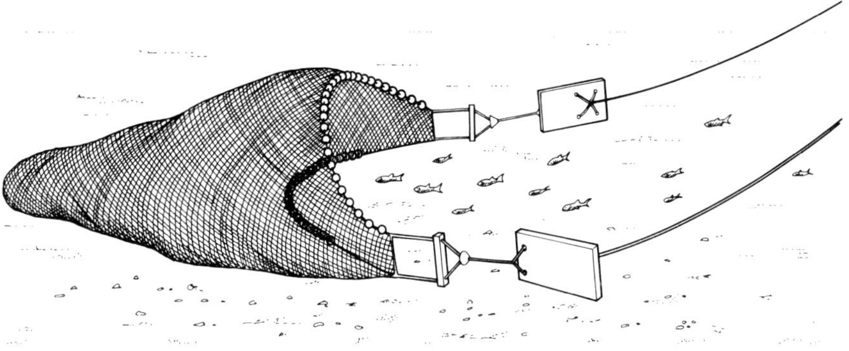

Fig. 3.50   The otter trawl.

square (top), belly (bottom), and cod-end which is the net-bag at the after end of the trawl and in which the fish are captured. The mesh size of the net decreases gradually from the mouth to the cod-end which has a very close mesh.

The square of the net extends forward above the forward selvedge of the belly in order to prevent fish from escaping from the net by swimming upwards. The net is kept open vertically by floats (either of cork or plastic materials) which are attached to the headline. The bottom forward edge is kept in contact with the bottom by weights and bobbins (heavy metal balls) attached to the ground rope.

The horizontal opening of the net is provided by the otter-boards or trawl doors. The traditional otter-board is a very strong rectangular board of wood or metal, but oval, V-shaped and slotted boards are also used. The trawl doors stand vertically, one attached to one trawl 'wing' and the other to the other 'wing'. Thus they lie at either side of the trawl mouth.

The boards are attached to the wings of the trawl via a swivel arrangement, comprising the 'butterfly' and 'dan-leno', and by the 'sweep' or 'bridle'. In turn, the doors are connected to the towing fishing vessel by the towing wires or 'warps'.

The bottom trawl is used for fishing all species of demersal fish like cod, haddock, etc.

The (Danish) seine. In bottom trawling, the essence of the catching system is to chase and overtake the fish with the net so that they are forced into the cod-end. The seine-net, however, is an encircling device which is designed to trap the fish prior to forcing them into the net. The net has a similar shape to that of the trawl net but is much larger and lighter and is fitted with longer wings. In fishing the gear is 'shot' by first attaching one end of the very long warps to a buoy. The buoy is placed over the side and the vessel then steams in a large circle paying out the warp as she goes. After travelling 180° the net itself goes into the water and the vessel continues steaming round the remaining 180° until she reaches the buoy. When this is reached the buoy is taken back on board and the warps are secured for hauling. The gear now encircles an area of the seabed with the net mid-way between the warps and directly behind the vessel.

The vessel now hauls in the warps which has the effect of rapidly reducing the encircled area of the sea bed. The fish within the circle are herded into the path of the net by their attempts to escape the encroaching vibration of the warps as they are hauled. When the warps are as close together as possible and all the fish have been herded into the path of the net, the net itself is hauled in, thus forcing the fish into its cod-end.

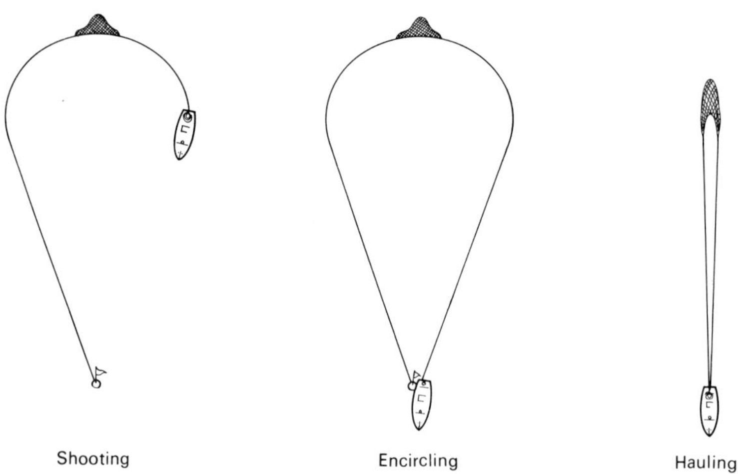

Fig. 3.51   Danish seining.

The beam trawl. This gear preceded the bottom trawl which was developed from it. In order to keep the mouth of the net open before boards were invented, a stout beam was used in place of the headline (rope from which the mouth of the net is suspended). Because of the beam, this gear was difficult and awkward to handle and this in turn placed a limitation on the size of net which could be used since a beam of too great a length would have been dangerous to use.

Fig. 3.52   Double-beam trawler.

The beam trawl is, however, still used for particular kinds of demersal fishing – shrimp fishing and flatfish fishing. The beam trawl is particularly advantageous in this kind of fishery because it digs into the bottom and forces those fish and shellfish which bury themselves in the bottom to swim up and into the net. Some beam trawls have a series of 'tickler' chains rigged between the bridles forward of the net mouth to do this more effectively.

Vessels employing this gear are often double-rigged, i.e. they carry one net one each side of the vessel and fish both simultaneously.

Long lines. Finally, some of the demersal catch is obtained from long lines. There is a variety of long-line arrangements and methods but basically several hooks (often hundreds) are connected to a single main line by shorter lines known as snoods. The hooks are baited and the line is shot by letting go the weighted end, the other end being attached to a surface buoy. The line is kept at the required depth by weights attached to the main line by weight lines of predetermined depth. The ship-board end of the main line is either secured on board or released attached to a marker buoy so that it may be easily picked up when fishing is done.

This gear can of course be rigged to fish pelagically, the line being held horizontally on the surface by floats with the snoods and hooks hanging vertically. This is in fact a popular method of poaching salmon, but its use is largely restricted to rivers, lakes etc.

a. Pelagic fishing gears

The mid-water trawl. A fish on the bottom can move in four basic directions, i.e. forwards, backwards, upwards or sideways. To escape the net, therefore, it must either swim faster than the net is moving, avoid the net square, or avoid the warps and wings. obviously if it swims backwards it will swim into the net. The bottom trawl, therefore fairly effectively blocks all avenues of escape. A pelagic fish, however, has an additional avenue of escape, it can swim down below the net. Because the pelagic fish has this additional freedom, mid-water trawling has to be supported by a sophisticated fish location and tracking system to enable the catchers to make continual adjustments to the depth and direction of the net so as to aim it at the escaping fish school.

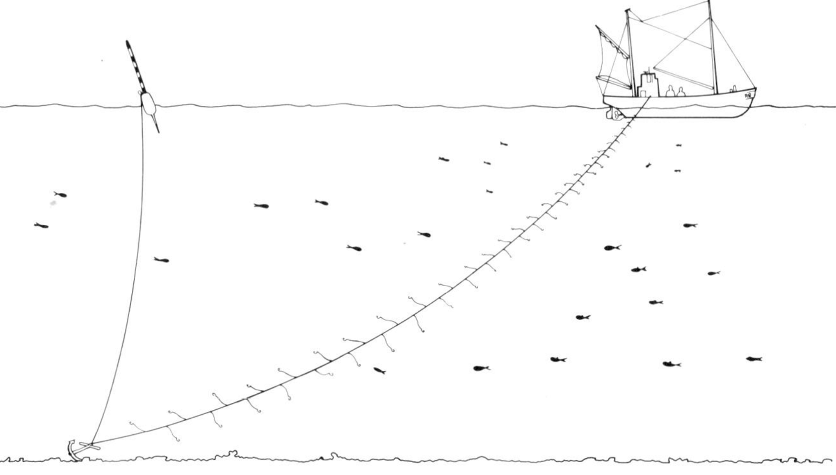

Fig. 3.53   Long lining.

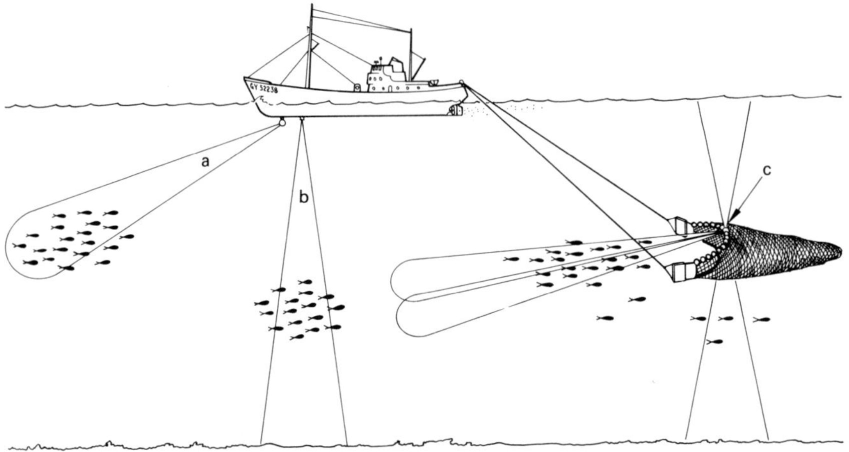

Fig. 3.54 One-boat mid-water trawl. Aimed trawling system. *a*: ahead scanning sonar to locate fish school and determine depth, speed and direction. *b*: echo sounder to calculate depth of bottom and depth and direction of the fish school located by *a*. *c*: net sounder which indicates depth of net from surface and/or height of net from bottom as well as position of net in relation to the advancing fish school.

The net itself is quite unlike the bottom trawl net and is of a much more conical shape. There are two main types of mid-water trawl: conical shaped nets made up of two similar panels, one above and one below, and fitted with small wings on each side; and box-shaped nets made up of four panels: top, bottom and sides.

The drift net. This is a wall of netting which is suspended by floats along the headline and vertically in the water by weights attached to the 'footrope' at the bottom. These nets can be several miles long from end to end and a number can be set by a single vessel.

Such nets are of two main types – gill nets and trammel nets.

In the gill net, the size of the mesh is so calculated as to permit the fish being hunted to pass its head but not its body through the net. On trying to move back out of the net, the fish catches its gills in the mesh and is thus held fast. This type of net can, therefore, be highly selective and has particular value in 'managed' fisheries like rivers and lakes where regular culling takes place. In northern Europe, it was until recently the principal gear used to fish herring, but the reduced stocks of this species which now obtain make this particular gear uneconomic in the majority of cases and it has now been largely superseded by the mid-water trawl and purse-seine.

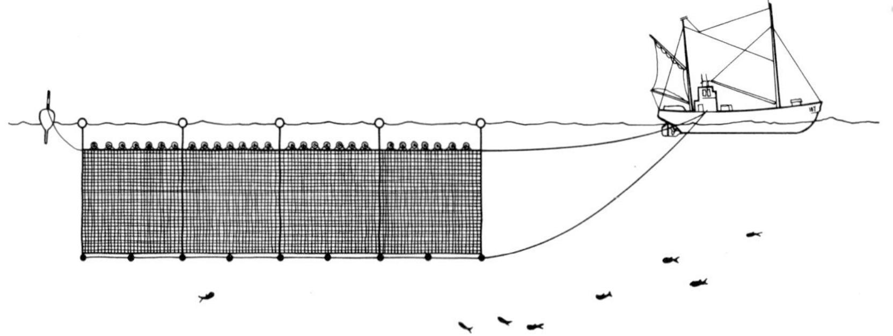

Fig. 3.55 Drift net.

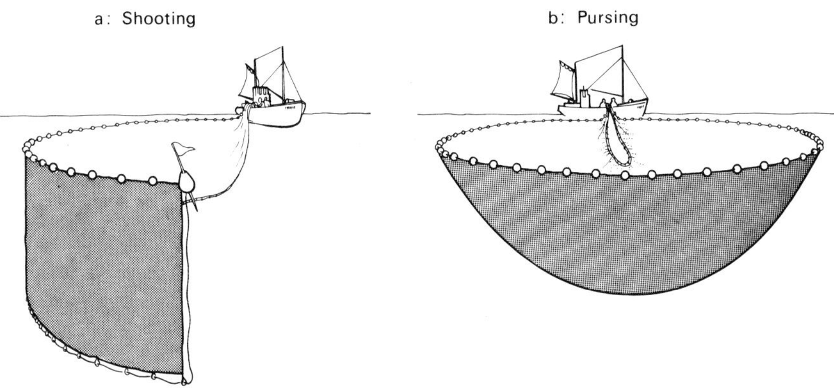

Fig. 3.56 Purse-seining.

Trammel nets comprise two walls of netting hanging side by side. The first has a large mesh which allows the fish to pass through unhindered, but the second has a fine mesh too small for it to negotiate. In trying to escape from between the two walls the fish become enmeshed. Obviously this type of net will only operate on one side and it is principally used in rivers and river estuaries to catch migratory fish whose direction of migration is known.

Both these types of net can of course be 'bottom-set' to act as demersal gears and can also be made stationary devices by anchoring them to the bottom or securing them to river banks, etc.

The purse-seine. Purse-seining is a particularly effective fish-catching method which combines the best features of the drift net and Danish seine. The gear itself is a large wall of netting which is shot in the same circular fashion as the Danish seine, resulting in the placement of a large escape-proof ring of netting around the fish school. When both ends of the net have been taken on board, the purse-line (foot-rope) is hauled in; this has the effect of drawing the bottom edges of the net together to form a large bag or purse. The fish within it now have no means of escape. The net is now hauled in in such a fashion as to reduce both the diameter and depth of the net simultaneously. When the size of the net has been reduced to its minimum in the water, the fish are removed from it by a brailer (wicker basket) or submersible fish pump and the net itself hauled on board when the last fish has been removed.

New fishing methods. In recent years a number of completely new non-net fishing techniques have been introduced which involve underwater electric lights to attract the fish, electric currents to narcotize them and submersible pumps to catch them and bring them on board. Of these approaches the combination of light and pump is the one which has found greatest commercial success and acceptance so far, and a very large part, for instance, of the sprat fishery in the Caspian Sea is caught in this fashion.

b. The vessels. The vessels used to catch and even to process fish at sea are as diverse as the gears they carry and range from small open boats to vessels like the huge 225-metre (738 ft) long, 43 400-ton fishing-fleet mother ship *Vostok* which carries its own fleet of 14 trawler-seiner fishing boats, each of which has a displacement of 69 tons and a length of 17 metres. Today's world fishing fleet comprises a total of 15 741 vessels of all kinds over 100 gross registered tons (see tables 3.6 and 3.7). But, of course, there are a multitude of highly-productive vessels of much less than 100 g.r.t.

Table 3.6 Number and size of world fishing vessel fleet (vessels over 100 g.r.t.)

			Gross registered tonnage			
Year	100–499	500–999	1000–1999	2000–3999	4000 + over	Total
1970	10135	1279	393	583	7	12397
1971	11144	1605	417	706	9	13881
1972	12182	1701	443	762	10	15098

Source: *Statistical Tables*, 1970, 1971, 1972, Lloyds Register of Shipping.

Table 3.7 Number and size of world fleet of fish carriers and fish factory fleet. (vessels over 100 g.r.t.)

			Gross registered tonnage			
Year	100–1999	2–3999	4000–5999	6000–9999	10000 + over	Total
1970	184	89	77	50	92	492
1971	211	118	91	55	112	587
1972	251	123	98	54	117	643

Source: *Statistical Tables*, 1970, 1971, 1972, Lloyds Register of Shipping.

Unlike other vessels, fishing craft are not only a means of transporting cargoes, but also a means of producing that cargo, and, furthermore, must produce it under the most adverse conditions, i.e. at sea. In order to operate effectively, therefore, fishing vessels must not only be equipped to navigate with safety but must also be built to operate for extended periods of time at sea. They must be capable of carrying or producing on board not only the essential consumable stores like provisions, oil, water, etc. which are necessary for the comfort of the crew and the operation of the vessel, but also the gear needed to catch fish and the consumables like ice and packaging materials necessary to preserve the catch.

In addition to all the above, the fishing vessel must also be capable of locating and maintaining contact with the fish she intends to catch. As a result all but the most primitive fishing boats are highly sophisticated and expensive vessels built to the very highest standards of specialized ship design and carrying on board several thousand pounds worth of the latest electronic navigational, communications and fishfinding equipment.

At this point it may be worth noting once again that fishing is a hunting industry. There is no certainty for any skipper when he sets out that he will in fact find sufficient fish of the right species and quality even to pay for his voyage, and even if his trip does in fact produce a large catch its value may be reduced by glut conditions at the market. It is only when one considers fishing in this light that it is possible to appreciate the true size and risk of the investment which every modern fishing vessel, of whatever size, represents.

Generally speaking, fishing vessels can be sub-divided into two basic groups; those which fish over the stern and those which fish over the side. The advantages and disadvantages of stern and side fishing are, even today, a matter of often heated debate between fishermen, but as a general observation it can be stated that a stern fishing configuration is particularly advantageous for dragged (trawled) gears, whereas side fishing is advantageous in the use of encircling and drift nets, and in pole and line

Fig. 3.57   Stern trawler (right) and two side trawlers in harbour.

fishing. In practice virtually all vessels today are equipped to operate a number of different gears and are capable of fishing with trawls and purse-seines.

In addition to their primary function of catching the fish, many vessels, like the freezer and freezer factory trawlers, are equipped to freeze the catch on board and even to process it into fillets, and other products, while other vessels do not fish at all, but are floating factories designed to take the catch of 'feeder' trawlers and to process it at sea.

3.9.3 Composition of fish

The body of a fish is made up of three basic constituents: water, which makes up from 56 to 79% of the total fish weight; protein, which accounts for 16 to 20%; and fat, which makes up 2 to 22%. In addition to these there are small amounts of minerals, sugars, amino acids and vitamins which together are known as extractives.

It can be seen, therefore, that the protein and extractive content are fairly stable whereas the fat and water content vary considerably. In fact, there is a definite relationship between the fat and water content: the greater the fat content, the smaller the water content and vice versa. However, their combined content is fairly constant at around 77 to 80% of the total weight. In any event, water always accounts for more than half the weight of the fish.

This variability in the proportions of the principal constituents is due not only to the differences between species but also to factors like sex, age, habitat and season of fish of the same species.

Generally speaking, as a fish grows older the proportion of water in its body decreases and fat increases. Sexual maturity also brings about compositional changes and the gonads may account for 25 to 30% of the total weight of the fish. Habitat obviously affects the composition of the fish for it is from its habitat that it gains its sustenance; i.e. where food is abundant the fish will grow more rapidly and lay down more fat.

Seasonal variations in the chemical composition of an adult fish can be considerable, since it undergoes a regular pattern of change throughout the year and it is this seasonal variation which determines the most propitious time for fishing. This pattern of change can be broken down into two basic periods: the bodily preparation for spawning and the recovery from spawning.

In the period preceding spawning there is a gradual redistribution of protein and fat in the body which is necessary for the development of the gonads. As spawning time approaches the fish tends to eat less as the gonads grow larger. At this time, however, the fish is often expending more energy than usual in migrating to the spawning grounds. As a result, fat content decreases to a very low level. Finally the spawning process itself uses up the fish's last reserves of energy and leaves it exhausted.

In the post-spawning period, the fish starts feeding intensively to replace the reserves lost and starts laying down fat again.

The commercial suitability of any particular fish species can be determined on the basis of its average fat content and fish generally can be divided into three groups for this purpose. Fish with a fat content of 4% or less are termed lean (cod, haddock, plaice and other demersal species), those with a fat content of between 4 and 8% are termed medium fatty (e.g. carp), and those with a fat content exceeding 8% (salmon, herring) are termed fatty.

These are, of course, only very general indications since the seasonal fluctuation of fat content and the effects of habitat may well increase or reduce fat content outside the limits given above.

The distribution of fat in the body of a fish depends upon whether or not it belongs to either the lean or fatty species. In the lean fish (cod, dogfish, skate, etc.) fat is laid down principally in the liver and other internal organs, or in the abdominal cavity in the form of fatty deposits around the viscera. In the fatty fish (herring and salmon), fat is laid down principally in the flesh either as fatty tissue between the myosepta and muscle fibres or in the subcutaneous layer.

The basic body parts are the head, skin, bones and fins, scales, flesh, roe, swim bladder, liver and digestive organs. The chemical composition of these various parts covers a wide range, which means that they can be utilized to produce quite different products.

The head of the fish is composed principally of protein, calcium phosphate and fat and is used to produce fish meal and oil which is consumed in animal feeds. Bones and fins, which comprise calcium phosphate and nitrogenous material, are also suitable for the production of fish meal. The skin, scales and swim bladder contain collagen which is used in the manufacture of glue. The flesh and roe comprise protein, fat and extractives and, of course, are principally used for human products. The liver contains nitrogenous material, oil, and vitamins A, D and B_{12} and is used for the production of medicines and as human and animal food. Finally the digestive organs containing nitrogenous material, fat and enzymes can be used for the production of fish meal and some technical products.

The size of the various parts relative to the overall size of the fish differs widely from species to species and determines, to a large extent, the way or ways in which the fish as a whole will be utilized and processed (tables 3.8 and 3.9). The overall size of the fish must be considered too, and the average size in commercial catches of some fish is given in table 3.10.

Table 3.8 Weight composition of some fish

Species	Weight (kg)	Percentage of weight of whole fish					
		Flesh	Head	Viscera	Bones	Fins	Scales
Bream	1.66	44.5	13.7	21.8	11.7	3.4	5.6
Carp	2.09	50.6	17.1	7.3	12.3	3.1	6.1

Table 3.9 Ratio of body parts of some fish

Species	Percentage of overall length		
	Body	Head	Trunk
Cod	90	26	57
Mackerel	88	23	61
Herring	89	19	66
Anchovy	89	22	63

Table 3.10 Commercial size of some fish

Species	Length, cm	Weight, g
Cod	35–80	400–5000
Haddock	30–60	300–2500
Herring	22–35	100–400
Redfish	25–65	200–3500

3.9.4 Handling

Fish are highly perishable and must be treated and handled with the greatest possible care from the time of death to their presentation as food. As with all living things, fish are subject to two principal kinds of spoilage after death – autolysis or 'self-digestion', and bacterial attack.

Autolysis is caused by the action of enzymes in the tissue of the fish. During life enzymes are responsible for the breaking down and conversion of the compounds like sugar and amino acids which are produced from digestion. However, since enzymes are not themselves 'alive', their activity continues after death and as the flow of food and energy has ceased with death they start to break down the protein and fat of the body. Most of the harmful bacteria in the fish are present on the skin and in the gut. Whilst the fish is alive its normal body defences prevent the bacterial invasion of the flesh. After death, however, the natural defence mechanisms collapse and invasion begins.

In addition to spoilage, two other important events take place in fish after death. The first is hyperaemia and the second rigor mortis.

Hyperaemia is a reaction of the dying fish to unfavourable surroundings in which mucus is released to the surface of the skin from glands within it. The mucus thus released is sometimes so heavy as to produce a thick coating of slime on the body which can account for as much as 2 to 2.5% of the total body weight. This slime is principally composed of the glucoprotein, mucin, which is an ideal habitat for bacteria. Because of this it begins to spoil and give off mal-odours very quickly. Hyperaemia does not mean that the fish is spoiling but, unless removed quickly, it will provide an ideal environment for bacterial invasion of the body.

Rigor mortis, or the stiffening of the muscles some time after death is due to a contraction of the muscles caused by the autolytic breakdown of the organic compounds in them. After some time has elapsed, the muscles relax and the fish becomes pliable again. The onset and duration of rigor mortis depends on many factors–the species of fish, its conditions at death (spent, exhausted), its manner of death (usually asphyxiation) and the temperature and conditions of storage. Knowing when rigor mortis is likely to set in is extremely important when freezing fish at sea.

The effects of fish spoilage due to autolysis and bacterial decay are similar and affect the flavour, texture and appearance of the fish.

It is autolysis which begins first and its gradual breakdown of the stomach wall and tissues assist the bacteria to penetrate the body. Within the fish the bacteria find ideal conditions for growth and multiplication. The products of autolysis and their own action provides them with the food and energy needed for rapid development which if not checked will completely digest the muscle within a few days.

Fish must, therefore, be handled in such a way as to control and reduce the rate of both autolytic and bacterial spoilage.

The first way of achieving this is the removal of those parts of the fish containing the greatest quantities of bacteria, i.e. the gut and other internal organs and the surface slime. The former is effected by 'gutting' the fish either manually or by machine. This entails slitting open the belly cavity from below the back of the head to the ventral opening and removing all internal organs. After this operation the fish is thoroughly washed to remove all the surface slime and any remains of the internal organs in the belly cavity. Most of the blood in the fish is also removed during these operations.

Throughout the whole handling process great care must be taken not to bruise or damage the skin of the fish not only because it may cause an unsightly blemish but also because it may also afford a possible entry for bacteria.

Enzyme action (autolysis) can be inactivated by irradiation, but the most practical control method available at present is the use of a low storage temperature since, generally speaking, the lower the temperature of the dead body, the slower the rate of autolysis. It must be remembered, however, that this does not inactivate enzyme activity and that autolysis still continues, albeit at a very low rate, at $-30°C$.

Autolysis is not directly proportional to temperature. Enzymes function most effectively at the normal body temperature of the fish which for most species in the North Atlantic, for example, is from $2°$ to $10°C$. Above this temperature, the enzymes become progressively more unstable and can be inactivated by heating the fish up quickly to a temperature above $40°C$ for a short time. This, however, would cause partial cooking which is undesirable. As the temperature falls to $0°C$ enzyme activity begins to fall, but from $0°C$ to $-4°C$ an increase is often observed which is probably due to the water in the fish freezing out as pure water and leaving enzymes and reacting substances in unnaturally high concentrations. The semi-frozen state must therefore be avoided as too must slow thawing and freezing. If fish are to be held in ice, it should be in melting ice which will ensure a temperature of exactly $0°C$. If frozen the fish should be kept at as low a temperature as possible and certainly at no higher than $-20°C$. For the rest of this section, we shall concentrate on the handling of fish stored in ice, as freezing is the subject of a later section.

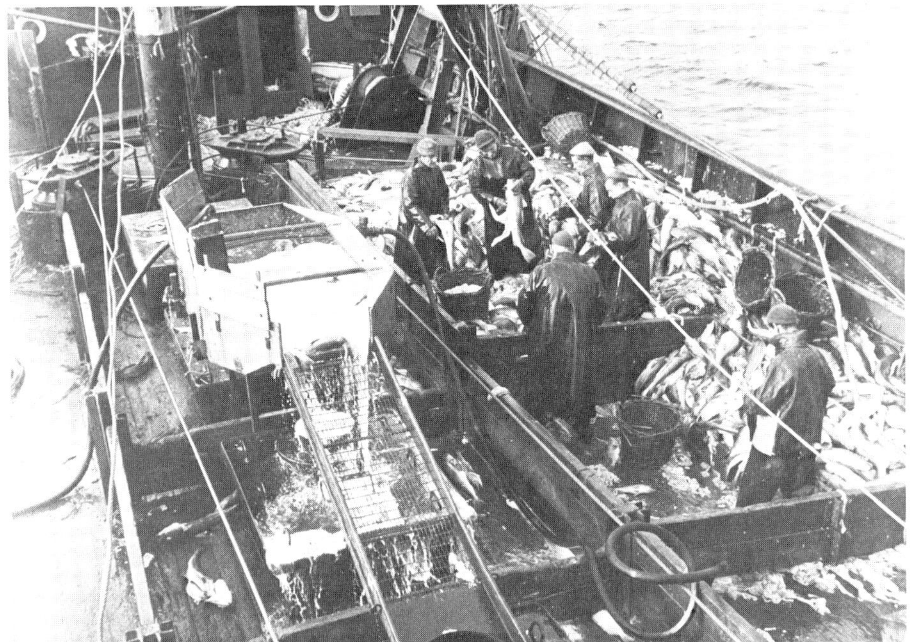

Fig. 3.58 Gutting and washing the catch preparatory to stowage in ice.

Once gutted, cleaned and washed the fish must be either landed immediately or stored at a low temperature, usually in ice. Fresh ice is generally loaded on board the fishing vessel just prior to its departure to the fishing grounds and is carried in the fish hold or fish room. When fishing begins, boards which have been scrubbed clean are arranged in the hold to form one of several methods of stowing the fish.

Bulk stowage is probably the oldest method and is still widely used. The boards are arranged to form 'pounds' (rectangular box structures). The interior bottom is covered with ice arranged so that ice is in contact with every surface and so that the surface of one fish does not come into contact with the surface of another or with the pound boards and their supports. The fish and ice are stacked in this fashion to a depth of around 457 mm (18 in) and then covered by boards which are supported not by the fish but by the structure of the pound. These boards serve as the bottom of the next section of the pounds and stowage continues as before.

Shelfing is an improvement on bulking. The same pound arrangement is employed but in this case only a single layer of fish are laid belly down and head to tail on a bed of crushed ice. No ice is placed on top of the fish, but an air space is left before the next shelf is added.

Boxing is the most recent improvement in iced fish stowage at sea, and in this case the fish and ice are packed directly into aluminium or plastic boxes and stored in the fish room. The advantages of this system are that these fish do not have to be handled again before reaching the shore processors and, of course, boxes, if of a standard size, lend themselves to mechanical handling in unloading and transportation.

The storage life of well-packed iced fish intended for human consumption is around 10 days from the time of death.

Unfortunately it is not always possible to gut the catch before landing. This is usually the case with herring which are caught in very large numbers, too great to be adequately handled on board. So too, there are some fish like redfish (*Sebastes*) which are difficult to handle on board because of their spiny fins and gill covers. In all these cases the very greatest care must be taken to ensure that the temperature of the fish is reduced as quickly as possible and that they are delivered as soon as conditions permit.

Either on board (in the case of a factory trawler) or ashore, the fish will usually be subject to further processing before reaching the consumer. Such processings may include one or more of the following: filleting, heading, skinning.

Filleting is the removal of the muscle from the back-bone of gutted fish so that the fillets are almost entirely free of bones. They are produced either skinned or unskinned depending on the species and condition of the fish.

Heading, as its name implies, is the removal of the head of the fish.

Skinning depends very much on the product and the condition of the fish. However elasmobranchei are almost always skinned because of the very rough nature of their skins.

All of the above processes can be carried out by machines as well as manually. Whether machines are used or not, hygiene is of critical importance and the production line must be so organized as to reduce processing time, temperature variation and handling to a minimum.

3.9.5 Canned fish

Of the world's total production of fish and fish products, canning accounts for some 6.3 m. tonnes, i.e. 9.1%. The principal species canned are pelagic, e.g. tuna, herring, sardine, anchovy, salmon, mullet, and mackerel, although some demersal species like cod and plaice are also processed in this manner. Crustacea (shrimps, crabs, crayfish, etc.) and mollusca (mussels, oyster, squid and scallops) are also canned.

Fresh, chilled (i.e. stored in ice) or frozen fish may be used for canning, but it must be of high quality because the canning process alters the nature of the raw material to some extent, and if autolysis has progressed too far, this will result in a poor quality pack.

Before the canning process itself is begun, the fish must be prepared by one or all of the following operations: washing, grading, scaling, boning, gutting, salting, heading, tail and fin removal, bleeding, and portioning. In the smaller fish (smelt, sprats, sardines, and some herring), the head and gut are removed in a single operation which does not entail slitting the belly. This process is commonly known as 'nobbing'.

In addition, canned fish products must contain 1 to 2% of salt which may be introduced directly into the fish by soaking in a concentrated brine solution, by adding measured dry quantities to the tins, or by adding salt to the liquid in which the fish will be sealed in the can.

The size and shape of the can employed depends basically on the market requirements, the species of fish being processed, and the type of product made. The processes involved in canning, including the preliminary preparation of the raw material, are often mechanized.

Once the fish have been prepared for canning they are packed into the can to a predetermined weight. The can is then filled with a measured amount of liquid or sauce (usually oil or tomato sauce) specially selected for that particular product. The sauce can, in fact, be added before or after packing and is sometimes added in two portions; the first before packing and the second afterwards.

After packing, the lids are fitted to the can in an operation termed clinching. The lids, though tightly fitted, do allow gases to escape from the interior of the can. The can is then moved on to the exhausting process, which creates a partial vacuum in the can. This is done in several ways. In heat-exhausting the can is passed, by conveyor, through an exhaust box which has an internal temperature of around 96°C maintained by steam sprays. The can remains within the box for a predetermined time long enough to exhaust the greater part of the gases within it. The lid is seamed immediately the can leaves the box. Vacuum-seaming machines are used instead of the exhaust box in some process lines. These machines withdraw the air from the can and seam the lids in a single operation. A variation of the exhaust box method of spraying the outside of the can with steam, is to inject the steam directly into the can itself. Additionally, in some canned products the cans are filled with pre-cooked fish or hot sauce.

The creation of a vacuum is necessary for two reasons. First, it reduces the increase in internal pressure when the can is subjected to heat processing. Secondly, a vacuum ensures that the ends of the can remain concave. This is important because if there were no vacuum, the expansion of gases in the can which would occur during heat processing would tend to blow out the ends of the can, making them convex. There are a number of reasons why this should be avoided: the blowing-out and resultant convex form would place the seam under stress so that it may well leak; cans with rounded ends would be difficult to stack and store; and, finally, if all cans are produced with concave ends a 'blown' (convex-ended) can is a good indicator of bacterial spoilage. If such spoilage should take place within the can after sealing, the bacterial action produces gases (e.g. hydrogen or carbon dioxide) which would force the ends of the can outwards.

After seaming, the cans are thoroughly washed in hot water to remove all traces of protein and fat which might discolour the tinplate. Following the washing stage, the cans are packed into a basket 'pack' and placed in a retort for cooking and sterilization.

As has been noted earlier, enzymes are inactivated at relatively low temperatures, but many bacteria (thermophiles) are resistant to high temperatures for varying lengths of time. One of the most dangerous of these organisms is the spore-producing bacterium *Clostridium botulinum* which can live at a temperature of 115°C for a considerable time. Many other thermophiles can withstand temperatures of 100°C for periods of up to 24 hours. The temperature and duration of the cooking-cum-sterilization stage must be sufficient, therefore, to ensure that the centre of the centre can in the pack (i.e. the point furthest away from the sources of heat) is held at at least 100°C for at least 24 hours. Sterilizing temperatures are usually in the region of 110° to 120°C and the duration of treatment depends on the size of pack being treated.

After sterilization the cans are washed in an alkaline solution at around 70° to 80°C to remove all foreign matter from the external surfaces and then cooled either within the retort (which is preferable as there is less risk of damage to the extremely hot can and its contents) or by cold air or water, to a temperature of about 30° to 40°C. When the cans are dry they are lacquered to prevent corrosion, labelled and packed.

Properly processed canned fish food products are a high-grade food of equal and sometimes of superior quality to similar non-canned products. The chemical composition and energy value of canned fish obviously varies widely from product to product. Where the fish has been cooked only by the heat processing operation and canned in its own liquor, the energy value depends upon the chemical composition of the flesh, but in delicatessen products which have been subjected to pre-cooking, the energy content

Table 3.11 Amino acid content of canned fish

Amino acids	Content (% of total nitrogen)	
	Canned fish	Canned beef
Arginine	5.0–8.9	6.2–8.5
Cystine	1.1–1.8	1.4–1.5
Histidine	1.6–1.8	1.7–2.3
Lysine	5.7–8.0	6.5–9.7
Methionine	1.7–3.1	2.5–2.9
Tyrosine	3.2–4.9	3.9–4.3
Tryptophane	0.8–1.1	1.1–1.7

depends on the kind of pre-cooking (scalding, frying, smoking) and the type and quantity of liquid used (i.e. oil, tomato sauce).

Table 3.11 compares the amino acid content of canned fish with canned beef; it will be seen that the protein content is similar.

3.9.6 Smoking

Smoking was originally carried out to extend the storage life of the fish. Today, however, this process is applied principally because of the pleasant and distinctive taste which smoking imparts to the fish.

There are two basic smoking processes – cold smoking and hot smoking. In the former the fish is smoked without cooking the flesh which means that the smoke temperature must not exceed around 29°C. In hot smoking, however, the intention is to cook the fish and smoke temperatures are consequently much higher, around 120°C. Generally speaking, the hot-smoked product does not store well and is not suitable for long-haul transportation, whereas cold-smoked fish form a more stable food product.

Fresh, frozen or chilled fish may be smoked and the preparation of the fish for processing is similar for both hot and cold smoking.

The fish must first be thawed (in the case of frozen fish), dressed and cleaned. The extent of the dressing operation will depend on the product. For instance, kippers, which are prepared from herring, are first washed to remove any loose scales or ice particles, and are then split along the backbone and the gut and roe removed. In the case of 'Finnon haddock', the fish is headed and the body split along the belly for the removal of the internal organs. After splitting the fish are thoroughly washed to remove traces of the gut, loose scales, gills etc. and are then soaked in a brine bath until the flesh of the fish has a salt content of around 2%. It is at this time that any permitted artificial dyes are introduced into the fish by adding them to the brine.

When the fish have been salted they are threaded onto rods for drying. This process, like the smoking stage which follows it, requires the fish to be hung on rods. The threading and hanging process is known as tentering and is the origin of the phrase 'to be on tenter hooks'.

The fish may be dried either inside or outside the kiln. As the suspended fish drip dry, protein dissolves into the brine and produces a sticky solution which dries on the cut surfaces of the fish to produce a smooth gloss which enhances the appearance of the product. When the fish are completely dry and the gloss evenly formed, they are ready for smoking.

Smoking is carried out in a kiln and the characteristic flavour and colour is conferred on the fish by the smoke of the fuel. Obviously the latter must be very carefully selected since it must not impart any unpleasant odour or taste to the fish. Experience has shown that wood, in the form of shavings or sawdust, is the only suitable fuel. The type of wood used is also important as not all woods are equally good. Those most often used are alder, beech, birch, oak and poplar.

Wood contains a large number of substances, some of which are combustible and some (moisture and ash) which are not. The former, which go to make up the smoke, include cellulose, lignin, pentosans, tannic acids, protein substances, resins and terpenes; water vapour is also present in the smoke.

The smoke produced by burning the wood is a typical aerosol, i.e. droplets dispersed in a vapour. Many of the organic substances produced by the combustion are present in varying proportions in both the vapour and the droplets, but those present in the vapour evaporate at lower temperatures than those in the droplets. In practice it is the chemicals from the vapour which are principally absorbed by the fish, the droplets playing only a minor role.

Combustion of the fuel is affected by a number of factors like the structure and thickness of the wood itself, the available air flow or draught, depth of the ash layer etc., and the kiln must be designed and operated so as to ensure the complete combustion of the fuel and the proper dispersion of the smoke around the fish.

There are three basic types of kiln: the traditional, the mechanical and the electrostatic.

The traditional kiln is basically a chimney in which the tentered fish are hung over a fire of smouldering sawdust. With such kilns the quality of the product depends absolutely on the skill and experience of the smoker, since there is very little control over the many factors affecting the combustion of the fuel and the actual smoking of the fish. A variety of mechanical kilns exist, and one of the most efficient is that developed by the Torry Research Institute in Scotland. In this machine the fires are laid in special hearths outside the kiln and the smoke, mixed with fresh air, is led into the interior of the kiln through ducts. The humidity of the smoke and the internal temperature of the kiln can be controlled; the former by the use of electric or steam heaters and the latter by increasing or reducing the amount of fresh air introduced into the smoke. The electrostatic process involves the removal of the droplets from the smoke by electrostatic precipitation, leaving the vapours alone to colour and taste the fish. The advantage of this system is that the smoking process itself is speeded up.

The time taken to smoke the fish properly depends on the species being processed, its fattiness and size, and whether or not the product is being cold or hot-smoked. Generally speaking herrings of up to 250 g (8 oz) weight can take up to around 36 hours whereas the same fish weighing around 1 kg (2 lb) can take some 44 hours smoking time. The colour, taste and odour of the finished product depends almost completely on the amount of phenols absorbed by the fish from the smoke.

After the fish have been properly smoked they have only to be cooled, removed from their tenterhooks and packed for distribution.

Another smoking method does exist and this is the use of smoking liquids which are derived from the distillation of wood. Earlier attempts at producing such liquid dips were not particularly successful because the flavour obtained differed markedly from that produced by natural smoking. However, the discovery that it was the smoke vapours rather than the droplets which gave smoked fish its characteristic properties provided the solution to the problem of producing an effective smoking liquid. In practice, however, smoking liquids are rarely used alone but, when used, are added to the fish at the brining stage.

A wide variety of fish and shellfish are smoked and include herring, haddock, cod, whiting, salmon, roe, trout, sprats, eels, and oysters. The fish may be smoked headed or heads-on, split or round or filleted. Handled properly, smoked fish will remain in good condition from around two to ten days, depending on whether it is hot or cold smoked.

3.9.7 Dried fish

Drying fish by deliberately exposing them to the dehydrating effects of sun and wind or by hanging them over open fires is probably the oldest method of preserving this highly perishable food. The objective of dehydration is, of course, to reduce the water content of the raw material to a point where it can no longer support the growth of spoilage micro-organisms. Of these organisms, bacteria and yeasts cease to multiply in the fish when the water content is less than 25% of the total body weight and moulds when it is less than 15%.

The time taken for this process to complete is highly variable and depends not only upon the size and thickness of the raw material and the geographical location of the processing, but also upon local variations in weather. Because local weather conditions are subject to changes which man cannot control, the finished product cannot always be prepared to a fixed quality standard. Moreover the process itself is a slow one taking up to several weeks and suffers to a greater or lesser extent from the uncertain behaviour of weather.

In many dried fish products it is common practice to salt the fish prior to the drying process. The reason for this is that although the drying process halts the development of bacteria and moulds, it has little or no effect on their spores which can continue to exist for decades. If, therefore, a dried product is allowed to absorb moisture after drying has been completed, the micro-organisms and particularly the fungi, which are less sensitive to moisture deficiency, will soon begin to develop. Increasing the concentration of salt in the fish by this means increases the osmotic pressure of the cellular fluid and thus disturbs the normal osmotic exchange between the cells of spoilage micro-organisms and their medium, i.e. the cellular fluid of the fish. The result is that moisture is transferred from the micro-organism protoplasm to the medium causing the micro-organism either to die or to pass into a state of anabiosis (lack of metabolic activity). If, therefore, the fish is salted prior to drying, the concentration of salt in its cellular fluid is significantly increased and extends the edible life of the product.

Natural drying produces a dessicated product which because of the complicated biochemical protein changes brought about by the process, has lost its raw flavour and has become permeated with fat. As such it is a highly nutritious food which can be eaten without cooking. It is, however, tough to eat and has a strong characteristic flavour which requires an acquired taste on the part of the potential consumer.

Two of the most common dried fish products are 'stockfish' (dried unsalted cod) and 'klipfish' (dried salted cod). The preparatory processing for these products, and indeed for all dried fish products, is the same. The fish are headed, gutted, thoroughly bled, and washed to remove all clots and extraneous matter.

To produce stockfish the cod is hung up in the open air and left for about six weeks by which time the process is complete and the moisture content reduced to the levels of acceptability mentioned earlier. Stockfish have been produced for centuries in Iceland and Norway and the product remains edible for several years.

Klipfish, on the other hand, are salted prior to drying. Usually the fish is laid flat on a bed of salt for about 12 days at a temperature of around 12°C. After salting, the fish are either exposed to the sun and wind or dried in a drier. In the first case, the salted fish are washed and then laid out in the open air, skin down, on wooden frames, known as 'flukes',

Fig. 3.59 'Klip fish' (dried salted cod) arranged on flukes for drying.

which allow the free circulation of air all around the product. From time to time during the process the fish are stacked into piles and weighted down. After some days in this condition, they are again spread out as before. The duration of the total drying process is highly variable depending as it does on the vagaries of the weather, but drying is usually completed within about 40 days.

Weather is obviously critical. If the sun is too hot or if it rains, the quality of the product is impaired and a great deal of time and effort has to be expended to protect the fish from the elements when such conditions obtain. Because of this, klipfish are often dried in a drier. The simplest driers are a plain room where the fish can be suspended over fires which both create the necessary draught and heat the air to a temperature of around 30°C. This approach, however, is still very slow and does not provide for enough control over the factors like air speed, temperature and humidity, which influence the effectiveness of the drying process and the quality of the dried fish.

Mechanical drying is usually carried out in one of two basic ways: air-drying and vacuum-freeze drying. In air drying, the fish are suspended in a controlled environment which provides exactly the right conditions of air speed, temperature and humidity to ensure the optimal rate of drying. The drying room itself is a ventilated space fed with warm air from a kiln (the fuel often being coke). Ideally, air speed should be around 80 m/min (250 ft/min) air temperature around 30°C, and relative humidity around 50 to 55%. If these conditions are not met, the quality of the product will be impaired.

Vacuum-freeze drying is a sublimation process in which the fish are dried while frozen. The end product is superior in many respects to air-dried fish. The fish are first frozen, then placed in a vacuum cabinet, laid between heated plates and subjected to a vacuum. The heat input is balanced against the rate of evaporative cooling to obtain the most rapid possible drying rate.

During natural and mechanical air drying, the flesh of the fish loses moisture and matures and the fat dissociates, oxidizes, becomes rancid and dries up. Such changes are obviously drastic and permanent and there is no way in which the finished product can be returned to its original state. In vacuum-freeze drying, however, the low temperature and absence of oxygen prevent oxidation. In addition, shrinkage is prevented by the fish being frozen. Because of this the vacuum-freeze dried product can be reconstituted by immersion in water.

3.9.8 Frozen fish

As noted earlier, fish will remain edible in ice for around 10 days from the time of death. This time is, therefore, critical since it limits the period during which a vessel can hunt for fish, the distance from the shore at which it can operate, and the distance over which its catch can be transported after landing.

In this respect the application of freezing technology to the particular problems of storage and transportation of fish has been the most important development in the industry this century. The successful introduction of freezing at sea has brought about fundamental changes in the design of vessels and the areas in which they fish, and in the handling, processing, distribution, and marketing of fish and the range of fish products available to the consumer.

Because of freezing, vessels can now spend several months at sea in their search for fish and can travel almost any distance from their base port to exploit new grounds or stocks. Landing can be more easily mechanized and the catch transported over any distance or be stored for extended periods of time in order to even out the seasonal peaks and valleys of marine fish production. All of this is possible without altering, to any marked degree, the appearance, flavour or other qualities of the fish on death, and this is the only method of processing in which all of the original qualities of the fish can be preserved.

Preservation of fish by freezing has a relatively long history and has been practised since the latter part of the eighteenth century on shore, and at sea since the early 1900s, though on a very small scale in both cases. It was not until the building of the British factory trawler *Fairtry I* in 1953, that freezing at sea became a practical commercial proposition capable of wide-spread application. Today, as table 3.12 shows, freezing at sea accounts for some 15% of all processed fish production.

As already noted, the major component of fish is water (some 60 to 80%). However, this water, which exists as the tissue fluid, could more accurately be described as a solution of several salts and other chemicals. Because of this, the water in the fish does not begin to freeze at $0°C$, but at around $-1°C$. As the temperature is lowered beyond this point, more and more of the water is frozen out and the remaining fluid becomes an increasingly stronger solution of salts and enzymes. As the water is frozen out it forms ice crystals within the muscle cells whose size and distribution are dependent upon the rate of freezing, i.e. the speed at which the temperature of the fish is lowered. Generally speaking, the faster the rate of freezing, the more numerous and smaller the ice crystals formed; the slower the rate of freezing, the fewer and larger the crystals formed, until there is just one large crystal in each cell. At very slow freezing rates, large crystals form between the cell walls and the walls themselves may be ruptured.

Freezing does not take place at a uniform rate because heat is only removed from the external surfaces of the fish. Heat in the centre of the fish (which is always the warmest part) has to travel outwards to the surface before it can be removed by the freezer. The fish therefore freezes from the surface inwards towards the centre. The fish cannot be considered 'frozen' until the temperature in its centre reaches at least $-5°C$ regardless of the temperature of the external surfaces. It is important to remember that although the freezing process is assumed to be completed when this temperature of $-5°C$ is reached, only about 80% of the total water content of the fish is actually frozen. In fact, around 5% of the water in a fish still remains unfrozen at a temperature of $-40°C$.

During the freezing period, when the water in the fish is converted to ice, cooling is relatively slow because the latent heat of solidification must be extracted. This period of slow cooling or thermal arrest ($0°$ to $-5°C$) is also known as the 'critical range' because during the time taken to reduce the temperature to this level the fish can suffer increased autolytic action which can cause irreversible spoilage. As we have already seen, the

Table 3.12 World catch by-product

Type or product	1966	1967	1968	1969	1970	1971
Fresh	18.3	18.0	18.1	17.7	19.3	19.6
Frozen	6.9	7.6	8.1	8.6	9.5	10.3
Cured	8.2	8.0	8.1	8.0	8.1	8.0
Canned	5.0	5.3	5.6	5.8	6.2	6.3
Reduction (conversion to fish meal etc.)	17.9	20.5	23.0	21.5	25.5	24.2
Other	1.0	1.0	1.0	1.0	1.0	1.0

freezing out of the water leaves the remaining fluid with progressively higher concentrations of enzymes and reacting substances whereas reduction of temperature to around $-4°C$ has virtually no affect on the activity of the enzyme themselves. The enzymes are therefore provided with a better environment and their activity increases. Below $-5°C$, however, enzyme activity is slowed down and the rate of autolytic spoilage reduced. Autolysis, however, cannot be completely halted as it still continues, albeit at a very slow rate, at temperatures below $-30°C$.

Freezing rate is, therefore, extremely important, as too slow a rate of cooling may well damage the cell walls of the fish and encourage the spoilage which the process is designed to retard. In addition, bacterial action, though slowed down by lower temperatures, does not cease until a temperature of around $-10°C$ is reached. A slow freezing rate will not only enable the bacteria to remain active but will also provide them with an opportunity to adjust themselves to the new low-temperature conditions, thus enabling them to remain active longer. At the same time, however, too rapid a rate of freezing, e.g. immersion in liquid air, must also be avoided since the uneven expansion caused by this particular freezing method makes the surface contract whilst the centre remains unfrozen, so that the flesh splits and cracks.

Although we have been discussing freezing using a single fish as an example, fish are normally frozen at sea in blocks measuring approximately a metre square by 10 cm thick. Ideally, quick-freezing should ensure that no part of the block of fish being frozen (which in practice means the centre part of the innermost fish) takes more than 2 hours to cool from 0 to $-5°C$.

Fig. 3.60 Glazed, frozen blocks of whole gutted fish being unloaded from a freezer trawler.

Even after the whole block has been reduced to a maximum of $-5°C$, its temperature must be further reduced in the freezer to make it suitable for cold storage and to halt bacterial action. The recommended cold storage temperature is $-30°C$ and the warmest part of the block being frozen must be reduced to $-20°C$ so as to ensure that when the temperature has become even throughout the block (several hours after cold storage has begun) its average temperature will be no higher than $-30°C$, since the external surfaces of the block will be at around the temperature of the refrigerant itself, i.e. $-35°$ to $-40°C$.

The total freezing time, therefore, is the time taken to reduce the temperature of the fish from its initial temperature to $-5°C$, plus the time taken to reduce its temperature from this level to $-20°C$. After the thermal arrest period ($0°$ to $-5°C$), the temperature falls rapidly and the total freezing time for a 10 cm-thick block is in the region of 4 hours.

Cold storage is as critically important as the freezing process itself and the greatest care must be taken to ensure that the temperature of the frozen fish is not permitted to rise at any time after freezing is completed. Any temperature rise will mean at the least an increase in autolytic action and, if high enough, will reactivate the dormant spoilage bacteria. This is possible because not all the bacteria are killed by reducing the temperature to below $-10°C$; some are merely rendered inactive and may become active again if the temperature is allowed to rise sufficiently. Of course, autolysis still continues even at the cold storage temperature of $-30°C$ but is then very slow.

In addition to slow autolytic action, frozen fish do suffer other kinds of spoilage or quality reduction in cold storage, like denaturation of proteins, oxidation of fat and dehydration.

Denaturation, like autolysis, is an unavoidable and irreversible process and causes changes in the colloidal structure and chemical composition of the protein. It is dependent on storage temperature and duration and its effects can therefore be limited by controlling these two factors. The rate of denaturation is greatest in the temperature range just below freezing (i.e. $0°$ to $-5°C$), but even at $-10°C$ the rate of denaturation is so rapid that a fish stored at this temperature would be unfit to eat after a few weeks. It is important, therefore, that the temperature of the fish is reduced to $-20°C$ and retained at that temperature as soon as possible. However, denaturation is not halted at these lower temperatures, and this together with the continuing slow process of autolysis means that fish cannot be stored indefinitely.

Oxidation is a problem particularly associated with the fatty fish like herring which, for this reason, have a more limited storage life than white fish such as cod or haddock. Because autolysis continues even at low storage temperatures, acid decomposition products form a process which is stimulated by oxidizing enzymes and the surrounding air. The result is that the fat in the fish slowly becomes rancid and the fish itself will change in flavour, become discoloured and eventually begin to smell unpleasantly. Dehydration also accelerates oxidation.

Dehydration, or loss of moisture from the surface of the fish, is caused by the difference in humidity between the air above the fish and the air in the cold store and cannot be completely avoided. If this process is allowed to continue unchecked it will not only cause the fish to lose weight, but will also impair the appearance, taste and texture of the fish in addition to accelerating the rate of denaturation and oxidation.

In order to reduce the effects of dehydration, and consequently the effects of denaturation and oxidation, the block of fish is 'glazed' immediately after removal from the freezer. This is done quite simply by immersing or coating the block in water. The water freezes immediately on contact with the fish thus forming a thin protective layer of ice. During cold storage, water will evaporate from the glaze without damaging the fish and storage life can be extended by renewing the glaze at appropriate intervals.

Because of these unavoidable quality changes in the fish owing to the conditions of cold storage, freezing cannot provide for the indefinite storage of the product. The actual storage life obtainable depends on a number of factors – the initial quality and condition of the fish; the completeness of the preparatory operations like gutting, bleeding, washing, etc.; the length of time occupied and ambient temperatures during these operations and during the feeding of the fish into the freezer from the processing area and from the freezer to the cold store; the freezing rate, degree of glazing, and the conditions in the cold store itself.

It is important to remember too, that freezing cannot improve on the initial quality of the fish at death and that from the moment of death to the time the fish is brought down to an average temperature of $-20°C$, the initial quality is being reduced by spoilage. It is essential therefore, that, in the case of a freezer trawler, ambient temperatures are kept constant and as low as comfort permits, and that all preparatory operations (gutting, etc.) are performed as quickly and as hygenically as possible.

If all of these conditions are met, the end product can be stored for several months or even years and after thawing still be virtually indistinguishable from the freshly caught fish. In fact, fish frozen at sea in this manner can be 'fresher' on thawing after several months in cold storage than fish stored in ice for only ten days.

At sea two principal freezing methods are used: contact plate freezing and air-blast freezing; of these, the former is the most common.

In contact plate freezing, the fish are frozen between pairs of metal plates through which a refrigerant is passed. The plates can be arranged either horizontally or vertically. Horizontal plate freezers are used almost exclusively on shore and usually for the freezing of packaged fish products (e.g. packs of smoked frozen fish, fish fingers, boil-in-bag fish dishes, etc.) whereas vertical plate freezers were developed specifically for the freezing of whole fish at sea.

In the vertical plate freezer, the fish are loaded (usually gutted, heads-on) head to tail one on top of the other between the plate so that when freezing is completed the fish will be welded into a single block of fish. The particular advantages of the vertical plate freezer are its compact size, ability to freeze fish into 'blocks', and rapid freezing rate. It is also an extremely robust and simple piece of equipment making it particularly suitable for the adverse conditions of an active fish vessel.

Air blast freezers occupy far more space than plate freezers and are usually employed to freeze single items, e.g. standard packs or very large fish. Such freezers are of two kinds: those in which the product remains stationary for a batch loading/freezing sequence, and those in which the product moves through the freezer for a continuous loading/freezing sequence. Freezing is effected by forcing cold air at relatively high speed over and around the product. Because of its obvious disadvantages for use at sea when compared with the vertical plate freezer, the air-blast freezer is usually found in shore processing plants, and particularly where mechanized production can benefit from a continuous loading/freezing sequence.

Other freezing methods are employed at sea and on shore to suit particular kinds of product. For example some shellfish (e.g. scampi) are frozen by sprinkling them with liquid nitrogen.

3.9.9 Manufactured products

Recent developments in food technology generally, and particularly the introduction of reliable freezing techniques, have made possible the production of a wide variety of fish products for the human consumer. The most important of these are the breaded-portion variety, which includes fish fingers (sticks) and fish 'steaks' which are marketed as easily-prepared quick-fry meals.

The production processes involved lend themselves easily to mechanization. The actual processing regime will vary from product to product, but generally speaking, it can be broken down into the following stages: preparation of the raw material; portioning; batter coating; breading; frying; packaging and freezing.

The raw material for these particular products is the 'laminated block'. This is a frozen block of fish fillets. The original fish used to produce the block must of course meet the same quality standards of frozen fish intended to be marketed directly. The fillets may be prepared from fresh, chilled or thawed frozen fish, and the laminated block itself is produced by laying the fillets one on top of the other in a horizontal plate freezer. When the block is frozen it is ready for portioning. This is done by cutting the block into the required sized portions with band saws and guillotine cutters.

After the portions have been correctly and accurately dimensioned and separated from the parent block, they are moved by conveyor to the batter coating process. This, like the breading and frying operations which follow, is usually completely automatic. The battering stage will ensure that each portion receives an even coating of the batter on all its surfaces. The batter mix itself varies from product to product and from manufacturer to manufacturer, but generally speaking it is starch-based, as this gives

Fig. 3.61 Inspection of fish fingers prior to packing.

better adhesion to fish products than does a flour-based mix. After battering the portion receives its coating of breadcrumbs in a similar fashion. Again the materials used will depend on the manufacturer and product, but generally they are designed to attain and retain an attractive colour (golden-brown) on cooking.

If the product is to be packed raw it now passes straight on to the packaging stage, if not it travels on, by conveyor, to the fryer. Usually the product remains on the conveyor and passes through a long fryer in which the oil is held at around 200°C. The speed of the conveyor and length of the fryer are so fixed as to provide a frying time, depending on the desired properties of the finished product, of around 20 to 45 seconds. As a rule frying is carried out as quickly as possible so as to reduce any possible weight loss due to evaporation of moisture from the product to a minimum.

After frying, the portions usually pass into an air-blast freezer and are machine-packed on exit from it. During the whole production process, the greatest attention is given to the quality of the individual portions and a careful check maintained on any changes in weight which occur at each stage.

After packing, the finished products ar held in cold storage until transportation. As with all frozen products the greatest care must be taken to keep any temperature rise to a minimum both during storage and during transportation and final display in retail cabinets.

3.9.10 By-products

The major by-products of fish are fish meal (5 330 000 tonnes in 1971) and fish oils and fats (1 300 000 tonnes in 1971). In addition, however, several other products like vitamins and fish glue are also produced.

a. Fish meal. Virtually any species of fish or fish offal (head, skin, scales, bones, etc.) can be used for the production of fish meal, although the greater part of the world output is produced from whole fish. A fishery based on the production of fish for reduction to meal and oil is termed an 'industrial fishery' and the fish which it catches, 'industrial fish'. This includes fish which are caught solely for their value as industrial fish (e.g. capelin, Norway Pout, and menhaden), and fish which are also used for human consumption (e.g. cod, haddock, hake, herring, anchovy, etc.).

The raw material can be divided into groups – fatty fish and lean fish. Of these two, the fatty fish provide by far the greater proportion of fish meal and it is from this group that most of the fish oil is produced.

The first stage in the process is cooking, in which the fish is heated to around 100°C in a long steam-jacketed cylinder through which the raw material is moved by a screw conveyor. This process coagulates the protein and liberates much of the oil and water, which can be screened off in a pre-strainer. The oil is then separated from the water and stored in tanks. After cooking, the fish are moved to the press which removes most of the remaining oil and water and produces a dry material known as 'press-cake'. The oil and water which are known as the 'press-liquid' are first screened by centrifuges or vibrating screens to remove the larger solid particles and are then passed to an oil/water separator. The oil produced by the separator is drawn off and, possibly after a further screening, is passed to the oil storage tanks. The water phase which remains after separation can contain as much as 20% of the total solids (mostly protein) in the original fish and is therefore valuable. This liquid phase is delivered to evaporators which produce from it a condensed, viscous liquid known as 'stickwater'. The stickwater can be sold as a separate

Fig. 3.62 Fish meal plant showing cooker before entering the press and, on the right, vibrating screens for treating the liquor before evaporation.

product (condensed fish solubles) but more often it is fed back into the press-cake in order to enrich the latter with additional protein. When the stickwater has been fed to the press-cake the new mixture (i.e. enriched press-cake) is fed to the drier.

Drying is an extremely important production stage as it plays a large part in the quality of the end product. If under drying occurs, the resultant meal is an ideal habitat for moulds and bacteria, whereas if it is over dried, its nutritional value may be impaired by scorching.

There are two principal drying methods; direct and indirect. In direct drying, very hot air (400 to 600°C) is passed over the material as it is tumbled in a rotating drum. This method is extremely rapid, with a total processing time of 10 minutes or less, but there is a greater possibility of heat damage. This system also generates rather more of the offensive odours associated with fish meal plants than does the indirect system.

Indirect drying is effected by placing the material in a steam-jacketed cylinder which is also able to tumble its contents. Because it operates at much lower temperatures, there is much less risk of scorching the meal, but the processing time is much longer, 30 minutes or more.

The final process is grinding the prepared meal to produce the required particle size, and bagging.

The adequate removal of the oil in the initial stages of the production process is extremely important and can have a decisive effect on the quality of the meal produced. Too high an oil content may give rise to oxidation which will result in the meal storing badly and in addition it will tend to build up in the tissues of the animals eating the meal, giving their flesh an unpleasant fishy taste. Properly produced white fish meal will have 60 to 65% protein, around 10% moisture, 20 to 25% ash and a fat content of less than 5%. Properly produced, fish meal is an extremely stable product which will store well at normal temperatures for long periods.

As already indicated, fish meal is used principally as an animal feedstuff, although some is also used as fertilizer. Fish meal is not fed to animals (chiefly pigs and poultry) in isolation, but is usually compounded with other feedstuffs so as to provide around 10% of the diet. It is rich in amino acids (table 3.13), though the amounts present depend to some extent on the species of fish.

As world production figures show, an increasing proportion of the world's fish catch is being utilized for fish-meal production. The reason for this is that the world's increasing meat production, particularly in pig and poultry farming, requires an abundant supply of highly nutritious feed-stuffs at an economic price.

b. Fish oils. As we have already seen, fish oils are an essential by-product of fish-meal production. Of course, most of the oil produced comes from the fatty fish like herring, anchovy, mackerel, sardine, pilchard, menhaden and capelin.

Fish oil is produced during the cooking and pressing stages of fish meal production. After draining-off and initial separation, the oil is usually subjected to further centrifuging to produce a more refined product. This process is known as 'polishing'. After polishing, further refining is necessary to remove some of the fatty acids. One process is 'winterizing' the oil in which the product is chilled to a temperature at which some of the fatty acids solidify. After separation the resulting oil will remain clear at low temperatures. As well as winterizing, the oil may also be subjected to alkali refining and washing. This further processing reduces the free fatty acids to less than 1% and removes some of the pigments from the oil, thereby improving its colour. The oil has now been refined to a point where it can be utilized either as a food-industry or other industrial product.

As a food, fish oils are used in the production of margarine and shortening. To be used for this purpose, the free fatty acid must be reduced to no more than 0.1%, and its melting point raised (i.e. the oil must be hardened). This requires the oil to undergo further processing known as catalytic hydrogenation.

The industrial uses of fish oils take advantage of either their unique type and high degree of unsaturation to produce elastic durable polymers, or of their long and diverse mixture of chain lengths to add lubricity, detergency or plasticity to the compound in which they are incorporated.

The principal applications to which the oil is put are in protective coatings – paints and varnishes – for a wide variety of materials (e.g. wood, metals, fibre and concrete). In addition, derivatives of the oils are used in lubricants, caulks and sealants, leather treatments, printing inks, insecticides, metal soaps and a number of other products.

c. Fish protein concentrate (F.P.C.). Our world faces a continuing shortfall between the production of animal protein for human consumption and the steadily rising population. This problem is at its most acute in the developing nations of Africa and Asia where, annually, millions of people, particularly children under the age of ten, either starve to death or, perhaps worse, suffer physically- and mentally-crippling protein deficiency diseases like quashiokor. There is no simple answer to this awful problem, but a partial solution has been made possible by the development of Fish Protein Concentrate.

Table 3.13   Mean value of total amino acid composition (g/16gN) of fish meals determined mainly by ion-exchange chromatography

	Herring meals	Anchovy meals	Pilchard and Maasbanker meals	Tuna (mixed species) offal meals	Menhaden meals processed by wet reduction	White fish meals
Lysine	7.73	7.75	7.94	7.30	7.56	6.90
Methionine	2.86	2.95	2.71	2.75	2.82	2.60
Cystine	0.97	0.94	0.95	0.79	0.90	0.93
Tryptophan	1.15	1.20	1.02	1.05	1.07	0.94
Histidine	2.41	2.43	3.02	3.41	2.32	2.01
Arginine	5.84	5.82	5.95	6.43	6.04	6.37
Threonine	4.26	4.31	4.38	4.34	3.97	3.85
Valine	5.41	5.29	5.41	5.31	5.10	4.47
Isoleucine	4.49	4.68	4.48	4.46	4.40	3.70
Leucine	7.50	7.62	7.30	7.20	7.14	6.48
Phenylalanine	3.91	4.21	3.91	4.10	3.95	3.29
Tyrosine	3.13	3.40	3.23	3.28	3.22	2.60
Aspartic acid	9.10	9.49	9.37	9.30	9.07	8.54
Serine	3.82	3.84	4.27	4.18	3.61	4.75
Glutamic acid	12.77	12.96	12.92	11.93	12.70	12.79
Preline	4.15	4.17	4.52	5.43	4.58	5.34
Glycine	5.97	5.62	6.92	8.15	6.78	9.92
Alanine	6.25	6.31	6.17	6.76	5.94	6.31
Crude protein %	73.6	65.4	65.4	53.24	62.01	65.01
Moisture %	6.93	8.01	9.0	6.20	8.25	8.49
Ash %						20.92

In its most refined form, F.P.C. is an inexpensive, bland, highly-nutritious light-coloured 'flour' whose protein content is significantly higher than that of the fish from which it is produced. Other forms of F.P.C. are produced, for instance a dark, coarse fish-tasting flour and a highly-flavoured dark-coloured paste or powder. These latter have been produced for centuries in the Far-East, using traditional methods, but the first type (refined F.P.C.) is of particular value because of its ability to be utilized as a tasteless food additive. It can therefore be added to traditional meals of rice or corn without impairing the flavour and so increase the protein content of a traditional diet without impairing its variety, taste, texture of other qualities.

The traditional, primitive methods of producing concentrated fish protein in the form of pastes or sauces rely on biological methods of rendering the proteins of the fish water-soluble and hence amenable to extraction, and they all produce relatively expensive products of low nutritive value and highly characteristic taste.

High grade F.P.C., however, is produced by solvent extraction; this removes not only the water from the raw material but also water-soluble odour-bearing compounds like ammonia, amines, and lipids. A number of methods have been developed using different solvents but the majority employ either ethyl alcohol or isopropyl alcohol.

Because the end product has to be of the highest quality and nutritive value, F.P.C. should ideally be produced only from, fresh, chilled or frozen fish known to be suitable and acceptable for human consumption. Some processes do, however, produce high-grade fish protein concentrate using conventionally processed fish meal as the raw material.

d. Fish glue. This is produced from fish skins and swim bladders, both of which contain collagen. The production process begins with washing the raw material to reduce the salt content to less than 0.1%. Fish scales contain guanine and any of this organic compound which is washed away at this stage is separated from the water to be further processed to produce 'pearl paste'. Scales are also directly processed for this purpose.

After washing, the skins and swim-bladders are cooked to extract the glue licquor. The extent and duration of the cooking process depends upon the required properties (e.g. purity) of the end product. When cooking is completed, excess water is evaporated and the glue blended to produce the required viscosity and density.

Fish glue has several excellent properties, being highly adhesive and having high initial tackiness. It is commonly used in the wood-working industries.

The swim-bladders of some fish can also be used to make isinglass, a high-grade product used in the food industries for preparing cooked dishes and for clarifying wine and beer.

In addition to the above, some fish livers, such as those of the halibut and cod, are processed to produce pharmaceutical (e.g. cod liver oil) and vitamin concentrates (e.g. vitamin A and D_3).

3.9.11 Shellfish

Shellfish is a general name given to all commercially important marine invertebrates and includes two main groups – molluscs (oysters, clams, mussels, abalones, etc.) and crustaceans: crabs, lobsters, nephrops (popularly known as 'Norway lobster' and marketed as 'scampi'), prawns and shrimps. Although not strictly shellfish, this term is also applied to the cephalopod molluscs like squid and octopus.

Shellfish are part of the benthos; that is they live on the sea bed and in comparatively shallow water. Of the whole group, only the cephalopods are capable of swimming, and so, therefore, catching methods for molluscs and crustaceans are limited to bottom operating gears. The principal catching methods for the molluscs is the use of a dredge or, in shallow water, a hand-held rake. Baited traps known as pots or creels are used to capture the larger crustaceans like lobsters and crabs. These traps are usually a wire or wood and netting cage into which the crustacean can climb to reach the bait but from which it cannot escape. The smaller crustaceans like shrimps and prawns are usually caught by beam trawl. Some species school in the upper reaches of the oceans and when this occurs, mid-water trawls are used.

The cephalopods are highly mobile and often swim in large groups in the upper reaches of the sea. The catching methods used, therefore, are essentially pelagic and vary from area to area. The principal method used is jigging. A jigger is a grapnel-shaped shaft having one or more circlets of barbs at one end and is attached to a long line at the other. When shot, the line is 'jigged' up and down in the water either mechanically or manually. After a short period has elapsed the line is hauled. If the operation has been successful, squid will have become entangled with the barbs by their tentacles. The jiggers themselves are brightly-coloured to attract the squid which are also often encouraged to school in the vicinity of the catcher boat by shining powerful lights onto the surface of the sea.

Virtually all shellfish can be considered to be part of the luxury food market and,

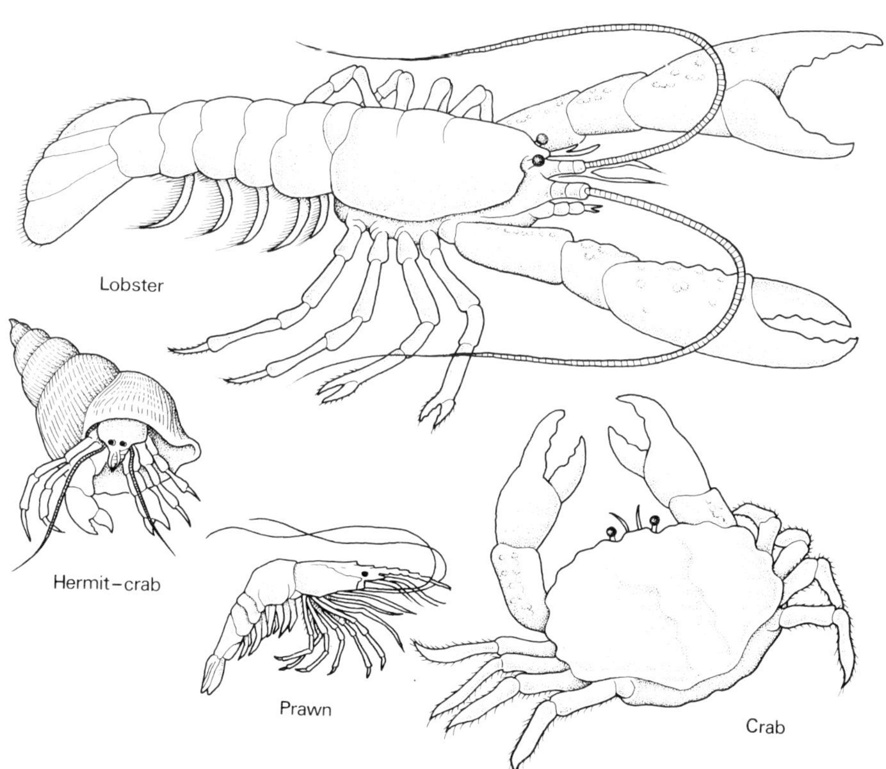

Fig. 3.63   Some common commercial shellfish.

generally speaking, they can be processed by any of the methods already described. There is one important difference, however, between the handling of fish and shellfish. Fish expire, due to asphyxiation, very quickly after their removal from the water. The very nature of the task of fishing means, therefore, that some fish will be dead for some time before even initial processing (gutting) is commenced. In the case of shellfish, however, many of these animals will not die immediately on leaving the sea and can survive for sometime in a damp or nearly dry environment. In addition, many of them, particularly lobsters and crabs, are marketed live. Because of this every effort is made to keep the catch alive until it can be processed or marketed. This is done by storing them in sea water on board until they can be landed.

If the shellfish are not to be marketed live, initial processing falls into three basic stages: shucking (shell removal); gutting and washing. Some shellfish, particularly shrimps and prawns are also deveined. When initial processing is completed, the raw material can be dried, smoked, canned, frozen, breaded or battered to meet the requirements of the market.

3.10 LITERATURE

3.10.1 Cattle, Pigs and Sheep

Bulletin No. 166: Sheep Breeding and Management. London, Ministry of Agriculture, Fisheries and Food, H.M.S.O., 1960.
W. BULLOCK. *The History of Bacteriology.* London, Oxford University Press, 1960.
F. M. BURNET. *Natural History of Infectious Disease,* 3rd ed. Cambridge, Cambridge University Press, 1962.
R. CRUICKSHANK. *Medical Microbiology.* Edinburgh, E. & S. Livingstone, 1968.
European Breeds of Cattle. Rome, Food and Agriculture Organization, 1966.
Pig Breeding, Recording and Progeny Testing in European Countries. Rome, Food and Agriculture Organization, 1958.
H. THORNTON. *Textbook of Meat Inspection.* London, Bailliere, Tindall and Cassell, 1968.
J. A. S. WATSON and J. A. MORE. *Agriculture: the Science and Practice of Farming.* Edinburgh and London, Oliver and Boyd, 1962.
W. R. WOOLBRIDGE. *Farm Animals in Health and Disease.* London, Crosby Lockwood, 1960.
The Yearbook of Agriculture: Power to Produce. Washington D.C., United States Department of Agriculture, 1960.

3.10.2 Poultry

G. SYKES. *Poultry – A Modern Agribusiness.* London, Crosby Lockwood, 1963.

3.10.3 Sausages and their production

F. GERRARD. *Sausage and Small Goods Production.* London, Leonard Hill, 1960.
E. KARMAS. *Sausage Processing.* New Jersey, Noyes Data Corp., 1972.
D. PEARSON. *The Chemical Analysis of Foods,* 6th ed. London, Churchill, 1970.
Recipe Leaflets. Swindon, Wiltshire, England. The Oppenheimer Casing Co. (UK) Ltd.
Symposium on the British Fresh Sausage – Present Day Manufacturing Techniques.

Leatherhead, Surrey, England, Symposium Proceedings No. 11, B.F.M.I.R.A., November 1971.

3.10.4 Meat Preservation

BIGELOW, BOHART, RICHARDSON and BALL. *Heat Penetration in Processed Canned Foods.* National Canners' Association Bulletin 16L, 1920.

Calculation of Processes for Canned Foods. American Can Company, 1952.

M. GUTTERSON. *Food Canning Techniques.* New Jersey, Noyes Data Corp., 1972.

S. W. F. HANSON. *The Accelerated Freeze Drying (AFD) Method of Food Preservation.* London, H.M.S.O., 1961.

R. A. LAWRIE. *Meat Science.* London, Pergamon, 1968.

Meat Chilling – Why and How. M.R.I. Symposium No. 2. Bristol, Meat Research Institute, 1972.

Meat Freezing – Why and How. M.R.I. Symposium No. 3 Bristol, Meat Research Institute, 1974.

M. J. MORLEY. *Tables of the Thermal Properties of Meat.* Bristol, Meat Research Institute.

The Science of Meat and Meat Products, 2nd ed. American Meat Institute Foundation. New York, Reinhold.

Statement of Policy on Commercial Sterilization of Foods in Hermetically Sealed Containers. Application of Emergency Permit Provision of Section 404 of the Food Drugs and Cosmetics Act as amended. National Canners' Association of America, 1972.

C. R. STUMBO. *Thermo-bacteriology in Food Processing.* New York, Academic Press, 1965.

3.10.5 Fish

G. THORSON. *Life in the Sea.* World University Library.

CHAPTER 4

Dairy Products

4.1 INTRODUCTORY

4.1.1 Milk

All dairy products are derived from milk. In temperate climates the cow is the main or only source, but in warm, dry climates the goat and the sheep may make a substantial contribution to the milk supply, especially in mountainous regions. In some hot countries such as India the buffalo may be a major source of milk, and in cold countries such as northern Sweden the reindeer may provide milk. Other mammals such as the horse (mare) may be used in a few regions on a limited scale, or for special purposes, e.g. the preparation of cultured milks.

In this section the word *milk* will mean cow milk unless otherwise stated.

The average or typical composition of the milk of various mammals is given in table 4.1. Like all biological properties, the composition of milk may vary considerably, especially when taken from a single animal. The chief factors affecting composition are: (1) breed, (2) stage of lactation, (3) method of feeding and general management, (4) age of animal, and (5) disease of the udder (mastitis). The typical composition of milk from two breeds of cows given in table 4.2.

4.1.2 Dairy products

Dairy products may be classified as:

1   Pure dairy products utilizing only the constituents of milk.
2   Traditional dairy products utilizing well established food ingredients such as salt.
3   Modern sophisticated products incorporating non-milk foods.

These are summarized in table 4.3. Milk and milk constituents are used to a considerable extent in the general food industry, so that many foods including processed meats such as sausages, soups, cakes, bread, confectionery and beverages, regularly contain such dairy products as whole and skim milk powder, sodium caseinate, co-precipitate, whey powder, cream, butter oil, butter, etc., but these foods are not considered as dairy products in the ordinary sense of the term because the proportion of milk ingredients is small.

For the purpose of this section we may consider dairy products to be foods made in a dairy from milk and consisting entirely or substantially of milk ingredients. Some materials, such as casein, may be used for non-food purposes.

Table 4.1 Typical composition of milk from various mammals

	Fat %	Lactose %	Casein %	Albumin etc. %	Minerals %
Ass	1.4	6.1	0.75	1.2	0.5
Buffalo	6.0	4.5	3.8	0.7	0.75
Camel	3.0	5.5	3.5	0.4	0.77
Cat	4.7	4.8	3.6	3.2	0.6
Cow	3.75	4.75	3.0	0.4	0.75
Dog	11.0	3.2	5.0	2.6	0.73
Goat	6.0	4.3	3.3	0.7	0.84
Human	3.7	6.4	0.9	1.2	0.3
Horse (mare)	1.1	5.8	1.3	0.7	0.3
Mule	1.7	5.3	2.1	2.1	0.4
Pig	6.5	3.3	3.7	1.5	1.0
Reindeer	17.1	2.4	8.4	1.7	1.5
Sheep	9.0	4.7	4.6	1.1	1.0

Table 4.2 Typical composition from Friesian and Guernsey cows

	Friesian	Guernsey
Fat %	3.50	4.65
Solids-not-fat %	8.65	9.10
Protein %	3.25	3.65
Lactose (anhydrous) %	4.60	4.70
Calcium %	0.115	0.13
Calories per 100 g	62.0	75.0
Vitamin A, activity i.u. per 100 g	125.0	190.0
Vitamin D, i.u. per 100 g	1.8	2.3
Vitamin B_1, μg per 100 g	40.0	40.0
Riboflavin, μg per 100 g	150.0	200.0
Nicotinic acid, μg per 100 g	80.0	80.0
Pantothenic acid, μg per 100 g	350.0	350.0
Vitamin B_6, μg per 100 g	35.0	35.0
Biotin, μg per 100 g	2.0	2.0
Vitamin B_{12}, μg per 100 g	0.5	0.5
Folic acid, μg per 100 g	0.1	0.1
Vitamin C, mg per 100 g	2.0	2.0

In considering the nomenclature of dairy products it must be borne in mind that names may have different connotations in various countries. For example, the line of demarcation between milk high in fat and what is legally cream may vary from country to country. In the USA the term *ice cream* can only be used for the pure dairy product, the term *mellorine* being used when vegetable fats are incorporated. In the United Kingdom the term *ice cream* can be used for both types.

  a. General observations on the dairy industry in developed countries. During the last 100 years the dairy industry has gradually changed from a crude, empirical farm-house industry to a highly sophisticated one employing modern and elaborate equipment, and scientific control of a high order. This development has escalated during the

Table 4.3 Classification of dairy products

---

1. *Pure dairy products (no additions)*
   Milk (various grades)
   Separated (skim) milk (fat free)
   Cream (various types)
   Evaporated milk (unsweetened)
   Butter (unsalted)
   Milk powder (dried milk)
   Yoghourt (natural) and other cultured milks
   Casein

2. *Dairy products utilizing accessory materials*

Product	*Other materials present*
Condensed milk (sweetened)	Sugar
Butter (salted)	Salt, (annatto)*
Cheese	Salt, rennet, (annatto)
Condensed and dried whey	

3. *Modern dairy products*

Milk (fortified)	Vitamin D. and other micronutrients
Cream (whipping)	Stabilizers
Margarine	Vegetable fats(mainly)
Dairy ice cream	Sugar, emulsifiers, flavourings, colouring matter
Ice cream	Vegetable fats, sugar, emulsifiers, flavouring, colouring matter
Fruit yoghourts	Fruit, sugar, flavouring, colouring matter
Chocolate crumb	Sugar, chocolate
Sodium caseinate	Sodium carbonate or other alkali
Co-precipitate	Calcium chloride

---

* Annatto is a vegetable colouring material

past 20 years. The factors mainly responsible for this change are:

1 Scientific research, mainly by chemists, bacteriologists and engineers.

2 The urbanization of populations leading to dense centres of population.

3 The perishability of milk, cream and certain other dairy products, necessitating research and methods for preserving and distributing these in a sound and safe condition.

4 The steady merging of companies leading to the establishment of large scale dairies, e.g., handling 200000 or more litres of milk a day. Such large units justify considerable expenditure and the employment of high grade executives which are beyond the means of small units.

5 The increasing shortage of labour, encouraging mechanization and automation.

6 Increase in the size of dairy farms and bulk collection of milk.

7 Stringent legal control of milk and milk products.

8 Progress in the chemical and engineering industries leading to new materials, processes and manufacturing methods.

b. Standards. All developed countries have standards, mainly for chemical composition, for all or most dairy products. There may be restriction on permitted ingredients, permitted names and descriptions (labelling), and sometimes there are microbiological standards. There are regulations governing fat, moisture and total solids for nearly all dairy products.

These standards vary in different countries and, as for food additives in general, the whole subject of food standards can be considered internationally as a 'jungle'. Standards throughout the world should be as uniform as possible for any one product, as trade would then be greatly facilitated. Efforts are already being made towards this end by the Food and Agriculture Organization, the E.E.C. and the International Dairy Federation.

There may on occasions be ample justification for varying standards for nutritional or climatic reasons or because of the availability of certain types of raw materials. For example, standards for the composition of evaporated milk may differ in temperate and hot countries because the physical stability may be influenced by atmospheric temperature. It is however, highly desirable that all rules for labelling food should be internationally standardized, and that there should be international agreement on what additives are permitted for particular foods. The present position is chaotic. Chemical and microbiological standards should be made as uniform as possible, and standard methods of analysis laid down.

Specific standards are considered for each dairy product as it is dealt with. The definitions and regulations quoted below are generally those current in the UK. The comparable standards applying in the USA are given in the Appendix, together with relevant statistics.

4.2 MILK

4.2.1 Definition

In the simplest terms milk may be defined as the 'normal post-parturition secretion of the healthy udder'. Legal definitions in the developed countries often attempt to be more precise, for example, to exclude colostrum (immediate post-parturition secretion), bad mastitis milk, grossly contaminated milk, etc., but this is difficult because there is no clear-cut demarcation between normal and abnormal milk. The USA definition states that 'milk is hereby defined to be the lacteal secretion practically free from colostrum, obtained by the complete milking of one or more healthy cows, which contains not less than $8\frac{1}{4}\%$ milk solids-not-fat and not less than $3\frac{1}{4}\%$ milk fat'.

Abnormal milks can most conveniently be excluded by requiring that milk shall not clot on boiling and shall clot with rennet. Compositional and other standards can be laid down; e.g., not less than 3% fat and 8.5% solids-not-fat; physico-chemical properties, e.g., pH 6.5 to 6.8; maximum bacterial count, e.g., 1 000 000 per ml; maximum of somatic cells, e.g., 500 000 per ml, but all such standards are controversial. Subjective assessments such as 'normal colour, odour and taste' have little meaning.

4.2.2 Production

Milk production is steadily increasing in most countries, largely because all governments now realise the exceptional nutritional value of milk and milk products, and because advancing technology is making it relatively cheaper to produce milk and its products by large-scale operation and mechanization. Progress in packaging and increase in keeping quality make dairy products more available and attractive to the housewife, and also facilitate the export trade from countries of high production when this exceeds the home requirement.

a. Control. The legal control of milk and cream in the United Kingdom is covered by the Food and Drugs Act 1955, Sections 28–48. Section 32 forbids the addition of water, colouring matter, dried milk, condensed milk or separated milk to milk intended for sale. More detailed provisions are given in the Milk and Dairies (General) Regulations 1959 which deal with production, distribution and health aspects. Various milk products such as cream, condensed and evaporated milk, dried milk or milk powder, butter and cheese are covered by specific regulations.

Ideally milk should be obtained from healthy cows free from all diseases, including tuberculosis, brucellosis and mastitis (infections of the udder), under near aseptic conditions so as to be almost germ-free, of long keeping quality and safe for consumption. This ideal is impossible to attain in practice, although cows in the UK, USA and many other countries are now tubercle-free, but mastitis and brucellosis are often more extensive now than 30 years ago. Hygiene in production and refrigerated transport and storage have increased the keeping quality (K.Q.) or 'life' of milk, but some form of heat-treatment is necessary, and probably always will be for the industry in general, to make milk safe and of adequate K.Q.

b. Grades or types of milk. Each country has its own grades of milk. These may be defined in terms of (i) type of cow, e.g., Jersey or Guernsey, (ii) production from cows free from tuberculosis, (iii) conforming to a specified chemical quality, usually fat content, and/or (iv) conforming to a specified bacteriological quality which may be based on the number of bacteria or on K.Q. Thus in the UK there is a Channel Islands grade of milk which must have at least 4% of fat. In addition there is a grade of pasteurized milk which may be homogenized, sterilized milk, ultra high temperature treated (U.H.T.) milk, and one grade of raw or untreated milk which must be bottled on the farm. In developed countries most of the milk is now pasteurized or otherwise heat treated. In the UK about 96% of all liquid milk is pasteurized or sterilized.

All milk in the UK must comply with the *presumptive* standard of at least 3% fat and 8.5% solids-not-fat (the latter consisting of lactose or milk sugar, casein, albumin and globulin (whey proteins)) and salts or minerals (see table 4.2).

With the entry of the UK into the E.E.C. the fat standard will be raised to 3.5% and in practice milk will be standardized at this level. There will be a half-fat milk which must contain 1.5 to 1.8% fat and separated or skim milk which must contain less than 0.3% fat.

4.2.3 The heat treatment of milk

a. History. It had been known for centuries that the keeping quality (K.Q.) of foods could be increased by heating them, as one does in ordinary cooking. This simple method, however, gave only a limited increase in K.Q. because immediately after cooling the food became contaminated by micro-organisms which are to be found almost everywhere.

In the early part of the nineteenth century Appert in France showed that if foods were heated in hermetically sealed containers the K.Q. could be greatly extended, sometimes almost indefinitely. At this time heating methods were quite empirical and many of them brought about chemical changes in the food which had an adverse effect on the flavour. In the 1860s Louis Pasteur, a French chemist, was asked to investigate the problems of undesirable fermentations in beers and wines. At that time our knowledge of microbiology was very limited, and although it was realized that micro-organisms existed and could cause decomposition of foods, it was thought that these arose after sterilization by spontaneous generation. In a series of experiments of classical importance and elegant simplicity Pasteur showed that micro-organisms always occured by contamination of a product from another source, and that in beers and wines the organisms bringing about these secondary or undesirable fermentations could be killed by a carefully controlled heating at a low temperature, such that it did not affect the flavour. Provided that they were protected from further contamination after cooling these liquids then had a good keeping quality. After some opposition Pasteur established his theory and his methods were subsequently applied to other liquid foods including milk. This method is now universally called 'pasteurization'.

Milk is an excellent medium for most types of micro-organisms so that it is not surprising that under ordinary commercial conditions it sours rapidly. In hot weather much of the milk supply was lost through this cause and had to be discarded. During the last two decades of the nineteenth century crude heating methods were used by dairymen to prevent the souring of milk, and at that time the dairy industry did not take any interest in the dangers arising from pathogenic bacteria in milk. By this same time many medical bacteriologists, particularly Koch in Germany, were working on bacteria as a cause of common diseases, and the importance of milk as a vehicle for many disease-causing bacteria gradually became realized. At the beginning of this century properly controlled experiments were carried out in the United Kingdom, the United States of America and other countries on the pasteurization of milk, not only for liquid consumption but for the manufacture of dairy products. In the second decade accurate bacteriological work was carried out on milk, particularly by North in the United States, and the necessary conditions (time and temperature) to kill the common pathogens in milk were elucidated. The most important of these bacteria were the streptococci of scarlet fever, septic sore throat, etc., the organisms of typhoid and paratyphoid fevers, the diphtheria organism, the staphylococci and the organism of bovine tuberculosis. It was found that the last was the most heat resistant of the pathogens commonly occurring in milk, and North's experiments enabled bacteriologists to recommend a minimum heat treatment for the certain killing of this bacterium. North showed that it could be killed in about 10 minutes at 63°C, and after further extensive work in many countries it was agreed that pasteurization would be defined as far as milk is concerned as heating to a temperature not less than 63°C and not more than 65°C for thirty minutes. Naturally there were slight differences in various countries but this treatment became generally accepted both by Government authorities and by the industry.

In the United Kingdom pasteurization became legally approved and defined from 1922 onwards. Fortunately this phase coincided with the introduction of glass bottles which gave excellent protection to the milk, and could moreover be heat sterilized. Previous to this, pasteurization was of limited value because the crude method of delivery by hand can and dipper obviously gave ample opportunity for recontamination of the milk. In fact experience showed that pasteurization resulted in a reduced K.Q. of the milk because at that time hygiene was practically non-existent in the dairy industry. The equipment was never cleaned properly so that the soured milk residues in the plant seeded the pasteurized milk with souring bacteria.

At this time also there was considerable opposition to the pasteurization of milk, partly for this reason. Pasteurized milk was criticized on account of its flavour (milk was often over-pasteurized) and it was also alleged that the process greatly reduced the nutritive value of the milk. This opposition lasted up to about 1930 but gradually died out and pasteurization is now universally accepted as a cheap, simple and effective way of making milk safe without affecting to any appreciable extent its nutritive value or flavour. Up to the time when it was introduced generally in Britain thousands of children, especially young babies, died every year from milk-borne diseases, particularly bovine tuberculosis, various fevers and especially summer diarrhoea. When pasteurization was introduced, at first in the cities, the reduction in these diseases was quite marked. Some medical officers of health commented on this without realizing at the time the reason for it. This general pattern of reduction in milk-borne diseases was followed all over the country as more and more milk was pasteurized. Occasionally an outbreak of milk-borne disease was caused by pasteurized milk, but in every case it was found that the reason was recontamination of the milk after pasteurization. To-day no informed person questions the value of pasteurization or alleges that more than very slight damage is done to the milk nutritionally. In her cooking processes in the home the housewife does far more damage to milk nutritionally than the dairyman.

b. Methods. Broadly speaking, there are three types of heat treatment applied to milk.
1   Pasteurization.
2   Commercial or traditional sterilization.
3   Ultra high temperature (U.H.T.) treatment.

Each country has its own laws controlling these treatments. In the UK the Milk (Special Designation) Regulations 1963 control (1) and (2) and special amendments control (3).

4.2.4 Pasteurized milk

Pasteurization is a controlled mild heat treatment which kills all common pathogens and most ordinary bacteria in milk. It does not kill the thermoduric bacteria, which are vegetative cells surviving pasteurization, or spores. In practice about 99% of bacteria in milk of good bacterial quality are killed, the milk is made safe and the K.Q. improved.

There are three legal methods for pasteurizing milk:
1   The holder method.
2   The high temperature–short time or HT–ST method.
3   Any other method approved by the appropriate licensing authority.

a. The holder process. When the experimental evidence for the value of pasteurization became so strong that governments were ready to enact legislation on the matter, a decision had to be made as to what temperature and time should be adopted for legal purposes. Most of the experimental work had been done with exposure times of half an hour or some similar period, and such a time is convenient to use in practice and can be measured accurately. For these reasons the method officially adopted by nearly all governments consisted of holding milk for 30 minutes in the range 63° to 65°C, and this process became known as the 'holder method'. It worked perfectly satisfactorily provided care was taken to prevent contamination of the milk after cooling. Recontamination was one of the most serious problems in the dairy industry over the period 1920–40 because at that time engineers had little or no knowledge of bacteriology and hygiene, and some equipment was designed in such a way that it could not be properly cleaned and sterilized. This difficulty was overcome in time but there remained certain disadvantages in the holder process which became more serious as dairies handled larger and larger quantities of milk. It was necessary to construct large tanks to hold volumes of several hundred gallons, which were expensive and took up much space, and in addition it meant a delay of half an hour before the milk could be bottled. Moreover, certain technical problems were gradually recognized. Among these was the possibility of recontamination as milk was pumped from the heating tank to the cooling equipment, and foam could form on the surface of the milk and in this way not receive the full heat treatment. During this time experimental work was going forward on another method which ultimately displaced the holder process.

b. The high temperature–short time (HT–ST) process. In 1923 Dr Richard Seligman, founder of The Aluminium Plant and Vessel Company (later to become A.P.V. Limited) began experimental and development work on the plate heat exchanger for milk and other liquids. This machine consists essentially of a stack of indented metal 'trays' provided with holes or ports at their extremities, and separated by rubber gaskets so that when the plates are compressed together there is a very narrow space between the metal surfaces. The holes at the extremities then form a channel which allows liquid to enter the space at one end and leave it at the other. Raw milk and hot water are passed through the alternate compartments formed by the trays. Originally made of tinned bronze or copper, the machine was later made entirely of stainless steel and in this way problems of corrosion and wear were overcome, and cleaning and sterilizing were simplified. Working in collaboration with the National Institute for Research in Dairying, near Reading, Seligman steadily improved the plate heat exchanger until today it can be regarded as near a perfect machine as can be made for the purpose.

The plate heat exchanger method gave many advantages over the holder method. The temperature of the milk could be raised to the necessary value in a few seconds, and similarly cooled again in a very short time. Experiments showed that milk could be properly pasteurized in a very short time if the temperature were raised a few degrees, and ultimately a holding time of fifteen seconds at 71°C was accepted by British public health authorities as equivalent to the holder treatment of thirty minutes at 63° to 65°C. As with the original form of pasteurization, some opposition to the HT–ST method was encountered, chiefly on the grounds that the fifteen-second period could not be measured accurately. This was a valid criticism because the temperature of milk cannot be altered instantaneously; it requires a few seconds to raise the temperature to 71°C and a similar time to cool it to 5° to 10°C. The fifteen-second holding period is in practice a minimum and a time of at least 17 to 18 seconds is usually given. These conditions have been based on experiments in which artificially infected milk has been subjected to various heat treatments. The time is measured by injecting a chemical into the milk and using a very sensitive instrument to detect the earliest time at which the chemical can be detected in the milk leaving the heating section of the machine.

The HT–ST process has many advantages. It eliminates the large tanks which were necessary for holding milk at the pasteurization temperature for half an hour. The machine itself is very compact and one dealing with 20 000 litres of milk an hour can be accommodated in a few square feet of dairy floor space.

The HT–ST process has another very important economic advantage. The hot

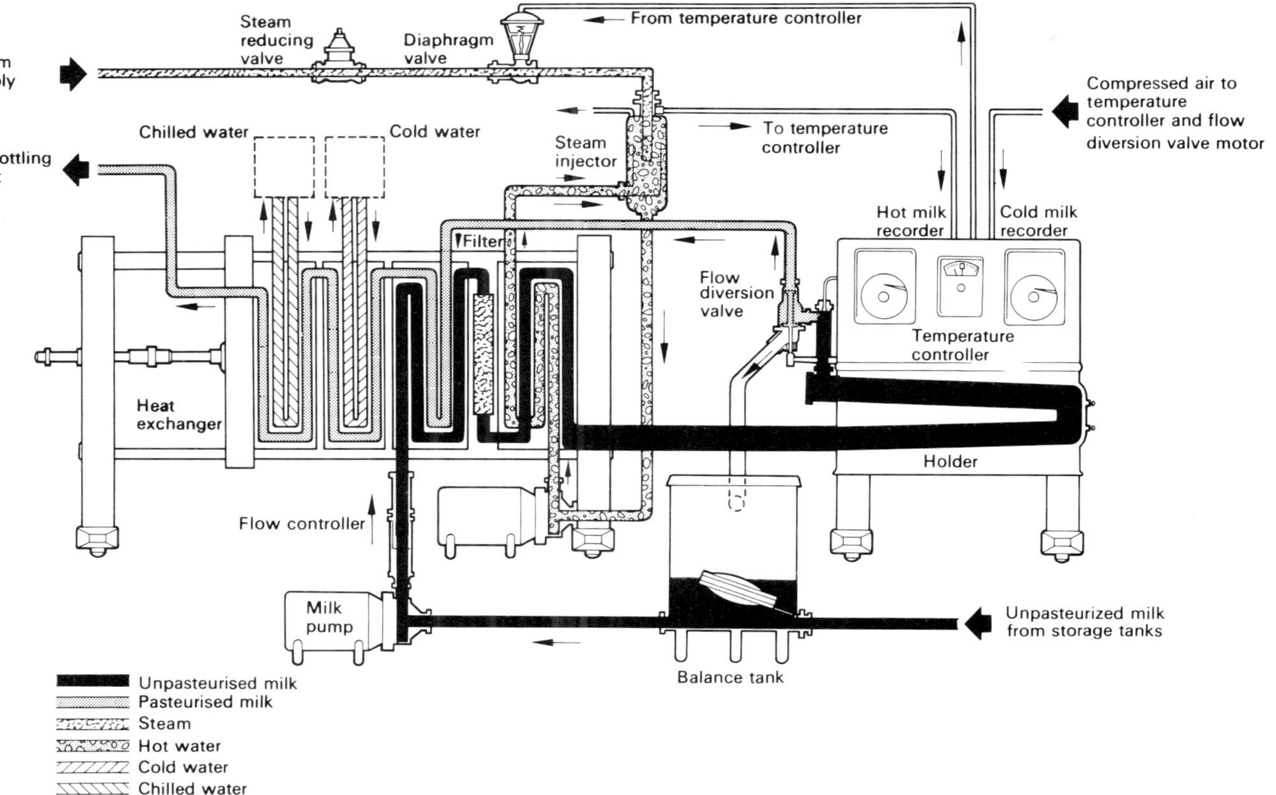

Fig. 4.1 Flow diagram for high temperature–short time pasteurizer.

pasteurized milk is used to warm the incoming cold milk, and conversely the hot pasteurized milk is cooled down by the incoming cold milk (regeneration). This affords a substantial economy in steam usage, since steam is used to heat the water to pasteurizing temperature, and a well-designed regeneration system can save up to 83% of fuel. Another advantage is that, since the heating chambers are totally enclosed, the milk is at all times protected from outside contamination.

Briefly the method is as follows. The raw milk is fed into a hopper or balance tank and from there is pumped through the first warming or regeneration stage, which raises it to about 50°C. Naturally this intermediate temperature varies with the design of any particular plant. The milk may be filtered through cloth at this stage, but it must be emphasized that such filtration removes only visible dirt and has no effect on the bacteriological quality of the milk. The filters are normally changed every twenty or thirty minutes. From this stage the milk passes to the heating section where the temperature is raised by hot water to the required level, usually 71°C. From the heating section the milk passes under pressure to the holding section which may consist of another 'pass' or series of plates or a holding tube, both of which have been designed to hold the milk at the pasteurizing temperature for at least 15 seconds. It is customary to design the equipment so that the holding time is 17 to 18 seconds, whilst the pipe-lines connecting the holding section to the other two sections add a further short time to the nominal holding time. From the holding section or tube the milk passes to the reverse side of the regeneration section where the hot milk is cooled by the incoming cold milk. Finally the milk passes to the chilled water or brine section. If the raw milk is not very cold a mains water section may be included at the cooling stage, but since all the raw milk is now refrigerated there is no longer any necessity for a mains water section. Formerly brine (a strong solution of calcium chloride in water) was used for the final cooling stage. It has the advantage that it can be used at very low temperatures, e.g., at $-10°C$, but brine is extremely corrosive and it is now customary to design HT–ST plants to use chilled water, which is simply ordinary water cooled to 1° to 2°C. A larger volume is required, but the difficulties associated with brine are eliminated.

The HT–ST method has virtually eliminated the old holder method, which is now to be found operating only on a very small scale in very remote districts. It has been universally accepted as an efficient and economic method for making milk safe and improving its keeping quality. In practice it extends the shelf life of ordinary retail milk for at least 24 hours and today with modern refrigeration in the home holding milk at 3° to 5°C a good pasteurized milk will keep in a good condition for up to seven days. The bacteria not killed by pasteurization, amounting normally to only about 1% of the original microflora, are biochemically sluggish and usually have little effect on souring and taint production. There is, however, one weak link in the chain, and that is that pasteurization does not kill bacterial spores. These may be regarded as the 'eggs' of bacteria and represent a resting stage in a form which is highly resistant to all adverse conditions such as heat, light, chemical disinfectants, absence of nutrients and radiation. Spores are extremely difficult to kill by ordinary heating methods, usually requiring at least 15 minutes at 121°C in a liquid at a neutral pH, or 160°C for 3 hours, or 170°C for 2 hours when embedded in powdery materials such as flour and chalk. Fortunately bacterial spores are not a health hazard in milk; if they were all milk would have to be autoclaved to make it safe. However, there is one type which can cause a keeping quality problem in milk supplies, and that is the organism producing sweet curdling and separation of fat, giving the condition known in the trade as 'bitty cream'. Recent research has led to the development of a method which overcomes this problem.

c. The ultra-high temperature (U.H.T.) method. The higher the temperature used for the killing of bacteria the shorter is the time required to achieve a given kill. The coefficients for the killing of bacteria differ from that for chemical changes in milk or any other liquid, and it has been established that by raising the temperature sufficiently it is possible to kill not only all the vegetative cells of bacteria but also any spores which may be present. For example, if milk is heated within the temperature range 135° to 140°C it is possible to kill all bacteria and all spores with a holding time of about two seconds. This represented a difficult technical problem but it has now been solved and there are available a number of methods which make this process technically efficient at a reasonable cost. The method has now become generally recognized as the ultra-high temperature or U.H.T. method: an ill-contrived and ungrammatical term, meaning literally 'beyond high temperature', but unfortunately it has come to stay.

An amendment to the Milk (Special Designations) Regulations 1963 now permits the method to be used legally in the United Kingdom. This requires milk to be raised to a temperature of at least 270°F (132°C) with a holding time of at least one second.

Two main methods have been developed for this process – the indirect and the direct. In the former the milk passes through a tubular or plate heat exchanger in which super-heated steam raises the temperature of the milk rapidly to the required level and then rapidly cools it after the very short holding period. In the direct method steam is injected into the milk or milk is injected into steam. The equipments which are now available for this process are summarized in table 4.4.

The essential feature of all these methods is that they are able to produce a sterile milk which is free from the usual disadvantages of ordinary sterilized milk, such as browning

Table 4.4 Equipment for the U.H.T. method

---

a Indirect heating methods
1. Using steam as a source of heat.
    (i) Tubular systems:
        1. Stork (Holland)
        2. Gerbig (Germany)
        3. Mallorizer (USA)
        4. Graves–Stambaugh (USA)
        5. Roswell (USA)
        6. Spiratherm (Cherry Burrell–USA)
    (ii) Plate systems:
        1. A.P.V. (England)
        2. V.T.S. (Vacu-Therm Sterilizer–Alfa Laval, Sweden)

2. Using electricity as a source of heat.
        1. Daveat (USA)

b Direct heating methods
1. Using steam as a source of heat.
    (i) Steam injected into milk:
        1. Uperization (Alpura–Switzerland)
        2. V.T.I.S. (Vacu-Therm Instant Sterilizer–Alfa Laval, Sweden)
        3. U.H.T. Aseptic Aro–Vac (Cherry–Burrell, USA)
        4. Grindrod (USA)
    (ii) Milk injected into steam:
        1. Laguilharre (France)
        2. Palarization (Denmark)
        3. Thermovac (Thimonnier, Breil et Martel–France)

---

and a cooked flavour. Properly carried out the method gives a milk which is almost indistinguishable from ordinary pasteurized milk. The chemical changes and loss of nutritive value are intermediate between those of pasteurized and ordinary sterilized milk, and are negligible in relation to the obvious commercial advantages of the method. Initially a cooked flavour is produced which may be described as 'cabbagy' and is caused by the production of volatile sulphur compounds; on keeping this gradually passes off and may be replaced by an oxidised or cardboard flavour.

If U.H.T. milk is merely bottled in the ordinary way, like pasteurized milk, there is no particular advantage to be gained because bacteria will gain access to the milk and the keeping quality will be little better than that of ordinary pasteurized milk. To obtain the greatest, and in fact only worthwhile benefit from the U.H.T. process – which costs more than ordinary pasteurization – it is necessary to protect the treated milk from subsequent contamination. The keeping quality of milk is determined solely by the numbers and types of bacteria which are present.

The systems are available for obtaining the greatest benefit from the U.H.T. process. The method may be used as a preliminary treatment for the traditional sterilization process in bottles (dealt with later), or the milk may be aseptically packaged in sterile containers in such a way that subsequent contamination is made impossible.

d. The indirect method. This was the only method legally permitted in the United Kingdom until recently. The equipment most commonly used is a specially strengthened type of plate heat exchanger in which milk is forced through at a high pressure, as otherwise it would boil when the temperature rose to above 100°C. The high temperature causes deposits to be formed on the plates to a greater extent than with pasteurization, so that the cleaning of the equipment requires special care. Unlike ordinary pasteurization

Fig. 4.2 The Steriplak UHT plant with cartoner.

equipment the U.H.T. plant must be *sterilized* after cleaning, and this is usually done by treating it with super-heated steam or with water at 140°C under pressure for about one hour. Naturally the U.H.T. process requires not only more sophisticated equipment but more elaborate scientific control than ordinary pasteurization because contamination by a single bacterium can undo all the benefit achieved by the U.H.T. process.

Tubular equipments are equally efficient, and simpler in construction than plate heat exchangers, but have the disadvantage that they cannot be taken to pieces as can the plate heat exchanger. Cleaning must therefore be done with the utmost care in order to prevent the build-up of deposits which might ultimately cause a complete blockage of the tube. If this occurs very drastic methods are necessary to clear it.

e. The direct method. In the direct method the temperature of the milk is raised with the necessary rapidity by the injection of steam into the milk, or of milk into an atmosphere of steam. The method depends essentially on the fact that when a substance in gaseous form condenses to the liquid form a considerable amount of heat is liberated. A simple example of this is the well-known experience that if the steam issuing from a boiling kettle is allowed to impinge one's hand a very bad scald can be sustained. The converse reaction takes place when a liquid changes into a gas by evaporation. Here again a simple example is afforded by allowing a few drops of ether to evaporate from the back of the hand.

This liberation or absorption of heat is called the *latent* heat of a substance, and is the heat liberated or absorbed when a solid changes to a liquid, and a liquid changes to a gas and *vice versa*. It is relatively much greater than the ordinary or intrinsic heat, which is the amount of heat required to raise or lower the temperature of a material. For example, 1 gram of ice at 0°C it will require 79 calories (330 J) to change it to water at the same temperature. It requires only 100 calories (420 J) to raise the temperature of the water from 0° to 100°C. If now we wish to change the 1 gram of water at 100°C to steam at the same temperature we shall have to provide 536 calories (2 300 J). This latent heat effect is utilized in the dairy industry for a number of purposes, of which the most important is the direct method for the U.H.T. treatment of milk.

f. Uperization. During the 1950s extensive research was carried out in Switzerland on the direct method for U.H.T. treatment and this resulted in the perfection of a method and ultimately the manufacture of a very ingenious piece of equipment for the purpose. The process was called *uperization*, and was based on the latent heat effect as described above. The first part of the process is basically the same as for the HT–ST method and either a plate heat exchanger or a tubular heater can be used to raise the temperature of the milk to nearly 100°C. Superheated steam is then forced under pressure into the milk with the result that its temperature is raised almost instantaneously to the uperization temperature of about 140°C. The milk then passes through a section of such length that it gives a holding period of about two seconds (it is not possible to measure this with great accuracy) and the milk is then ejected into a vacuum or expansion chamber which has the effect of producing a rapid evaporation of water from the milk. Just as the sudden contact with superheated steam brought about a rapid increase of temperature by the condensation of the steam to form water which was immediately mixed with the milk, so now the evaporation of water from the milk in the vacuum chamber produces a cooling effect and so lowers the temperature of the milk to what it was immediately before the injection of the steam. The ingenuity of this equipment essentially consists in such an accurate control of both the injection of steam and a few seconds later the evaporation of the condensed steam from the milk, that the two stages are balanced. If there is no loss of heat then the amount of water evaporated should correspond exactly to the amount of steam condensed.

Naturally it is impossible to ensure that it is the actual water from the condensed steam which is evaporated in the expansion chamber, and for this reason the method could not be used legally in the United Kingdom because it was a contravention of the Food and Drugs Act. Recently an amendment to the Milk (Special Designations) Regulations was passed in order to make this method legal in the United Kingdom. Previously it had been used in other branches of the food industry.

Of the other methods the commonest is a reverse procedure, in which milk is sprayed into superheated steam, but from the physical point of view the methods can be regarded as identical.

Some claims are made for the superiority of the direct over the indirect method for

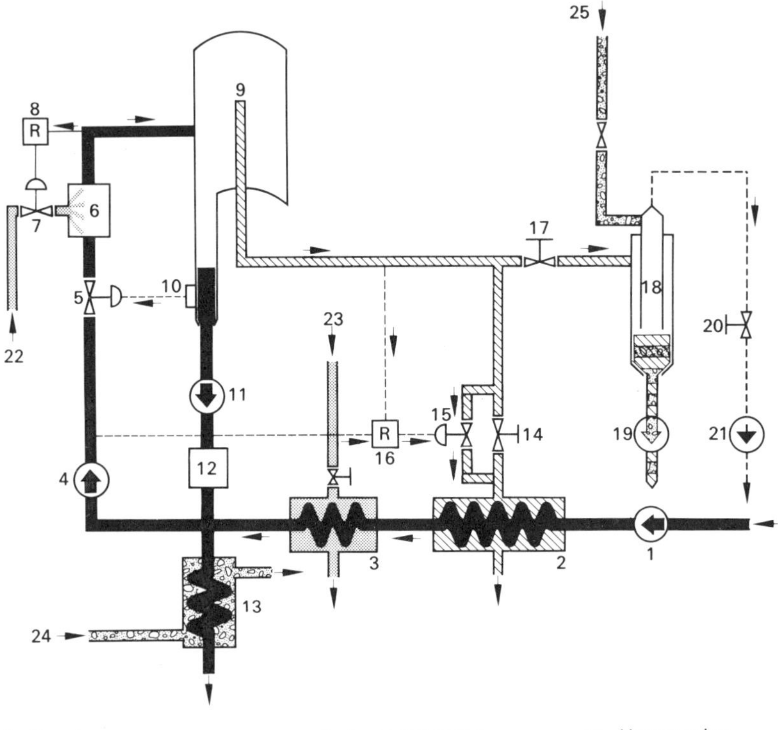

1	Transfer pump	9	Expansion vessel	17	Vapour valve
2	Preheater	10	Cushioning vessel with level sensor	18	Injection condenser
3	Second preheater	11	Sterile milk pump	19	Condensate pump
4	Milk pump	12	Homogeniser	20	Vacuum regulating valve
5	Pneumatic flow regulating valve	13	Sterile milk cooler	21	Vacuum pump
6	Uperisation unit	14	Vapour valve	22	HP steam
7	Pneumatic steam regulating valve	15	Pneumatic vapour regulating valve	23	LP steam
8	Uperisation temperature regulator	16	Specific gravity regulator	24	Cooling water
				25	Cooling water

Fig. 4.3 Flow diagram for Uperizer. Milk flow shown solid; water and steam shaded.

U.H.T. Probably the most important of these claims is that the steam treatment removes any taint or volatile obnoxious substances in the milk. This principle has been utilized in dairy manufacture for many years, as for example in the *vacreation* of cream. However, it is most important in using the direct method to ensure that the steam used is of a very high degree of purity. In commercial practice steam can easily be contaminated with a minute proportion of lubricating oil, and very small traces of this in milk can produce a very disagreeable taint.

      g. The cream line on milk. Whenever ordinary milk is allowed to stand for a few hours a layer of cream (commonly called 'cream line' in the trade) forms on the top of the milk. This is more noticeable with Jersey and Guernsey milk because more of the vitamin A activity is present as the coloured carotene in the milk of these breeds. It has been commonly thought by the housewife that this cream line is a measure of the richness of milk in fat, but this is not a reliable index of the fat content.

    The rate of formation and the depth of this cream layer is controlled by the percentage of fat in the milk, the size of the fat globules, the temperature at which the milk is held, whether the milk has been overheated during pasteurization, and by any agitation which the milk has received, especially when very cold or very hot.

    Milk fat has a relative density of 0.91 compared with a figure of 1.035 for skim milk, in which the fat globules are dispersed. They rise because they are lighter and the rate of rise, or the speed of formation of the cream line, is described by Stokes' law as follows:

$$R = \frac{r^2 2(d_1 - d_2)}{9v}$$

where R = the rate of rise of the fat globules, r = the radius of the fat globule, d_1 = relative density of skim milk, d_2 = density of fat globule and v = viscosity of the skim milk.

    As all these values except r are practically constant we can write $R = kr^2$, or in other words the rate of formation of the cream line is proportional to the square of the radius of the globules. In the Channel Islands breeds the fat globules are appreciably larger (about 5 μm) than those in Friesian and Shorthorn breeds (3 to 4 μm). It is for this reason that Jersey and Guernsey milk form a deeper cream line, and more rapidly than Friesian milk. With the larger globules there is more space between them, and another important factor is the clustering of the globules which in effect is equivalent to making the radius of the globules larger, so that by Stokes' law clusters rise faster than single globules. Clustering is due to the presence of agglutinins (a special type of protein) on the surface of the fat globules, and overheating and agitation damage the cream line by destroying these agglutinins.

4.2.5 Homogenization

    The formation of the cream line or layer on milk leads to technical and legal difficulties. The fatty material on the surface may, on storage, adhere to equipment and thus make for difficulties in cleaning and sterilizing, and when the milk is transferred from one vessel to another, or filled into bottles or cartons, the first milk through may be deficient in fat, and so possibly lead to a prosecution of the dairy. For these reasons it is most important that the fat in milk should be kept evenly distributed at all times. Mixing of the cream layer into the bulk of milk in a 3 000-gallon tanker which has stood overnight is not an easy problem, requiring steady agitation for at least 20 minutes.

If the milk is so treated as to break down the fat globules to a very small size, e.g. 1 μm or less, then these problems disappear because the viscosity of the milk prevents such 'tiny globules from rising to the top. This reduction in the size of the fat globules is brought about by the process known as homogenization.

A homogenizer is a machine in which the milk, usually at about 70°C, is forced by a system of reciprocating piston pumps through special valves at pressures up to 3 000 p.s.i. (200 bar) the valves being returned to their seatings by very strong springs. All homogenizers used for milk are based on this principle. The force required to pass the milk through the machine is very high because the lift is only a few thousandths of an inch and the milk is forced through this very small annular space at a very high velocity. The stress on all parts of the machine in this section is high, and the valves and seating must be made of a very strong material. The pumps must be extremely efficient, and both suction and discharge valves must open quickly and close tightly. A pressure gauge and safety valve are always included in the machine for obvious reasons.

The type of homogenizer shown below is the established machine in all branches of the dairy industry. Other types of machine and other methods have been tried, but they are either very expensive or inefficient.

a. Dairy products for which homogenized milk is used. On account of the capital expenditure involved and the substantial cost of operating a homogenizer, dairies only employ this process when the advantages of homogenized milk justify it. These advantages may be summarized as follows:

1   As the fat remains evenly distributed throughout the milk during the whole of its life, the dairyman has no anxiety about inequalities in fat content and possible prosecution by the local authority and conviction in a court for selling milk below the legal presumptive limit of 3% fat.

2   Homogenization makes milk more creamy and smooth in appearance and feel on the

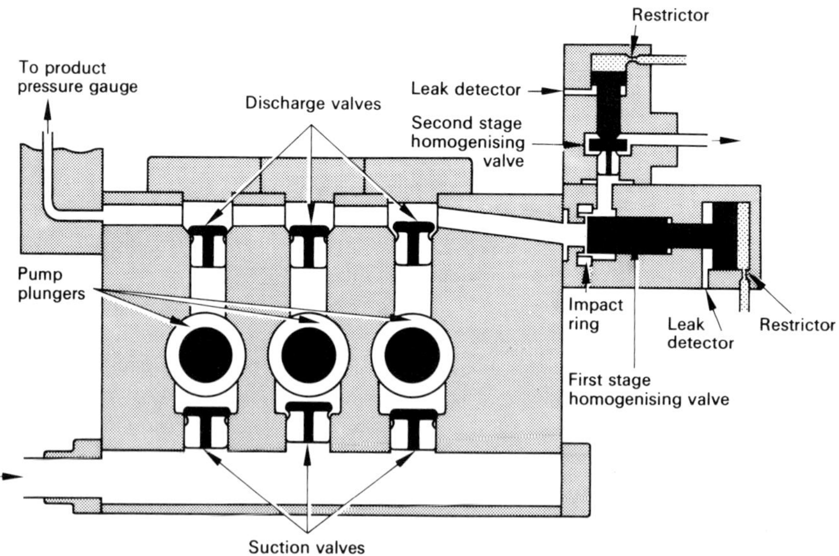

Fig. 4.4   Sectional diagram of homogenizer.

tongue. For example, when used for tea or coffee the degree of creaminess or whiteness in the beverage can be obtained with slightly over half the amount of ordinary milk required.

3   The division of the fat globules into particles of very small size makes homogenized milk more easily digestible. It is found that persons who cannot take ordinary milk can sometimes take homogenized milk without digestive troubles.

4   Homogenization brings about a subtle change in the protein-fat structure of milk, so that when it is acted on by rennet under weakly acid conditions it no longer forms a tough clot, as in cheese-making. Thus when homogenized milk is taken into the stomach, which is normally in an acid condition on account of the hydrochloric acid secreted there, the gastric juice converts the homogenized milk into a very fine 'grainy' precipitate instead of a clot which makes bigger demands on the digestive mechanism of the alimentary tract.

5   Although homogenized milk is not normally employed for cheese-making, the manufacture of blue veined (Roquefort type) cheese in the United States of America is often carried out using homogenized cream which is then re-incorporated in the skim milk. This sub-division of the fat particles enables the blue-green mould which develops in the cheese to bring about a faster hydrolysis by lipolytic enzymes. In this way the production of the typical flavour which is largely caused by the break-down of the fat by the growth of the mould is accelerated.

The use of homogenized milk is not without its disadvantages. The sub-division of the fat globules not only greatly increases the surface area of the fat, but it also disrupts the mobile lipid-protein membrane or skin which is present on all the fat globules. This membrane material is re-adsorbed to some extent but the protection afforded to the fat against chemical and enzymic attack is greatly reduced. The fat in milk is readily oxidized in the presence of free oxygen and especially if trace amounts (e.g. 1 ppm) of copper are present. This oxidation is catalysed not only by copper and some other heavy metals, but also to a large extent by light. In practice it is impossible to store milk in such a manner that oxygen is completely absent, and although with modern stainless steel equipment contamination by copper is very rare today, milk in glass bottles nearly always gets exposed to some light, with the result that by the time the homogenized milk reaches the customer it has usually acquired a characteristic taste, which is variously described as cardboard, oxidized or oily. If a bottle of homogenized milk is exposed to sunlight a peculiar burnt or feathery flavour may be induced. This is more common in the USA than in the UK. Persons who habitually consume homogenized milk become accustomed to the flavour and do not notice it. In effect all pasteurized milk is slightly oxidized but as the taste is familiar this is not detected.

Homogenized milk is popular in catering establishments, the even distribution of fat being an advantage when milk has to be held in a vessel for several hours and used from time to time. Incidentally, when milk is kept hot the production of sulphydryl substances induces reducing conditions in the milk which exert an anti-oxidant effect, so that the milk may have a cooked flavour but the oxidized flavour does not appear.

Milk used for the manufacture of sterilized milk and evaporated milk (see later) is always homogenized. Milk used for yoghourt and other 'fancy' dairy products is also frequently homogenized.

With the entry of the United Kingdom into the European Common Market the legal position of the fat content of milk will be fundamentally changed. Instead of requiring that milk shall be sold 'as it comes from the cow' and with all its fat, having only as a supplementary requirement a presumptive legal standard of 3% fat, milk will in future be required to have a fat content of 3.5%. Under the new system it will be an offence to sell milk containing less than 3.5% fat, and correspondingly the dairyman will be allowed to abstract fat from milk provided he maintains this level of 3.5% fat. In view of the ease with which fat in ordinary milk rises to the top, it will become a virtual necessity for all dairies to homogenize milk for ordinary retail sale. The British public will, therefore, in future be drinking milk homogenized and standardized at 3.5% fat instead of ordinary un-homogenized milk with a fat content varying from 3.3 to 3.9%.

4.2.6 Sterilized milk

Several years elapsed from the time of Appert's work on the preservation of foods by heating in a closed container before systematic work started on the preservation of milk by heating it in a bottle. This seems to have begun independently in Liverpool and in Germany about 1894. The stimulus came from the scientific work of Soxhlet on the heating of milk to destroy the organisms of tuberculosis, as by now it had been realized that milk was a dangerous source of this organism. The process was rapidly adopted in London and other big towns in Britain and in the USA and Europe generally. The advantages of this heated milk were quickly appreciated at a time when ordinary milk rapidly became sour and clotted, or developed taints. The medical profession soon realized that this heated milk was a safe milk and its easier digestion was soon appreciated by those with weak stomachs. The invention of the homogenizer by Gaulin in 1899 resulted in a great improvement in the quality and appearance of this heated milk. The fat remained evenly distributed and there was no longer the adhesion of the objectionable gelatinous mass of fat on the neck of the bottle. A further improvement was the substitution of the Crown Cork seal in place of the rubber-gasketed swing-type porcelain stopper which was in general use until about 1930.

The term 'sterilized milk' needs some explanation. Strictly speaking the term 'sterile' means the complete absence of life. Thus when a food is sterile it will keep indefinitely from the microbial point of view unless it becomes recontaminated. The success of milk heated in a bottle was primarily due to the fact that recontamination was prevented. The earliest procedures involved either heating the bottles to 100°C for about one hour or under pressure to a temperature between 105 and 110°C for 20 to 30 minutes. This process was sufficient to destroy all vegetative cells of bacteria and a certain proportion of spores, but a few spores survived. Nevertheless such milk remained fit for human consumption for several days, and sometimes several weeks. The term 'sterilized' has come to be used in the food industry to mean 'almost sterile' or in other words containing only very few bacteria, usually spores. Thus it will be seen that the term is used rather loosely; bacteriologists only use it in the sense that no living organisms are present.

In England sterilized milk rapidly became very popular, particularly amongst the poorer people of the midlands and the north. In addition to the advantages of homogenized milk it had the merit of good keeping quality.

a. Definition. In the British Milk (Special Designation) Regulations 1963 the product is required to be 'sterilized', that is to say, filtered or clarified, homogenized and thereafter heated to and maintained at such a temperature, not less than 100°C, for such a period as to ensure that it will comply with the turbidity test. '... The milk shall be heated as aforesaid in bottles and in such a manner that on or before completion of the treatment the bottle shall be sealed with an airtight seal. The term 'Bottle' in this context means any container which is approved by the licensing authority'.

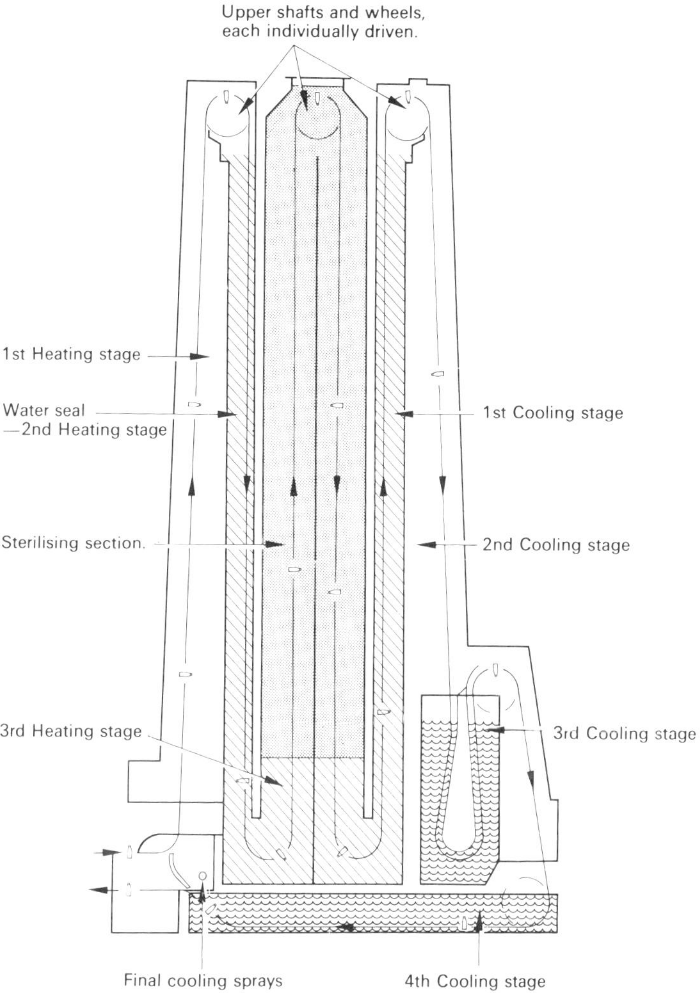

Fig. 4.5 Sectional diagram of hydrostatic sterilizer.

b. Turbidity test. This is a very simple test which merely indicates that the milk has been held for at least a few minutes at 100°C, to ensure the complete precipitation of all the albumin and globulin in the milk. It is intended only to differentiate sterilized milk from pasteurized milk, and is not in any sense a measure of sterility or keeping quality.

c. Methods. Initially the bottles of milk were merely heated in an atmosphere of steam or immersed in boiling water for about 1 hour, but this method was found to have certain disadvantages. The fat rose to the top and keeping quality was very variable. It was found that the separation of fat and irregular keeping qualities could be eliminated by agitation of the bottles during the heating process. Experience also showed that by raising the temperature through the use of an autoclave (or retort as it is commonly called in the food industry) and a shorter time it was possible to get a superior product with good keeping quality and less cooked flavour. Agitation was brought about by a cradle inside the autoclave which could be rocked during the heating process, and later a more sophisticated equipment, the rotating autoclave, came into use. This method gave good results but from the point of view of production had some serious disadvantages. It was necessary to load the frame inside the autoclave by hand, and on completion of the process some time elapsed before the autoclave could be opened and the bottles removed, again manually. Thus the method was expensive and time consuming.

A great advance was made in the technology of sterilized milk by the invention of the continuous mechanical or hydrostatic sterilizer. This equipment had practical and technical advantages over that previously used. Perhaps the greatest advantage consisted in the fact that it completely eliminated the handling of the bottles. The equipment consists essentially of a very long chain conveyor holding cradles into which the sealed bottles can be put and which takes them successively through towers containing water at specific temperatures. Water can only be heated to above the boiling point if it is held under pressure, and this fact is the essential principle on which the process is based.

Milk for sterilizing is always clarified, homogenized and filled into narrow neck bottles at a high temperature, usually about 80°C. The bottle is then sealed and passes into the sterilizer on a chain conveyor holding twenty or more bottles in a row. Here the bottles undergo a sequence of heat treatments. They pass first through warm water, then hot water whose temperature is gradually raised as the bottle proceeds through the machine, until it finally enters the sterilizing zone through a short water section. The actual sterilization takes place in an atmosphere of steam at a temperature of 108° to 112°C. To hold steam at this temperature requires a pressure of 5 to 8 p.s.i. (0.35–0.4 bar), and this pressure is maintained by the column of water in the machine. As atmospheric pressure corresponds to a 34-ft column of water, there must be a difference in water level of 14 to 20 ft (4–6 m) between the columns to maintain this pressure. At the completion of the sterilizing period the bottles descend through another water sealed compartment at about 100°C and subsequently through reverse stages so that the bottles are slowly cooled and finally leave the machine after passing through a section containing water at ordinary temperature.

The development of the continuous sterilizer marked a great advance in the technology of sterilized milk, and the product is today of very high quality. There were at one time two types in production: (1) a milk receiving a low heat sterilization and consequently having a comparatively short K.Q., but having a good flavour and white colour not very different from ordinary pasteurized milk, and (2) milk receiving a more drastic heat treatment and correspondingly longer K.Q. The latter type had a marked cooked flavour and was distinctly brown in colour. However, it found acceptance where K.Q. was the major requirement, and the question of flavour in any food is largely a matter of habituation.

Recently another important advance has been made in the technology of sterilized milk. The U.H.T. process effectively kills all bacteria and spores, but does not of itself protect the milk from subsequent contamination. If milk is first subjected to the U.H.T. process and then filled hot into bottles, which are sealed and sterilized in the continuous sterilizer, it is possible to get a milk of long keeping quality, of attractive flavour and white in colour. This is simply achieved by using a lower temperature and sometimes a shorter time for the sterilization process. This combined method, namely U.H.T. followed by continuous in-bottle sterilization, is now nearly universal for the production of this type of milk.

d. Aseptic filling. Research work during the past 30 years, particularly in the USA, has demonstrated the advantages of sterilizing liquid foods by a continuous process at a high temperature for a short time, and then filling the sterile product under aseptic conditions into sterile containers and sealing these containers aseptically so that it is impossible for any micro-organism to obtain access to the product. This process has reached a high degree of reliability with the Martin–Dole equipment which is now widely used in the food industry. It involves the use of super-heated steam to maintain a sterile atmosphere and is therefore unsuitable for the filling of milk into glass or plastic bottles or simple paper cartons. Realizing the advantages of the aseptic filling of U.H.T. milk in this way, a number of firms have tackled the container problem. The most successful to date has been the Tetrapak system (see below) which is based on the now well-known tetrahedral carton. The aseptic filling of milk into any container is a difficult process because it needs only one weak link in the chain–i.e., one place where asepsis is not maintained–to permit contamination of the milk.

The ingenious Tetrapak system is an elaboration of the ordinary Tetrapak filler. The sterile milk passes from the U.H.T. equipment into the Tetrapak machine and is delivered into the carton as it is being formed. The carton is made from a strip of special paper which is lined with polythene and aluminium foil, or alternatively with black polythene, and this is sterilized immediately before being shaped into a carton by passing it through a bath of 50% hydrogen peroxide solution. This exerts a drastic germicidal effect on the paper which then passes in front of heating lamps which destroy the hydrogen peroxide so that none remains on the strip of paper as it is formed into a carton. The method is now successfully employed in many dairies throughout the world and can be accepted as the most reliable aseptic filling system for milk at the present time. However, other carton manufacturers are now bringing out aseptic filling equipment, and development work is also proceeding with plastic bottles and sachets. It is likely that during the next decade satisfactory equipment will be evolved for the aseptic filling of all types of containers, including glass bottles. However, all these systems naturally cost more than the filling of pasteurized milk into ordinary glass bottles, and the market for this special type of milk, which must necessarily cost appreciably more than ordinary milk, remains problematic. The extra charge can be justified for such purposes as casual sales in shops, milk required on sea voyages, holiday camps and other places where sales are erratic and long keeping quality without refrigeration is an obvious advantage for the seller. Pasteurized milk is today of such good keeping quality that it will keep for several days in the ordinary domestic refrigerator, and now that about 70% of households in the more affluent countries possess refrigerators, it is a satisfactory product for all ordinary purposes. However, it may be anticipated that aseptically filled U.H.T. milk will in time displace the product of the ordinary commercial sterilizer, because it has the advantages of a better flavour, whiter colour and less loss of nutritive value.

e. Filtration. During milking a certain amount of fortuitous dirt, straw, hair, etc., gains access to the milk, and this is always strained through a metal strainer on receipt at the creamery. During pasteurization or other process milk is filtered through a cloth which removes the coarser particles of dirt and a certain amount of body cells etc., but does not remove bacteria or most of the cells which are always present in milk. A certain amount of fine dust also passes through the filter cloth and if a bottle of milk is held for some time this foreign matter falls to the bottom and forms an objectionable sediment.

Fig. 4.6 Aseptic Tetrapak cartoner.

Fig. 4.6a Aseptic Purepak unit.

There is no danger to health if the milk has been pasteurized but the consumer is naturally apprehensive. It is therefore standard practice, when making sterilized milk, to remove this foreign matter by clarification. Some dairies clarify milk for pasteurization, but because of the extra cost it is not generally used.

f. Clarification. This is a centrifugal process in which milk is rotated between discs revolving at a very high speed on a vertical shaft. The clarifier, which is fundamentally the same as a cream separator, consists of a set of conically-shaped thin stainless steel discs which revolve on a central shaft at a very high speed, e.g., 7000 rev/min. All the matter which is heavier than the milk, e.g., leucocytes or white blood cells, red blood cells, epithelial tissue fragments from the udder, fine dust, etc., is flung outwards and collects in the bowl of the clarifier, from which it is removed at intervals. The fat globules or cream collects in the centre of the discs because it is lighter than the skim milk; it is not drawn off separately but is recombined with the skim milk. What the clarifier does, therefore, is to pass the milk through a process which removes dirt etc., but otherwise does not alter the composition.

g. Bactofugation. The use of preservatives to increase the K.Q. of milk is forbidden in most countries, so that the only practicable method possible for making milk safe and prolonging its K.Q. is by heating. Research in the dairy industry has therefore been directed towards making heat treatments more efficient, and to reduce their effect on flavour and nutritive value. In addition, ancillary methods have been explored with a view to reducing the intensity of heat required. Bacteria are slightly heavier than milk but because of their size are not readily removed by ordinary centrifugation, as in a clarifier or separator. Single cells are virtually unaffected but clumps of bacteria are more easily removed by centrifugal force. A drastic heat treatment is necessary to kill bacterial spores in milk, but these bacteria often form clumps and it is possible to remove many of them by high-speed centrifugation, therefore making it possible to obtain a milk of good K.Q. without drastic heating. The method has been developed extensively by Simonart in Belgium who has found that about 90% of these bacteria can be removed by high-speed centrifugation. A double process naturally removes a much higher proportion. The process has been termed *bactofugation*, and is sometimes used in the preparation of sterilized milk. It does, however, involve the employment of an expensive piece of equipment.

4.3 CREAM

Milk is an emulsion of fat globules in fat-free milk, the globules occupying between 3 and 5% of the volume. Cream is simply a concentration of these fat globules to give an emulsion in which the fat globules form up to about 60% of the volume. Cream may be made for sale as retail cream, for making butter, for the manufacture of cream cheese or as a convenient product in the dairy for bringing milk to a given fat content.

Cream was probably the first dairy product ever made by primitive man. Whenever milk is allowed to stand the fat globules rise to the top and, until the invention of the cream separator by de Laval, this was the only method for making cream. Gravity creaming may be employed on farms where no separator is available, and is still in use today in creameries to obtain low fat milk for the manufacture of certain cheese such as

Parmesan and Blue Vinney. Apart from these minor exceptions, all cream is made today by the use of the mechanical separator.

4.3.1 Definition
The British Cream Regulations 1970 define seven types of cream for legal purposes. These are as follows:

Type of cream	Minimum fat percentage
Half or coffee cream	12
Single cream	18
Sterilized or canned cream	23
Cream for whipping	35
Whipped cream	35
Double cream	48
Clotted cream	55

4.3.2 Half and single creams
These are intended for use where a product intermediate between milk and the thicker double cream is suitable, e.g. in coffee and other beverages, for making puddings, etc. This type of cream is always homogenized using a high pressure such as 2500 or 3000 p.s.i. (170 or 210 bar). Homogenization confers the same advantages on this thinner cream as it does on milk.

4.3.3 Sterilized or canned cream
This product is analogous to sterilized milk and to evaporated milk. The standard of 23% minimum fat is set in order to allow the manufacture of a stable cream which will not 'butter' or go oily. Sterilized cream, like evaporated milk, is always homogenized. The general method of processing is virtually identical with that for evaporated milk; the same problems are found, namely how to achieve a sufficient heat treatment to ensure sterility without affecting flavour unduly and without destabilizing the cream. Bacteriological quality is most important because if the milk has soured this will have a marked destabilizing effect, and the presence of any appreciable number of heat resistant spores will shorten the K.Q. of the canned cream. The phenomenon known as 'age thickening' occurs with sterilized cream as it does with evaporated milk, and is caused by the instability and precipitation of the protein. This problem may be overcome to some extent by adding stabilizing salts such as sodium bicarbonate and sodium citrate to the cream. If too much is added the cream darkens on sterilization and flavour may be affected. It is therefore necessary to make tests beforehand with varying amounts of stabilizing salts in order to find out the minimum concentration required to give a satisfactory product.

After stabilization the cream is homogenized at 70° to 75°C, and the lowest pressure necessary should be used, i.e. 1000 to 1500 p.s.i. (70–100 bar), as homogenization itself can have a slight destabilizing effect. The viscosity of the product can be controlled to some extent by the method of homogenization.

After filling, the cans are sealed and then pass to the sterilizer which may be of any type–static retort, rotating autoclave or continuous sterilizer. With the first two it is customary to cool the cans in water, and this presents a hazard as bacteria in the water

Fig. 4.7 Cream separators.

may penetrate the cans at the seams during cooling. This type of contamination can cause bitterness and thinning of the cream, and may be due to spore forming organisms such as *Bacillus subtilis* or to *Proteus* and similar bacteria which are common in water. It is important, therefore, that hygienic precautions are observed and the most important of these is to ensure sterility of the cooling water by chlorination.

Overheating the cream may result not only in cooked flavour and loss in stability, but may also produce black or purple discolouration on the inside of the can, which is caused by the liberation of volatile sulphur compounds from the protein in the milk.

4.3.4 Cream for whipping or whipping cream

The manufacture of cream for whipping, whether used in the household or by bakers for cakes etc., is an important part of the cream industry. In the UK the legal standard is 35% fat but the general preference in the baking industry is for a cream containing about 40% fat. The mechanism of whipping lies in the fact that the globules of fat are surrounded by a skin which enables them to cluster together and entrap air. Cream may be whipped simply by beating or using a high-speed food mixer of the propellor type. Bakeries commonly use a more sophisticated machine which blows air into the cream. Whipped cream occupies a much larger volume than the original cream, and the increase in volume expressed as a percentage is called the *over-run*. A satisfactory over-run is about 120%. This enables the baker to employ a cream which remains stable and does not leak liquid (serum), and at the same time gives him good 'value for money'. It is possible to obtain a higher over-run, sometimes up to 140 to 150% but a whipped cream containing this amount of air is unlikely to be stable and may not only leak serum but may collapse on standing.

a. Manufacture. It is advisable to separate cream for whipping at a rather lower temperature than usual, e.g. at about 30°C, rather than the usual 40° to 45°C. Cream for whipping must not be homogenized in the ordinary way, although some dairies employ a very low pressure treatment, e.g. 300 p.s.i. (20 bar) for whipping cream. It appears that the temperature of pasteurization does not affect the properties of this cream.

Certain treatments are of the greatest importance in manufacture. After separation and standardization of fat content the cream should be cooled to 4° to 5°C and maintained at this temperature for at least 24 hours. Some makers age the cream for up to 48 hours but it appears that the extra ageing does not have much effect. It is important to keep the cream at this low temperature during transport to the bakery. If not whipped immediately the cream should be held in a cold store at 0° to 5°C and whipped at 4° to 5°C. It is important that whipping stops as soon as the desired over-run has been obtained. Whether the over-run increases or not, further agitation will lead to lack of stability in the whipped cream.

Bakers often express dissatisfaction with the quality of their whipping cream, but often the fault lies with their own procedures. Lack of stability is due to one or more of the factors discussed above, and development of taints in the cream is due to the lack of hygiene, either in the manufacture of the cream or in the treatment in the bakery. Cream differs from milk in that it is usually older when used, and further time may elapse between the utilization in the bakery and consumption in the home. For this reason cream may develop a high bacterial count and have a poor flavour if not an actual taint. Both the dairyman and the baker are therefore involved in ensuring that the consumer receives a good quality whipped cream.

The British Cream Regulations 1970 permit the incorporation of certain stabilizers for whipping cream, and the use of these depends upon the contract between maker and buyer. The stabilizers include sodium alginate, sodium carboxymethyl cellulose, carrageenan and gelatine, but the total present in the cream must not exceed 0.3%. Whipping cream may also contain not more than 13% of sugar, and where the cream is dispensed from an aerosol container, nitrous oxide may be employed as the propellant gas.

4.3.5 Double cream

Double cream having a minimum fat content of 48% is by far the most popular type of cream for the ordinary consumer. Sales are higher during the soft fruit season and at weekends, and cream has now become a common luxury in the home and one of the most profitable products for the town dairyman. The method of distribution differs appreciably from that of ordinary milk. Whereas milk is produced one day, pasteurized the next and delivered to the housewife the following day, so that it is nearly always consumed within two days of pasteurization, cream is manufactured, distributed and utilized much more spasmodically. In practice this means that cream may be some days old before it is finally consumed, and this inevitably affects its bacteriological or hygienic quality. The dairyman, therefore, has to take special precautions in the manufacture and distribution of cream. A more severe heat treatment is given to cream than to milk; this is partly because of the greater viscosity of cream passing through the heat exchanger and partly because heat transfer in cream is poorer than in milk because of the greatly increased fat content. Cream has to undergo more operations than milk and every piece of equipment with which it comes into contact can contaminate it unless the most stringent hygienic precautions are observed. Cream has first to be separated, filled into containers, aged and finally distributed, often through a wholesaler. Thus the more complicated sequence of operations and delays in distribution all contribute to increase the hazards affecting the quality of the cream.

One of the curious properties of cream is its viscosity. In spite of all the research which has been done on the subject, the dairy industry cannot guarantee to produce a cream of a given viscosity without adopting procedures which may be uneconomical or impossible in practice. There appear to be seasonal and regional differences in the fat globules of milk which affect this property. A common method for increasing the viscosity of cream is the so-called 'rebodying' of cream. It consists of cooling cream after heat treatment to a temperature in the range 20° to 30°C and then allowing it to cool slowly over a period of several hours down to the normal storage temperature of 5°C. The mechanism of this process lies in the fact that the fat in milk has a melting point of about 30°C (although there is no definite value and the melting and solidification of milk fat take place slowly over a range of temperature) and if the fat in the cream is allowed to solidify slowly *without agitation* the fat globules become 'tacky' and adhere to each other and in this way increase the viscosity of the product. It is possible to produce cream of almost any viscosity by this method; some firms may even use surprisingly high temperatures and long periods of cooling to achieve a very high viscosity, but the higher the temperature and the longer the period of cooling the more opportunity there is for bacteria to grow and produce souring and taints. The method is therefore attended by risk and requires the greatest care in use.

Some dairies make a 'long life' cream in which nearly all the bacteria have been killed by some kind of heat treatment. The U.H.T. method cannot be used for cream much above about 35% fat because this increases the danger of physical instability or 'buttering'. One method which has been used for high fat creams is to heat the product in a bottle hermetically sealed at 100°C for two or even three periods of 1 hour at 90° to 100°C. This is a traditionally used method for sterilizing liquids without having recourse to pressures above ordinary atmospheric conditions, and was formerly known as 'Tyndallisation'. Even with this method the cream can in time become unstable with separation of serum at the bottom of the bottle. Modern methods have now been improved to overcome this difficulty.

4.3.6 Clotted cream

This is a famous traditional luxury product made chiefly in Devon and Cornwall for the tourist trade in the summer. It was formerly only made in farmhouses by a very simple and crude method. Milk was placed in pans in front of the farmhouse fire for several hours, which permitted a number of changes to take place. The milk would naturally contain a number of bacteria which would at first be encouraged to grow by the warmth of the fire. These bacteria would produce not only lactic acid (the ordinary souring process) but also aromatic substances such as ketones, aldehydes, volatile fatty acids, etc. which are responsible for the traditional flavour of this product. During this time the fat globules would rise to the surface and, as in earlier times Jersey and Guernsey cows were common in this region, the large fat globules with their yellow colour would rise more quickly than those in ordinary Shorthorn and Friesian milk. The increasing heat would gradually coagulate or precipitate the 'whey proteins' (albumin and globulin) and form a matrix with the entangled fat globules, thus giving ultimately a solid mass on the surface of the milk. Finally the heat would kill off nearly all the bacteria so that the product would have an enhanced keeping quality.

Clotted cream is now made mainly in factories or creameries. The same basic method is used but the cream is separated in the ordinary way and filled into glass bottles. These are then placed in a shallow water-bath at about 90°C and held there for an hour or longer. Exactly the same sequence of changes takes place as described above but with better control and more hygienic conditions. A cream of about 50% fat content is used and the heating process results in the loss of a certain amount of moisture. The treatment must be controlled so as to ensure a fat content of at least 55% in the product as sold, in order to meet legal requirements.

4.3.7 The standardization of milk and cream

Unless otherwise qualified, in the dairy industry the term 'standardization' means the adjustment of the fat content of milk or cream to a certain concentration. Milk is normally standardized for the manufacture of condensed and evaporated milk, dried milk or milk powder and yoghourt and cultured milks. Cream is standardized at concentrations from 12% to nearly 55% according to the type of cream which is being manufactured.

Although sophisticated equipments are now being put on the market for the standardization of milk, these are expensive and most dairies standardize by calculating the volume of skim milk (or cream) which is necessary to add to a given volume of milk, skim milk or cream to give a product having the desired fat content. The process consists of two parts: 1. the calculation of the proportion of basic ingredients, and 2. the mechanical operations involved in mixing two liquids. The simplest method of calculation is to use the Pearson Square. To do this a square is drawn with diagonal lines and the fat percentages of the two ingredients are placed at the left-hand corners of the square. The desired fat percentage is then placed at the intersection of the diagonal lines and simple subtraction between these figures gives the necessary quantities to use. An example is given in the diagram below; the quantities appear in the right-hand corners.

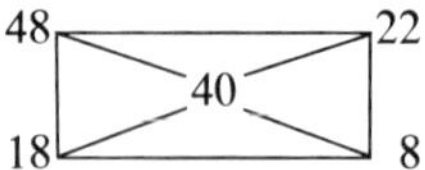

This very simple method of calculation is quite satisfactory for all purposes but the manipulation of the liquids may present some difficulty. The usual procedure for milk is to run a given volume, e.g. 5000 litres, into a tank and then to calculate the necessary amount of skim milk to be added. The volume may be metered in from another tank, or alternatively if the first tank has the required capacity and is fitted with a gauge the skim milk can be run in up to the required volume mark.

When standardizing cream it is usual to separate a higher fat content than required, and then to reduce this by adding skim milk. Two common faults in this process are the failure to observe hygienic precautions throughout the entire process, and failure to mix the two liquids thoroughly, resulting in 'pockets' of higher or lower fat content.

4.4 BUTTER

Butter was without doubt the first solid dairy product made by man as soon as he had acquired the art of obtaining milk from domesticated animals. Under the primitive conditions then holding, the milk would rapidly sour and with agitation under warm conditions it would quickly 'butter', i.e. the fat globules would become destabilized, shedding their membranes, leaving free fat which would then congeal into a solid mass. In course of time the advantages of using cream obtained by gravity creaming would have been realized, and the whole process of butter making was revolutionized when the cream separator was invented by de Laval in the nineteenth century. This enabled dairymen to separate cream at any desired fat content, and butter-making in large cylindrical wooden churns became a well-developed technology. In more recent years the wooden churn has been replaced by stainless steel churns, and the latest invention in this field is the continuous butter-making machine.

4.4.1 Definition
Formerly butter was only defined as 'the substance usually known as butter' and was required to be made from milk and have a moisture content not greater than 16%. The British Butter Regulations 1966 give more precise definitions and legal requirements for chemical composition. Butter is defined as the fatty substance intended for sale for human consumption and derived exclusively from cow's milk, the pH of which may have been adjusted by the addition of an alkali carbonate. It may contain salt, lactic acid cultures, and annatto, carotene and/or turmeric as colouring matter. The definition includes whey butter which is butter made from cream separated from cheese whey.

The chemical composition requirements are now that butter shall contain not less than 80% milk fat, not more than 2% milk solids other than fat, and nor more than 16% water. There is however, a qualification to meet the slight complication introduced by the presence of salt in butter. This provides that butter may contain less than 80% but not less than 78% milk fat if the amount by which the milk fat content percentage falls below 80% does not exceed the amount by which the percentage of salt in such butter exceeds 3%, and provided that the words 'salted butter' appear clearly on the label.

4.4.2 Manufacture
Today only very small amounts of butter are made on farms and virtually all butter is manufactured in large creameries with very advanced stainless steel equipment and detailed and scientific hygienic control. Both cream and butter are powerful absorbants of odours and these odours are strongly retained by cream and butter, and impossible to remove from the latter. It is therefore of the greatest importance that the cream used for butter-making, and also the milk from which the cream is separated, shall be entirely free

from any objectionable odour, whether this has been produced by micro-organisms or has been absorbed from the environment. The quality control system therefore starts with the milk itself and the organoleptic examination of the milk is more important when it is to be used for butter-making than when it is to be used for pasteurization and ordinary retail sale. As the yield of butter depends upon the fat content of the milk, such milk is commonly purchased on a quality basis calculated either from the fat content or the estimated quantity of butter which the milk would yield.

Unless the milk is of good hygienic quality it may be neutralized if conditions of collection have permitted the growth of souring bacteria, and hence the production of acid. It is a standard practice in countries where butter-making is a major part of the dairy industry, as for example in Australia and Ireland, for the milk to be separated on the farm and the cream collected from the farm only every two or three days. The system naturally depends upon the distance of the farms from the creamery. Under such conditions souring is inevitable and the cream will be neutralized, usually with sodium bicarbonate or sodium sesquicarbonate.

It is now universal practice to pasteurize or otherwise heat treat cream which is to be used for butter-making. Any system of heat treatment, if efficient, will destroy all pathogens and reduce the numbers of bacteria, yeasts and moulds to a negligible proportion, but ordinary pasteurization will not necessarily eliminate 'feed' or absorbed flavours from the cream. It has therefore become a common practice to treat cream by a steam or vacuum process which not only heats the cream but removes any obnoxious flavour substances which may be present. Equipments of this type include the Vacreator (New Zealand), the volatilizer (USA) and the counter-current cream treatment unit (A.P.V., England).

Immediately after heat treatment the cream must be cooled to 4° to 5°C and held at this temperature for several hours in order to allow the fat to solidify or 'crystallize'. This is an essential procedure if a good body and texture are to be obtained.

The churning of butter is brought about by some form of agitation, resulting in an inversion of the cream emulsion. A number of theories have been proposed to explain this phenomenon but it is now generally accepted that there is a sequence of the destruction of the fat globule membrane, a close packing of fat globules in the foam which is formed as the cream is agitated in the presence of air and finally the inversion of the emulsion such that the fat forms the continuous phase and water the disperse phase. As with the problem of the viscosity of cream, the churning process is governed by many factors. Sometimes it may require a long time and this was formerly called 'sleepy butter' or 'sleepy cream'. The optimum fat content is about 37% fat; this gives a good churning time and also a minimum loss of fat in the buttermilk. Temperature of churning is probably the next most important factor and the skilled butter maker will vary this according to the season of the year, which is also linked with the stage of lactation of the cows, the nature of the feed (grass or artificials), the breed of cow, the acidity of the cream, the intensity of agitation (speed of rotation of the churn), the extent to which the churn is filled with cream, and the period of holding the cream between heat treatment and churning. Under the present conditions of the dairy industry, where large bulks of milk are manufactured every day, there is little change in these factors. Cream derived from Jersey and Guernsey cows will churn more easily and give a softer butter. An acid cream will churn more readily than sweet cream and two types of butter are recognized – sweet (non-acid) cream butter and soured or ripened cream butter. Cream churns most rapidly when the churn is about one third full, and if a churn is over two-thirds filled churning will be delayed because adequate agitation then becomes impossible. The churn should be operated at the speed laid down by the manufacturer of the equipment. Finally if the cream is churned too quickly after heat treatment a soft butter of poor texture will be obtained and there will be a higher loss of fat in the buttermilk. When the cream 'breaks' or becomes changed into butter, a large solid mass of butter granules becomes visible and the excess milky liquid from the cream is evident as buttermilk in the bottom of the churn. A simple calculation will show how much crude butter containing 80% of fat can be obtained from a given weight of 37% fat cream. Buttermilk contains only about 1% fat and is commonly mixed in with the separated milk obtained when cream is made, but it is rich in the lipid-protein substances of which the fat globule membrane is composed, and for this reason buttermilk has valuable emulsifying properties. It is especially useful in the manufacture of ice-cream and for the reconstitution of skim milk powder. On account of the relatively small quantities involved, however, it is usually not an economic proposition to dry it separately from the skim milk.

4.4.3 The churning process

Originally butter churns were constructed of wood, and a few wooden churns are still in use. They were formerly fitted with internal rollers so that the butter could be worked in the churn, but simple vanes are sometimes fitted now. Stainless steel churns are now a standard piece of equipment, and the trend is to construct them without rollers or vanes, the churning being achieved solely by concussion.

Fig. 4.8 Stainless steel butter churns.

Particularly with wooden churns, it is most important to sanitize them immediately before use. This is usually done with boiling water or sometimes a cold chlorine preparation. After the usual laboratory tests have been made on the cream, this is pumped into the churn so that it is slightly over one-third filled. When the churning begins the cream expands in volume and gases escape from it, and these are released through a vent after a short time. The churning time is usually 30 to 40 minutes, and the course of the churning can be estimated by a change in the concussive sound. Finally, when the fat globules coalesce to become discrete granules the break occurs and can be detected in addition by the clearing of the sight glass in the side of the churn. This is then opened and water at a temperature 2 to 3°C below the churning temperature, and to the extent of about 5% of the volume of the cream, is added to improve the condition of the buttergrains. When these have reached the size of 3 to 5 mm the buttermilk is run off through a strainer. The correct procedure in the churning of cream must be most carefully observed because otherwise there will be an unnecessary loss of fat in the buttermilk and the butter may have a poor body and texture.

After the removal of the buttermilk a further quantity of water at slightly below the churning temperature is worked over the butter granules and then run off. Further water is then added and the churn revolved several times in order to wash and harden the butter.

a. Working and salting. It is necessary next to change the numerous small granules into the continuous compacted solid form of butter. This process is termed 'working' and also brings about the dispersal of the moisture into minute droplets only a few microns in diameter. Thus cream, which is an emulsion of fat in water, has been converted to butter, which is an emulsion of water in fat. This is why the microbiological keeping quality of butter is superior to that of milk and cream.

Working is carried out in the churn, if this is provided with internal rollers, or it may be done in a separate machine. If the butter is to be salted, the salt is incorporated at the same stage. Broadly speaking there are two types of butter on the market, salted and unsalted, and the amount of salt added depends simply on the market requirement. The amount of salt and of water required to bring the moisture content as near as possible to the 16% limit is weighed or measured out, put into the churn, and worked in by rollers or the mixing arms if a separate machine is used.

b. Over-run. Just as the over-run in whipped cream is a measure of the increase in volume when air is incorporated in the cream, so the over-run in butter is a measure of the difference between the final quantity of butter compared with the original amount of fat which has been obtained from the milk. Thus over-run is controlled by the amount of moisture and other non-fat substances in the butter, and the loss of fat during the butter-making process, i.e. that lost in the buttermilk.

c. Grading. Grading or judging of butter is always carried out by specialist graders when the butter is at least some days old. Butter is always stored at a low temperature, e.g., − 10°C or lower according to its method of marketing, but should be raised to 12° to 13°C for grading, because otherwise the aroma and taste of the butter cannot readily be assessed.

Each country has its own system of awarding marks for the grading of butter. For example, in the UK 50 points are awarded for flavour, 20 for body and texture, 20 for general appearance and colour and 10 for absence of free moisture, i.e. visible droplets on the surface of the butter.

From the consumers' point of view, flavour is the most important property of butter and faults are variously described as bitter, cheesey, cooked, cowy, feed flavour, fishy, flat, musty, rancid, metallic, sour, stale and unclean. The skilled butter-maker and grader can relate these defects to certain faults in the quality of the cream or in the method of manufacture.

Body and texture are considered together. Body may be defined as the firmness of the butter, and is highly dependent upon temperature, hence the necessity for grading butter at a defined temperature. Texture is the appearance of the butter apart from colour and is particularly concerned with homogeneity and holes or spaces in the butter. Body and texture may be described as greasy, gritty, leaky, mealy (or crumbly), sticky or weak.

Colour is considered primarily in relation to the breed of cow concerned, and the amount of colouring matter such as annatto or carotene which may have been added. Homogeneity of colour is obviously of the greatest importance. Defects in colour may be described as dull, mottled or uneven, and streaky.

4.4.4 Modern methods of butter manufacture

As in all branches of food technology, the classical butter-making equipment is now being rapidly superseded by modern machines which have great advantages over the orthodox churn. These all permit the conversion of cream into butter by a continuous process and so the machines are commonly referred to as continuous butter-making machines. In the early years of this new development (from 1940 onwards) several types of machine involving different principles were developed. Broadly these fell into two groups – one in which a simple phase inversion (from a fat-in-water to a water-in-fat emulsion) was precipitated by a sudden temperature shock, and the other which simulated what happens in the orthodox churning process. The former type was never popular with the graders, and it suffered from two main disadvantages. The structure of the butter was not considered to be analogous to that of traditional butter, and the simple phase inversion resulted in the retention of the non-fatty ingredients of the cream. These included small amounts of protein (curd) which acted as a nutrient for bacteria so that the microbiological quality of the butter was affected.

Practically all continuous butter-making machines are now based upon a standard working principle. The milk is first warmed to about 45°C and is then passed to a hermetic separator to yield a cream of about 35% fat. This is then pasteurized or treated by a vacuum-steam process which raises the temperature of the cream to about 95°C, from which it is cooled down to about 45°C and further separated to give a cream of a suitable fat content for the butter machine. The cream then passes through a mixing tank where colour, salt and water can be added according to the type of butter required. The cream is pumped through a special machine (a 'transmutator' or a 'votator') which consists essentially of jacketed stainless steel cylinders (usually three) about 2 m long and 37 cm diameter. The cream passes through the cylinders in turn and is cooled by brine (strong calcium chloride solution) or other suitable refrigerant to successively lower temperatures, and when the butter reaches a temperature of about 10°C the phase inversion takes place and the cream is converted into butter. The agitation of the cream inside the cylinders may be brought about by various methods, for example by raised spiral strips, or by a votator knife or blade which is a strip of metal held on two projecting flanges from a central shaft and which exert a scraping action on the internal surface of the cylinder.

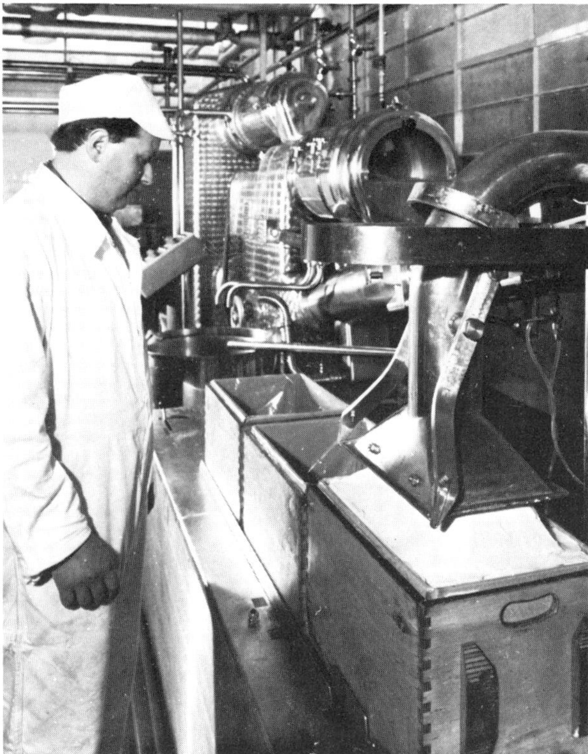

Fig. 4.9 Continuous butter-making machine.

Finally the butter is extruded from the machine and packed directly into boxes lined with greaseproof paper.

4.4.5 Sweet cream and ripened cream butter

Although all types of butter are basically the same, there are two broad divisions into which they can be grouped. Sweet cream butter is made from cream which has not been allowed to sour or 'ripen'. It is produced under the most hygienic conditions and pasteurized before conversion into butter. It was realized long ago that if milk or cream was allowed to go sour, one obtained not only quicker churning, because the acidity helped to destabilize the fat globules, but also a more highly flavoured butter. Without modern scientific control such souring is uncontrolled and may easily lead to defects, consequently if a highly flavoured butter is desired today the cream is inoculated with a special type of souring organism called a butter culture or butter starter (cf. cheese, section 4.7). It was discovered accidently that if such cultures were held at a low temperature for several days a more delicate and refined butter aroma was obtained, and it was ultimately found that this aroma was due to the production of diacetyl by *Streptococcus paracitrovorus* and related organisms. Later a faster growing type named *Streptococcus diaceti-lactis* was discovered which was capable of growing quickly and producing large quantities of diacetyl, and this organism is often used today in butter cultures. Thus there are in the market two types of butter–the sweet cream butter as commonly made in the UK, Australia and New Zealand, and the highly flavoured ripened cream butter as made in France and Denmark. There is a relation between the extent of flavour produced in the butter by the use of ripened cream and the amount of salt incorporated. For a sweet cream butter which is relatively flavourless, salt naturally improves the taste, whereas flavoured butters may not contain so much salt.

Another very important aspect of the two main groups of butter is keeping quality. Butter may deteriorate through the growth of micro-organisms, particularly moulds and yeasts and also by chemical oxidation leading to rancidity and tallowy flavours. This oxidation is catalysed or accelerated by acidity (low pH value) and also by traces of heavy metals of which copper is by far the most important, the next most important in practice being iron. Formerly when copper and bronze were in common use for dairy equipment, cream contained traces of copper which passed into the butter, but today when virtually all dairy equipment is made of stainless steel this problem no longer exists.

For obvious reasons butter which has to be transported long distances, as from Australia and New Zealand to the UK, is made from sweet cream. Provided that such butter is held at from $-10°$ to $-15°C$ it can be kept in good condition for several months.

4.4.6 Renovated butter

The keeping quality and marketing of butter, as of all dairy products, is one of the most important problems with which the dairy industry throughout the world has to contend. In periods of high production of milk, which are largely controlled by temperature and rainfall in different countries throughout the world, there is naturally an accumulation or stock-piling of dairy products such as butter and milk powder. If butter has to be stored too long it will inevitably deteriorate and the industry then has the problem of utilization. It is possible to remake such butter by melting it, clarifying and washing the fat which separates, and then re-emulsifying this fat with fresh or ripened milk and so preparing a reconstituted cream which can then be churned into butter again. This type of butter is called 'renovated', 'processed' or 'rechurned' butter and although not of the highest quality can be readily used for human consumption either as butter or in manufactured foods.

Butter may also be reworked or blended. The object of this may be to produce a butter of a desired flavour or body, or sometimes to control the water content of the butter. Such working tends to increase the size of the water globules in the butter and so increases the bacterial count.

4.4.7 Buttermilk

This is the milky liquid which separates when cream is churned into butter. It is rich in lipoproteins, which contain lecithin, and for this reason has a characteristic flavour. Although considered rather a by-product and of little value, it is a pity that its valuable emulsifying powers are not better utilized than by merely re-incorporating it in skim milk for drying. In the USA a product termed 'cultured buttermilk' has been on the market for many years, and this is now sold in the UK. The name is a misnomer because the product is simply a culture of common starter organisms, such as *Streptococcus lactis* and *Streptococcus cremoris*, in ordinary skim or separated milk. The demand for this is increasing, partly because such cultured milks are now becoming more fashionable and also because of the present tendency to regard animal fats as dangerous from the point of view of 'heart disease'. The term should strictly speaking be confined to real buttermilk.

Fig. 4.10   Multiple effect evaporators.

major part in the evaporating process. Stainless steel is now universal because of ease of cleaning and sterilization and freedom from corrosion. The milk is heated either by steam coils at the bottom of the pan or by a calandria. When used, coils are installed in sections, each with its own steam inlet valve and condensate trap in order to give accurate control of the rate of evaporation. Coils are now becoming obsolete and replaced by calandrias. A calandria may be regarded as a cylinder through which passes a number of tubes, similar to the tubular heater used for sterilizing milk. The milk is forced through the tubes by pressure or by thermal effect and the tubes are surrounded by an atmosphere of steam in the cylinder. Calandrias are superior to coils not only thermodynamically but because there is less formation of deposit and so cleaning is easier.

Two or more evaporators may be joined together to provide great economy in steam consumption, although a single pan is generally used for the manufacture of sweetened condensed milk. The vapour from the first-effect evaporator is used to heat the milk in the second-effect, the latent heat of this steam increasing its effectiveness. For the greatest efficiency a temperature difference of about 27 deg. C is required between the working temperatures of the two evaporators. Typical temperatures and vacua for a triple effect plant might be 200°F (93°C) and 89 mm, 180°F (82°C) and 354 mm, and 49°C and 660 mm of vacuum. Multiple effect evaporation may be unsuitable for products which are very heat sensitive because the temperature in the first effect may be too high for the quality of product required, but is generally used in making unsweetened evaporated milk where a continuous process is desired. Increased efficiency may be obtained by the process known as thermo-compression, in which part of the vapour from the milk is heated and compressed by high pressure steam, and then this supplies the steam for the heating surface of the next evaporator.

The rate of evaporation depends upon the temperature difference between the steam and the milk, the area of the heating surface, the speed of circulation of the milk over the heating surface and the degree of vacuum. Also important is a large diameter vapour off-take pipe because small pipes mean high vapour velocities with consequent high losses of milk by entrainment or the carry-over of spray from the boiling milk. This obviously leads to inefficiency and incidentally increases the polluting effect of the condenser effluent, a problem which is attracting the attention of the river and hygiene authorities today. An interceptor chamber on the top of each pan traps most of the droplets of milk spray which are then returned to the pan, although unless such material receives a separate heat treatment bacteriological difficulties may be encountered. Part of the skill in condensing is to reduce entrainment to the minimum.

It is desirable not only to produce a satisfactory vacuum, but to maintain it as constant as possible throughout the entire process. Not only water vapour but incondensible gases, mainly air, have to be removed continuously during the evaporation. All the vapour is passed through a condenser where it is cooled by a spray of cold water. The vacuum may be produced by a vacuum pump or by a steam injector. The effluent temperature should be from 3 to 8 deg. C lower than the working temperature of the evaporator (about $120°F$ ($50°C$)), and from 20 to 30 kg of water are required to condense 1 kg of vapour.

Basically the same method is used irrespective of the final product (sweetened or unsweetened). The milk is first pre-heated to between $75°$ and $140°C$ for suitable times in order to confer stability on the milk and then is drawn into the pan by vacuum from a hot well. The pre-heat treatment not only stabilizes the milk by bringing about changes in the constitution of the proteins, but kills the bacteria, yeasts and moulds which may be present in the milk, and also inactivates enzymes which are always present in raw milk. All these are capable of bringing about faults in the final product.

The hot milk is steadily drawn in to maintain a certain level in the evaporator and the process is controlled by the operator watching this level, the pan and condenser effluent temperatures, and the vacuum gauge. When all the milk has been drawn in the intake valve is shut and evaporation proceeds until the desired concentration has been achieved, which is usually from 10 to 15 minutes after the last ingress of milk. The most important stage in the whole operation is the 'striking' of the batch and this is where experience can play an important role, because the operator can base his judgment not only on certain tests on samples withdrawn from the evaporator, but also the appearance of the boiling milk.

The tests which are made are usually based on specific gravity, viscosity and/or refractive index. Of these, specific gravity is more reliable because this is linearly related to the total solids content of the concentrated milk, whereas viscosity is controlled not only by composition but by other factors which are variable and not fully understood. Tests must be made under carefully standardized conditions and, as in cheese making, the operator relies on experience as well as on the results of the analytical tests.

4.5.4 Evaporated or unsweetened concentrated milk

This type of milk has to be sterilized in a hermetically sealed container. In the UK regulations require a concentration about $2\frac{1}{2}$ times and the raw milk has to be appropriately standardized. It is then pre-heated, evaporated or concentrated, homogenized and filled into cans; then the cans are sealed and sterilized. One of the main problems in the manufacture of evaporated milk is the heat stability of the product. Unless this is managed efficiently the sterilization may produce lumps which make the product unmarketable. When the concentrated milk is heated all the albumin and globulin are coagulated and in an unstable milk some of the casein is also precipitated. The extent of coagulation is controlled by the salt balance in the milk, the degree of souring, the pre-heat treatment, the extent of concentration and the sterilizing temperature. The conditions for pre-heating have been studied extensively and one of the best appears to be heating the milk to a temperature of about 118°C with a holding time of 3 to 4 minutes.

Instability owing to incipient acidity can be corrected by the addition of sodium bicarbonate, but this must be done most carefully because over-neutralization leads to a darkening of the colour. Instability caused by salt imbalance (which occurs most generally in the spring and autumn when the method of feeding the cows changes) is usually corrected by adding sodium citrate. This is done after concentration, using experimental pilot batches containing different concentrations of sodium citrate and heating them at the sterilizing temperature.

The milk is standardized by adding skim milk or cream to the milk in a tank, making the calculation to adjust the fat and solids-not-fat values to give a ratio of 1:2.44 so that the product complies with the legal requirement of 9% fat and 22% solids-not-fat. Government regulations require both fat and total milk solids standards for the full cream product, but only total milk solids for the skim milk product.

When the concentrated milk leaves the evaporator at about 50°C it is homogenized to prevent fat separation during and after sterilization, and also to improve viscosity. As this process usually reduces heat stability the minimum pressure should be used.

The use of a stationary autoclave is not recommended for the production of evaporated milk because an even temperature rise and fall cannot be obtained with all the cans in such a machine. A rotating autoclave in which the cans are loaded into crates and then the crates held in a frame inside the autoclave which rotates continuously during the sterilizing process can be used successfully. The autoclave is first filled with hot water at about 55°C so that some of the cans are submerged, and then the temperature is raised by steam over a period of about 20 minutes. The frame should rotate at 10 to 15 rev/min. The actual time-temperature conditions to give a satisfactory product must be found by trial and error and will vary with the season of the year. Usually a temperature of 112° to 118°C for 15 to 20 minutes is suitable. With inadequate heat treatment spores of *Bacillus subtilis* will survive and cause bitterness and thinning in the sterilized product on storage. Over-sterilization will lead to a darkening of the product and the production of lumps in storage. On the completion of the sterilizing stage the cans are cooled by water which should be chlorinated, as on cooling bacteria can sometimes be drawn into the cans through the seams.

Ordinary autoclaves are only used today for small-scale production. In large factories continuous sterilizers are almost universal. The conventional equipment consists of three tanks or vessels which carry out the three stages—raising the temperature, holding the cans at the sterilizing temperature for the required time, and finally cooling. The cans pass on a rotating spiral trackway through each vessel in turn and are transferred mechanically from one vessel to the next through a steam trap.

Another type of continuous sterilizer is constructed upon the same principles as the equipment for sterilizing milk in bottles already described. The cans are operated in the same way and are passed up into a pressure steam chamber with a water seal on either side. This type of equipment has the advantage that the cans can be heated and cooled gradually and with ease of control. Finally the cans issue from the machine at about 30°C, a temperature which allows them to dry off in the air.

4.5.5 Sweetened condensed milk

Although the method of concentrating the milk is the same, sweetened condensed milk differs fundamentally from evaporated milk in that the microbiological stability is obtained by the incorporation of 42 to 45% sugar (sucrose), and the product is subsequently not sterilized. The organisms which are most likely to cause faults in condensed milk are yeasts and moulds, and as these are commonly present in sugar the sugar syrup must be heated sufficiently to destroy any such organisms which may be present. After suitable heat treatment the sugar syrup is drawn into the pan by vacuum at such times and in such manner as the operator feels is the best method.

Like evaporated milk, condensed milk may suffer in quality because of the heat instability of the milk, but for a different reason. Instead of forming lumps the whole mass may become more viscous and ultimately solid over a period of several months, and this form of deterioration is called 'age-thickening'. The phenomenon is not fully understood but it appears to be influenced by changes in the salt balance of the milk, variations in which give an unstable period during April and May, and a recurrence to some extent in the autumn in the northern hemisphere. Salt imbalance may be corrected by sodium citrate, and disodium phosphate is also used. These salts raise the pH slightly but their chief effect is probably to increase the ratio of monobasic to dibasic cations, and the phosphate and citrate radicals also enhance stability. Incipient souring also acts adversely but probably the most important factor in respect of this fault is the proportion of colostrum or early lactation milk in the bulk. This may largely explain the periods of instability in the spring and in the autumn. Mastitis milk also has an adverse effect.

In addition to correcting the salt imbalance, the choice of a correct pre-heat treatment is a most important device for the manufacturer. It appears that the best results are obtained by heating to about 82°C for 15 minutes or alternatively to a temperature above 110°C for a few minutes.

In condensed milk the inclusion of sugar introduces a new problem. The concentration of the sugar in the water phase is very high (60 to 65%) and the lactose concentration is also increased in proportion, so that when the product is cooled to atmospheric temperature the concentration of lactose is greater than the solubility, and it crystallizes. If crystallization is slow the crystals of lactose may be large enough to be detectable on the tongue, and in this way the product produces a feeling of sandiness or grittiness. This can be detected when the crystals are greater in size than about 15 microns, so that it is necessary to ensure that the lactose crystals are kept down to a size of about 10 microns. In order to achieve this the product is 'seeded' by addition of finely powdered lactose which accelerates the crystallization of the lactose in the condensed milk, and the product must be cooled sufficiently quickly to encourage rapid crystallization and so the formation of very small crystals. The seeding is usually carried out at 30° to 33°C; the condensed milk is then agitated and cooled to 15°C. The crystallization should be completed in about 3 hours.

4.6 DRIED MILK OR MILK POWDER

Milk is an almost perfect food for children and adults but it has one serious defect. It contains a large amount of water so that when a tanker conveys milk for hundreds of miles, as it does in the UK to supply the big cities, seven-eighths of the load is water. From the point of view of cost it would be much more economical to remove the water in regions of production so that only one-eighth of the load would require transport, in the form of milk powder. In spite of the technological advances in the drying of milk, it is either impossible or very expensive to produce a powder which is free from the characteristic 'cardboardy' flavour and which will keep well. But for this fact it is likely that by now the distribution system for liquid milk, which is an outstanding feature of the dairy industry in the UK, would have disappeared. Instead of the housewife being supplied with bottles or cartons of milk on the doorstep every morning she would buy cans or packets of milk powder at the grocers or supermarket.

4.6.1 Regulations

All milk powder sold for human consumption must comply with the Dried Milk Regulations 1965. In these regulations, dried milk 'means milk, partly skimmed milk or skimmed milk, intended for sale for human consumption, which has been concentrated to the form of powder or solid by the removal of water, and includes any such milk which has been sweetened, modified or compounded'. These regulations apply to dried milk to which no other substance has been added and to the dried milk contained in any powder or solid of which not less than 70% consists of dried milk.

Dried milk must contain not more than 5.0% of moisture, and have milk fat contents as given in table 4.7.

The label must give the equivalent volume of liquid milk based on the composition in table 4.8.

The label must also include the following words as may be appropriate:

1   The word 'sweetened' where only sugar has been added.
2   The word 'modified' where only a constituent of milk has been added.
3   The word 'compounded' in every other case.

Table 4.7   Description of dried milk

Description	Percentage of milk fat
Dried full cream milk	Not less than 26
Dried three-quarter cream milk	Less than 26 and more than 17
Dried half-cream milk	Not more than 17 and not less than 14
Dried quarter-cream milk	Less than 14 and not less than 8
Dried partly skimmed milk	Less than 8 and not less than 1.5
Dried skimmed milk or dried low-fat skimmed milk	Less than 1.5

Table 4.8   Composition of dried milk

Description of milk	Minimum amount of milk fat %	Minimum amount of milk solids including fat %
Milk	3.6	12.4
Three-quarter cream milk	2.7	11.6
Half-cream milk	1.8	10.8
Quarter-cream milk	0.9	9.9

With dried skimmed milk the label must bear the words 'unfit for babies' or 'not to be used for babies'. With all other reduced fat dried milks the label must bear the words 'should not be used for babies except under medical advice'.

4.5.2 Manufacture

Although many methods have been proposed and used for the drying of various materials, only two are used to any extent for milk. These are the roller process, which is the older and cheaper method, and the spray process, which costs more but gives a powder of superior quality. Other methods which have been used to a very limited extent are freeze-drying or accelerated freeze-drying, the foam spray method, and some esoteric methods such as the Birs method, which is based upon drying in a very tall tower in air of very low humidity at a low temperature, for example 30°C. Such methods give a virtually perfect product of high solubility and an absence of the usual milk powder flavour, but are quite uneconomic for the ordinary type of product. They can only be justified when the product commands a high price.

a. The roller method. The equipment for this method consists of one or more usually two hollow steel cylinders rotating on a horizontal axis and heated internally by steam under pressure. The cylinders are usually from 3 to 4 ft (about 1 metre) in diameter and may be up to 12 ft (about 4 metres) in length. The milk, which is always concentrated by evaporation, is spread in a very thin film on the roller, or in the narrow gap between a pair of rollers, by dropping from small holes in a horizontal pipe. The milk immediately dries on the hot surface of the roller which is at about 140°C and after about three-quarters of a revolution the dried film of milk is shaved off by a very carefully adjusted scraper or knife. The rollers turn at a speed of about 15 rev/min, so giving a contact time of about 3 seconds.

This method has many advantages over the spray method. The equipment is comparatively cheap, it does not occupy much space and it requires only steam at a moderate pressure. The disadvantages are that unless the knife is very carefully adjusted the film of milk is not removed completely and consequently becomes burned on the rollers, so giving a very inferior quality product. Even when the equipment is operating properly, roller powder is much less soluble than spray powder because it receives a more drastic heating, resulting in appreciable denaturation or precipitation of the proteins. Where economy is the most important factor to be considered, space is limited and the manufacturer has a ready market for this type of powder, then the roller process is the method of choice. However, it is becoming less and less used and nearly all plants which have been established since 1950 adopt the spray method.

Nearly all factories employ the double roller method, with the rollers separated by a distance of about 0.7 mm. The adjustment of this gap is crucial, and this distance must be employed at the working temperature. Not only must it be accurate but the rollers must be most carefully aligned so that they are exactly parallel. The milk is always concentrated, the degree of concentration depending upon the temperature and other conditions employed, and forms a miniature reservoir between the two rollers. These rotate inwards so that a film of milk is carried down between the rollers and is scraped off by a knife pointing downwards when viewed from the side. Usually the milk may be condensed up to about 17% total solids, sometimes higher. It is possible to operate a roller dryer in a vacuum chamber, and then a much higher concentration of milk solids may be used to produce a product of superior quality. Such a process is naturally much more expensive.

The quality of roller powder is controlled by the following factors:

1   Accuracy in alignment of the rollers.

2   Condition of the knives. These must not only be very sharp but accurately set. Knives are usually reground after about 100 hours working, and the rollers require resurfacing or being made true again after about 2500 hours.

3   The speed of rotation and the temperature at the surface of the rollers must be carefully controlled.

4   The rapid and efficient removal of steam and the humid air over the powder.

5   The temperature of the milk as it is fed on to the rollers. High temperatures increase efficiency but above 70°C they may damage the quality of the powder.

6   The quality of the milk; for example, any developed acidity or the presence of colostrum or mastitis milk will affect the quality of the powder.

7   Clarification of the milk, which removes dirt, cells, etc. This usually improves the quality of the powder to some extent.

8   The speed with which the powder is cooled on coming off the rollers.

It is common practice to neutralize milk for roller drying. This may be done with calcium hydroxide or a mixture of calcium hydroxide and sodium hydroxide. Solubility is greatly improved, but over-neutralization will lead to a soapy flavour and darkening of the powder.

b. The spray method. The drying of milk by the spray process has well-marked advantages and disadvantages compared with the roller method. It gives a much superior product from nearly all points of view, but the equipment is much more bulky and much more expensive, both in respect to capital expenditure and running costs. It is normally not economic today to set up a milk-drying plant unless the throughput is large, and for the spray-drying equipment it is virtually essential to construct a special building for the purpose on account of its size and especially its height.

In the spray process hot concentrated milk, either whole, partially skimmed or skimmed is converted to a fine mist which is then dried by contact with hot air in the drying chamber. In this way the milk is given a very large superficial area in the form of minute drops so that the evaporation of water from the milk is very rapid. This evaporation has a cooling effect on the droplet so that the temperature of the solids never rises to that of the hot air, and only approaches it closely after the drying has been almost completed. For this reason the complex colloidal structure of the milk protein etc. is not greatly affected. Spray powder consists of very small spherical or egg-shaped particles which entrain a considerable amount of air so that it has a much lower bulk density (i.e., a lower specific gravity) than roller powder. In order to produce a uniform particle size it is essential to obtain a rapid and even mixing of the milk mist and hot air coming into the chamber. This can be controlled mainly by two factors, the concentration and temperature of the milk when sprayed, and the mechanism used for producing the mist. There are three main types for this purpose: 1. the centrifugal disc, 2. the high pressure milk jet, and 3. the use of compressed air jets impinging on a low-pressure milk jet. As the speed of the air increases, the size of the drop decreases up to a speed of about 400 ft/sec (121 m/sec), but no further reduction in size is obtained with higher speeds.

In the centrifugal method the milk impinges on a disc, usually of saucer shape, rotating at speeds in the range 10000 to 16000 rev/min according to the diameter of the disc. The disc is placed in a cool zone in order to prevent drying of the liquid milk on the edge of the disc, which leads to a loss of efficiency. The disc method permits easy cleaning and by the avoidance of high pressures and small apertures, blockage is unlikely to occur. On the other hand some high-pressure jet systems may work at such high pressures, e.g. 2000 p.s.i. (140 bar), that they may become blocked by particles of solid milk or incrustation.

In the earlier plants the chamber was cylindrical in shape with a flat floor from which the deposited powder was continuously removed by a mechanical sweeper. The rate of deposition of the powder particles is controlled not only by their size but by the velocity of the hot air flow and the way in which it enters and leaves the chamber. In earlier equipments fairly low temperatures and air speed were used, but modern equipments depend upon a short exposure of the milk mist to higher temperatures, e.g. 150° to 170°C. A downward current of air accelerates the fall of the powder and the modern drying chamber is shaped as a cylinder with an inverted conical base so that the powder can be continually removed. The heavier particles leave through a pipe at the bottom, and the lighter particles, which get carried over as the air passes out of the chamber, are recovered by a system of cyclone separators and cloth filters.

As with the roller process, milk is always pre-condensed, usually to a value of 42 to 45%. This not only effects a considerable thermal economy but reduces the amount of air held within the particle. This gives increased bulk density, improves keeping quality and gives quicker reconstitution or solution in water. Because of the milder heating conditions, bacteria and enzymes, which are important factors in the keeping quality or shelf life of milk powder, are not destroyed to the same extent as in the roller process.

Fig. 4.11 Spray-drying plant.

The pre-heating associated with pre-condensing largely destroys both enzymes and bacteria and so greatly improves keeping quality.

c. The Birs tower method. Even the best orthodox spray drying method involves the heating, first of the liquid milk before concentration, and subsequently exposure of the spray droplets to very hot air, and so results in at least some denaturation of the soluble proteins, which affects dispersibility and solubility to some extent.

The Birs method eliminates this heating by allowing the spray droplets to fall gently from the top of a tower about 70 m high and 15 m in diameter. The air in the tower is held at 30°C and the efficiency of drying depends upon an extremely low relative humidity in this air which accelerates the evaporation of water from the droplets. In this way it is possible to obtain a powder of satisfactory moisture content in which none of the soluble protein fraction has been denatured, apart from any denaturation which may have been brought about by the preheating of the milk, as for example by pasteurization to destroy pathogens, which results in only about 5% denaturation. The method gives a powder which is much superior to ordinary spray-dried powder.

Unfortunately the heavy capital expenditure involved (about £500000 for the tower and necessary equipment) and the high running cost arising from the fact that the air must be dried by chemical means in order to obtain the required low relative humidity, mean that the procedure is much more costly than the ordinary spray method. Only three towers have so far been built in France, Switzerland and Italy. As an economic proposition, the method depends largely on whether the equipment can be used for drying fruit juices, vegetable extracts and similar materials of high value during the season of low milk production.

d. The foam spray drying method. In this method gas, which may be air or nitrogen, is injected under high pressure into the concentrated milk between the pump and the atomizer at the top of the spray drying chamber. This method, which has only been developed during the last ten years, shows great promise and may possibly ultimately displace ordinary spray drying. The throughput is increased for a given sized plant and it produces a rapidly dispersible product without the necessity for a subsequent ag-glomerating procedure which is required for converting the ordinary spray powder into an instant powder. Skim milk can be concentrated up to 60% solids and dried satisfactorily. The cost is little more than that of the orthodox method as only very slight changes are necessary in the standard equipment. As might be expected, the particles contain gas vacuoles, and on reconstitution these cause the powder to form more foam on the surface of the milk than is usual before it dissolves.

In a typical procedure whole milk is standardized to 3.3% fat, heated to 73°C and held for 15 seconds in a suitable tubular or plate heat exchanger, homogenized at 2500 p.s.i. (170 bar) and concentrated to 50% solids in a falling film evaporator processing the milk at the rate of 500 litres per hour. The concentrate, adjusted to a temperature of 32°C, is then pumped into a pipeline containing a specially designed pressure nozzle which injects nitrogen gas at very high pressure into the feed line. A suitable pressure is 2000 p.s.i. (150 bar). This method gives a powder of individual particle size from 60 to 100 microns, and with average clump size from 150 to 300 microns. The powder has a bulk density varying from 0.2 to 0.5, a residual moisture from 2.5 to 5.0% and contains a highly variable proportion of free fat, ranging from 3 to 20%. The dispersibility can vary from 70 to 95%. The physical properties of the powder can be controlled by varying the drier feed rate, the nitrogen injection rate and the diameter of the nozzle orifice.

e. Freeze-drying. Any heating of milk to a temperature above 70°C will result in the initiation of denaturation or precipitation of the soluble whey proteins, and consequently reduce the solubility of the dried milk. In the freeze-drying process the liquid food is not heated, apart from any pasteurization beforehand in order to destroy pathogens. The liquid is frozen in shallow layers and then subjected to vacuum in order to sublime the ice. The sublimation, like evaporation, lowers the temperature of the product and consequently the rate of sublimation, and in the accelerated freeze-drying process (A.F.D.) heat is applied, for example, by circulating warm water through the hollow base of the trays in which the frozen product is held.

This process, which was developed in Britain during and after the Second World War, has been used for many foods and gives a product with perfect reconstitutibility when water is added to the dried product. Freeze-dried milk powder is without question the best possible product of its type but the expense of producing it is so high that it is unlikely ever to be a commercial proposition for milk. It has, however, been used for yoghourt and soft cheese in Germany and other countries.

4.6.3 Instant milk powder

On account of their low bulk density and minute particle size, spray dried powders easily form a dust when manipulated in the food factory, and although they have a high solubility it requires a certain time and a special technique to prepare a homogeneous reconstituted milk free from lumps. Instant powders are those which have received a treatment after drying so that when added to water they dissolve almost instantly to give a homogeneous milk. Various methods are available for producing powders of this type, some of them combined with the drying procedure. A popular subsidiary method is to treat

Fig. 4.12 Instantizer for spray-dried milk.

the powder issuing from the drying chamber with hot moist air to humidify it and cause the minute particles to agglomerate, thus forming a granular structure containing numerous capillary channels within the granules. These are then treated with hot air to reduce the moisture content to a level only slightly above that of the normal powder. Instant powders have an appreciably higher bulk density than normal powders.

4.6.4 Properties of milk powder

The two main methods for drying milk (roller and spray) yield products having very different properties. The more drastic heating given in the roller process produces appreciable denaturation of the protein, so that roller powder only dissolves to the extent of about 85% in water. The fat globule structure is completely destroyed, with the result that the fat becomes freed and on reconstitution rises to the surface as a yellowish oily layer. Both the odour and appearance of reconstituted roller powder are less attractive than those of spray powder.

Spray powder consists of particles ranging in size from about 5 to 150 microns, the size being controlled by the particular atomizer used. The structure is broadly a shell having an outer layer of amorphous lactose glass, the particle enclosing the fat and also a certain amount of entrapped air. This not only controls the bulk density but is an important factor in the rate of oxidation and hence the keeping quality of the powder. A good spray powder should have a solubility of about 99%. The milder heat treatment results not only in a much higher solubility but also a higher survival rate of bacteria, so that hygiene in operation is more important than with the roller process. Heating of the milk on the roller to temperatures above 75°C has a drastic killing effect on bacteria but results in some loss of solubility.

a. Flavour. All milk powders have a characteristic flavour, which may be described as tallowy, cardboardy or oxidized. This flavour is produced by the oxidation of the fat, which loses its protective fat globule membrane when milk is dried. This flavour is much more pronounced in the full cream powder, for obvious reasons, and ultimately the powder becomes very tallowy and distasteful. It is mainly for this reason that the use of milk powder by itself as a food has not been generally adopted. The characteristic flavour is impossible to disguise, and anti-oxidants, even when legally permitted, are not very effective in preventing it. The characteristic oxidized flavour is brought about by oxidation in the first place of the double bonds of the unsaturated fatty acids in the fat to form peroxides, which then decompose to aldehydes and other compounds which markedly affect the flavour of any food, even in extremely minute concentrations. Apart from the amount of oxygen present in the powder, the rate of oxidation is increased by trace amounts of heavy metals, particularly copper. Formerly when tin-plated copper and bronze equipment was used in the dairy industry two or three parts per million of copper might be present in the powder, and this had a powerful catalytic effect in accelerating oxidation and so the production of the characteristic taint. Today all dairy equipment is made of stainless steel, which has resulted in a much lower amount of copper in milk. Storage at high temperatures naturally accelerates the oxidation, but the moisture content of the powder does not affect it.

Many methods have been tried to reduce the development of the unpleasant taint. One method is to pack the powder in cans, first evacuating the can and then filling it with an inert gas such as nitrogen or carbon dioxide. This process is very efficient and may give a powder with a life of up to seven years, but it is obviously expensive. To be effective this form of packing must be carried out within three days of manufacture, and it is advantageous also to repeat the evacuation and gas filling process in order to remove the last traces of oxygen from the powder, some of which is, of course, still entrapped within the particles and can only slowly be replaced by the inert gas.

A simple alternative method is to use a high preheating temperature before spray drying, e.g. 90°C. When milk is heated above 80°C some of the albumin decomposes to yield sulphydryl compounds; these have reducing properties and act as antioxidants. In this way not only is the development of off-flavour curtailed but there is less loss of vitamins A and C during storage. This method is not so effective as gas packing but is very much cheaper and has now become a standard method throughout the world.

b. Hydrolytic rancidity. The taints produced in milk powder and other foods by oxidation are commonly referred to as 'oxidative rancidity' but there is another type of rancidity which is common to butter and may occur in milk powder. If the heat treatment given to the milk or during the processing of the food is not sufficient to destroy all the lipase present, this enzyme can then act upon the fat to split it into free fatty acids and glycerol. The free fatty acids then produce a characteristic off-flavour commonly called hydrolytic or true rancidity. The lipases may be derived from micro-organisms growing in the milk before drying but also occur in milk as it leaves the udder. This type of taint can be prevented by preheating the milk to at least 75°C and by thorough hygienic control of the equipment and the processing method.

c. The deterioration of protein. In addition to the taints associated with fat, changes can take place in protein during the storage of milk powder, and these changes lead to the development of various off-flavours which are called stale, gluey and/or cardboardy. There is also a loss of solubility which occurs at the same time, and these changes are more common in skim milk powder which contains more protein and has no fat to exert some protective action on the protein. This deterioration is not only accelerated by high temperatures of storage but also by high moisture content in the powder. For this reason the moisture content of the powder should be kept as low as practicable, and if possible at about 3%. If it exceeds 5% the defect may become quite prominent.

The main chemical mechanism involved in protein deterioration is a slow combination between the lactose or milk sugar aldehyde group and certain amino groups in the protein, particularly the ϵ-amino group of lysine, an amino acid in which milk is particularly rich. This and other chemical changes slowly lead not only to off-flavours but to a darkening of the colour and a loss of solubility of the milk powder. Moreover, as lysine is an essential amino acid largely responsible for the nutritive value of milk protein, this chemical reaction can lead to as much as a 40% loss of the nutritive value of the protein. Gas packing can reduce this deterioration to some extent.

4.6.5 Microbiology

From the microbiological point of view milk powder is the most stable of all dairy products. Bacteria require food, moisture and warmth to grow and if one of these factors is absent they cannot grow and will die out at different rates according to their species. This basic fact accounts for the great usefulness of drying as a method for the preservation of food. However, once the powder is reconstituted or dissolved in water the surviving bacteria can grow as easily as they would do in the original milk. It is therefore necessary to exert some control on the bacterial content of milk powder.

The pre-heating which is now a standard process in manufacture is more severe than ordinary pasteurization so that the micro-flora are reduced very considerably and only the more heat-resistant types such as spore formers, micrococci, corynebacteria and certain streptococci survive. Usually the bacterial quality of roller powder is superior to that of powder made by the spray process. In some methods the pre-condensed milk may be held in a 'hot well' at about 45°C, a temperature at which certain bacteria can grow very rapidly and may not later be destroyed in the spray drying process. This is therefore a vulnerable stage in the process and even if such bacteria are killed in the drying chamber they may have grown sufficiently to produce an off-flavour in the powder. If killed they will not be detected by the ordinary plating or colony count tests for viable bacteria, and it is advisable therefore to make counts not only of living bacteria but of the total living and dead by a microscopic count method. This sometimes gives extremely high values running into many millions per gram. A good powder should not contain more than 5000 living bacteria per gram.

4.7 CHEESE

4.7.1 Historical

It is virtually certain that butter and cheese were made by primitive man long before recorded history. They are both mentioned in the earliest known writings and frequent references to them are made in the classics of world literature.

It has been suggested that cheese-making was discovered accidentally by the carrying of milk from domesticated animals in the stomach of an animal under hot conditions, when the milk would sour rapidly and be acted on by the milk-clotting enzymes of the stomach. The clots of milk so formed would be broken up by agitation and in this way moisture would be expelled and the curd would contract to give a primitive type of cheese. It is interesting to note that this basic action is identical with what happens when a calf suckles a cow, and it is also basically the method used today even in the most sophisticated types of equipment for making cheese in modern factories.

Definition. In this chapter the term 'cheese' is used to include both singular and plural forms. Many definitions have been proposed for cheese. In the United Kingdom the Cheese Regulations 1970 define cheese as the fresh or matured product intended for sale for human consumption, which is obtained as follows:
(a) in the case of any cheese other than whey cheese–1. by coagulating any or a combination of any of the following substances, namely milk, cream, skimmed milk, partly skimmed milk, concentrated skimmed milk, reconstituted dried milk and buttermilk, and 2. partially draining the whey resulting from any such coagulation;
(b) in the case of whey cheese–1. by concentrating whey with or without the addition of milk and milk fat, moulding such concentrated whey or 2. by coagulating whey with or without the addition of milk and milk fat. This definition would not be universally accepted. The important features are that it permits the use of seven different kinds of milk for cheese-making. It requires the draining of whey from the coagulated product, and this aspect is accepted in all definitions. It makes no reference to the mammal from which the milk is obtained, although elsewhere the British regulations state that, unless otherwise mentioned, 'milk' means cow milk. It makes no reference to the use of rennet or other coagulating enzymes, which some authorities regard as an essential material

for the manufacture of cheese. Finally it includes whey cheese which is a very different product from ordinary cheese, such as Cheddar.

A more detailed and scientific definition has been suggested as follows: Cheese is the curd or substance formed by the coagulation of the milk of certain mammals by rennet or similar enzymes in the presence of lactic acid produced by added or adventitious micro-organisms, from which part of the moisture has been removed by cutting, warming and/or pressing, which has been shaped in a mould, and then ripened by holding for some time at suitable temperatures and humidities. The Shorter Oxford English dictionary defines cheese as the curd of milk (coagulated by rennet) separated from the whey and pressed into a solid mass.

The word cheese itself is derived through the Old English cese from the Latin *caseus*. Although most cheese throughout the world is made from the milk of cows in temperate regions such as western Europe, the USA, Australia and New Zealand, sheep and goats supply a considerable amount of the milk used for making cheese in sub-tropical countries and particularly in mountainous regions. In India and other countries the milk of the water buffalo may be used, in northern Scandinavia the milk of the reindeer and in the USSR mares milk may be used for making cheese and cultured milks; there is often no clear-cut difference between these types of dairy products.

4.7.2 Regulations

The United Kingdom Regulations classify cheese as hard cheese, soft cheese and whey cheese, and the standards are as follows.

1 'Full fat hard cheese' must contain not less than 48% milk fat in the dry matter (f.d.m.) and not more than 48% water.

2 'Medium fat hard cheese' contains less than 48% but not less than 10% f.d.m. and not more than 48% water.

3 'Skimmed milk hard cheese' contains less than 10% f.d.m. and not more than 48% water.

For soft cheese a different system is used in that the fat content is expressed as the percentage of the whole cheese in the following manner.

1 'Full fat soft cheese' must contain not less than 20% milk fat and not more than 60% water.

2 'Medium fat soft cheese' contains less than 20% but not less than 10% milk fat, and not more than 70% water.

3 'Low fat soft cheese' contains less than 10% but not less than 2% milk fat, and not more than 80% water.

4 'Skimmed milk soft cheese' contains less than 2% milk fat and not more than 80% water.

Separate standards are laid down for cream cheese as follows:

1 If described as 'cream cheese' it must contain not less than 45% milk fat.

2 If described as 'double cream cheese' it must contain not less than 65% milk fat.

Whey cheese must comply with the following standards

1 'Full fat whey cheese' must contain not less than 33% f.d.m.

2 'Whey cheese' contains less than 33% but not less than 10% f.d.m.

3 'Skimmed whey cheese' contains less than 10% f.d.m.

Special regulations are laid down for processed cheese and cheese spread.

digestion of the mother's milk in the stomach of the calf, and its peculiar property is that it has a very high ratio of clotting to proteolytic (protein digesting) action. The enzyme produced by the adult animal is pepsin, which is sometimes used for cheese-making, but this has a lower ratio and is less satisfactory, its use tending to result in the production of bitter cheese. Owing to the steady increase in cheese-making throughout the world and the diminishing number of milk-fed calves available, there is now a world shortage of true rennet, and a 50:50 mixture of rennet and pepsin solution is now in common use. Milk-clotting enzymes can also be obtained from plants, and these have been used in many countries, including England, over the centuries. In general they are much less satisfactory than rennet but find a place in the manufacture of vegetarian cheese. Very active research is now in progress on the manufacture of milk-clotting enzymes from fungi, although here again troubles have been experienced with the production of off-flavours and reduced yield, both attributable to the comparatively strong proteolytic activities of the fungal preparations. Rennet is an extremely powerful enzyme and is normally used at the rate of about one part in 5000 parts of milk. Its action has a very high temperature coefficient, and this is an important factor in cheese-making.

d. Treatment of the curd. The combined action of the acid produced by the starter organisms and of the rennet enzyme at a temperature of about 30°C converts the liquid milk into a soft jelly or junket, which is technically called the coagulum. This change takes about half an hour and the next stage is to break up this coagulum by cutting with special knives or ladling out with a flat spoon or dipper. In early times the coagulum was broken by hand. As soon as this breakage takes place moisture immediately begins to leave the curd and as this contains the soluble proteins (albumin and globulin) and the yellow-green riboflavin (vitamin B_2) it has the characteristic appearance which we recognize as cheese whey. The higher the acidity and the finer the degree of cutting or division of the curd the faster is the whey expelled.

e. Cooking or scalding the curd in the whey. The expulsion of whey from the curd is greatly influenced by temperature and with many types of cheese there is a stage called 'scalding' in which the particles of curd floating in the whey are heated to a temperature which is commonly in the range of 36° to 40°C, but for some varieties such as Parmesan, Emmenthal and Gruyère it may be as high as 56°C. This scalding process is one of the most important in controlling the character of the final cheese.

f. Separation of the curd from the whey. When the scalding or cooking of the curd has been completed, the whey is removed. The curd is allowed to 'pitch' or settle on the bottom of the vat, or it may be removed by straining, or in some modern methods by a centrifugal process. Whatever method is used, this is a crucial step in manufacture because it is no longer possible to influence the behaviour of the curd from the point of view of its equilibrium with the whey. Treatment from this stage onwards varies greatly with the type of cheese. For example, the curd may be filled immediately into moulds, or it may be allowed to mature or 'cheddar' for two hours.

g. Salting the curd. The essential ingredients of practically all cheese are milk, souring organisms, rennet and salt. Without salt, which acts as a preservative, all cheese would deteriorate fairly rapidly. The amount of salt used varies according to the type of cheese, and different varieties are salted in a number of ways. The simplest method is to grind up or mill the curd into small fragments in the cheese vat, and then to sprinkle the required amount of salt over the curd–about 2% for Cheddar and similar cheese. A second method is to mould the curd into the shape characteristic for the cheese, e.g., a cylinder or a block, and then to dry-salt or rub salt on to the cheese by hand, usually over a period of two or three days. The third common method is by immersion in a solution of salt (sodium chloride) at a concentration of about 20% and a temperature in the range 10° to 15°C. It will be obvious that in each of these methods the salt mixes with the curd at different rates, and this is a factor in the initial chemical changes taking place in the young cheese, as salt has a repressive effect on the activity of the bacteria present. It is important that the salt should attain an even concentration throughout the cheese in a fairly short time.

h. Pressing the cheese. The pressure which is applied to a cheese to help form it in its earliest stage of ripening may vary from nothing (apart from the weight of the cheese itself) up to very high pressures, in some cases 120 p.s.i. (18 bar). In general the lower the moisture the firmer becomes the cheese and the higher is the pressure applied. Early presses were of the simple lever type but in the modern factory the cheese (in their moulds) are placed in a horizontal gang press and the pressure applied by pneumatic or hydraulic rams.

i. Maturing the cheese. The maturing or ripening of cheese is essentially a slow microbiological-biochemical process in which the lactose or milk sugar remaining in the watery part of the cheese is completely fermented to lactic acid, and the protein (mainly casein) is gradually digested by the enzymes in the rennet and liberated from the starter and other bacteria in the cheese. There is also a variable amount of digestion or breakdown of the fat. In hard cheese such as Cheddar the fat breakdown or lipolysis is not very extensive, but in other types such as Parmesan lipase preparations are deliberately added with the rennet in order to bring about an extensive breakdown of the fat which is largely responsible for the strong flavour of the cheese. Moulds produce powerful proteolytic and lipolytic enzymes, so that those varieties in which moulds are an essential feature, such as the blue-green penicillium which grows in the interior of Stilton, Roquefort and Gorgonzola cheese, and the white fluffy penicillium and other moulds which grow on the surface of Camembert and Brie cheese, undergo considerable breakdown of both protein and fat. This breakdown is mainly responsible for the characteristic flavours of these varieties, and the differences in the location of growth, the type of mould and the conditions under which the mould grows are all responsible for the subtle differences between the flavours of the cheese. Formerly cheese were allowed to ripen at ordinary atmospheric temperature and without any particular control of conditions, apart from such varieties as the blue veined Roquefort which have been traditionally ripened in caves in an atmosphere of very high humidity to encourage mould growth. In modern cheese-making, however, particular attention is paid to the conditions of ripening, especially as it is realized that higher temperatures are more likely to result in tainted cheese. When the cheese is not treated by a surface process or by film wrapping in the early stages of ripening, it is now standard practice to ripen cheese at a controlled temperature (usually 10° to 12°C) and humidity. The latter varies somewhat but for ordinary cheese such as Cheddar and Gouda may be in the range 75 to 85% relative humidity. The time of maturing may vary, from 2 or 3 days, as with unripened soft cheese, to as much as two years for Parmesan.

j. Special treatments. During the period of ripening some types of cheese receive special treatments. These are mostly given to produce a certain type of micro-flora characteristic for the variety. For blue-veined cheese such as Stilton it is necessary to allow air to obtain access to the interior of the cheese, because the characteristic moulds can only grow in the presence of a certain amount of air. Formerly this was allowed to take place by chance, the cheese-maker relying on cracks developing in the cheese as the result of distortion in the cheese mass produced by different rates of drying out. For many years, however, it has been standard procedure to stab or prick cheese of this type by inserting stainless steel wires into the cheese mass, with a special machine. It is also a standard practice to inoculate the curd during manufacture, or alternatively the milk initially, with a suspension of mould spores. For the surface mould ripened type, such as Camembert, the cheese may be sprayed with a suspension of the mould spores, although in practice the air in the ripening room becomes so charged with the spores, which grow easily on the surface of the cheese as they have plenty of air and moisture, that this artificial addition is unnecessary. With cheese of this type there may be a sequence of different types of bacteria, yeasts and moulds growing on the surface and with modern methods of manufacture there may be a sequence of as many as three maturing rooms, each with their characteristic temperature and relative humidity. Other varieties of cheese may rely on a surface growth of certain slime-forming bacteria, such as *B. linens*, for the production of their own type of flavour. This is typical for certain semi-hard varieties such as Limburg, and this group of cheese is commonly referred to as 'smear-ripened'.

Not all cheese receive the treatments described above. In general the soft high-moisture unripened types are made by a simple method and consumed quickly after making, whereas the long-maturing, highly flavoured types receive highly specialized treatment.

4.7.4 The development of cheese types

From the early beginnings cheese-making has developed into a very varied process, producing many types and varieties. A number of classifications have been suggested for cheese, and broadly speaking all varieties can be placed in about a dozen groups according to the method of manufacture, the types of micro-organism responsible for the ripening, and the ultimate analysis of the cheese. The most important single factor in the last is the moisture content, and this is the main factor controlling keeping quality. Cheese are available not only in an amazing number of different varieties, but also with a large range of keeping quality. The various factors controlling the type of cheese are summarized in tables 4.9 and 4.10. Every variety can be placed in one of these groups or types.

An interesting point is why certain cheese have survived from early times, because some of the famous cheese of history have records going back for about 1000 years. The factors which have controlled this are, apart from general attractiveness and acceptibility by a local population, ease of manufacture, keeping quality and ease of transport. Thus hard (hard pressed) varieties such as Cheddar and Parmesan have become famous for their keeping quality and ease of marketing, and it is this factor which is probably responsible for the fact that Cheddar cheese is the most popular variety in the world. Parmesan keeps even better and is commonly used as a grating cheese. The soft unripened varieties are usually consumed locally apart from the special types sold in

Table 4.9 Classifications of cheese types and methods of manufacture

Type	Cheese variety	Characteristics	Milk — Skimmed	Milk — Ripened	Coagulation — Acid	Coagulation — Rennet	Cutting — Ladled	Cutting — Large	Cutting — Small	Scalding — None	Scalding — Low	Scalding — Medium	Scalding — High	Draining — Vat	Draining — Hoop	Salting — Curd	Salting — Cheese	Salting — Brine	Shaping — Hoop	Shaping — Hand	Shaping — Pressure
Hard	Parmesan	Extremely hard	+	+		+			+				+				+				+
	Emmenthal	Large gas holes		+		+			+				+		+		+				+
	Cheddar	No gas holes		+		+						+		+		+					+
Semi-hard	Port du Salut	Fairly firm, mild flavour	(+)	+		+			+			+					+				+
	Brick	Fairly strong, sweetish flavour				+			+				+		+		+				+
	Pecorino	Sheep's milk				+			+				+			+	+		+		
	Edam	Fairly firm				+		+			+				+		+	(or +)			+
	Gouda	Mellow	+			+		+				+			+			+			+
	Cacciocavallo	Full flavour, long keeping	(+)			+			+				+	+				+		+	
Soft	Limburg	Strong flavour, bacterial ripening	(+)			+			+	+					+		+		+		
	Camembert	Strong flavour, external mould ripening				+	+								+		+		+		
Mould-ripened (blue veined)	Roquefort	Peppery flavour, internal mould ripening		+		+	−	+									+		+		
Acid coagulated	Cottage	Soft-lactic flavour	+	+	+			+			+			+		+			+		
	Sapsago	flavoured by herbs	+	+	+							+				+			+		
Cream	Cream	Made from cream		+		+								+		+			+		

Table 4.10 Rheological properties of different varieties of cheese.

Variety	Type	Moisture %	pV (Viscosity factor)	pM (Elasticity factor)	pS ('Springiness' factor)	Keeping quality
Convalli	Soft	59.06	7.3	5.6	1.7	Few months
Little Dutch	Soft	46.27	–	5.7	–	Few months
Port du Salut	Soft	44.98	7.1	5.7	1.4	Few months
Bel Paese	Soft	42.72	–	5.6	–	Few months
Lancashire	Semi-hard	–	7.3	5.5	1.8	Few months
Leicester	Semi-hard	39.62	7.5	5.7	1.8	Few months
Pont l'Evêque	Semi-hard	39.24	–	5.7	–	Few months
Cheshire	Semi-hard	38.24	7.8	5.9	1.9	Few months
Caerphilly	Semi-hard	37.11	–	5.9	–	Few months
Derby Double	Hard	35.61	8.0	6.0	2.0	Several months
Gloucester	Hard	35.77	8.2	6.1	2.1	Up to 1 year or longer
Dunlop	Hard	34.45	8.2	6.1	2.1	Up to 1 year or longer
Cheddar	Hard	34.08	8.2	6.1	2.1	Up to 1 year or longer
Gruyère	Hard	31.25	8.6	5.9	2.7	Up to 1 year or longer
Parmesan	Very hard	19.97	9.5	7.0	2.5	Several years

Paris and other large cities where there is a gourmet market prepared to pay the necessary high price.

4.7.5 The manufacture of cheese

The greatest advances in the dairy industry since the Second World War have been concerned with the heat treatment and packaging of liquid milk, and with the manufacture of cheese. So great have been the changes in the latter that a cheese-maker of the 1930s would hardly recognize the latest types of cheese-making equipment as being concerned in any way with cheese. Mechanization and automation have transformed the whole conception of conventional cheese-making, although the basic principles remain the same.

When milk production was seasonal – that is, the cows calved in early spring and produced a quantity of milk rising to a peak in May–June (in the northern hemisphere) with the supply fading out towards the end of the year – cheese-making like other dairy manufacturing activities was entirely seasonal. However, the policy of the Milk Marketing Board (M.M.B.) in the United Kingdom has resulted in a fairly level production of milk throughout the year, because nearly 70% of the milk production in the United Kingdom is required for liquid consumption. In some countries, such as New Zealand, cheese-making may be almost entirely seasonal. This has some important advantages. The cheapest way of producing milk is by the cows grazing on good pasture, and it is well recognized that milk produced in this way is the best for cheese-making, grass being the natural food of the cow. Seasonal production demands intensive use of the equipment, which modern factories can manage by repeated use of the vats, but it also gives a short period of non-production which allows the workers to have holidays and gives the management an opportunity to overhaul the equipment and carry out any alterations to the buildings, repairs and replacements which may be necessary.

It is impossible to describe in detail the methods for making the innumerable varieties of cheese throughout the world and we shall describe in outline the manufacturing method for a typical cheese in each of the main categories as follows.

Unripened, soft cheese and cream cheese. These represent the simplest possible methods of making cheese, involving nothing more than the souring of the milk or cream and draining whey from the clotted material. A very little rennet may or may not be used.

Cheddar cheese. This, probably the most famous cheese in the world, includes nearly all the various manipulations which are involved in cheese-making.

Gouda. This famous Dutch cheese is similar to Cheddar in some ways but has a higher moisture content and is therefore softer. The curd receives slightly different treatment.

Provolone. This well-known Italian variety includes a characteristic stage in which the curd is plasticized by treatment with very hot water.

Grana (Parmesan). This differs markedly from the three types above in that the curd is scalded at a high temperature and enzymes other than rennet are generally added to increase the breakdown of fat. Grana cheese is matured for a long time, has a very low moisture content and a characteristic flavour or 'bite'.

Emmenthal. This is the commonest Swiss type of cheese and is easily recognized by the large holes or 'eyes' in the body of the cheese. It is a very firm cheese, has a mild flavour and the holes are produced by the carbon dioxide resulting from a propionic acid fermentation.

Limburg. This is a very pungently flavoured, semi-hard variety in which the ripening is partly brought about by bacteria in the slime or 'smear' on the surface of the cheese.

Stilton. This is one of the three most famous blue-veined cheese in the world. To permit the growth of the characteristic blue-green mould, this type is stabbed or pierced a few weeks after making in order to allow air to penetrate the cheese. This is necessary for the growth of the mould.

Cottage Cheese. This is the most recently developed well-established variety originating from the USA. The soft curd is washed to remove much of the lactic acid and the cheese is consequently fragile and of short-keeping quality. Its bland flavour allows it to be used with many other foods, and the very low fat content makes it a popular cheese for slimming purposes.

a. Soft and cream cheese. These are probably the oldest types of cheese ever made by man. Basically milk or cream is allowed to sour, in old days by natural means and today with the aid of a properly cultured cheese starter, and methods may vary from the very simplest type which can be practised by the housewife in her kitchen to the most modern mechanized operations of the cheese factory. The method is essentially as follows. The milk or cream may be used raw (as commonly in farmhouse making) or given a heat treatment equivalent to or slightly less than the normal pasteurization. In either case some starter (0.1–5%) should be added. The amount depends upon whether an overnight incubation period is allowed or not. Milk of any fat content may be used; whole milk gives the best quality cheese but naturally it is more expensive. Often separated or skim milk is used, and the manufacture of soft cheese is a common way of utilizing skim milk, especially on the farms, and in small butter-making dairies. The greater the fat content the higher will be the yield and the better the quality of the cheese.

Skim milk cheese are always of inferior quality, and are tougher and more leathery than a whole milk cheese, other factors being the same.

Soft cheese may be made without any rennet at all, in which case the cheese is really nothing more than a cheese starter which has lost a certain amount of whey. A very small amount of rennet, for example, from 1 to 5 ml per 500 litres of milk is often used and gives rather better clotting and a better flavour in the cheese. The temperature of the milk should be adjusted to 32° to 34°C, before the addition of the rennet, which must be diluted with about five times its volume of water and thoroughly mixed into the milk. In soft cheese-making the curd is not cut fine, nor is it scalded in the whey, and it is important therefore that the ambient temperature should be maintained at 20° to 25°C. Cheese knives are not used but the coagulum, which forms within two or three hours according to the methods used, is sliced or ladled into the cloth or mould (hoop). This operation must be carried out carefully in order not to damage the curd. In the simplest possible method, as generally practised in farmhouses, the curd is scooped into a coarse cloth, the four corners tied together and the bundle suspended on a hook to drain. In modern factories highly mechanized methods are used for making soft cheese, particularly for the well developed varieties such as Gervais and Camembert.

There is no set period for drainage of the curd; this depends upon the particular variety to be made and the equipment used. With the simple cloth method, the pores may become choked and the drainage hindered; the curd is then taken out and either transferred to a fresh cloth, or the old cloth may be scraped and cleaned, and then the curd resuspended as before. Drainage does not usually last beyond 24 hours, and the curd is then taken out and transferred to moulds. Alternatively, the curd may initially be scooped into a suitable mould, which is often now made of plastics, and is somewhat taller than the final height of the cheese to allow for contraction of the curd. The mould is usually perforated with small holes and stands on a plastic or rush mat in order to permit continued drainage.

Soft cheese should have a clean and delicate lactic flavour, and the characteristic flavours of ripened cheese such as Cheddar are entirely absent. Unripened cheese are consumed within a few days and the ripened varieties usually in two to three weeks.

The same basic method is used for cream cheese but the rate of drainage is slower because of the large amount of fat which is entangled with the clotted casein.

b. Cheddar. It appears likely that Cheddar cheese did not originate from the village of that name, but early records show that the area covered by the Mendip Hills in Somerset produced the best Cheddar cheese. Its fame was probably due to the fact that the cheese was sold to tourists visiting Cheddar to see the famous caves, so that in this way it became known generally throughout the country as Cheddar cheese. As with Stilton (see below) it obtained its name through marketing and not manufacturing activities. It is certainly the cheese made in the greatest quantity, being the predominant variety in the USA, Canada, the UK, South Africa, Australia and New Zealand. Its manufacture is now being taken up in many other countries, and countries in the far east such as Japan are now importing large amounts. It owes its importance in that it is a straightforward cheese to make, has a keeping quality of several months and is a firm, strong cheese which is easy to transport.

The method of manufacture of Cheddar cheese used in cheese factories in the United Kingdom and elsewhere until quite recently is as follows. The general trend has been to design premises on a two- or three-tier principle. The tanks, holding 2000 to 3000 gallons (10000–15000 litres) of milk, are placed on the highest floor and the milk allowed to run by gravity to the next floor containing the cheese vats holding from 1000 to 2000 gallons (4500–9000 litres). The curd-whey mixture can then be run from these vats on to drainers on the ground floor where the curd is separated from the whey and further manipulated.

The essential ingredients for Cheddar (and in fact, most varieties of cheese) are milk, starter, rennet and salt. It is essential that these all be of good quality; if any one of these products is unsatisfactory it can easily result in a defective cheese although the other ingredients and the method of manufacture are sound. Factories commonly convert any quantity of up to 100000 gallons (450000 litres) of milk a day into cheese. Small factories have virtually disappeared because the economics of manufacture are progressively improved as the scale increases. Farmhouse making, which nearly disappeared during the Second World War, has received a fillip in the United Kingdom through the previously mentioned enlightened policy of the M.M.B. so that an appreciable quantity of farmhouse cheese is now on the market (table 4.11). Factories receive their milk from a large number of farms, often nearly 1000, and this is carried in churns (cans) holding 45 litres, transported by lorries. This is a cumbersome method and milk is now being increasingly collected by the bulk tanker. The farmer pours his milk directly into a refrigerated tank near the cowhouse where it is rapidly cooled to 5°C. A special road tanker visits the farm, takes a sample of the milk for control purposes, and then the milk is pumped from the farmer's tank into the tanker through a hose. In this way from 1000 to 2000 gallons (4500–9000 litres) of farm milk can be delivered to the creamery in one vehicle with obvious advantages of economy and convenience. After checking, the milk is pumped to a storage tank from which it goes to the pasteurizer. With a few exceptions in some countries, milk for cheese-making is now always heat treated but the full legal pasteurization is not always given because this kills about 99% of the bacteria in the milk and consequently reduces flavour production and the rate of ripening. For this reason a

Table 4.11   Sales of milk for the manufacture of cheese (million gallons – April–March)

	England and Wales			United Kingdom		
Variety	*1969–70*	*1970–71*	*1971–72*	*1969–70*	*1970–71*	*1971–72*
Creamery						
Caerphilly	5	5	5	5	5	5
Cheddar	86	104	136	144	167	207
Cheshire	65	68	68	66	68	68
Double Gloucester	6	7	8	7	8	10
Dunlop	–	–	–	2	2	1
Lancashire	12	12	12	13	13	13
Leicester	5	5	7	6	7	10
Stilton	6	7	8	6	7	8
Wensleydale	6	7	8	6	7	8
Others	2	4	7	2	4	8
	193	219	259	257	289	338
Farmhouse						
Caerphilly	2	2	2	2	2	2
Cheddar	15	16	20	15	16	20
Cheshire	5	5	5	5	5	5
Others	1	2	3	1	2	3
	23	25	30	23	25	30
Grand total	216	244	289	281	314	368

somewhat milder treatment is often given, for example 68°C with 15 seconds holding. Where it is desired to produce a full-flavoured cheese with a 'bite', still lower temperatures may be used, but naturally the less the heat treatment the greater is the possibility of a bacteriological defect in the cheese.

Pasteurization is so adjusted that the milk can flow into the cheese vat at the required temperature for ripening (acidification by the starter organisms) and renneting. For Cheddar this is usually 30° to 31°C. Formerly a period of ripening was allowed and for this purpose the starter, usually 1 to 2%, was added and souring allowed to proceed until the acidity had increased by about 0.02% lactic acid. However, today the starter and rennet are added almost simultaneously. Using a standard rennet, about 22 ml are added for each 100 litres of milk. The rennet must be diluted about six times with water before addition and thoroughly mixed into the milk in order to obtain a homogeneous coagulum. After mixing in, the milk is surface-stirred for about 3 minutes, and it is then important that it be not disturbed until the coagulum is ready for cutting. This usually takes about 35 minutes and the cheese-maker determines the cutting time by examining the firmness of the coagulum or junket. Thus must break cleanly when a finger is pushed through the surface horizontally, and it must come cleanly away from the side of the vat when gentle oblique pressure is applied.

The cutting operation is carried out using 'American knives'; these form a framework, rather like a grid-iron, with either sharp steel blades or steel wires. Knives with vertical and horizontal blades may be used, and cutting in this way was formerly done by hand. Today practically all the stages of cheese manufacture are mechanized, although this does not affect the fundamentals of the process. The vats are now equipped with an overhead gantry with attachments to which the knives can be fitted in such a way that they can be moved up and down the vat and at the same time undergo a circular motion. When the curd has been cut to a sufficiently small size, usually to pieces of 3 to 4 mm, the scalding process is started. This term is perhaps misleading because it consists of raising the temperature of the curd-whey mixture to a temperature in the range 38° to 40°C. It is important that this operation be carried out smoothly with vigorous agitation of the curd particles as otherwise these may acquire a skin on the outside which would interfere with drainage. The changes taking place in the curd are complex but the most important are a steady loss of moisture or whey, and a subtle change in the constitution of the protein or casein-calcium phosphate complex which constitutes the basic structure of the curd; the fat globules are entangled with the protein particles but otherwise take no part in the process at this stage. The temperature is raised at the rate of about 1 deg. C every 5 minutes and the curd particles contract and harden slowly during this treatment. When the maximum scald temperature is reached, the heat is turned off and after a further period of stirring the curd attains a condition such that it is ready for 'pitching'; that is, the stirring is stopped and the curd particles are allowed to fall to the bottom of the vat. The cheese-maker determines this by certain tests, a popular one being to move the hand briskly through the curd-whey mixture when the curd particles produce a peculiar tingling or 'shotty' sensation on the back of the hand. Another test is to gather a handful of curd and compress it in the hand, testing its hardness and resilience. It is important to determine the pitch point with accuracy. The curd now falls to the bottom of the vat and forms a 'mattress'. The chemical and physical changes continue steadily and a short time later the whey is run off from the vat. This stage is perhaps the most crucial of all because once the whey has been removed the cheese-maker is severely limited in his control of the curd; he can no longer apply heat or stir the curd in the whey.

Fig. 4.13 Cheddar cheese manufacture.

Formerly, after the whey had been run off a channel was cut down the middle of the mattress of the curd in order to form a gulley to permit the residual whey to escape from the curd, but with the modern two-tier method the pitching operation is allowed to take place by running the curd–whey mixture out of the vat onto a drainer – a shallow vat – on the floor beneath. A weir and strainer at the end of the drainer holds back the curd and allows the whey to run off. The curd now forms a mattress on the drainer and receives the traditional 'cheddaring' treatment which is characteristic for this variety. Conventionally it consists of cutting the mattress of the curd into blocks of about $30 \times 20 \times 20$ cm and piling them one on top of the other. Every quarter of an hour or so the blocks are re-piled and gradually flatten out. In the conventional method this cheddaring stage might take up to 2 hours and the curd would be slowly transformed from a rather tough, rubbery material into a mellow and slightly softer product. When this stage is completed the curd can be torn into strips and develops an attractive somewhat buttery aroma.

Up to this stage the cheese-maker uses acidity tests for the purpose of following the development of the process and the rate of working of the curd. Conventionally the acidity test is carried out by titrating the whey with caustic soda solution, as in the chemical laboratory, but in modern methods the pH value or true acidity is now increasingly determined, usually by inserting a glass electrode into the whey or curd. An additional test may be used to determine the end of the cheddaring stage. This is the 'hot iron test' and is carried out by pressing a piece of iron at black heat against a piece of the curd very firmly and then slowly drawing it away. Threads of the molten curd are formed and the more mature the curd the longer are the threads. The cheese-maker expects to obtain threads of about 4.5 cm as an indication that the curd has been cheddared satisfactorily.

The next stage is milling and salting. The mill consists of two shafts bearing short rods or pegs, or circular sharp steel discs, rotating inwardly inside a box. Pieces of curd are fed into this mill and the small pieces so formed fall back into the vat or, in modern practice, on to a conveyor belt. Formerly the curd was salted by hand but with modern equipment the salting is carried out mechanically and the rate and amount added are proportional to the amount of curd and the speed with which it passes along the conveyor belt. Cheddar cheese is salted at the rate of 2% which gives a final concentration of just over 5% in the aqueous phase of the cheese. The pieces of curd are then fed into the moulds and compressed with the aid of a pneumatic ram. The well-known cylindrical moulds formerly used to give a Cheddar cheese of 60 to 80 lb (30–40 kg) are now rapidly being replaced by three-piece square moulds to give a block cheese of 40 lb (20 kg) because they are much more convenient to handle, can be packed with greater economy of space, and can be more easily wrapped. The moulds packed with curd are then pressed. The latest development is a horizontal gang-press in which a pressure of 80 to 120 p.s.i. (5.5–8 bar) is applied pneumatically or hydraulically. One of the faults in Cheddar cheese is

Fig. 4.14 Cheddar cheese: salting and filling of curd into moulds.

mechanical openness which is caused not through gas production by bacteria, but by occlusion of air inside the pieces of curd. The simplest way of avoiding this fault is to carry out the pressing of the cheese in a vacuum chamber, a method developed some years ago in New Zealand and now used with great success. Formerly pressing occupied three days, but with modern equipment and higher pressures the process can be completed in 18 to 24 hours. Block cheese are today almost invariably wrapped in plastics film, and this may be done 24 hours after manufacture or at any convenient time thereafter.

c. Gouda. This variety, somewhat softer than Cheddar, is the most important type made in the Netherlands and, like Cheddar, its manufacture is now rapidly developing all over the world. Cheddar and Gouda are probably the most important 'table cheese' throughout the world today.

Gouda cheese takes its name from a town near Rotterdam, and is made in large quantity in the Netherlands, both in factories and on farms. Whole milk is used so that Gouda cheese is more waxy and less firm and rubbery than the other well known Dutch cheese, Edam. Normal manufacture is as follows. The milk is pasteurized at 75°C without holding, cooled to 31°C and 0.5 to 0.75% starter added. Sodium nitrate (0.005%) may be added to prevent gas formation by coli bacteria, and calcium chloride (up to 0.02%) may be used to restore the rennet-coagulating properties of the milk, which are slightly reduced by the heat treatment. A common practice is also to add 22 ml of annatto solution to 100 litres of milk in order to produce the characteristic yellow-orange colour of the cheese. After a short ripening period 25 to 30 ml rennet are added and in about half an hour the coagulum is cut to pieces of 0.5 to 1.0 cm cube. After stirring for about 20 minutes about half the whey is run off over a 10-minute period and then hot water added to raise the temperature to 38° to 41°C. After a further period of about half an hour the curd is pressed under the whey by a metal tray with weights and then cut into blocks. These are removed from the whey, packed into moulds and pressed for up to 5 hours. Thus Gouda manufacture differs from Cheddar by this partial hot-water treatment which has the effect of reducing the acidity and removing some of the calcium salts, thus giving a softer body than that of Cheddar.

Gouda cheese is salted by immersion in saturated brine at 15°C for 3 to 5 days. The pH value of the cheese should now be 5.2, and after drying off for a day the cheese are wrapped in foil or coated with a plastic.

As with Cheddar and all major varieties of cheese, the procedure for making Gouda cheese can now be fully mechanized. This may involve some modifications (shorter time etc.) but does not affect the fundamentals of the cheese making process.

d. Provolone. This fairly firm, semi-hard cheese originated in southern Italy and is now made not only all over that country but in many other countries, particularly the USA. It is the best known variety involving the plastic curd (pasta filata) or 'spun curd' technique in its manufacture. Under the difficult (hot) conditions in southern Italy for cheese-making it is possible that this particular system, demanding treatment of the curd with water or whey at over 90°C for about 5 minutes, pasteurized the product to produce a cheese of good keeping quality and safe to eat. Two main types are made; the Dolce or milder flavoured, using ordinary liquid rennet, and the Piccanti, piquant or strong flavoured type, using a rennet paste prepared from the stomachs of kids or lambs. Provolone and Caciocavallo are variants of the same basic type, but differ in shape.

The basic method of manufacture is as follows. About $1\frac{1}{2}\%$ starter, which may contain thermoduric streptococci and *L. bulgaricus*, is added to whole milk which is generally first pasteurized and then cooled to 35° to 38°C. Liquid rennet may be added at the rate of 25 ml per 100 litres of milk or, as commonly with Italian cheese, rennet paste containing protein-splitting and fat-splitting enzymes may be used. The milk clots in 7 to 10 minutes and the coagulum is cut with revolving knives to give a small size curd. The whey is then run off to the level of the curd, heated to about 53°C and then poured back over the curd which has been heaped at one end of the vat. The temperature is maintained for 4 to 10 hours according to the season, and the curd tested at intervals by dipping a piece into water at about 82°C and working it by slowly twisting and stretching. When it is possible to draw out the curd to a thread 1 metre long it is considered ready for the next stage. The curd is now cut into strips and put on a drainer for about half an hour during which time the acidity increases. For the next stage a lump of curd, usually about 3 kg, is weighed out and immersed in water at about 55°C. It is then worked and squeezed with the aid of a paddle when the curd is in the water and removed to work with the hands at intervals.

When ripened the rind has a smooth shiny appearance and is yellow or brown-golden in colour. The interior is white or pale straw in colour. The weight is usually 4 to 5 kg, but many forms and a wide range of weights are now in use. The common pear- or

Fig. 4.15 Provolone cheese manufacture.

Fig. 4.16 Provolone cheese in store.

cone-shaped Provolone cheese is usually 36 to 45 cm high. The flavour may be described as creamy and delicate after two or three months ripening but becomes steadily sharper with age, and the use of porcine rennet gives a strong flavoured cheese. If ripened for over 6 months it becomes a grating cheese. The minimum fat content is 45% in the dry matter.

e. Parmesan or Grana. The term 'Grana', which means granular, refers to a number of varieties which are made, originally in northern Italy but now all over the world, and have become famous as grating cheese. They are usually matured for at least 1 year, and sometimes up to 2 years, are very hard, have a low moisture content and consequently a very long keeping quality. If cheese-making is regarded as the best simple method for conserving milk nutrients in a safe and palatable form, then the Grana type cheese must be recognized as the most successful variety ever invented. It is certainly the best known long-keeping cheese, and the most commonly used for grating and incorporation in soups and other foods. It is known in English-speaking countries as Parmesan, this word being derived from Parmigiano or Parma.

The essential features in manufacture are a partly skimmed milk, a very high scalding temperature for the curd, a long immersion of the young cheese in brine, and a long maturing period. Techniques vary somewhat from locality to locality and from season to season, but all conform to the following general pattern. Evening milk is held overnight in cans, and a certain amount of cream removed by gravity separation. This primitive method is now sometimes replaced by centrifugation to leave about 2% fat in the milk. Traditionally raw milk was used (this would give a stronger flavoured cheese) but pasteurization is now commonly practised. The very high scalding temperature makes pasteurization less important than for other varieties. Cow milk is used and this may be bleached by benzoyl peroxide or coloured yellow-orange by annatto or other vegetable colouring, according to season and market requirements.

The milk is warmed to 32° to 35°C in conically shaped copper vessels (or kettles) holding up to about 700 litres. Formerly whey from the previous day's make was used, but now a controlled starter consisting of *Str. thermophilus* and *L. bulgaricus* or similar thermoduric lactic acid bacteria is used. After a short ripening period about 25 ml rennet are added per 100 litres milk; a special lipase-rich type of rennet, or a separate lipase preparation may be used. In about 20 minutes the coagulum is ready for cutting with wire

Fig. 4.17 Parmesan cheese manufacture.

knives, the cutting and stirring continuing until particles of curd about 3 mm in size are obtained. The temperature is then raised to 43°C over about 15 minutes, held there for 15 minutes, and then raised to 54° to 58°C over 15 to 30 minutes, according to the rate of acid development. The mixture of curd and whey must be stirred very vigorously. When the particles are firm enough they are allowed to settle and a weight may be applied to help firm the curd together. After about 10 minutes the curd is manoeuvred into a submerged cloth, the corners tied together and the mass lifted out of the kettle and held on hooks to drain for about 30 minutes. The mass is then placed in a mould, the cloth folded over the top, a wooden follower (disc) put into position and the cheese put to press overnight.

After removal from press the cheese is transferred to the brining chamber where, after holding for about 3 days at 15°C they are immersed in brine (26% salt) at 8° to 10°C for about 14 days. They are then removed and allowed to dry off at 15° to 20°C for a few days.

The first stage of maturing takes place at 10°C and 85% relative humidity for 6 to 12 months. During this time the cheese must be turned frequently to maintain a regular cylindrical shape and avoid the development of soft patches. They are also washed, scraped and rubbed with oil, with or without a dark colouring matter. For the second stage the cheese is coated with a mixture of lamp black or similar material dispersed in dextrin and a vegetable oil. The cheese is usually sold when 1 year old and is normally consumed from about $1\frac{1}{2}$ years onwards. The yield is from 5 to 6% according to the fat content of the milk and the age of the cheese. A typical composition is 31% protein, 28% fat and 30% moisture. Grana cheese is thus very rich in protein and may be regarded as the most concentrated and nutritious food in common use.

The very strong flavour or 'bite' of Parmesan cheese is largely due to the free fatty acids which result from the breakdown of the fat by the lipases added with the rennet. This flavour may be regarded as hydrolytic rancidity; as with well-matured Canadian Cheddar cheese there is no firm line of demarcation between rancidity as a defect and the strong flavour of this type of cheese. Flavour is entirely a matter of what the customer likes and requires in a cheese.

f. Emmenthal. Throughout the world the phrase 'Swiss cheese' has come to mean a very hard, somewhat rubbery cheese with large holes or 'eyes'. The most important variety in this group is Emmenthal cheese and, like Cheddar, it is made in large quantities all over the world, and in German-speaking countries is the main table cheese. The name is derived from the Emmenthal valley in Switzerland. It has a long history and, although originally a farmhouse cheese, it is now made only to any extent in the larger dairies in the valleys. The method of manufacture is as follows.

The vat, formerly of copper but now often of stainless steel, holds about 800 litres of milk. The evening milk may be inoculated with a minute amount of starter and a certain amount of hand skimming carried out the following morning. The morning milk is then added uncooled to the evening milk, and in factories the mixed milk may be standardized to 2.8 to 3.1% fat to give a cheese with 45% fat in the dry matter. In the morning the milk is inoculated with half a litre of a culture of *Str. thermophilus*, 1 litre of a culture of *L. helveticus* (both being lactic acid bacteria capable of growing at high temperatures) and a small quantity of a culture of propionic acid bacteria. The temperature is then raised to 31° to 32°C and sufficient rennet added to clot the milk in 25 to 30 minutes. A small quantity of calcium chloride may be added if experience shows that a weak coagulum is likely to be obtained. This is cut with wire knives until the curd particles are 3 mm in size.

The temperature is then raised at the rate of 1 deg. C every 2 minutes with constant stirring. When this reaches 45°C the rate is increased to 1 deg. C per minute up to the maximum scald temperature of 53° to 57°C. The temperature is very critical because the non-thermoduric types of bacteria are killed in this range, and the temperature selected will vary with the properties of the milk, and particularly with the rate of development of acidity. The total time of heating and stirring is about 45 minutes. When the firmness of the curd indicates that the pitching point has arrived, stirring is stopped. Some whey is then drawn off for the preparation of the starter for the next day. A little cold water may be added and the curd collected in a large coarse cloth and lifted out of the vat. It is then lowered into a round mould made of wood or stainless steel, the cloth folded over the curd, a wooden follower inserted and pressure applied. This is increased three stages over two days up to 15 kg per kg cheese. The cloth is then taken away and the cheese removed to a salting room being maintained at 10°C. In modern manufacture the cheese is first dry-salted for a day or two and then immersed in strong brine at 8° to 12°C, the cheese just floating in the brine. The strength is critical because if it is too weak the cheese will sink to the bottom and become slimy and develop a bad flavour. The cheese are turned over at least once a day and salt sprinkled over the exposed surface. The cheese are left in the brine for 1 to 2 days and then removed, drained and placed on a clean, dry board in a room at 10° to 12°C and 80 to 85% relative humidity for 8 to 10 days. During this time they are brushed, dry salted and turned every day. Subsequent stages of maturing take place at higher temperatures, the times varying according to the season and the rate at which the 'eyes' develop. The temperatures and relative humidities are important factors in the maturing of Emmenthal cheese and must be most carefully controlled.

The holes in cheese of this type are produced a few weeks after manufacture by the development of the propionic bacteria. The lactose or milk sugar is first fermented to lactic acid by the ordinary lactic streptococci in the starter, and this is then attacked by the propionic acid bacteria to produce propionic acid, acetic acid and carbon dioxide gas which is responsible for the holes.

Gruyère cheese is a separate variety and has a somewhat more mellow and fatty body, with holes which are fewer and smaller in size. It is identical with the French Comté. However, the term 'Gruyère' may be applied to any Swiss cheese with holes, and Emmenthal, which is made in the greater quantity, is often sold as Gruyère in English-speaking countries.

g. Limburg. This rather soft, semi-hard cheese is easily recognized by its block shape and rather small size, its brownish red smeary or greasy coat, and its penetrating odour. It originated in Belgium and takes its name from Limburg, a town in the Flemish province of Liège. The method of manufacture is as follows. Whole milk, now usually pasteurized, is adjusted to 30° to 34°C and starter and rennet added to give a coagulum ready for cutting in 30 minutes or slightly longer. The coagulum is cut into pieces 1 cm in size, and then the temperature gradually raised to 36° to 37°C. Stirring continues until the curd is sufficiently firm and then the whey is almost completely removed. The pieces of curd are removed by a ladle into multiple rectangular block moulds, formerly made of wood but now usually of aluminium or stainless steel. The total time from renneting up to this stage is $1\frac{1}{2}$ to 2 hours. Slight pressure may be applied to the curd in these moulds by resting a board on them, and the young cheese must be turned frequently while draining in this way.

Salting is carried out in various ways. The cheese may be immersed in brine at 13°C for one or two days or it may be dry salted by rubbing daily with salt over the whole surface, or the cheese may be simply smothered in salt. Ripening takes place at about 15°C and a relative humidity of 90 to 95%. A characteristic brown-red, slimey coat soon forms, and the cheese are rubbed frequently with a cloth soaked in brine. This not only keeps the coat in the desired condition but transfers the typical smear organisms (*Bacterium linens*) from cheese to cheese. These bacteria are responsible for the characteristic and powerful odour and taste of Limburg and similar varieties of cheese. Ammonia, fatty acids and other volatile organic compounds are the main constituents of this flavour complex, which is unmistakable. Ripening is completed in one month and the yield is about 12%. The cheese is usually made in the form of a cube or rectangular block $15 \times 15 \times 8$ cm and weighs about 1 kg, but a number of shapes are made. A typical composition is: fat in the dry matter 40%, moisture 52%. A half-fat cheese may also be made and the corresponding figures are 20% and 59%.

h. Stilton. Stilton cheese is not only world famous, being the best-known blue veined cheese in the United Kingdom, but is one of the few major varieties of which the origin is reliably known. The cheese was invented (if one may use this word) by Mrs Paulet about the year 1730 in Wimondham in the Melton quarter of Leicestershire. A claim has also been made for Mrs Orton at Little Dalby at the same time, and it is possible that Stilton cheese was invented by both independently. Whatever the true history may be, the fame of this cheese soon spread and it was sent to Cooper Thornhill, a relative of Mrs Paulet, who kept the Bell Inn at Stilton in Huntingdonshire. This was a well-known coaching inn and the reputation of the cheese was carried far and wide by the travellers calling there. Thus, as with Cheddar, the name of the cheese was adopted from the marketing and not the manufacturing background.

For some time it was claimed that true Stilton cheese could only be made around the Melton Mowbray area, but this has since been found to be unjustified. Good quality Stilton cheese are not only made in Leicestershire but also in Nottinghamshire and Derbyshire. Many cheese connoisseurs consider that Stilton is the king of all cheese, but naturally this is a matter of opinion. The French may well consider that Roquefort and the Italians that Gorgonzola is the best of the blue veined cheese.

As far as can be known, Stilton cheese today is similar to that made by Mrs Paulet and Mrs Orton in 1730. The chief characteristics of this variety are as follows.

1   It is made from whole milk or milk enriched with cream. It is considered that the best Stiltons are made from evening milk (which has more fat in it than morning milk) in September. Cool temperatures and the 'second bite' of grass are undoubtedly factors of importance. In factory making, which is the only method in use today, mixed evening and morning milk are normally used.

2   The absence of scalding and pressing leads to slow drainage of whey and to a curd with a high moisture content. This results in a high acidity on account of the lactose in the whey. Later the loss of whey and shrinkage of the curd leads to distortion and cracks, so letting air into the cheese. The acid curd and the presence of oxygen create conditions favourable for mould growth.

3   Until some years ago Stilton cheese was made from raw milk and without pricking or stabbing the cheese. This resulted in a full flavoured cheese but with wide variations in extent of mould growth, and in cheese quality. About 25% of the cheese were excellent,

but the remainder of inconsistent quality. Nearly all milk is now heat treated and all Stilton cheese, like Roquefort Gorgonzola, are stabbed a few weeks after making.

4   The adoption of pasteurization of the milk and stabbing of the cheese has led to more consistent quality, but the 13 factories each still tend to produce a characteristic cheese, and this is especially true so long as raw milk is used. An expert can usually identify the factory from the cheese.

Method of manufacture. The earliest cheese were made in round vats or pans, sometimes of wood. At a later stage ceramic sinks were introduced. In modern factory practice ordinary rectangular vats are used and the method is as follows.

Mixed milk is pasteurized at 71°C in the summer and 68°C or even lower in winter, and run into the vat at 29° to 31°C. Very little starter (about 0.01%) is added and then rennet at the rate of about 200 ml to 1 000 litres of milk, according to the acidity. About 1 hour is allowed for ripening or souring of the milk, and then the coagulum is cut 1 to $1\frac{1}{2}$ hours after rennetting to give pieces 1 to 1.5 cm in size, and the pieces of curd are allowed to settle for about $1\frac{1}{2}$ hours. The whey is then slowly drawn off over a period of about 2 hours. The treatment of the curd varies from factory to factory but a typical method is to cut it into blocks from 10 to 15 cm in size and to make a channel for drainage down the middle of the vat. The blocks are turned from time to time and may be cut again. This drainage may continue overnight but the temperature must not be allowed to fall below 21°C.

When the acidity has reached about 1.1% lactic acid, the curd is broken by hand or very coarsely milled into pieces about 3 cm cube. Then 13 kg (or a weight according to the size of the mould) is weighed out, salted to the extent of about 2%, and carefully filled into the mould by hand. The curd must not be pressed at this or any other stage. The moulds are allowed to stand on a cloth-covered table or shelf for a few days with a temperature maintained at about 16°C. The cheese is turned every day and no follower is used. It is then taken out of the mould and trimmed, the surface being scraped very lightly with a broad bladed knife and any fissures filled by working in the curd to give a continuous smooth surface. A calico binder may be fitted round the cheese at this stage, using pins to secure it. When a satisfactory coat has formed, usually after about a week, the cheese is transferred to the ripening room where it stands on a shelf at a temperature of 13°C and a relative humidity of about 96%. Formerly this atmospheric control was rather empirical, but today factories have instrumental control for both conditions. It is important that the temperature be maintained constant because fluctuations lead to uneven shrinkage and distortion of the cheese.

A characteristic stage in Stilton manufacture, as for all blue veined cheese today, is the pricking or stabbing of the cheese, usually after 5 to 6 weeks. This was first done by simple metal skewers but today a machine having stainless steel rods 1 to 2 mm in diameter and 2 cm apart is used to make air channels into the cheese. This machine was developed for making Stilton about 1953 and may be operated by hand or foot or sometimes mechanically. The admission of oxygen into the cheese in this way permits growth of the *Penicillium roqueforti* (or *glaucum*) which confers the characteristic blue-veined appearance of the cheese. The cheese is rotated between stabbings so that it may have as many as 160 holes made in it. The cheese may be stabbed again once or twice at weekly intervals. Stilton cheese should be 'blue' when 10 weeks old and are then ready for wholesale distribution.

In early days cheese-makers left the inoculation of the cheese by the spores of the mould to chance; at that time there was no knowledge of microbiology. Today, with elaborate scientific control in the cheese industry it is general practice to add a

Fig. 4.18   Stabbing of Stilton cheese.

suspension of mould spores either to the milk or to the curd when placing this in the mould.

The low yield (about 7.5%) and care necessary during ripening, and the uncertainty in obtaining a well-blued cheese, are the factors responsible for the high price of Stilton, which may be nearly double that of Cheddar cheese. The best supplies go traditionally to the good clubs and restaurants.

i. Camembert. Camembert and Brie (these two varieties are basically similar) are the most famous surface mould ripened soft cheese in the world. They originated in Normandy, France, probably from an early type of cheese which was mentioned as far back as the thirteenth century. In 1791 Madame Harel, a farmer's wife in Camembert, perfected a method for making this type of cheese. It remained a family secret for some time but its manufacture and fame steadily grew until now it is made all over the world.

Manufacture. Camembert cheese made by the traditional method, using raw milk and with no cutting of the curd, ripens faster than the modern type made from pasteurized milk, and develops brown patches on the coat due to the growth of *B. linens* etc. The coat wrinkles and the odour of the uncut cheese is characteristic and strong. The modern method now used in factories involves the use of pasteurized milk and a simple method of cutting the coagulum. The cheese ripens more slowly and the coat is unwrinkled, becoming covered with a pure white growth of *Penicillium candidum*. The odour of the uncut cheese is faintly ammoniacal. Methods vary for Camembert as for all varieties, and a certain amount of raw milk cheese is still made for the connoisseur market by a slightly different procedure.

Fig. 4.19 Camembert cheese manufacture.

A typical contemporary method is as follows. Good quality fresh milk is pasteurized at 69°C and cooled to 33° to 34°C, the cheese room being maintained at about 25°C. From 1.5 to 2.5% starter is added and an hour later 16 to 17 ml rennet per 100 litres milk. This clots in 25 to 30 minutes and the coagulum is ready for cutting in about $1\frac{1}{4}$ hours. When sufficiently firm the pieces of curd, which are 3 to 4 cm in size, are put into open cylindrical moulds, made of plastic, aluminium or stainless steel, standing on draining mats. The moulds are normally about 10 cm in diameter and about 14 cm high, and are arranged in groups to permit easier and quicker filling. In the larger factories the curd-whey mixture is poured or pumped into a large tray or distributor fitted over several ranks of moulds. The curd drains rapidly and in 5 to 6 hours sinks to about half its original height, the acidity of the whey by this time being 0.6 to 0.7%. The temperature of the room must not fall below 22°C during the night, and early in the morning the moulds are removed and the cheese left on the drainers. They are trimmed before being transferred to the saloir or salting room at 18° to 20°C where they are dry salted on both flat surfaces and sprayed with a penicillium culture about an hour later. This process is repeated after three hours, and an hour later the salting is carried out again. Next day the mats are removed and the cheese turned over on to stainless steel drainers and placed in another room at 11° to 14°C for 10 to 12 days. They are then wrapped and are considered to be fit for eating in 4 to 6 weeks.

About two litres of milk are required to produce one Camembert cheese weighing 210 to 260 grams. The moisture should not exceed 48% and at least the cheese must contain 110 grams of total solids and 45% fat in the dry matter. A smaller size is also made.

The microbiological stages in the ripening of a Camembert cheese are complex and interesting, involving in sequence lactic streptococci, lactobacilli, yeasts, oidium, penicillium and finally the red-brown pigment-forming bacteria. The secret of making a good Camembert cheese lies in the control of this sequence of micro-organisms.

Examination of a cut Camembert cheese at different ages readily reveals the course of ripening. First, the early soft lactic curd becomes firmer as it dries out and then, as the mould develops on the coat, it gradually changes from snow-white to a slightly greyish creamy colour, and the body becomes softer and more velvety or waxy. The central white 'chalky' layer steadily becomes thinner as the powerful proteolytic and lipolytic enzymes produced by *Penicillium candidum* (*camemberti*) diffuse inwards. When this chalky layer has disappeared the cheese is fully matured and fit for eating. It will subsequently become over-ripe if held at ordinary temperatures and ultimately liquefy.

j. Cottage cheese. Cottage cheese is interesting from several points of view. It is of recent origin and probably the most important and original contribution of the USA to the cheese industry. It is now popular not only in the USA but in the United Kingdom and in many other countries. It is a distinct variety and should not be confused with the old 'cottager's cheese' which was formerly made in Britain (and in other countries with the appropriate name) by the farmer's wife from any milk, whole or skimmed, which had been left over from other operations.

Definition. In the United States and Canada Cottage cheese is defined as a product containing not more than 80% moisture, and creamed Cottage cheese must contain not less than 4% of fat. The latter is by far the more popular form.

Only one type (creamed Cottage cheese) is produced to any extent in Britain, but several variants are recognized in the USA. The best known of these are:

1 Fat free, salt free (for people suffering from high blood pressure).
2 Pot Cottage cheese, made from 0.75% fat milk.
3 Philadelphia or flake Cottage cheese, also made from 0.75% fat milk, but the curd is dressed with 18% homogenized cream.
4 Californian style, curd dressed with 14 to 20% homogenized cream.
5 Popcorn (home-made), curd dressed with 40% cream.
6 Country style (regular), curd dressed with 18% ripened or soured cream.
7 Whipped cream, curd carefully blended with whipped cream to give 8% fat in the cheese.
8 Vegetable or salad types, (chives, carrots, etc.).
9 Vegetarian type, without calf rennet.

The characteristic properties of Cottage cheese are a high moisture content, cream added without loss of identity of curd particles, absence of ripening or maturing and consequently a very mild, bland flavour, a fragile structure and a short keeping quality necessitating refrigerated storage and distribution. The most characteristic feature in the making process is repeated washing of a soft and delicate curd in water. These properties and the method of manufacture make Cottage cheese very liable to attack and deterioration by bacteria and it is important, therefore, that very good hygienic conditions are observed in manufacture.

Manufacture. Methods of manufacture for Cottage cheese vary greatly, even for the same type. Usually separated milk or milk containing 0.75% fat is used. The amount of starter and of rennet can vary considerably and the times required for the stages in the making process also differ. In outline, the basic method is as follows.

Milk, which must be of good bacteriological quality, is heat treated at 71°C for 12 seconds, separated, cooled to 21°C and about 7% starter added. After about 1 hour a very small amount of rennet, e.g. 2 ml for 500 litres of milk, may be added. The coagulum is ready for cutting when the acidity has reached nearly 0.5% lactic acid, with a pH of about

Fig. 4.20 Cottage cheese manufacture.

4.6, and this stage is reached nearly 4 hours after adding the starter. A knife with 1 cm spaced blades is used for cutting, and after a few minutes the temperature is carefully raised to 49°C over 25 minutes. As soon as the curd has become sufficiently firm the whey is run off and the curd washed in cold water. The washing process performs two functions – it removes whey (with its acid and calcium salts) from the inside of the curd, and it also cools the curd. A common procedure is to use three washes at about 27°, 17° and 3°C at intervals of about 20 minutes, the curd finally remaining in the third wash for 30 to 60 minutes. The third wash may be adjusted to pH 4.8 by addition of phosphoric acid. Even if of good bacteriological quality, the water may with advantage be very lightly chlorinated as a safeguard against contamination.

After washing, the curd is drained for 30 to 60 minutes, the vat being kept cold. The curd is then dressed or creamed, for example by adding about a third of the curd weight of 18% fat cream. The mixing may be done in the vat or by using a machine of the dough mixing type. Sufficient salt should be added to give a final concentration of about 1%.

Since it is a high-moisture product of low acidity, Cottage cheese is especially vulnerable to spoilage by bacteria of the type found in water. Not only must the greatest care in hygiene be observed during manufacture, but the cheese must be held at a low temperature, e.g. below 5°C, during distribution and sale.

4.8 YOGHOURT AND OTHER CULTURED MILKS

4.8.1 History
Cultured or fermented milks of the yoghourt type, which are cultures consisting entirely or mainly of lactobacilli in milk, have been made in the warm countries of the world for centuries. These were prepared quite empirically and constituted a common

method for the consumption of milk, because this would be contaminated with souring bacteria as soon as it left the cow or other mammal, and under warm atmospheric conditions would rapidly go sour. Souring in this way is an effective method under primitive conditions for keeping the milk free from pathogenic organisms, or at least of reducing their number.

As far as the Western world is concerned, the history of yoghourt goes back to the turn of the century when Mechtnikoff, working at the Pasteur Institute in Paris, studied the bacteria in the common fermented milk of the Balkan countries. He ascribed the health and longevity of the peasants in these countries to the consumption of yoghourt, and to the beneficial effect of the characteristic lactobacillus in yoghourt on the conditions in the human bowel. He described his researches and theories in his book *The prolongation of life* and this achieved wide publicity, the organism supposedly responsible becoming known as 'the bacillus of long life'. He claimed that the therapeutic effect was achieved through this organism repressing the growth of putrefactive bacteria in the bowel, and thus preventing the absorption by the body of the associated toxins. Subsequent research, particularly in the USA, showed that this bacillus, which became known as *Lactobacillus bulgaricus*, could not proliferate in the bowel, and that for therapeutic purposes a similar organism, *L. acidophilus*, should be used.

4.8.2 Regulations

There are no official regulations for yoghourt or other cultured milks in the United Kingdom but it is anticipated that they will be introduced in the near future. The legal presumptive standards of 3% fat and 8.5% solids-not-fat apply. Yoghourt has now become an established product in the Western world and milk for its manufacture is commonly standardized to 3% fat. To give firmness to the very fragile curd it has become a common practice to add skim milk powder to the extent of 4 to 5%. When no other material is added the product is generally known as 'natural yoghourt', but many variations have been introduced during the past 25 years in Western countries. To meet recent medical prejudice against animal fats, and also for inclusion in slimming diets, yoghourt may be made with milk of reduced fat content or even from skim milk. About 1960 yoghourt incorporating real fruit came on the market and this type of product rapidly attained great favour with the public, and now constitutes nearly 90% of the market in the United Kingdom. The development of these products has constituted a problem for health authorities responsible for controlling the quality of food, and in the absence of specific regulations only the Labelling of Food Regulations 1970 can be applied for this purpose. It is generally accepted that a fruit yoghourt should contain at least 80% of yoghourt, the balance being made up of cooked fruit and sugar syrup.

4.8.3 Manufacture

The manufacture of yoghourt is essentially the same as that for cheese starter, but because different types of lactic acid bacteria are employed, a different temperature is used. Yoghourt today is milk soured by a culture consisting of equal parts of *L. bulgaricus* and *Streptococcus thermophilus*. This method has been developed because it has been found that these two organisms form a symbiotic relationship, each enhancing the growth of the other. The streptococcus helps to initiate the growth of the lactobacillus by removing oxygen, and it also produces diacetyl which confers a pleasant buttery aroma on the product. The lactobacillus initiates the breakdown of the protein to liberate amino acids which stimulate the growth of the streptococcus and

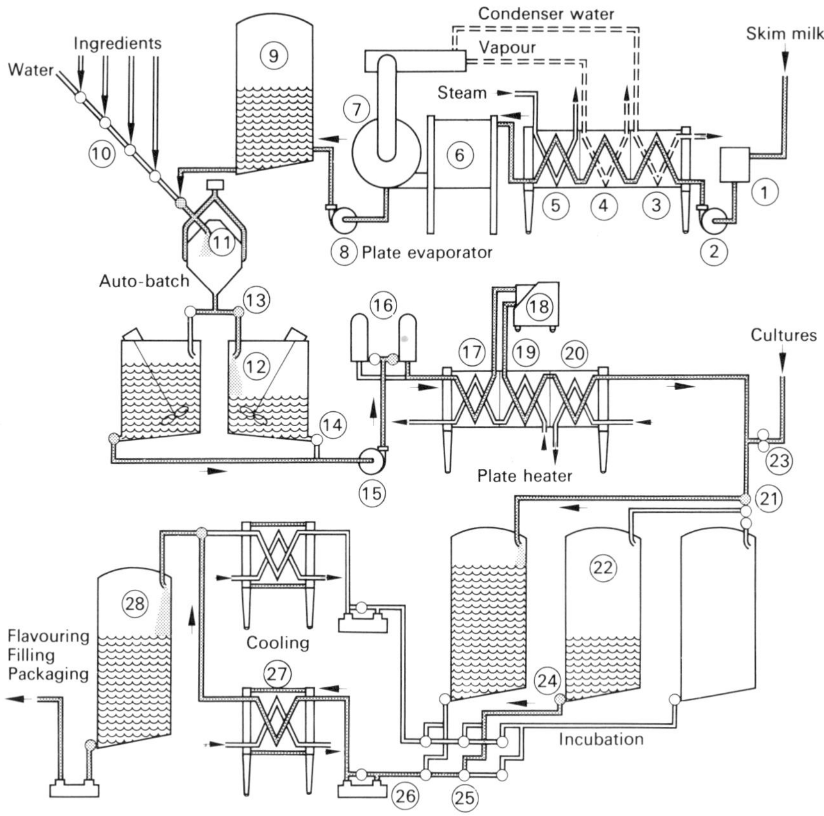

1 Feed balance tank
2 Feed pump
3 Paraflow preheater: 1st stage
4 Paraflow preheater: 2nd stage
5 Paraflow preheater: 3rd stage
6 Plate evaporator
7 Separator
8 Concentrate pump
9 Concentrate tank
10 Ingredient inlet valves
11 Weigh hopper
12 Mix vat
13 Mix vat changeover valve
14 Mix vat outlet valve
15 Mix pump
16 Duplex filter
17 Paraflow 1st stage —preheating
18 Homogeniser
19 Paraflow 2nd stage —U.H.T.
20 Paraflow 3rd stage —rapid cooling
21 Tank inlet valves
22 Incubation tank
23 Cultures metering pump
24 Tank outlet valve
25 Valve manifold
26 Positive pump
27 Paraflow cooler
28 Yoghourt tank

Fig. 4.21 Flow diagram for yoghourt manufacture.

also produces the characteristic yoghourt flavour by forming acetaldehyde. The effect of the two organisms together is to produce a highly soured milk having the specific yoghourt type of flavour.

The details of manufacture are as follows. The milk is standardized to 3% fat, or about 1.7% fat if a low-fat yoghourt is to be made. It is then homogenized in order to increase the viscosity of the final product. The milk, or skim milk if this is used, is then heated to about 90°C for 15 to 30 minutes, or to a higher temperature for a shorter time. This heating process not only 'sterilizes' the milk and destroys bacteria which would otherwise contaminate the yoghourt, but makes the milk a better medium for the growth of the lactic organisms by removing oxygen and producing reducing substances and growth factors (bacterial vitamins). The milk is cooled to 43° to 45°C and 1 to 2% of the yoghourt culture (equal proportions of *Str. thermophilus* and *L. bulgaricus*) added and carefully mixed in. Incubation takes place at this temperature until the desired acidity is reached. This may vary according to the country, a higher acidity being preferred by the public in the Middle East. In western Europe a figure of about 1% lactic acid may be required. It is preferable to measure the acidity and control it on the basis of pH measurements, the final pH usually being in the range of 4.0 to 4.2. For the natural yoghourt (i.e. without the addition of fruit) the incubation may take place in the cartons or bottles. When this is done the milk is filled out into the container immediately after inoculation, the cartons or bottles placed on trays which are transferred to trolleys and then wheeled into large incubators at the required temperature. The trolleys are then withdrawn shortly before the desired acidity is reached, and transferred to a cold store at about 0°C. Yoghourt organisms stop growing when the temperature has fallen to about 10°C.

A different procedure is followed for the manufacture of fruit yoghourt. The milk is incubated in bulk and then cooled to 5°C or slightly lower and held at this temperature overnight in order to firm the curd. The following morning it is pumped out and mixed with the appropriate fruit syrup in the proportion of 80 to 90% yoghourt to 10 to 20% fruit syrup. As fruit yoghourt constitutes an ideal medium for the growth of yeasts and moulds it is essential that the syrup be free from these organisms. Today the fruit syrup is usually bought by dairies in cans or polythene bottles already sterilized. After mixing, the fruit yoghourt is distributed by a suitable machine into cartons or bottles, the container sealed, and then held at 0° to 5°C during distribution and up to the point of sale.

The time of incubation is usually about three hours so that the inoculated milk rapidly becomes soured. As all ordinary bacteria, including pathogens, are unable to grow at pH 4 yoghourt is a perfectly safe product from the health point of view, and natural yoghourt without sugar (sucrose) has a good microbiological stability. However, some manufacturers now add a little sugar to natural yoghourt to reduce the sharpness of the taste, and this makes the product more vulnerable to yeast growth. Fruit yoghourts contain an appreciable amount of sugar (5% or more) and form an excellent medium for yeasts. Both types must be regarded as perishable foods, natural yoghourt having a shelf life of 10 to 14 days, and fruit yoghourts a shelf life of 7 to 10 days. The most popular fruits are strawberry, raspberry, blackcurrant and mandarin in the United Kingdom. The popularity of the fruit varies from region to region and apricot, peach, banana and mixed fruits are fairly common. Other foods such as honey and nuts may be incorporated. In the UK all the ingredients must be given on the label in decreasing order of concentration. For example, a typical list might read: separated milk, whole milk (or cream), fruit, sugar, flavouring and colour. The last two ingredients come from the fruit syrup, being used to 'top up' the natural flavour and colour of the fruit.

4.8.4 Acidophilus milk

The investigation of Mechtnikoff's claims by bacteriologists in the USA and other countries during the 1920–30s showed that although *L. bulgaricus* could not proliferate in the bowel another lactobacillus, *L. acidophilus*, could do so under certain conditions. This bacillus was therefore recommended for those who wished to undergo 'lactobacillus therapy,' which became popular for a short time.

Acidophilus milk differs from yoghourt in containing a pure culture of a single organism, *L. acidophilus.* The method of manufacture is much the same, and is based on the inoculation of heated milk with 1 to 2% of a culture of *L. acidophilus* and incubation at 37°C until the milk becomes clotted. In the absence of *Str. thermophilus,* or any similar streptococcus, no diacetyl is produced so that acidophilus milk has a sharp, rather astringent flavour. Mainly for this reason it is not as popular as yoghourt, and is only taken by those people who suffer from such complaints as constipation, offensive stools resulting from excessive putrefaction in the bowel, unnatural fatigue, migraine, bad breath and similar disorders which have been found to be capable of being cured or at least reduced by the regular consumption of acidophilus milk. For such purposes it is logical to take acidophilus milk rather than yoghourt, but some people who take yoghourt regularly claim to gain benefit from it. The whole subject is highly controversial and must be left to the experience of the individual. Some firms have put a special type of yoghourt on the market, claiming that the organism they use can proliferate in the bowel. Nevertheless, for serious therapeutic purposes acidophilus milk should be taken, preferably using a culture bought from a reputable firm which will advise on the correct method for taking acidophilus milk.

4.8.5 Cultured buttermilk

This is by far the most popular cultured milk in the USA. It was originally prepared from buttermilk, but is now universally made from skim or separated milk. The method of manufacture is basically the same as that for a cheese starter. Either skim milk or a spray-dried skim milk powder reconstituted at a 10% level in water is heated to 80° to 90°C for about 30 minutes, cooled to about 21°C and inoculated at about 1% level with a culture of a butter starter. This type of culture contains a streptococcus which produces a buttery aroma. Incubation takes place at 21°C for about 14 hours after which time an acidity of 0.75 to 0.9% lactic acid is reached. A method which is often used to increase the aroma is to add 0.1 to 0.2% citric acid before inoculation; this acts as a precursor of diacetyl which is the major constituent of the butter aroma. As with yoghourt, various additions such as fruit flavours and fruit may be made.

4.8.6 Kefir

If a yeast is incorporated in a cultured milk, in addition to the streptococcus and the lactobacillus, a fermented milk is obtained. The most famous preparation of this type is kefir which probably originated from the Caucasus Mountains in Russia. It is now made all over the world but principally in the Middle East. A characteristic feature is that it is necessary to use kefir grains for its manufacture. These resemble small cauliflower heads when moist but can be stored in a dry condition. They are normally about 1 cm in diameter and consist of clotted milk containing a mixed population of a streptococcus (usually *lactis*), a lactobacillus (usually *bulgaricus*) and a lactose fermenting yeast.

If dried grains are used these are soaked in water at about 30°C for a few hours and then used for inoculating sterilized milk. Repeated incubation and inoculation at about 20°C may be necessary in order to get the culture working efficiently. For the preparation of the kefir proper, whole, low-fat or skim milk is pasteurized, cooled to about 15°C and 2 to 3% of the moist grains added. Only a loose cover should be fitted to the container in order to allow gas to escape. After about 8 hours the culture is strained through a clean cloth and then the clear serum bottled and held at about 0°C in order to check the fermentation and keep the product in a sparkling condition. Kefir contains about 1% lactic acid and from 0.5 to 1% alcohol, but skilled manufacturers can obtain higher alcohol values.

4.8.7 Kumys (Kumiss)

Kumiss is basically the same as kefir but is made from mare's milk. As this contains more lactose than cow milk it is possible to obtain more alcohol and values of up to 3% have been reported. This is a favourite drink in the Russian steppes. Incubation of the product may last for several days, and it is sometimes distilled to produce a highly intoxicating beverage.

4.8.8 Urda and Skuta

These are similar to kumiss and kefir and are prepared from whey derived from sheep milk in the Carpathian Mountains.

4.8.9 Kaeldermaelk (cellar milk)

This is the most popular cultured milk in Norway. Milk is boiled, cooled and inoculated with a previous sample. Microbiologically it is similar to kefir, with an alcohol content of about 0.5% and a lactic acid value of up to 2.5%. As in other countries this is a primitive means of storing milk.

4.8.10 Taette

Taette and Lange melk are popular cultured milks in Sweden. They are viscous or ropy products made with a slime-forming strain of *Str. lactis.*

4.8.11 Leben

Leben is the traditional fermented milk in Egypt and surrounding countries and is analogous to *Kumiss.* It may be made from the milk of the cow, goat or buffalo.

4.8.12 Dahi and Lassi

These are the traditional fermented milks in India. The microbiology is very variable but lactobacilli of the bulgaricus and related types usually predominate. Streptococci and yeasts are sometimes present.

4.9 WHEY

In cheese-making 100 units of milk are converted to 10 to 15 units of cheese and 85 to 90 units of whey. The latter contains only roughly 7% of total solids and until a few years ago the primary concern of the cheese industry was to dispose of it as easily as possible. The traditional method was to feed it to pigs, and where practicable feeding whey to livestock is the simplest and best method of disposal. Although whey can be condensed and dried the high content of water (93%) makes the processing costly in relation to the amount of solid material produced. However, the steadily rising prices of skim milk powder and whey powder during recent years makes the drying of whey far more profitable than formerly. Even in 1972 millions of gallons of whey were poured into quarries etc., because there was no local method for processing it, and the necessary transport to convey it to a drying centre would have been unjustifiably expensive. Today virtually all whey is either condensed to a paste containing 20 to 60% total solids or dried, usually by the spray method.

A typical analysis of whey powder is given in table 4.12. From the nutritional point of view the greatest drawback is the high proportion (73%) of lactose. Whey powder can be

Table 4.12   Composition of whey powders and paste

	Spray dried (A) %	Spray dried (B) %	Roller dried %	As fraction of solids in whey paste %
Lactose	73.7	70.0	69.0	71.5
Protein	13.0	12.5	14.0	14.7
Minerals	8.0	10.0	7.5	10.3
Lactic acid	1.8	1.3	2.5	2.0
Fat	0.8	1–1.5	1.7	1.5
Water	3.0	4–5.0	5.0	–

Whey paste may contain 33% solids

used as an ingredient in dairy products such as cheese spread, and in many compounded foods. The disadvantage of lactose is that it may crystallize and make a product gritty and it can also cause adverse symptoms in people who suffer from lactose intolerance. However, the protein content, although less than in skim milk powder, is marginally superior nutritionally to casein, the chief protein of milk. Large amounts of whey are used for the manufacture of lactose which finds application in the pharmaceutical industry.

4.10 CASEIN

Casein is the principal protein in milk, in which it occurs to the extent of about 2.6%. It was formerly used for making plastics, adhesives, etc. but now that its high nutritional value is generally realized it is used almost exclusively as a protein in the food industry.

Casein is prepared by adding an acid (usually hydrochloric) solution slowly and with vigorous agitation to skim milk at a temperature of about 50°C until a pH of 4.6 is reached. This is the isoelectric point of casein at which it has minimum solubility in water and so is almost all precipitated. It can be purified by redispersing in a weak alkaline solution and then reprecipitating with acid. It is used mainly as a protein supplement in special foods and for the manufacture of sodium caseinate.

4.10.1 Sodium caseinate

Casein is practically insoluble in water but can be solubilized by conversion to the sodium salt. This is effected by dispersing it in a mildly alkaline solution, using a final pH end-point of 6.2 to 6.4. In this form the protein is more easily managed as a food additive. It is employed to a considerable extent in processed meats, such as sausages, and also in certain other foods. It is particularly valuable as an emulsifying agent and for conferring water binding properties on the food in addition to increasing the protein content. These properties can be improved by addition of polyphosphates.

4.10.2 Co-precipitate

Milk contains nearly 80% of its protein as casein, the remainder being the soluble whey proteins. Thus nearly 80% of the protein can be isolated by the simple method described above, but the remainder, which is of marginally superior nutritive value, stays in solution and so would normally be lost. An ingenious method devised in Australia gives a combined precipitation of all the proteins in milk by the addition of an acid and a calcium salt under controlled heating conditions. This permits the utilization of all the protein in milk for compounding other foods.

4.11 ICE CREAM

Various types of frozen milk and cream confections have been known for centuries. The first record of an 'ice cream' appeared in the reign of Charles I. Water ices were made in southern Europe in the fifteenth century and since then ice-cream manufacture has developed until today it is a highly organized branch of the food industry. The first example of bulk manufacture was in the USA in 1851 and today the most elaborate and technically developed ice cream industry is in America. Pasteurization was introduced just before 1900 and the homogenizer, invented in France in 1899, was rapidly brought into use in the American ice-cream industry. The circulating brine freezer was invented in 1902 and the direct expansion freezer in 1913. The Vogt continuous freezer was introduced in 1928 and in 1950 vegetable fat was utilized in the ice-cream industry. In 1953 the HT–ST method (section 4.2.4 *b*) for the pasteurization of ice-cream mix was approved in the USA and definitions and standards were drawn up by the Food and Drugs Administration in 1960.

4.11.1 Definitions

In general, ice cream may be regarded as a frozen confection based on milk or cream, sugar, and emulsifying and stabilizing agents. In the modern industry a wide variety of ingredients is used.

In the United Kingdom the Ice Cream Regulations 1967 define *ice cream* as 'the frozen product intended for sale for human consumption which is obtained by subjecting an emulsion of fat, milk solids and sugar, with or without the addition of other substances, to heat treatment and either to subsequent freezing or to evaporation, addition of water and subsequent freezing, whether or not fruit, fruit pulp, fruit purée, fruit juice, sugar, flavouring or colouring materials, nuts, chocolate or other similar substances have been added before or after freezing, and includes any ice cream present as an ingredient of any composite article of food, but does not include any sherbet, sorbet, water ice or ice lolly...'

In the USA the ice-cream type of product is classified into about 28 different types of confection. One interesting and important difference from the UK is that ice cream can only contain milk fat, whereas in the UK vegetable fats can be and are largely used. A product which contains only milk fat can be designated as 'dairy ice cream'. In the USA a product containing non-milk fat cannot be called ice cream but is designated 'mellorine'. This type is illegal in many States, but in general the recipes for mixes and methods for processing and distribution are practically identical for ice cream and mellorine.

4.11.2 Regulations

In the United Kingdom the Ice Cream Regulations 1967 require that all ice creams shall contain not less than 5% fat and not less than 7.5% solids-not-fat (s.n.f.) from milk. If the ice cream contains fruit etc., the total content of fat and milk s.n.f. must be not less than 12.5% and not less than 7.5% must be fat and not less than 2% must be milk s.n.f. The requirements for 'dairy ice cream' are identical except that all the fat must be milk fat.

Regulations also exist in the United Kingdom for the heat treatment of the ice-cream mix. Formerly ice-cream products sold in the street and on stalls were easily liable to contamination and on occasions caused food poisoning. The Ice Cream (Heat Treatment) Regulations 1952 permit three methods of heat treatment for the mix: (i) 30 minutes at 150°F (66°C), (ii) 10 minutes at 160°F (71°C), and (iii) 15 seconds at 175°F (79°C). These are somewhat more severe than the requirements for milk on account of the higher amount of solids in the ice cream mix.

In the USA compositional standards are laid down for ice cream, ice milk, fruit sherbets and water ices. In the United Kingdom ice cream is exempted from the requirements of the Labelling of Food Regulations 1970 so that the ingredients need not be specified.

4.11.3 Manufacture

Typical recipes are given in table 4.13. The most commonly used ingredients today are butter or vegetable fat, skimmed milk powder, sugar, emulsifier, stabilizer, flavouring and colour. Only permitted emulsifiers, stabilizers and colours can be used as designated in the respective regulations. There is no restriction on the type of flavouring or other ingredients such as fruit, nuts, etc. Originally milk and cream were used but it is more convenient to use the concentrated or dried products for large-scale production.

Common emulsifiers are egg yolk, whole egg and other egg products, usually dried.

Table 4.13 Recipes and formulas for ice cream

(a) Basic types

	Custard ice (Inferior texture) %	Ice cream (Genuine) %
Fat	3.0	12.0
Milk solids-not-fat	7.5	10.5
Sugar	13.0	13.0
Stabilizer	3.0 (starch)	0.5 (gelatine)
Total solids	26.5	36.0

(b) Typical formulas

Fat %	10.0	12.0	14.0	16.0	18.0
Milk solids-not-fat %	11.5	11.0	10.0	8.5	7.0
Sugar %	15.0	15.0	15.0	17.0	18.0
Stabilizer/emulsifier %	0.3	0.3	0.3	0.25	0.2
Total solids %	36.8	38.3	39.3	41.75	43.2

(c) Typical recipe and formula

Ingredient	Weight	Fat	Milk s.n.f.	Sugar	Egg solids	Stabilizer	Total solids
Cream, 30% fat	145.0	43.5	9.1				52.6
Sugar	63.0			63.0			63.0
Skim milk powder	41.7		40.5				40.5
Stabilizer	2.5					2.25	2.25
Dried egg yolk	2.4	1.5			2.25		2.25
Water	195.4						
Total	450.0	45.0	49.5	63.0	2.25	2.25	160.5
Calculated %		10.0	11.0	14.0	0.5	0.5	35.7

Quantitities are in lb or kg (or other units of weight).

Lecithin and monoglyceryl stearate are also commonly used. The emulsifier accelerates the formation of the emulsion when the ingredients are mixed and the product homogenized. Stabilizers act in two ways; they help to form a stable emulsion and they also ensure a permanent smooth texture in the frozen and hardened ice cream. Common stabilizers are gelatine (traditionally the most popular), sodium alginate, and various gums such as carrageenan and agar. Improvers and anti-oxidants are generally not used. Sugar is normally the only sweetening agent, and there should be no need to use a synthetic sweetener except for special dietary purposes. Invert sugar, which is a mixture of glucose and fructose produced by the hydrolysis of sucrose, has on occasions been used, but there seems to be no advantage to be gained thereby.

4.11.4 Calculation of recipe

The method of calculating the amounts of ingredients required for any given recipe can best be illustrated by giving a simple example. Let us assume that the ice cream is required to have 12% fat, 10.5% milk solids-not-fat, 13% sugar and 0.5% gelatine. This gives a total solids content of 36%, the water content therefore being 64%. Let us also assume that is convenient for the manufacturer to use whole milk, butter and condensed skim milk having the following composition:

	Whole milk (a) %	Butter (b) %	Condensed skim milk (c) %
Fat	3.7	84	0.3
Milk solids-not-fat	3.8	1	27.5
Water	87.5	15	72.2

To calculate the proportion of these ingredients we formulate equations to relate the amount of each of these ingredients (fat, milk s.n.f. and water) to the amount in the final mix. In this way we obtain three equations from which we can calculate the weight of each ingredient to make up a given amount of mix. For example, if a, b and c are the weights of milk, butter and skim milk required the following equation is obtained for the fat equivalent

$$\frac{3.7}{100}\,a + \frac{84}{100}\,b + \frac{0.3}{100}\,c = 12$$

Similar equations can be formed for milk s.n.f. and water. From these it can readily be calculated that the mix should contain, say, 54.3 lb of whole milk, 11.82 lb of butter and 20.38 lb of condensed skim milk. When we add to these quantities 13 lb of sugar and 0.5 lb of gelatine we obtain 100 lb of mix.

The colloid chemistry of ice cream is a complicated subject for which reference should be made to the standard works. Here we shall only deal with the main aspects of the problem, of which one of the most important is *over-run*. This may be defined as the increase in volume which takes place as a result of beating in air during the freezing process. For example, if 100 litres or gallons of mix after freezing yield 175 litres or gallons of ice cream, then the over-run is said to be 75%. The optimum is considered to be in the range 70 to 100%. If the over-run is less than 50% it will yield a soggy, heavy product, while if the over-run appreciably exceeds 100% the ice cream will be too light and fluffy.

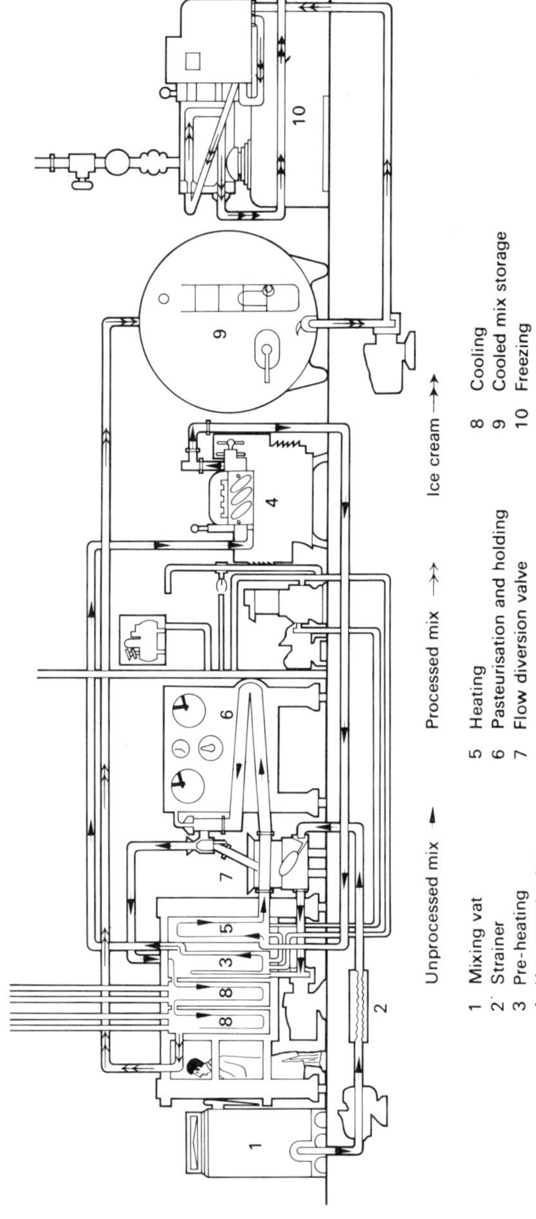

Fig. 4.22 Flow diagram for ice cream manufacture.

Another important aspect of ice cream manufacture is the possibility of sandiness or grittiness developing in the product. This is caused by the crystallization of the lactose in the milk s.n.f., lactose not being a very soluble sugar. The size of crystal which can be detected on the tongue varies with the individual but is considered to be in the range 15 to 30 microns. At the upper limit sandiness would probably de detected by nearly all consumers.

After blending in the ingredients, the mix is pasteurized, usually now by the HT–ST process, and is then homogenized at about the same temperature. The heat treatment not only kills nearly all the micro-organisms present in the various ingredients but also destroys or greatly reduces the content of any enzymes which may be present. Lipase is the most important of these as this enzyme can break down fat to produce characteristic off-flavours.

In the early days of the industry freezing was carried out in very primitive equipment, using a mixture of ice and salt to produce the necessary low temperature. Today ice cream is nearly all made by very large companies with a vast output from a single large factory. Continuous freezers are universally used. The low temperature is produced by direct expansion of liquid ammonia and the mix, after preliminary cooling, is forced through the freezer which consists essentially of a barrel inside which a rod bearing a scraper blade or knife rotates and so scrapes the internal surface of the barrel. In this way the mix is both frozen and aerated to give the desired over-run. The time required for this in a modern plant is only about 15 seconds, during which time the mix is cooled down to about $-6°C$. The frozen product is then extruded and passes immediately to a packaging machine which fills it into tubs, cartons or a brick shape which

Fig. 4.23 Processing room for ice cream manufacture.

Fig. 4.24 Small ice cream plant.

is then suitably wrapped. From this the ice cream passes through a hardening tunnel in which forced air circulation at a temperature in the range $-30°$ to $-40°C$ freezes the ice cream still further to harden it so that it can be readily transported and no chemical or microbiological changes take place.

4.11.5 Quality control

Any quality control system must take into consideration the fat and milk s.n.f. contents in order to ensure that the ice cream meets the legal requirements of 5 and 7.5% respectively. There are no legal microbiological standards for ice cream, but the methylene blue test for bacterial quality is used by the public health laboratory service in the United Kingdom and has thus acquired a semi-official status.

4.12   THE NUTRITIVE VALUE OF DAIRY PRODUCTS

Milk is sometimes spoken of as a perfect or complete food, but this is an exaggeration. There is no perfect or complete single food for children or adults, but milk and milk products (except butter which is virtually all fat) rank as some of the best foods we have and are nearly complete in themselves. The milk of any mammal *if normal* can be considered the perfect food for the suckling of that mammal for the first few weeks or months of its existence, although cow milk, for example, is deficient in certain nutrients such as copper, iron and vitamin D.

Fig. 4.25 Hardening tunnel for ice cream.

The simplest way of assessing the value of milk for human beings is to consider the contribution of 1 pint (568 ml) of average milk to the recommended daily allowances for a child and an adult. This is illustrated in table 4.14.

Milk and all its products normally receive varying types of heat treatment which result in a certain destruction of some of the nutrients, particularly the heat labile

Table 4.14 Contribution of 1 pint of average milk to recommended dietary allowances

| | 2–6 year old child | | Man (medium work) | |
	Recommended daily allowance	Contribution %	Recommended daily allowance	Contribution %
Riboflavin	0.9 mg	97	1.8 mg	49
Calcium	1.0 g	67	0.8 g	84
Thiamine (vitamin B₁)	0.6 mg	39	1.2 mg	20
Protein	56.0 g	34	87.0 g	22
Vitamin A activity	3 000 i.u.	24	5 000 i.u.	15
Nicotinic acid	6.0 mg	8	12.0 mg	4
Vitamin D	400 i.u.	3	–	–
Vitamin C	15.0 mg	78*	20.0 mg	58*
Calories	1 500 kcal	24	3 000 kcal	12

* In milk as it leaves the cow. Vitamin C is destroyed by light and heat so that the values in milk by the time it reaches the consumer are appreciably lower than these.

vitamins such as C, B_1, and some of the other water-soluble vitamins. However it should be appreciated that the treatment which milk, etc., receives in the home is nearly always more damaging to the nutritive value than that given by modern food technology. Those constituents of milk which are of the greatest importance in relation to dietary requirements are practically unaltered by pasteurization, and only slightly damaged by the most drastic heat treatments given in the dairy industry.

The emphasis in world nutrition today is on the protein content of foods, and dairy products play a very important role in this respect. The biological value of a protein depends upon the content of essential amino acids; milk, meat and eggs are rich

Table 4.15   Milk production in certain countries (million gallons)

	1964	*1967*	*1970*
United Kingdom[a]	2338	2499	2647
Australia[b]	1521	1497	1600
New Zealand[c]	1282	1317	1275
Canada	1797	1768	1775
United States	12327	11531	11402
Austria	669	719	712
Belgium	788	826	814[e]
Denmark	1120	1112	993
Finland	817	762	707
France[d]	5598	6493	6063[g]
West Germany	4458	4648	4678
Greece[d]	229	271	
Irish Republic	641	740	770[e]
Italy[d]	2056	2234	2290[e]
Netherlands	1492	1613	1763
Norway	357	371	370
Portugal[d]	104	103	127
Spain[d]	829	929	1077
Sweden	779	710	632
Switzerland	650	701	686
Yugoslavia	478	556	548
Bulgaria	198	258	267
Czechoslovakia	803	925	1023
East Germany	1231	1478	1518
Hungary	396	422	410
Poland	2721	3092	3189
Rumania	585	818	
Soviet Union	13500	17100	17700
Argentina	967	932	927[e]
Brazil	1353	1475	1519[e]
Israel[f]	57	84	97
Venezuela	129	153	
Japan	650	763	1020

[a] Total sales off farms.
[b] Twelve months ending 30th June of following year.
[c] Twelve months ending 31st May of following year.
[d] Includes milk from sheep and goats.
[e] Unofficial estimate.
[f] Twelve months ended September.
[g] Revised basis (estimate on previous basis gives 6586 m. galls).

in these substances, whereas cereals are comparatively poor. When a dairy product is eaten at or about the same time as the cereal food the high content of the essential amino acids in the milk protein raises the biological value of the poor protein in the cereal so that it becomes almost equal to that of the milk protein.

In the UK, dairy products contribute about 20% of the energy value and protein intake, over 60% of the calcium and about 30% of the vitamin A, as a proportion of the total consumption of food nutrients. Dairy products account for 18% of the average household food expenditure, liquid milk being responsible for 10%, and butter and cheese 3% each.

4.13 STATISTICS

Milk production for 1964, 1967 and 1970 of the developed countries of the world is shown in table 4.15. Figures for the production and consumption of milk and other dairy products in the UK are given in table 4.16 (statistics for USA are given in the Appendix). Great Britain is not self-sufficient in regard to milk products; figures for imports are shown in table 4.17. Data for production and utilization over the period 1938–73 are given in table 4.18 onwards.

Table 4.18 shows that sales have progressively increased over the years. In 1972–73 liquid sales for the whole of the UK formed 56% of the total (the remainder going into manufactured products). The distribution of milk for manufacture is shown in table 4.19, and the steady increase in per capita milk consumption (now levelling off) is apparent

Table 4.16 Dairy statistics for the UK, 1971–72 (million gallons)

	England and Wales	Scotland	Northern Ireland	UK total
Milk production	2336	251	170	2757
Average fat content %	3.78	3.82	3.73	
Average solids-not-fat %	8.68	8.73	8.73	
Liquid sales	1441	132	44	1618
Liquid consumption (pints per head per week)	4.71	4.12	5.02	4.66
Milk made into: butter	291	33	51	375
cheese	289	51	28	368
condensed milk	93	24	14	131
whole milk powder	49	–	3	52
fresh cream	154	11	3	168
sterilized cream	4	–	17	21
other products	15	–	10	25
Total milk made into dairy products	895	119	126	1140
Total money received by producers (£ million)	459	51	29	539
Price per gallon (pence)	19.6	20.1	17.2	19.5
Average price realized for manufactured milk (pence per gallon)	13.7	13.2	12.0	13.5

Table 4.17 Main imports of dairy products into the UK, 1971 (thousand tons)

Country	Butter	Cheese	Condensed and evaporated milk	Milk powder	Cream
Australia	32.3	6.6			
Canada	3.1	12.2			
New Zealand	144.5	70.3		5.6	
Austria	2.0	1.0		6.9	
Belgium	10.3	1.2		0.8	
Denmark	69.5	9.7		0.5	6.7
Finland	13.4	1.4		0.8	
France	9.0	9.6		0.6	
Irish Republic	30.4	25.3	12.0	19.0	4.3
Netherlands	21.3	17.4	12.5	0.6	
Norway	0.4	4.1			
Poland	5.6	0.7			
Rumania	5.1	0.6			
Sweden	2.5				
USA	26.0				

Source: Dairy Facts and Figures, 1973, Federation of Milk Marketing Boards.

Table 4.18 Total sales of milk off farms (million gallons)

April to March	England and Wales	Scotland	Northern Ireland	United Kingdom
1938–39	1 118.7	134.1	40.2	1 293.0
1944–45	1 187.3	134.8	60.6	1 382.7
1949–50	1 521.5	174.7	88.9	1 785.1
1954–55	1 653.4	209.6	96.0	1 959.0
1959–60	1 798.4	229.8	100.3	2 128.5
1964–65	1 989.5	238.9	127.5	2 355.9
1969–70	2 204.6	249.1	155.9	2 609.6
1972–73	2 492.3	263.5	186.5	2 942.3

Table 4.19 Sales of milk in the United Kingdom for manufacture (million gallons)

April to March	Butter	Cheese	Condensed milk [b]	Utilization Whole milk powder	Fresh cream	Sterilized cream	Other products	Total
1938–39	131	99	91	19	66	12	4	422
1944–45	49	44	44	34	–	–	2	173
1949–50	62	93	60	41	–	–	1	257
1954–55	103	162	110	39	16	12	3	445
1959–60	111	218	131	46	48	17	10	581
1964–65[a]	124	269	139	46	99	17	21	715
1969–70	303	281	134	46	155	22	23	964
1972–73[a]	503	404	127	46	183	18	26	1 307

[a] Owing to rounding, the figures for individual countries do not always add up to these totals.
[b] Includes any milk for chocolate crumb.

from table 4.20. Data for E.E.C. countries are given in tables 4.21, 4.22 and 4.24. From the tables it will be observed that liquid sales in the UK form a much higher proportion of the total utilization than in the E.E.C. countries. (In these tables the E.E.C. countries are those forming the community up to the end of 1972.) Broadly speaking, the UK, West Germany, Italy and Belgium and Luxembourg are strong net importers of manufactured milk products (chiefly butter and cheese), the other E.E.C. countries being exporters. Figures for consumption of butter and cheese in certain countries are shown in table 4.23. The source for statistics given in table 4.15 is the Commonwealth Secretariat, 1972. Tables 4.16–4.24 are derived from *Dairy Facts and Figures*, 1973, Federation of Milk Marketing Boards, Thames Ditton.

Table 4.20   Estimated liquid milk consumption per head (for non-farm population) (pints per week)

April to March	England and Wales	Scotland	Great Britain	Northern Ireland	United Kingdom
1938–39	3.01	2.88	2.99		
1944–45	4.00	4.13	4.02		
1949–50	4.96	4.81	4.95	Not available	
1959–60	4.85	4.35	4.80		
1964–65	4.94	4.39	4.89	5.05	4.89
1969–70	4.80	4.25	4.75	5.09	4.76
1972–73	4.75	4.20	4.70	5.03*	4.71

Note: With the exception of the figure marked with an asterisk, the above figures have been calculated by deducting an estimate of the farm population from the total population and dividing the resulting population into quantity of milk sold liquid to consumers through commercial marketing channels. It is assumed that the farm population, i.e. farmers and farm workers and their families, obtain their milk from supplies retained on farms and not sold to the Boards.

Table 4.21   Number of dairy* cows in the E.E.C. (thousands)

Country	December 1965	December 1970	December 1972
Germany	5854	5593	5511
France	8471[a]	8490	8507
Italy	3432	3214	N.A.
Netherlands	1695[b]	1874	1998[c]
Belgium[d]	963	997	960
Luxembourg[d]	55	62	63
Total–The Six	20470	20230	N.A.
United Kingdom	3270	3337	3482
Denmark[e]	1350	1153	1122
Irish Republic[e]	1547	1713	1894

* The definition of 'dairy cows' in the E.E.C. statistics varies as between countries and, therefore, such statistics may not be fully comparable. (Also applicable to table 4.22.)
[a] At October.
[b] Not strictly comparable with figures from 1970.
[c] Excludes cows on holdings of minor economic importance.
[d] At May.
[e] At June.
N.A. = not available.

Table 4.22 Annual average milk yield per dairy* cow in the E.E.C. (gallons)

Country	1965	1970	1972
Germany	778	812	844
France[a]	540	607	632
Italy	N.A.	N.A.	N.A.
Belgium	789	768	777
Netherlands	899	925	973
Luxembourg	724	747	799
Average of Five	664	714	739
United Kingdom	801	839	894
Denmark	843	842	909
Irish Republic	493	535	540

[a] Average of dairy and beef; for example, in 1971 yields were 682 gallons for dairy cows and 410 gallons for beef.

Table 4.23 Estimated consumption per head of butter and cheese in certain countries in 1972

Country	Butter	Cheese
United Kingdom	15.9[a]	11.9
Australia[b]	19.2 (1971)	9.2 (1971)
Belgium	22.2[e]	18.1[e,f]
Canada	14.9	13.2
Denmark	19.2	23.7
France[c]	19.1[e]	31.1[e]
Italy	4.2	24.0
Netherlands	4.2	20.1
New Zealand[d]	39.2 (1971)	9.7 (1971)
Sweden	10.7[e]	19.2[e]
United States	5.0	13.1
West Germany	15.7	13.4

[a] Including butter oil.
[b] Twelve months ending 30th June of following year.
[c] Twelve months ending 31st May of following year.
[d] Commonwealth Secretariat estimate. Including varieties previously excluded.

4.14 APPENDIX

Dairy products in USA

The USA is probably the greatest producer of milk (table 4.15) and of dairy products in the world (tables 4.25, 4.26, and 4.27). Full details are given in *Production of manufactured dairy products 1971* (US Department of Agriculture) and *Milk facts 1972* (Milk Industry Foundation). Very detailed regulations have been promulgated for their legal control, and these are summarized as follows. They should be read in conjunction with the definitions and descriptions for various dairy products already given above (section 4.2.1 *et seq*).

Table 4.24 Utilization of whole milk by dairies in the E.E.C. in 1971 (million gallons)

Country	Total utilization	Liquid sales	For cream[a]	Other fresh products	For manufacture into					
					Butter	Cheese	Condensed milk	Milk powder	Other[b]	Total
Germany	3855[c]	600[d]	351	48	2161	367	222	65	41	2856
France	4122	415[d]	96	34	2141	1170	98	91	77	3577
Italy	1595	657	–	–	297	635	3	3	–	938
Netherlands	1689	237[d]	50	9	577	450	214	98	54	1393
Belgium	568[e]	103[d]	22	5	339	50	2	52	– 5	438
Luxembourg	43	6[d]	4	…	33	–	–	–	–	33
Total–The Six	11872	2018	523	96	5548	2672	539	309	167	9235
United Kingdom	2624	1555	189	–	336	331	135[f]	53	25	880
Denmark	899	78[d]	50	20	524	158	65		4	751
Irish Republic	616[g]	89	17	…	371	71	31[f]	22	15	510

[a] For fresh and sterilized cream except that milk used in Denmark for sterilized cream is included in condensed milk and milk powder.

[b] Includes exports and residuals.

[c] Includes 11 million gallons of imported milk and fresh products.

[d] Consumption of liquid milk is higher than this gallonage because dairies standardize milk.

[e] Includes 10 million gallons of milk from an unspecified source.

[f] Includes milk used for chocolate crumb.

[g] Includes producer retailer sales from farms.

Table 4.25 Fluid milk product sales 1950–71* in USA

| | *Total product weight* | | | | *Per capita*[c] | | | |
Year	*Fluid whole milk* m.lb	*Low fat milk*[a] m.lb	*Cream*[b] m.lb	*Total* m.lb	*Fluid whole milk* lb	*Low fat milk* lb	*Cream*[b] lb	*Total* lb
1950	37280	2100	1490	40870	278	15.6	11.1	304
1955	43710	3020	1450	48180	290	20.0	9.6	320
1960	47360	4080	1560	53000	276	23.8	9.1	309
1965	49750	6390	1430	57570	264	34.0	7.6	306
1971[d]	45275	12800	1125	59200	223	63.0	5.5	291

* Excludes milk used on farms where produced; includes consumption of Armed Forces stationed in the United States.

[a] Includes skim milk, buttermilk and flavoured milk drinks.

[b] Includes milk and cream mixtures.

[c] Based on estimated population using fluid products from purchased sources.

[d] Preliminary.

Source: Milk Industry Foundation, USA, 1972.

Table 4.26 Proportion of the total milk supply used in the United States in various dairy products, 1950–71*

| | *Fluid milk and cream* | | | | | | | *Used on farms* |
Year	*Sold by dealers* %	*Sold by producers directly to consumers* %	*Butter creamery* %	*Cheese* %	*Evaporated and condensed* %	*Frozen* %	*Other*[a] %	*where produced*[b] %
1950	33.1	3.4	23.8	10.2	6.0	5.9	2.0	15.6
1955	37.8	2.2	22.8	11.0	5.1	6.6	2.6	11.9
1960	41.4	1.7	23.9	10.9	4.4	7.7	2.6	7.4
1965	43.1	1.5	23.0	12.6	3.7	8.5	2.8	4.8
1971	42.2	1.4	20.2	17.7	2.6	9.4	3.3	3.2

* Includes in addition to US milk production, total imports of milk fat and solids from sources outside the US and adjustments for the net change in inventories of storage cream.

[a] Dry whole milk, creamed cottage cheese and other miscellaneous uses.

[b] Milk fed to calves, consumed on farms as milk and cream, and used for farm-churned butter. Computations made by the Milk Industry Foundation based on data from the US Department of Agriculture.

Source: Milk Industry Foundation, USA, 1972.

4.14.1 Milk and cream

The recommendations of the US Public Health Services 1965 are given in considerable detail in Grade *A* Pasteurized Milk Ordinance (US Department of Health, Education and Welfare).

Milk is defined as 'the lacteal secretion, practically free from colostrum, obtained by the complete milking of one or more healthy cows, which contains not less than $8\frac{1}{4}\%$ milk solids-not-fat and not less than $3\frac{1}{4}\%$ milk fat.'

Table 4.27 Per capita sales of fluid and manufactured dairy products, United States, 1950, 1955, and 1960–71

| | Fluid products | | | | | Manufactured products | | | | | | | Frozen desserts | |
| | Total fluid milk products | Fresh whole milk | Cream | Skim milk or low fat items | Evaporated and condensed whole milk | Butter | American cheese | Other cheese | Cottage cheese | Dry whole milk | Non-fat dry milk | Evaporated and condensed skim milk | Net milk used | Ice-cream products |
Year	US quarts	US quarts	US quarts	US quarts	lb	lb	lb	lb	lb	lb	lb	lb	lb	lb
1950	141.7	129.3	5.2	7.2	20.5	9.1	5.4	2.2	3.1	0.3	3.5	5.1	45.6	17.6
1955	148.7	134.9	4.5	9.3	16.5	7.5	4.8	2.5	3.9	0.3	4.9	4.6	49.7	18.4
1960	143.7	128.4	4.3	11.0	13.9	6.8	5.3	2.9	4.8	0.3	5.6	4.4	52.5	18.6
1965	142.1	122.8	3.6	15.8	10.8	5.8	5.7	3.4	4.7	0.3	4.8	5.0	54.6	18.8
1971	135.5	103.7	2.6	29.2	6.2	4.2	7.0	4.8	5.2	0.2	4.8	4.8	54.3	18.0

Source: Milk Industry Foundation, USA, 1972

Cream is the sweet, fatty liquid separated from milk, with or without the addition of milk or skim milk, which contains not less than 18% milk fat. Light or coffee cream contains 18 to 30%, whipping cream not less than 30%, light whipping cream 30–36%, and heavy whipping cream not less than 36% fat.

Concentrated milk is a fluid product, unsterilized and unsweetened, resulting from the removal of a considerable proportion of the water from milk, which, when combined with potable water, results in a product conforming with the standards for milk fat and milk solids-not-fat as defined above.

Concentrated milk products such as homogenized, skim, etc. milk are defined analogously.

Frozen milk concentrate is similarly defined.

Skimmed milk contains less than 0.50% fat, and

Low fat milk contains 0.50 to 2.0% fat.

Vitamin D milk contains at least 400 U.S.P. units of vitamin D per US quart. (1 US quart = 2.15 lb = approx. 1 kg; 1 US gallon = 0.8326 Imperial or British gallon = 3.785 litres.)

Homogenized milk must give no visible cream separation after 48 hours storage at 45°F (7.2°C) and the fat percentage of the top 100 ml in a quart bottle must not differ from that in the remainder by more than 10%.

Buttermilk is a fluid product resulting from the manufacture of butter from milk or cream containing not less than $8\frac{1}{4}$% milk solids-not-fat.

Cultured buttermilk is a fluid product resulting from the souring, by lactic-acid-producing bacteria or similar culture, of pasteurized skim or low fat milk. (Note the inconsistency.)

Acidified milk and milk products are obtained by the addition of food grade acids to pasteurized milk, cream, etc. resulting in a product acidity of not less than 0.20% expressed as lactic acid.

Milk products include numerous types of milk and cream but not sterilized milk and milk products processed in a hermetically sealed container, condensed milk, ice cream and other frozen desserts, butter and cheese.

Pasteurization is for all practical purposes defined as in the UK.

4.14.2 Examination of milk and milk products

The official methods are given in *Standard methods for the examination of dairy products (American Public Health Association)* and *Official methods of analysis (Association of Official Agricultural Chemists.)* The basic tests for hygiene and safety are:

1   Standard plate (colony) count at 32°C (35°C) for raw and pasteurized milk.
2   Coliform test on solid media at 32°C.
3   Disc assay test for antibiotics.
4   Phosphatase test.

Grade A raw milk must be cooled to 50°F (10°C) or less and so kept until pasteurized.

Individual producer milks must not have more than 100000 and commingled milk not more than 300000 bacteria per ml before pasteurization. Antibiotics must be not detectable by the *B. subtilis* zone test.

Grade A pasteurized milk and products (except cultured milks) must be cooled to 45°F (7.2°C) or less and not contain more than 20000 bacteria and not more than 10 coliform organisms per ml. The phosphatase standard is less than 1 mg per ml by the

Scharer Rapid Method. The standards for cultured milk products are the same with the exception of the bacterial count test.

Very detailed requirements are prescribed for sanitation.

4.14.3 Cheese and cheese products

Very detailed definitions and standards are laid down in the *Code of Federal Regulations, Part 19, Title 21*, under the Federal Food, Drug and Cosmetic Act. These cover all the indigenous and many imported varieties of cheese and include labelling requirements and permitted ingredients.

4.14.4 Frozen desserts

Ice cream, ice milk, frozen custard, sherbet, water ices and related foods are dealt with in *Code of Federal Regulations, Part 20, Title 21*, under the same Act. This gives definitions and standards of identity.

4.14.5 Other milk products

Other regulations give definitions and standards for dry buttermilk, dry whole milk, dry whey, instant non-fat dry milk, non-fat dry milk (spray and roller process), edible dry casein (acid) and for grades of butter and some cheese varieties.

4.14.6 Statistics

Some statistics for the USA dairy industry are given in the tables above.

4.15 LITERATURE

4.15.1 Books on dairy products in general

C. ALAIS. *Science du Lait et Principes des Techniques Laitières.* Paris, S.E.P., 1961.

Annuaire National du Lait 1969. Paris, Editions Comindus, 1969.

J. CASALIS, E. MANN and M. E. SCHULZ. *Dairy Dictionary* (English, French and German). Nürnburg, Carl, 1963.

Chiffres-Clé de l'économie laitière Française. Paris, Fédération Nationale de l'Industrie Laitière, 1971.

J. G. DAVIS. *A Dictionary of Dairying.* London, International Textbook, 1955.

J. G. DAVIS. *A Dictionary of Dairying: Supplementary Volume.* London, International Textbook, 1965.

J. G. DAVIS. *Dairy Products* in *Quality Control in the Food Industry* (Ed. Herschdoerfer). London, Academic Press, 1968.

J. G. DAVIS and F. J. MACDONALD. *Dairy Chemistry.* London, Griffin, 1953.

K. J. DEMETER. *Bakteriologische Untersuchungsmethoden der Milchwirtschaft.* Stuttgart, Ulmer, 1967.

K. J. DEMETER and H. ELBERTZHAGEN. *Grundriss der Milchwirtschaftliche Mikrobiologie.* Hildesheim, Mann, 1968.

E. ESCHE and M. DREWS. *Der Europäische Milchmarkt.* Hamburg, Parey, 1963.

E. M. FOSTER *et al. Dairy Microbiology.* New Jersey, Prentice Hall, 1957.

A. M. GUERAULT. *Les techniques nouvelles dans l'industrie laitière.* Paris, S.E.O., 1960.

B. W. HAMMER and F. J. BABEL. *Dairy Bacteriology.* New York, Wiley, 1957.

D. S. HSU. *Ultra High Temperature (u.h.t.) Processing and Aseptic Packaging of Dairy Products.* New York, Damana Tech. Inc., 1970.

J. JACQUET and R. THEVENOT. *Le lait et le froid. Les produits laitiers (laits, crêmes, beurres, fromages, crêmes glacées et leur traitement frigorifique).* Paris, Baillière, 1961.

Dairy Facts and Figures: U.K. Thames Ditton, Milk Marketing Board.

E.E.C. Dairy Facts and Figures. Thames Ditton, Milk Marketing Board, 1971.

R. JENNESS and S. PATTON. *Principles of Dairy Chemistry.* New York, Wiley, 1959.

H. F. JUDKINS and H. A. KEENER. *Milk Production and Processing.* New York, Wiley, 1960.

R. KELLERMANN. *Milchwirtschaftliche Mikrobiologie.* Hildesheim, Heinrichs.

Laboratory Manual: Methods of Analysis of Milk and its Products. Washington, D.C., Milk Industry Foundation, 1959.

L. M. LAMPERT. *Modern Dairy Products.* New York, Chemical Publishing, 1970.

R. A. MCCANCE and E. M. WIDDOWSON. *The Composition of Foods.* London, H.M.S.O., 1960.

J. A. NELSON and G. M. TROUT. *Judging Dairy Products.* Milwaukee, Olsen, 1965.

M. E. SCHULZ and E. VOSS. *Das grosse Molkerei-Lexikon.* Kempten/Allgau, Volkswirtschaftlicher, Verlag, 1965.

E. SPEER. *Technologie der Milchverarbeitung.* Leipzig, 1968.

M. R. SRINIVASAN and C. P. ANANTAKRISHNAN. *Milk Products of India.* Delhi, India Council of Agricultural Research, 1964.

W. STORK and H. HARTWIG. *Qualitätsfehler bei Milch und Milcherzeugnissen. Vermeidung und Abstellung.* Hildesheim, Mann, 1965.

G. THIEULIN and R. VUILLAUME. *Elements pratiques d'analyse et d'inspection du lait, des produits laitiers et des oeufs.* Paris, Rev. Gen. Quest. Lait, 1967.

R. VEISSEYRE. *Techniques laitières.* Paris, Maison Rustique, 1966.

W. G. WALTER (Ed.). *Standard Methods for the Examination of Dairy Products.* New York, American Public Health Association, 1967.

B. H. WEBB and A. H. JOHNSON (Ed.). *Fundamentals of Dairy Chemistry.* Westport, A.V.I., 1965.

B. H. WEBB and E. O. WHITTIER (Ed.). *Byproducts from Milk.* Westport, A.V.I., 1970.

G. WILCOX. *Milk, Cream and Butter Technology.* New Jersey, Noyes Data Corp., 1971.

4.15.2 Abstract journal

Dairy Science Abstracts. Farnham Royal, Bucks, England, Commonwealth Bureau of Dairy Science (Monthly).

4.15.3 Milk

H. BURTON (Ed.). *Pasteurizing Plant Manual.* London, Society of Dairy Technology, 1966.

H. BURTON, J. PIEN and G. THIEULIN. *Milk Sterilization.* F.A.O. Agricultural Study 65, Rome, Food and Agriculture Organization, 1965.

J. G. DAVIS. *Milk Testing.* London, Dairy Industries, 1959.

A. GODED-Y-MUR. *Técnicas modernas aplicadas al análisis de la leche.* Madrid, Ed. Dossat, 1966.

C. W. HALL and G. M. TROUT. *Milk Pasteurization.* Westport, A.V.I., 1968.

W. C. HARVEY and H. HILL. *Milk Production and Control.* London, Leurs, 1967.

J. L. HENDERSON. *Fluid Milk Industry.* Westport, A.V.I., 1971.

Milk Hygiene. Monograph Series No. 48. Geneva, World Health Organization, 1962.
G. S. WILSON. *The Pasteurization of Milk.* London, Arnold, 1942.

4.15.4 Cream
W. STORCK. *Schlagsahne-Sterilmilch.* Hildesheim, Mann, 1961.

4.15.5 Condensed and Evaporated Milk
O. HUNZIKER. *Condensed Milk and Milk Powder.* La Grange, USA, Hunziker, 1949.

4.15.6 Milk Powder or Dried Milk
C. W. HALL and T. I. HEDRICK. *Drying of Milk and Milk Products.* Westport, A.V.I., 1971.
O. HUNZIKER. *Condensed Milk and Milk Powder.* La Grange, USA, Hunziker, 1949.
M. E. SCHULZ. *Zerstaubungstroknen und Instantisieren.* Nürnberg, Carl, 1963.

4.15.7 Butter
O. HUNZIKER. *The Butter Industry.* La Grange, USA, Hunziker, 1940.
F. H. MACDOWALL. *The Buttermaker's Manual.* Wellington, University of New Zealand Press, 1953.
W. MOHR and K. KOENEN. *Die Butter.* Hildesheim, 1958.
H. POINTURIER and J. ADDA. *Beurrerie industrielle – Science et Technique de la fabrication du beurre.* Paris, Maison Rustique, 1969.

4.15.8 Cheese
ANDROUET. *Guide de Fromage.* Nancy, Berger-Levrault, 1971.
Cheesemaking. London, H.M.S.O., 1959.
V. CHEKE. *The Story of Cheesemaking in Britain.* London, Routledge & Kegan Paul, 1959.
J. G. DAVIS. *Cheese, Vol. I, Basic Technology.* Edinburgh, Churchill-Livingstone, 1965.
J. G. DAVIS. *Cheese, Vol. II, Annotated Bibliography to 1961.* Edinburgh, Churchill-Livingstone, 1965.
J. G. DAVIS. *Cheese, Vol. III, Manufacturing Methods.* Edinburgh, Churchill-Livingstone, 1974.
J. G. DAVIS. *Cheese, Vol. IV, Annotated Bibliography 1962–1970.* Edinburgh, Churchill-Livingstone, 1974.
O. K. HALTENBERGER and J. KAMMERLEHNER. *Der Schnittkäse. Herstellung – Behandlung – Verpackung.* Hildesheim, Mann, 1965.
F. V. KOSIKOWSKI. *Cheese and Milk Foods.* Ithaca, New York, Kosikowski, 1966.
T. A. LAYTON. *The Wine and Food Society's Guide to Cheese and Cheese Cookery.* London, Wine and Food Society, 1967.
J. L. SAMMIS. *Cheesemaking.* Madison, USA, Cheesemaker Book, 1946.
G. P. SANDERS. *Cheese Varieties.* USA Department of Agriculture, 1953.
G. WILCOX. *Eggs, Cheese and Yoghourt Processing.* New Jersey, Noyes Data Corp., 1971.
G. H. WILSTER. *Practical Cheesemaking.* Oregon, Corvallis, 1959.

4.15.9 Yoghourt and other cultured milks
H. ELLER. *The Technology of Sour Milk Products.* Estonia, Tallin, 1971.
H. J. KLUPSCH. *Sauermilcherzeugnisse.* Hildesheim, Mann, 1968.

F. V. KOSIKOWSKI. *Cheese and Fermented Milk Foods.* Ithaca, New York, Kosikowski, 1966.

W. STORCK. *Milchmischgetränke und Sauermilchgetränke einschliesslich Joghurt und Bioghurt.* Hildesheim, Mann, 1959.

4.15.10 Ice cream and related products

W. S. ARBUCKLE. *Ice Cream.* Westport, A.V.I., 1972.

A. H. BAYER. *Modern Ice Cream Plant Management.* New York, Donnelly, 1963.

4.15.11 Dairy equipment and control

J. G. DAVIS. *Laboratory Control of Dairy Plant.* London, Dairy Industries, 1956.

A. W. FARRALL. *Engineering for Dairy and Food Products.* New York, Wiley, 1963.

H. S. HALL, Y. ROSEN and H. BLONBERGSSON. *Milk Plant Lay-Out.* F.A.O. Agricultural Studies 59, Reading, Food and Agriculture Organization, 1963.

4.15.12 Legal

Food & Drugs Act 1955 (and subsequent regulations on dairy products). London, H.M.S.O., 1955.

F. L. GUNDERSON, H. W. GUNDERSON and E. R. FERGUSON Jr. *Food Standards and Definitions in the United States. A guide book.* New York, Academic Press, 1963.

CHAPTER 5

Material Resources and their Conservation

5.1 INTRODUCTION

The size of the Earth's population is such that we now depend directly on a continuing supply of material resources to provide us with the means of maintaining our present civilization. Even more important, this flow of resources has to increase ever rapidly with the passage of time because of three major factors. The world's population is growing rapidly – it may well have doubled by the end of the century or thereabouts. All over the developed countries of the world, material standards of living, already high, are rising steadily. And finally, over 60% of the world's population are to be found in what we call the 'developing world', and these peoples are, naturally, demanding continuing improvements in a standard of living which is well below that of the Western World – they want to catch us up, even as we continue to press for more economic growth for ourselves. The total effect of all these factors is to cause the demand for resources to rise *exponentially*. This phenomenon is clearly illustrated by a curve showing the growth in demand for a typical non-ferrous metal such as copper. Clearly, this exponential process cannot go on for ever unless we have an infinite supply of resources to draw on, even with extensive recycling. Since the world is of finite size, and since the recycling of material resources entails the use of further resources – notably energy – it is obvious that we have a major problem facing us if we are to give consideration to the resource needs of future generations. The problem becomes even more complex when we realize that many of our natural resources are unevenly distributed over the Earth's land mass, in its crust, and in the ocean beds. Nations do not have an equal and direct access to all the natural resources their society requires. Here lie the seeds of international strife, either economic or military, coming in advance of total world scarcity of resources.

What, precisely, is meant by 'natural resources'? These are the supplies of food, building and clothing materials, minerals, energy and water that we draw from the planet. So much is obvious, but less familiar is the concept that whilst some resources such as food can be replaced year in and year out by the seasonal growth of plants and the perpetuation of animal life, and are thus continuously *renewable resources*, others like minerals are depleting or *non-renewable resources* because the Earth only has fixed supplies which are being steadily consumed. It is true that meteorites add to the total, but the amount is so small as to be disregarded. As regards renewable resources, perhaps the most important question is, at what maximum rate can food resources be produced

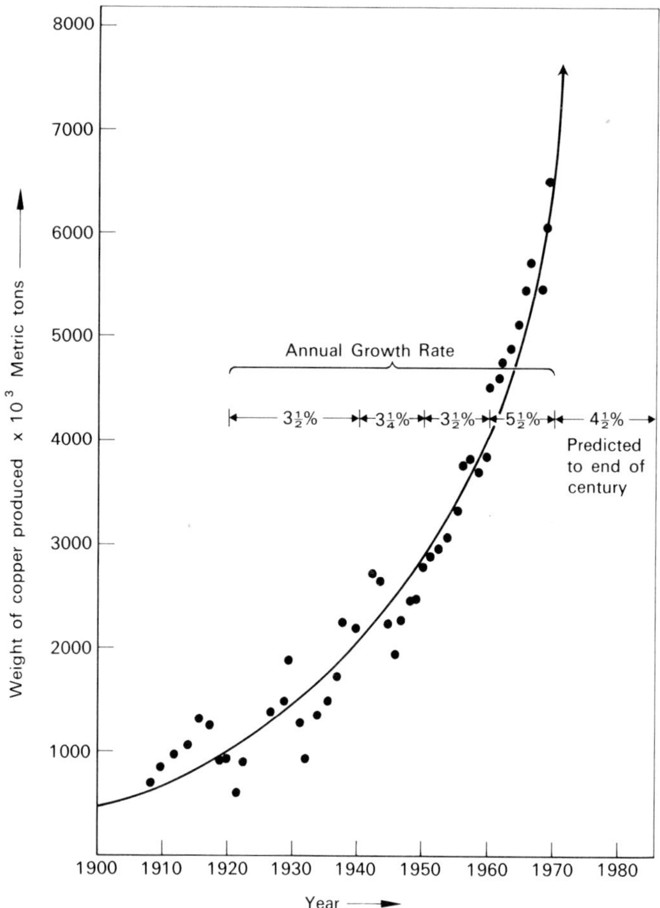

Fig. 5.1 Exponential growth in world production of copper.

without irretrievably eroding the surface of the Earth and ruining its productivity? The key question with regard to non-renewable resources is whether supplies will last until we have achieved a stable world population, and a world order where everyone receives the materials to maintain an adequate quality of life.

It is essential that we try to understand the limits of our natural resources because some of the important ones for our present way of life may already be approaching exhaustion, within a decade or two–for example, petroleum, helium gas and mercury. We also need to appreciate how we can take conservation measures, how to manage our natural resources over a long time-scale.

5.2 RENEWABLE RESOURCES

Food, natural fibres like cotton, and forest products such as timber and paper pulp are derived from plants, and because plants are seasonally regenerated, all the resources that they supply are renewable. The growing plant removes carbon dioxide from the atmosphere and by the process of photosynthesis uses the sun's energy to combine the carbon dioxide with water from the soil to make carbohydrates and oxygen which it releases to the atmosphere. Plant carbohydrates are the basic food supply for animals, including man, either by direct consumption or indirectly by consumption of plant-eating animals. The consumption of plants by animals is essentially a decay process in which carbohydrates are again reduced to water and carbon dioxide by oxygen from the atmosphere; thus, the chemical reason why animals breathe oxygen is to burn their carbohydrate food supply. The energy trapped by the growing plant during photosynthesis is released in the decay process and provides the energy needed for animal life. The growth and consumption process is cyclic; similarly, the same carbon dioxide and water can be recycled endlessly. The only component of the cycle that is not returned to its original state is the sun's energy, and this is limitless. Provided the soils and waters in which we grow our crops are given adequate care, the supply of food, fibre, and forest products that we draw from them need not be exhausted. Although production of food resources can be endless in principle, there is a limit to the annual rate of production. Plants grow only in the narrow zone where the atmosphere meets the surface of the Earth–the zone where the essential ingredients, carbon dioxide, water, and light energy are all readily available in correct proportions. The Earth's surface has a fixed area, and the rate at which the sun's energy falls on it is constant, so that these two fixed quantities limit the ultimate rate at which plant food can be produced. Though we may learn to exploit the surface more effectively and may even develop plants that use the sun's energy more efficiently than present plants, limits fixed by considerations of area and energy will remain. Although the population of the Earth has increased continuously since man's first appearance, it is now increasing at a much faster rate than our increase in plant food. Even with enormous efforts to expand food production from all sources, including sea foods from the oceans, it is inevitable that the population must cease growing at some point or else food demand will exceed production. A critical study of resources and man sponsored by the National Academy of Sciences in the USA estimated the Earth's *theoretical* carrying capacity at approximately 33×10^9 people–ten times the world's present population. If all possible means to increase food productivity were used and strict regimentation of diet were practised, the estimated maximum number of people could only be fed at a level of chronic near-starvation. It was considered possible that this doleful day could be reached about 100 years from now. Clearly, we should be foolish to reach such a disaster point, and a more practical population limit, by which we can live in equilibrium with a stable rate of food production, must be reached. The definition of this practical limit, and the means of holding it in stable balance, are two of the greatest problems which mankind faces today. The urgency of the problem is not fully appreciated as yet in all quarters of society; some people feel that the growth of technology will automatically lead to the finding of solutions in the future.

5.3 NON-RENEWABLE RESOURCES

5.3.1 Re-usable and depleting resources
The present population level on Earth could not have been reached without the incredibly complex system of health control, power distribution, transportation, and communication that underlies our civilization. But civilization is totally dependent on the

minerals and mineral fuels needed for complex technology, and these, unlike plant products, are not formed by rapid cyclic processes and cannot be regenerated once they have been mined or pumped from the Earth. The geological processes which caused their formation are virtually infinitely slow in comparison with man's growing rate of usage. The Earth contains fixed amounts of all the non-renewable resources. The so-called 'fossil fuels', coal and oil, once burnt, cannot be recycled. Metals like iron, lead, and copper, which can of course be recovered in scrap form, are only partially recovered after each cycle of use. A few of the Earth's non-renewable resources such as water are not permanently consumed. We may locally redistribute water but it soon returns to the evaporation–transpiration–precipitation cycle that governed its original distribution. No water is lost in the continuous cycle, and in contrast to the depleting resources, water is a *re-usable resource.* It is important to realize that we are discussing water on a *global* scale, because water pumped from a lake or underground supply at a rate faster than natural replenishment would certainly be a case of depleting *local* resources. For example, the south and south-east of Britain is going to find increasing difficulty in meeting its water needs in the future, although the problem will be less severe in the north. Pumping water south to even out the resource distribution could be very costly; a canal running at a fixed contour level has been suggested as one answer to the problem. Provided we do not irretrievably pollute the waters of the Earth by needlessly distributing the debris from our usage of the depleting resources, we should be able to re-use water endlessly. However, the position with the Earth's supply of *minerals* is different, in the majority of cases, from that of water, in that they are our *depleting* non-renewable resources. These are the resources of prime concern for the future.

5.3.2 Classification of depleting non-renewable resources

Having identified the area of depleting non-renewable resources as of prime concern, it is now necessary to identify the actual mineral resources involved. We have two major groupings, 'metallic mineral resources' and 'non-metallic mineral resources'. The latter group may be subdivided further on the basis of use, namely: (i) minerals for chemical and fertiliser production, (ii) building materials, (iii) fossil fuels. The former group, metallic minerals, can usefully be subdivided on a basis of abundance and scarcity; for this purpose we can use the geochemists' criterion of average concentration in the Earth's crust, usually called the 'crustal abundance', or 'Clarke'. Table 5.1, showing crustal abundances, has been divided into two parts; one containing the abundant metallic elements, and the other, the main scarce metallic elements in the crust. It will be noticed, perhaps with some surprise, that although the common industrial metals iron and aluminium appear in the abundant group, other familiar industrial metals, such as copper, lead and zinc appear in the scarce group (*vide infra*). It is for this reason that 'scarce' appears in quotes in table 5.1.

Returning to the three subdivisions of the non-metallic minerals, it can be shown that with the single exception of the fossil fuels, we are essentially concerned with compounds based on the abundant elements of the Earth's crust – e.g. sodium chloride, calcium phosphate, sulphur, gypsum, alkaline earth oxides, silica (sand) and clay. With perhaps the exception of the phosphates, any preliminary examination of non-renewable resource conservation problems can therefore omit the non-metallic minerals from detailed consideration. The resources of minerals upon which conservationists are now focussing particular attention – and will increasingly do so in the future – are those from which we extract many of the important metals of commerce, as well as the fossil fuels.

Table 5.1 Relative abundance of mineral resources

Element	Crustal abundance weight %
The abundant metallic elements of the earth's crust	
Silicon	27.0
Aluminium	8.0
Calcium	5.1
Iron (Fe^{++} and Fe^{+++})	5.8
Sodium and potassium	4.0
Magnesium	2.8
Titanium	0.86
Manganese*	0.1
The 'Scarce' metallic elements of the earth's crust	
Chromium*	0.02
Vanadium*	0.017
Zinc	0.008
Nickel	0.007
Copper	0.006
Cobalt	0.0028
Lead	0.001
Molybdenum	0.0001
Uranium	0.00016
Mercury	0.00002
Tin	0.00015
Beryllium	0.0002
Tungsten	0.0001
Silver	0.00001

* These three elements may be considered marginal in each category.
Note. The weight % figures vary in the literature, but the data in these tables are from the work of K. K. Turekian (1969).

5.4 SCARCITY ASPECTS OF METALLIC MINERALS

5.4.1 Crustal abundance and ore grade

Examination of table 5.1 reveals the striking fact that there are only ten metallic elements (silicon, aluminium, iron, calcium, magnesium, sodium, potassium, titanium, manganese and chromium) whose individual abundance by weight in the Earth's crust exceeds 0.02% of the total mass. All metallic elements with crustal abundances of 0.01% and below have been defined as geochemically scarce by Professor Skinner of Yale University. Even the element carbon, the basis of our fossil fuels, coal and petroleum, and hence of our rapidly growing plastics industry, only constitutes 0.02% of the Earth's crust; a few other materials, such as chromium and vanadium, are of similar concentration. Now if one takes into consideration the total mass of the Earth's crust, at approximately 2.5×10^{19} tonnes, it will be obvious that the total amounts present of the geochemically scarce metals must be truly enormous. For instance, a crustal abundance figure of 0.01% would correspond to a total quantity of 2.5×10^{15} tonnes of the metal element in the crust. But it is important to appreciate that the background or crustal abundance figure is of no direct *economic* importance at the present time, because man

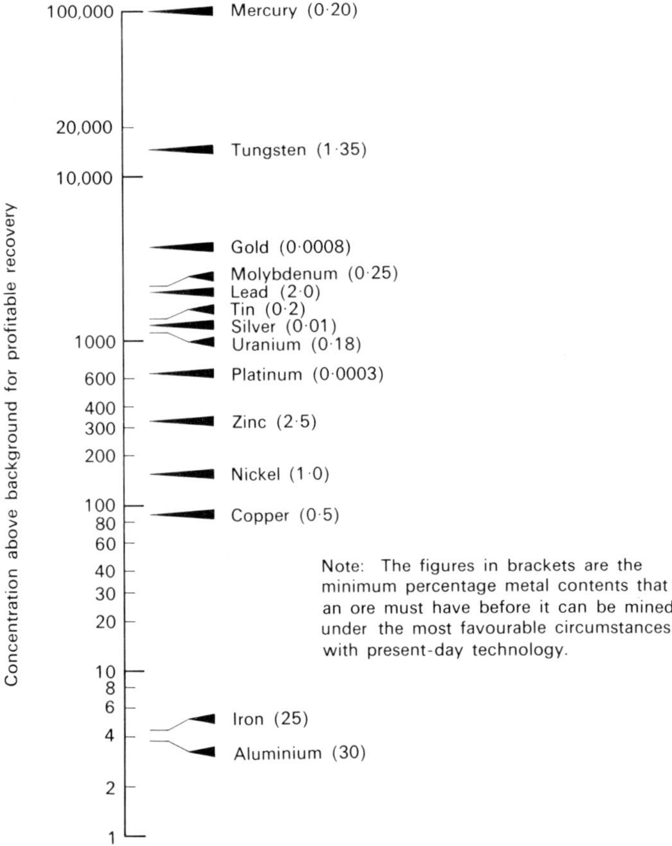

Fig. 5.2 The minimum concentrations above background needed for metals to be profitably mined on present-day technology.

only extracts metallic minerals from what are termed workable ores; that is, rock deposits in which geological processes in the past have concentrated out a mineral to a very high degree. The minimum concentrations needed for metallic minerals to be profitably mined and smelted, using present-day technology and against present-day market demands for these materials are often many times – even hundreds or thousands of times – above background concentration. When the world's reserves of various minerals are quoted in the technical press, it is the reserves of *ores* to which reference is made.

Now in actuality, geologists know that the total volume of workable mineral ore deposits is an insignificant fraction of one per cent of the Earth's crust and each deposit represents some geological accident, as it were, in the very remote past. Deposits have to be mined where they occur – often far from centres of consumption – and each site eventually reaches the stage where it becomes uneconomic to work owing to increased operating costs in relation to world prices for that particular mineral. So prospecting constantly takes place for locating the sites of new mines.

One might expect that when all the easily won ore reserves become worked out, man will move on to extract lower grade ores, but at an increased cost, provided that the consumers are prepared to pay higher prices. If the exponential increase in demand referred to earlier continues, it is not difficult to imagine the metal producers eventually obtaining these higher prices and thus working out the more difficult ore deposits. The important point to be made is that when mining experts talk about the presently available world reserves of a mineral, this is really dependent on the market price. To take a fairly extreme example, in 1963 the United States Atomic Energy Commission quoted uranium ore reserves of 650 000 tons in the USA as corresponding to the market price of U_3O_8 of 10 dollars a pound (22 \$/kg). However, at 30 to 100 dollars market price, lower concentration ores could be economically extracted as the higher operating costs could be sustained and the reserves figure would increase to 20 million tons. At 100 to 500 dollars, the reserves might be increased to 1700 million tons.

We started this section by discussing the geochemist's crustal abundance table of the elements, and then showed from a superficial examination that this bore little relationship to the concentrations of metals in the ores being commercially worked (but see below). We have then gone on to discuss the concept of ore reserves, but we now need to consider various aspects of the difficult problem of identifying the *ultimate limits* to the world's reserves of metallic ores. To state the problem in a more dramatic form, for how many years will the total reserves of the various minerals last?

5.4.2 Assessing the life of mineral reserves

a. The reserve life index. In order to express the remaining lifetime in years for a mineral resource, it is necessary to adopt some form of reserve life index.

A reserve life index is a measure of the long-term availability implied by a resource's reserve and consumption characteristics. Commonly used is the Static Reserve Life Index (SI), obtained by dividing the estimated total reserves by the *present* annual rate of consumption:

$$SI = Reserves/Annual\ consumption$$

Now this index will, on a first examination, appear to give an unduly high forecast of reserve life, when one recalls that all the metallic minerals have an exponential growth in rate of consumption. In 1971, *Mining Journal* predicted that the world consumption of almost all metals and minerals would double within the next ten to fifteen years and by the end of the century the demand for the base metals would increase by three to six times. The US Bureau of Mines foresees a 7% annual growth rate (AGR) for non-ferrous metals over the period of time from 1970 to 2000. Thus, a more useful reserve life index would now appear to be the Exponential Index (EI) given by:

$$EI = \frac{\log_e\ [(AGR \times SI) + 1]}{AGR}.$$

The SI and EI indices can be related in the form of a family of curves which can be used for reference purposes in comparing various minerals, each of which has its own reserves and also its own annual growth rate. A simple example will serve to illustrate the significance of a high annual growth rate in consumption: a static reserve life of 200 years is reduced to just less than 50 years with a growth rate in consumption of 5% per annum. However, this will *only* apply if this average annual growth rate is fully sustained over a period of time as long as nearly 50 years; this is a long time ahead for which to give any reliable forecast for an individual mineral because of such processes as substitution in use by other materials and changes in the fraction of metal recycled. Also, it does not

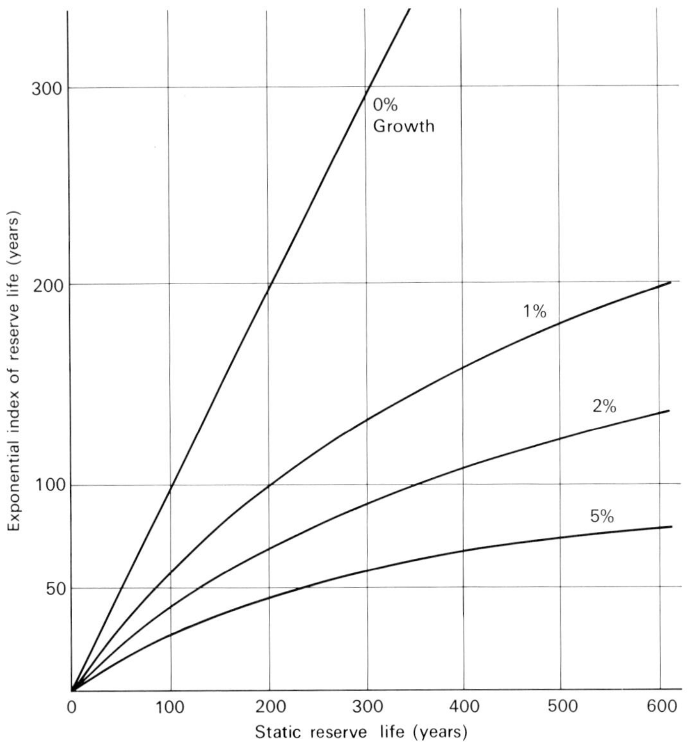

Fig. 5.3 Exponential Index from Static Index.

apply to a situation where new reserves are constantly being found, which has really applied in the real world to date.

So far, we have only examined the denominator in the basic reserve life relationship. The numerator is a more complex quantity to assess. A mining organization would use a 'reserves' figure, being concerned with thinking in terms of the economic present or the economic near future. However, in carrying out an ultimate reserves appraisal, we must think of the more distant future and take into account deposits of ore well beyond the miner's present interest; we need to set arbitrary limits more far reaching than the short-to-medium-term appraisals of importance to the mining industry. This is a difficult problem. The whole aspect of what we mean by 'reserves' of minerals was discussed by Blondel and Lasky some years ago and they proposed the simple relationship:

$$\text{Resources} = \text{reserves} + \text{potential ores}$$

and that this be further broken down as follows:

$$\text{Resources} = \text{reserves} + (\text{marginal resources} + \text{submarginal resources} + \text{latent resources}).$$

They also reaffirmed that the 'reserves' term should itself be broken down into 'proved', 'probable' and 'possible', but this is not very important in the present context, where we are concerned with ultimate resources. The full equation of Blondel and Lasky explains the fact that mineral reserve lives, in general, stay constant, since marginal resources eventually become reserves and submarginal resources will become marginal resources and so on, depending upon the economic conditions prevailing. Now the mining industry only requires to have a limited fixed quantity of 'reserves' available at any given point in time, since it is wasteful to carry out unnecessary, expensive prospecting. With an infinite supply of mineral resources of fixed grade to draw on, reserves could in principle go on being added to *ad infinitum*. Speaking generally, it has been fairly common practice for the non-ferrous mining industries to operate with a constant reserve life of 20 years or more, based on the Static Index.

Thus, it is clear that mineral reserves will continue to increase, through new discoveries, in response to rising demand, because all the exploitable ore-bodies in the Earth's crust have not yet been found. The problem of identifying the *ultimate* limits to probable additional reserves is difficult; it will depend on the state of mining and mineral extraction technology in the future, and on future demand for the various minerals.

    b. Ultimate limit to reserves. It has been pointed out by various workers in both the US and Japan that a linear relationship exists between the crustal abundance of a mineral and its probable world reserves. For example, Sekine (1963) deduced the following approximate relationship:

$$R = (1.7 \text{ to } 36.5) \cdot A \cdot 10^{10} \text{ to } 10^{11}$$

where R is the probable world reserves in tons and A is the crustal abundance in weight per cent. Thus we have returned to the geochemists' abundance index and have seen a relationship to the world's ore reserves, despite the statement in section 5.4.1 that superficially, there would not seem to be any connection between the two. Very recently, Govett and Govett in Canada have plotted the known reserves of world minerals as at 1970, and obtained the relationship:

$$R = A \cdot 10^{10} \text{ to } 10^{11}$$

which is similar to Sekine's relationship. The crucial thing in assessing likely world total reserves is to determine the reasonable upper limit to the factor. However, Govett and Govett suggest that in the foreseeable future of mining and extraction technology, the crustal abundance multiplier (10^{10} to 10^{11}) is unlikely to be increased by more than a factor of five. This, of course, is a very subjective opinion; they point out that such a small increase, however, would make little difference to depletion dates if exponential increase in consumption continues. But it would require a major increase in exploration success – for example, for every deposit of copper now known, five new deposits of the same size, and grade, would have to be found.

Crustal reserves can be plotted against world crustal abundance in weight per cent, for some well-known metals. The crustal abundance data is supplied by the geochemists and other Earth-science specialists. Thus a straight line can be drawn representing the mean of Govett's relationship given above ($0.5 \times A \times 10^{11}$), and an upper line representing the mean of the Govett relationship increased by a factor of five (i.e., $R = 2.5 \times A \times 10^{11}$). Normally the crustal reserves increase progressively with time, as exploration proceeds, although in certain cases there appears to have been a reduction in the amount of known reserves for certain metals, as with tin and manganese. A special comment must be made about aluminium; the reserves data with respect to this element are based on bauxite reserves. These are probably very limited in extent, whereas if aluminium were to be

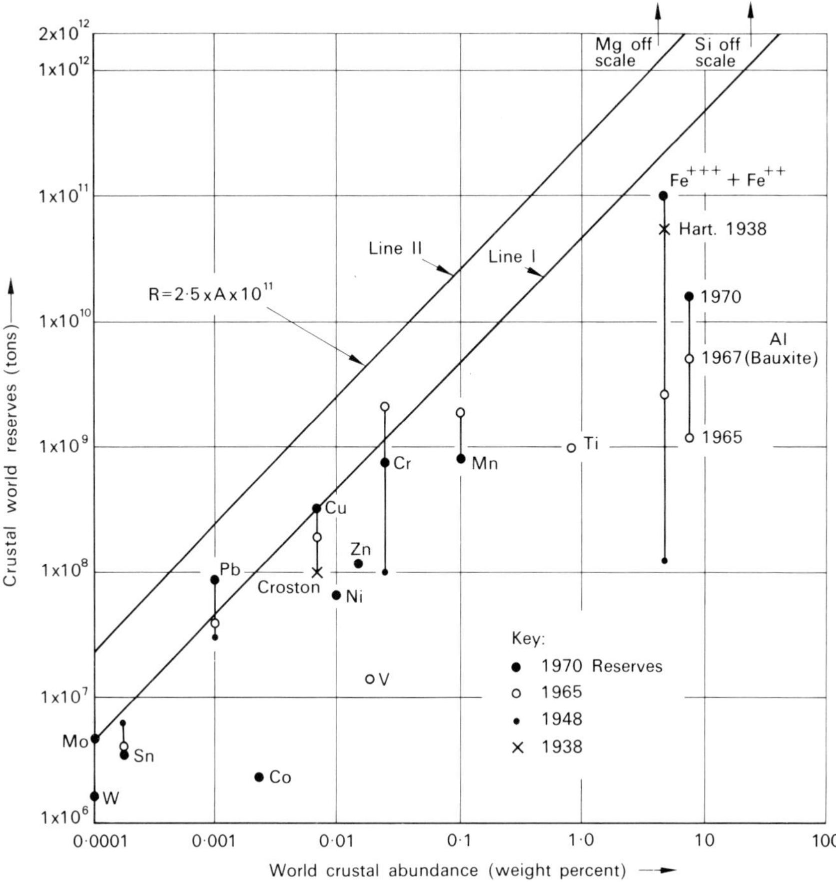

Fig. 5.4 Relationship between crustal reserves and crustal abundance.

extracted from clays, which are widespread in many types of sedimentary rock, then there are undoubtedly huge reserves, as for iron and magnesium.

It must be emphasized that the ultimate crustal reserves of mineral ores are unknown, and the Govetts' view that we may well reach our ultimate limits for many of the geochemically scarce metals at Line II on Fig. 5.4 is merely a suggestion. Some geologists and mining experts incline to the view that the ultimate ore reserves will turn out to be much greater. But at least there does seem support for the general thesis that the ultimate reserves will turn out to bear some relation to the crustal abundance data. In other words, the geochemically-scarce minerals listed in table 5.1 would seem to be the ones to concern ourselves with first in any investigation into long-term scarcity. Table 5.1 only gives the average percentage weight of the elements as metals; the main chemical constituents of the crust may be expressed as follows: 62% silica, 15.6% alumina, 6.5% iron oxides, 3% magnesium oxide, 0.8% titanium oxide and 0.1% manganese oxide, alkali and alkaline-earth metal oxides being neglected. The geochemically scarce metals in the table are in general only present as a few parts per million; they appear at the bottom of the Govett curves whereas the abundant metals of importance in the crust – silicon, aluminium, iron, magnesium, titanium and manganese – appear near the top.

It is a fairly well established fact that with several minerals the average grade of ore being mined has steadily fallen over the years. The classical case is perhaps that of copper, where 2.0% ores were the minimum grade economically exploitable several decades ago; now, 0.5% is generally agreed as being an exploitable concentration, provided the *total* reserves on the site are large. Some years ago, Lasky postulated that with certain minerals, reserves would increase *geometrically* as the average grade to be mined decreased *arithmetically*. It is unfortunate that this relationship has been taken to be more widely applicable than Lasky intended, and there are sound geological reasons why caution should be used in its application; it can lead to serious optimism about ultimate reserves, as Lovering (1969) has stressed. Many ore deposits in fact do not exhibit the Lasky Ratio, as it is sometimes called, and this includes those of mercury, zinc, lead, gold, silver, tungsten, antimony, tantalum and molybdenum. However, the porphyry copper deposits (*vide infra*) are a good example of the applicability of the rule. But from the environmental aspect, we have to bear in mind that when extracting a metal from an ore deposit where its concentration is 0.5%, we will have 199 tons of waste material to dispose of when winning one ton of metal.

c. Dynamic modelling of the life-cycle of a non-renewable resource. It seems obvious that in practice, exponential growth in consumption of a mineral will not go on right up to its ultimate depletion. The failure to go on identifying reserves will eventually have a feed-back to industry, and as the available reserve life begins to shrink, real scarcity will become apparent, prices will rise more steeply and substitutes – if there are any available – will rapidly penetrate the market (or society will have to do without). M. K. Hubbert in the USA developed a dynamic model, along this line of approach, to explain the life-cycle of natural gas and petroleum reserves for that country. Hubbert's treatment is worthy of examination; it is based on S-type growth curves for the cumulative annual production of the resource over its life-cycle. He pointed out that the discovery and production of crude oil in the continental United States (excluding Alaska) had grown exponentially from the beginning of the present century until about 1959, at which time the rate of increase in discovery, had begun to level out. He plotted out the *cumulative* discoveries Q_d and *cumulative* production Q_p up to 1970 and presented a model for the exploitation cycle of the resource. This assumes that the relation between resource and resource life can be described by a perfectly symmetrical S-shaped curve

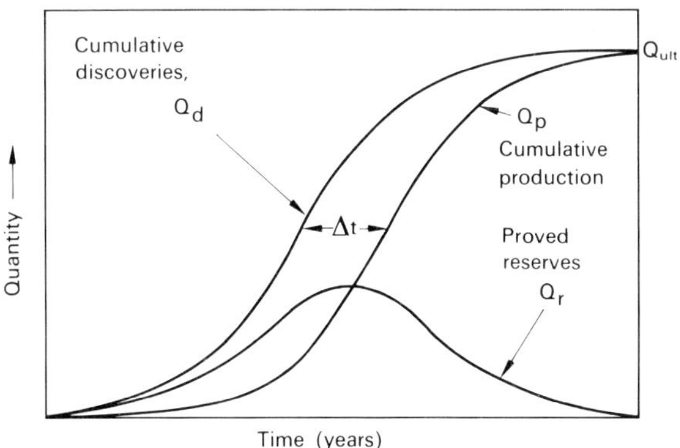

Fig. 5.5   Dynamic approach to resource life: integral curves.

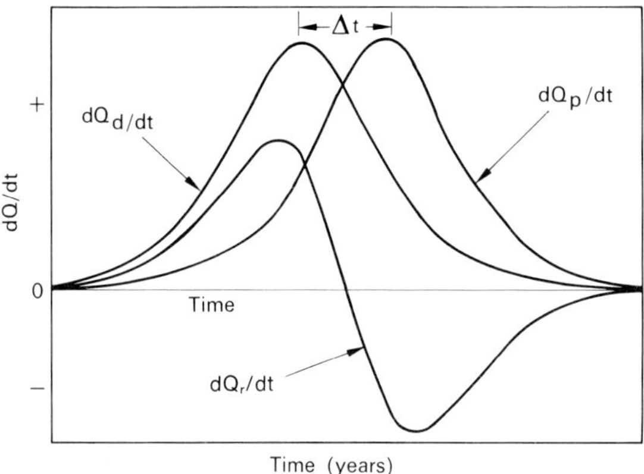

Fig. 5.6   Dynamic approach to resource life: differential curves.

approaching an ultimate total quantity of the resource, Q_{ult}. The vertical difference between the Q_d and Q_p curves at any time is the *proved* reserves Q_r at that time. By assuming that the Q_d curve passed through its point of inflexion in 1959 and the Q_p curve approximately 12 years later he estimated the total quantity of crude oil available and the approximate depletion dates. If the three parameters, Q_d, Q_p, and Q_r are differentiated with respect to time, and the respective dQ/dt data is plotted against time, then a graphic picture of the life-cycle is obtained.

The general conclusions to be drawn from this analysis are that firstly, if normal market forces are relied upon to control the pattern of use of a non-renewable mineral, such as oil, society will actually continue to use the material at an increasing rate when the rate of discovery of new reserves is actually declining. Thus, a crisis situation will arise if conservation measures are not adopted by society at an early enough date in the life-cycle of the mineral. Secondly, although this may not be fully apparent, in fact the slowing of demand growth in the cycle as early as possible has a greater effect on the *total* lifetime than merely increasing the value of Q_{ult}, i.e. the ultimate reserves figure, within limits.

An application of the Hubbert-type modelling to the life-cycle of certain non-ferrous metallic minerals by the present author seems to lead to the same general conclusions.

5.4.3 Regional and political factors affect material resource availability

It is the considered view of the Forecasting Team of the Science Policy Research Unit, directed by Professor Freeman at Sussex University, that economic, social and political considerations, rather than physical, may well turn out to be the critical factors limiting our non-renewable resources. Most of the geochemically 'scarce' metallic minerals referred to in table 5.1, as well as petroleum, are very unevenly distributed in geographic terms. It is well known how the OPEC countries have been able to force up crude oil prices in the world market, thus exploiting the fact that the majority of the known world oil reserves occur in the Middle East and across North Africa, i.e. in developing countries. Speaking very generally, in the case of the non-ferrous metallic minerals, although traditionally the USA has so far had home supplies of these, as has the USSR, Western Europe has had to depend on imports to a great extent to supplement indigenous supplies. A good deal of nickel, lead, zinc and silver has come to Britain from

Commonwealth countries. But economic growth has meant that the developed world has had to keep increasing its imports of raw materials, including especially minerals, from the developing countries. Much copper now comes from Zambia, Chile, New Guinea, tin ores from south-east Asia and Latin America, and a huge proportion of the world's forward reserves of tungsten are in China, to take just a few examples. It would be very surprising if the developing countries did not in the future begin to exploit growing opportunities for exerting monopolistic control over world prices of these minerals. Ayres and Kneese have pointed out that such a trend could result in an accelerated redistribution of wealth between the developed and underdeveloped worlds. Thus, it will become increasingly important for those concerned with mineral resources availability, to consider regional distribution factors as well as the total world supply available.

5.5 A BRIEF REVIEW OF SOME INDIVIDUAL METALS

In section 5.4 we were concerned with developing general principles regarding the continued availability of mineral resources for the future. Individual metals were only mentioned by way of illustrating a point. It will now be appropriate to go on to consider individually some potentially scarce materials from the standpoint of resources, trends in consumption, environmental problems etc. Such a review must necessarily be brief and incomplete in each case. The objective is merely to give the reader a 'feel' of the materials conservation problem; to indicate present trends, and perhaps help him in making decisions regarding the use of certain resources which are liable to become of increasing scarcity and hence increasing price. Consideration is given first to the four base metals, copper, lead, zinc and tin, in that order. Silver and mercury are briefly discussed together. Then some ferro-alloy minerals likely to become scarce are discussed, and finally a few comments made regarding some of the newer metals of industry such as cadmium, bismuth, selenium and indium.

5.5.1 Copper

World consumption of copper during the period from 1900 to 1970 increased exponentially. From 1951 to 1970, world mine production increased by 130%. The US Bureau of Mines has forecast a world demand (primary metal plus recycled scrap) of somewhere between 2.1 and 4.6×10^7 tons for the year 2000, with a most probable figure of 3.2×10^7 tons, nearly five times the present demand. The Bureau has forecast that this metal will be nearly double its price in real terms by the year 2000, mainly due to the need for working ever lower grade ores, increasing exploration, and rising costs of capital investment. Present reserves of copper total about 3.4×10^8 tons and discoveries have been expanding quickly. However, one of the most significant factors affecting the exploitation of this mineral has been the steady decline in grade of ore worked, and new discoveries are essentially of lower grade minerals. The minimum economic level is now down to about 0.5%. From the environmental standpoint, this figure of 0.5% means that for every ton of copper extracted, 199 tons of waste must be handled, and disposed of, in a non-polluting manner; clearly, by the time we reach the year 2000, if the US Bureau of Mines prediction were to be correct, this would mean about 4×10^9 tons a year of waste accumulating from copper extraction operations. Land-use conflicts are increasingly envisaged because an increasing amount of copper will be extracted by surface or opencast mining, which involves removal of overburden, thus adding to the tonnage of rock being disturbed. Such operations are only profitable if the reserves are large. Economies of scale are of course obtained by working such projects, helping to keep down costs.

Copper has been used since prehistoric times, although its average crustal abundance is less than 60 parts per million. Originally, deposits in the form of veins or contact metamorphic deposits of a million tons or less were widespread and commonly worked. But such quantities are small for purposes of modern mining, even though the ore grade of these deposits often exceeded 10% of copper. During this century, production from vein deposits has continuously declined in importance as the discovery and exploitation of the large, easily mined deposits such as the porphyry copper has proceeded. These are the deposits which permit the workable economic grade at the present time to be as low as 0.5%, as previously stated, by using surface mining techniques. The copper mineral here is usually chalcopyrite, $CuFeS_2$, and the associated non-copper containing constituent is generally silicic in character. The main geographical areas for this type of copper deposit are to be found principally around the Pacific, especially the Americas, but have also been discovered in the USSR, in Yugoslavia and in the Bougainville area of New Guinea. About 50% of the world's copper now comes from porphyry-type deposits. A second type of large local copper deposit is commonly called stratiform, and these reserves probably yield about 25% of the world's production of copper. Each deposit has its own characteristics. The longest-worked deposits, best known geologically, are those in the Kupferschiefer shales of Central Europe, especially those in East Germany and Poland, and again the copper is present as a sulphide. But the bulk of the world's stratiform copper deposits currently worked are those of the Zambian–Katangan belt in Africa. Because of the modern importance of these porphyry and stratiform deposits, seven countries now account for fully 83% of the world's production. Chile and Zambia each produce about 750000 tons a year, although the USA and USSR produce rather more, at nearly a million tons each. Canada produces somewhat less than Chile or Zambia, the remaining two out of the seven principal producer countries being the Congo (300000 tons) and Peru (250000 tons). As regards the location of world reserves of copper, about half of these lie in Chile, the USSR and the USA.

If one calculates the Static Index for the life of the known and inferred reserves in various major producer countries, as described above, Congo, Chile and Peru come out highest at about 60 years, then comes the USSR with 40 years, then USA and Zambia with 33 years and lowest, Canada at 14 years. The average of these is just under 40. However, if we calculate the Exponential Index, this is seen to be somewhat less than 25 years, even in the case of Congo, Chile and Peru. As discussed above, in practice the *actual* life of the reserves of copper will be determined by complex factors and is very difficult to predict with any accuracy, as indeed is the case for other minerals. The purpose of quoting SI and EI values here is merely to obtain a rough comparison of reserves and their future as between different producer countries, by using simple indices.

It will be obvious that the smelting of all these deposits of essentially copper sulphide minerals must necessarily produce sulphur either as a pollutant from stacks, or in the form of aqueous or solid effluents, or even as a marketable commodity. As regards the latter, observers have predicted, however, that an unsaleable glut of sulphur, or its compounds, could result in the future, owing to the imbalance in demand for copper and sulphur when compared with their stoichiometry in the original mineral. The form in which future anti-pollution legislation is framed could have a considerable effect on the choice and development of smelting processes, e.g. pyrometallurgy versus hydrometallurgy; depending on whether air pollution or ground pollution is deemed to be the most serious threat.

Turning to the user side, copper consumption by industry varies somewhat from one developed country to another. Japan seems to use about two-thirds of its copper in electrical applications, West Germany and Italy about half, and in the United Kingdom, a little less than half goes into electrical industries (this includes all wire and cable applications). The next major use of copper is in 'general engineering', much of this in the form of alloys with other base metals and aluminium. Uses in construction and transport, although together small in the case of Japan, account for between 20 and 30% of the copper consumed in the USA and Western European countries. Rising prices of copper during the last decade have led to a more economical use of the material in the form of pipe inasmuch as thinner wall tubing has been introduced in building; moreover, a good deal of substitution by plastic tubing has also been taking place.

The copper used in industry is made up of *primary* material, coming from refiners and smelters which process ore from the mine, together with *secondary* material, based on

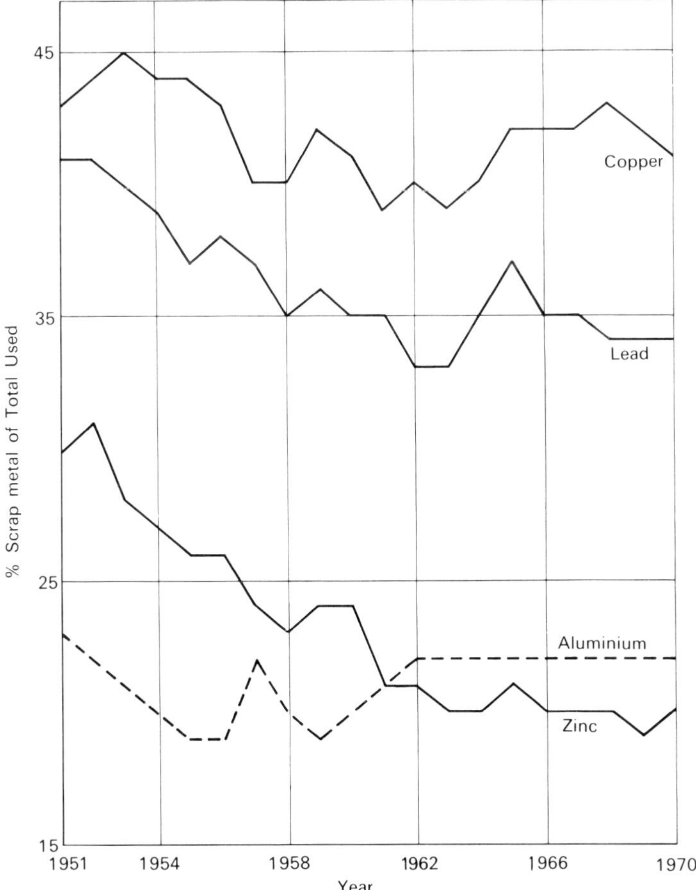

Fig. 5.7 Western World scrap/consumption ratio, 1951 to 1970.

the smelting of recycled scrap. In 1970, scrap formed about 40% of the *total* copper consumption in the Western World. However, there were country-wide variations. For example, the figure for the USA was about 50%, comparing with about 34% for Japan, and 40% for the UK. A most important point to make here is that although there seems to be a high proportion of recycled copper scrap circulating in the world, as a percentage of the annual copper consumption it has been virtually constant at around 40% in the Western World over the last 15 years. Over this same period of time, *total* annual consumption of copper has gone up by 75%. Thus, subtracting a *constant* fraction recycled leaves the *net* annual percentage growth rate in consumption of primary mineral copper virtually the same as for total copper. So the exponential growth phenomenon persists. To significantly help the long-term situation with regard to the conservation of copper resources, a much higher recycle fraction would seem to be needed, or at least a steadily increasing fraction.

Finally, it is worth recording that the world's copper-in-use resources have been estimated to be 9×10^7 tons at the present time, this being made up of 4×10^7 in the USA, 4×10^7 in Europe and 1×10^7 tons for all other countries combined. Compared to the developed world, the developing countries are thus at a serious disadvantage in this respect, although they do possess most of the unworked reserves of copper, as mentioned above.

5.5.2 Lead and zinc

It is useful to consider these two base metals together since their ore minerals often, but not always, tend to occur together. In each case, a single mineral species, galena (PbS) and sphalerite (ZnS) respectively, accounts for much of the world's production of these metals. They are potentially scarce metals in the sense that zinc has a crustal abundance of 80 parts per million, and lead only 10. Like copper ores, the main reserves occur either as contact metamorphic deposits or as sedimentary deposits, but the main difference from copper is that the ores tend to be richer, though the workable deposits are confined in size, and there are more underground mining sites. So far, low-grade deposits of lead and zinc analogous to the porphyry copper have not been discovered, and the Lasky Ratio referred to in section 5.4.2b is hardly of help in resource prediction studies.

It is noteworthy that zinc concentrates are currently our main source of some important new minerals of advanced technologies, e.g. cadmium, thallium, indium and germanium; these materials should therefore be considered as *potentially* scarce, or liable to become very costly, if lead and zinc ever ceased to be demanded by industry and their ores were only to be worked for cadmium etc.

The world's annual production of both lead and zinc is dominated by four countries, the USSR, Australia, Canada and the USA, in that order. World production of lead is around 3 million tons, and in the early 1960s total reserves were indicated to be about 50 million tons. Now since 1967, there have been extensive new discoveries of the mineral, giving increases in reserves, notably in Missouri, USA, Canada and Queensland, Australia. Hence in 1971, the world's lead reserves were stated to be approaching 100 million tons. The Static Index thus increased from about 17 years in 1967, to 30 years in 1971, for the Western World; the corresponding Static Index figures for the Eastern Block were only 9 years in 1967, but 16 years in 1971. However, turning to the more pessimistic index, the Exponential Index, for the Western and Eastern Blocks the figures are now estimated at about 23 years and 12 years respectively, the world average being 21 years. Of course, more reserves of lead and zinc ores will undoubtedly be discovered in the future but it is thought that some of these may well have a higher zinc-to-lead ratio than applies with many currently worked ore fields. This may not seem surprising when it is noted from table 5.1 that the crustal abundance ratio for zinc/lead is about 8:1. However, the present-known world reserves ratio for zinc/lead is *much* lower, roughly 1:1, perhaps reflecting the annual consumption ratio of these metals, which is about 1.5:1. But it is important to realize that world consumption of zinc is rising at a faster rate than lead; the world consumption of slab zinc overtook that of refined lead for the first time in 1940, and since 1945 has been at a consistently higher level. Since 1960, world average annual rates of growth in consumption for these metals have been about 4% for lead and 6% for zinc.

Why are these two metals so important to us? Will they continue to be in the future, and for how long? In the case of lead, 53% was destined for the transport industries by the end of the 1960s; 35% of all lead used currently goes into batteries, mainly for automotive purposes, and between 10 and 20% is made into anti-knock compounds for use in motor spirit. These are average values, the actual figure depending upon the country concerned and its car population. But lead is also used in general building and construction in the form of sheet and piping, for printing, for cable covering, in various solders and alloys, and in small arms ammunition. Zinc goes extensively into galvanizing–that is, for the protection of iron and steel from corrosion. On average, this use accounts for about 30–34%. About 20–25% goes into the manufacture of brasses and alloys, and about the same fraction into zinc die-casting; the rest is mainly used for making paints and chemicals. As with lead, there are individual country variations from these quoted averages. Also, patterns of usage are changing with the passage of time. Previously, the major outlets were for brass, for sheet zinc in building and for galvanizing, but by the early 1950s galvanizing and other protective zinc coatings had replaced brass as the most important end-use of this metal. Furthermore, developments in the USA about the same time led to pressure die-casting becoming more significant than sheet zinc, although it is still a traditional building material in many parts of the world. Thus, the degree and type of industrial development in any particular country may well be reflected in different balances between the various consuming sectors for zinc, and this generally applies to lead. So it is fair to assume that these are key metals and although use patterns change, their overall demand will continue to increase throughout the world for the foreseeable future.

The fraction of lead recycled after use in the UK showed a dramatic rise between 1960 and 1970, from 49% to 66%. This is in contrast to the USA, where the figure has stayed more or less static at about 42% scrap recycle over this period of time. France and Germany also maintained a roughly constant recycle. Japan fell from 38% lead recycle to 25%, and figures for Italy were similar. The average lead scrap recovery for the Western World as a whole has not changed very much between 1960 and 1970, falling from 35% to 34%. It is less than applies for copper, at 41%. This may reflect the greater degree of difficulty in recovering lead from several of its uses, if one excludes battery plates. Also, the world price of lead metal tends to be about a quarter of that for copper and since the cost of scrap recovery is about the same, there is perhaps less incentive to collect scrap for re-use.

The picture for world zinc recycle is worse on average, having been only 24% in 1960 and this had fallen to 19% by 1970. Corresponding figures for Britain were 24.7 and 25.0 over this period of time, showing the situation to be fairly static and somewhat better than average. This lower percentage recovery for zinc compared with lead and copper may well reflect the nature of its main end-uses; namely, as a protective layer on steel, as small components in motor cars, and in alloys. The low scrap recovery for zinc (which does not appear to be increasing, but rather the reverse) gives cause for concern from the conservation viewpoint, particularly in view of the 6% continued exponential growth in demand for this metal.

5.5.3 Tin

The crustal abundance of tin is now believed to be not greater than one and a half parts per million.

Tin differs from copper, zinc and lead in that it mainly, though not entirely, occurs in nature as the oxide rather than the sulphide. It has tended to form in the same geological environment as tungsten, but its most common mineral, cassiterite (SnO_2), has tended to migrate from its source of origin to become concentrated in what are called 'placer' deposits. These secondary deposits have been the source of much, though by no means all, of the world's supplies of tin to date. Much of the world's tin reserves are concentrated in two narrow belts, one running along the Malaysian Peninsula south-east to Java and Indonesia and another running along the eastern side of the Andes in Bolivia and Peru. Now whereas lead and zinc cost much less than copper, tin is approximately four times as costly as the latter metal. And as regards the future, it is not difficult to envisage the price of tin rising still further in view of the fact that the major reserves are concentrated in a few of the developing countries and these countries may begin to realize they have a strong bargaining position. Thus, political and economic pressures may well cause scarcity problems in advance of actual depletion of total world reserves of economically-minable ore deposits.

The US Bureau of Mines have estimated world demand for tin by the year 2000 to be somewhere between 321 000 and 530 000 tons per annum, the most probable figure being 440 000 tons. Of the four base metals considered, tin is predicted to be going to have the lowest annual growth rate in future, the US Bureau of Mines suggesting this might average only 1%.

The USA imports 65% of its primary tin, being very short of indigenous resources, and strenuous efforts have already been made in recycle technology in that country. In fact, the US leads the world in tin recycling, secondary production of tin in that country being about a quarter of total tin production, compared with a world average of only about 5%. This is a harder metal to recycle economically than copper, lead or zinc because of the make-up of its various applications. For example, just under half of the market for tin is in tin plating. The rest goes into various alloys, notably 10% into bronzes, 10 to 30% into solders and 10% into bearing metals. About 5% goes into tinning. However, as with the other metals, the use pattern of tin varies between countries. An important point about tin is that small, but increasing quantities nowadays go into a variety of applications related to advanced technology. For example, molten tin is used for high-strength titanium alloys, and also to produce plate glass; dioctyl tin is used to stabilize p.v.c. and impart crystal clarity, shatter resistance and low oxygen permeability. Organo-tin compounds are used as fungicides, insecticides and anti-slime agents with the great advantage from the environmental viewpoint that these agents degrade to non-toxic stannic oxide. Tin is also used in powder metallurgy to improve the sinterability of powder compacts. Thus, it will be very difficult in the future to increase the fraction of the metal recycled. Although the view has been expressed that the exponential increase in demand for tin may only be around 1%, and this might seem advantageously low from a conservation standpoint, present world reserves are low at just over 4 million tons, giving a static index of about 17 years.

In conclusion, the four base metals copper, lead, zinc and tin have been used by man since prehistoric times, have well-established use patterns throughout the world, and in all cases we could be heading for serious resource problems within several decades.

5.5.4 Silver and mercury

Concern about future resources for these two metals has been expressed in many quarters. As with the base metals, they usually occur as sulphides in the Earth's crust, and their abundance is very low indeed, being of the order of 0.1 to 0.2 parts per million. Both minerals occur in hydrothermal vein deposits and in the case of silver, the mineral is usually associated with lead, zinc or copper ore deposits. Only a very small fraction of silver-producing deposits are rich enough to be worked for that metal alone; thus, although it is a valuable by-product from base metal extraction operations, its production rate is controlled strictly by the production rates of these metals. Few new silver deposits have been found in recent years and silver production will hardly meet the needs of the future. In 1965, the US Government withdrew silver from currency and the price has been rising in real terms. Some old mines might thus become economic to work again in the future, and there will gradually be a greater incentive to exploit marginal sources of silver; but in general, future uses may well have to be curtailed to meet a restricted supply. Suppose the known reserves were to be increased by a factor of five, then the Exponential Index for reserve life would still only be 42 years. As regards geographical distribution, over half the world's supply of silver comes from the Americas, the rest coming mainly from the USSR and Australia.

In the case of mercury, the future resource position may well give cause for concern. Most of our mercury comes from cinnabar (HgS), which has tended to occur only in small quantities at any given point in the crust and these sites can get worked out fairly quickly. The largest supply of mercury, about 32% of world consumption, comes from Europe (Spain and Italy), the original deposits in the USA being nearly exhausted. Known reserves are small, despite the price increases which have taken place over the last 20 years; the price of mercury is said to have risen by over 500% in real terms during that period. The Static Index reserve life currently lies somewhere between 15 and 45 years and the Exponential Index is less than 25 years. There seems little point in speculating on what the life might be if the reserves were somehow to be increased by a factor of five, since experts believe there is very little hope of this happening.

The greatest demand, and probably the most important use of silver at the present stage of technology, comes in photography (about 25% of total demand for this metal in 1968). A similar amount goes into silverware and plated ware, somewhat less into coinage, with electrical and electronic components and accessories coming into fourth place with 12% of the demand. Overall, 20–25% of silver is recycled after use at the present time, although one might hope for an improvement on this figure in the future. Turning to the question of substitutes, although these are possible in some of the electronic applications, there is as yet no satisfactory substitute for silver in most of its key photographic uses, which are in black and white photography and X-rays. There *are* some known substitutes which can reproduce a photographic image but they all operate too slowly. Companies like Messrs. Kodak spend a good deal of effort in finding a sound substitute for silver although to date they have not been successful. At some point in the not-too-distant future it may be necessary to contemplate action by governments to ban the use of silver for ornaments and similar less-essential applications in order to conserve the metal for the all-important uses in photography and other technical fields.

Turning to mercury, although we could find substitutes for some uses in face of growing scarcity and cost, certain applications depend upon mercury's unusual combination of physical and chemical properties, which include very high specific gravity, fluidity at normal temperatures and good electrical conductivity. The main uses of mercury are in the cells for manufacturing alkalies and chlorine, in electrical applications, and in instruments. Diaphragm cells, which do not use mercury, can be operated successfully commercially, and indeed may gain popularity in the heavy chemical industry as mercury costs rise and concern grows over the dissipation of mercury due to losses from plant. The battery industry alone accounts for more than half of the total mercury consumed in electrical applications, especially for military use, and substitution is possible here. Mercury is also used in chemical compounds, both organo-metallic and inorganic, as well as in certain catalysts, typically in the acetylene-consuming plants for the production of vinyl chloride. But ethylene can be used in its place, indeed it is probably cheaper than acetylene. Fungicides based on mercury may become less popular than those based on organotin compounds, for environmental reasons; the toxic quality of mercury and its compounds is responsible for some of mercury's uses, but also presents us with health problems which are of growing public concern. Also, because of considerable fluctuations which occur in the price of mercury, this is an added incentive to consumers to find substitutes. Recycling is costly with many mercury uses owing to the large ratio of extraneous materials to the small quantity of mercury.

5.5.5 The iron and steel industry

a. Introduction. Iron and the several mineral elements which are alloyed with iron to produce varieties of iron products and steel belong to what one might call the family of 'ferrous minerals'. Iron is itself indisputably an abundant element in the earth's crust being present to an extent exceeding 5%. It accounts for over 95% of all metals consumed at the present time. The US Bureau of Mines in 1970 quoted a Static Index of 240 and an Exponential Index of 93 years for the life of known iron reserves, and a factor five increase in discoveries would raise even the Exponential Index to 173 years. It could, of course, be argued that we do not have infinite reserves of this element in the planet, but iron deposits are so plentiful that the world is going to have iron and steel industries, and all that that implies, for very many years to come, provided that

(i) the non-ferrous alloying elements remain available, and

(ii) the necessary raw materials for processing, expecially the reducing agents, continue to be in good supply.

In the latter case, there is unease about future supplies of coking coals for producing coke for the blast furnace, which is used to smelt the ore, and the supply situation could be critical in several developed countries, especially the UK, by the turn of the century. The iron industry is faced with supporting a probable annual growth rate of iron and steel production of around 2% for some time to come. Although somewhere between 45 and 50% of iron and steel products are eventually recycled, the reprocessing necessarily consumes energy and materials, and the actual fraction of many of the *alloying* constituents recovered and recycled is well under this figure owing to losses because of dilution.

When the blast furnace process for making iron begins its decline as a result of growing coking coal and oil shortages – oil is used in the blast – new types of plant will be developed in which the reductant is quite likely to be a gas which could be generated from coal or other sources. The new direct-reduction processes are then likely to gain ground rapidly.

Another processing raw material problem in iron and steel making could conceivably arise from *flux* shortages. Perhaps the most important flux used is fluorspar (CaF_2); it is needed not only in the open-hearth process, which is of course gradually becoming obsolete, but also in the newer electric arc and basic oxygen (BOF) processes, as well as in electro-slag refining for making high-grade steel and certain alloys. The mineral gives fluidity to the slag, giving a better steel recovery and lower fuel cost; it also enables the maximum amount of lime to be made available for effective removal of sulphur and phosphorus from the steel. However, fluorspar has many other uses; it is the only major source of fluorine and is of course used in making hydrofluoric acid and other fluorine chemicals, but it is also important as an opacifier and flux in the ceramics industries. It is even used in brick making to prevent staining and as a bonding material in abrasive wheels. Although the UK is fortunate at the present time in being able to mine indigenous supplies, notably in Derbyshire, the US Bureau of Mines predicts that world growth in fluorspar demand could be over 5% per annum between now and the end of the century, and known world reserves as a whole could be used up within 18 to 19 years. Consumption of fluorine in aluminium metallurgy, which only accounted for 18% of the total demand in 1968, is generally predicted to become the major end use by the year 2000, owing to the high growth in demand for aluminium metal. However, improvements in the technology of the supply of fluorspar are expected, which will increase reserves, but no doubt the cost will rise. Of course, technological advances may well come about which result in the elimination of fluorspar for fluxing purposes in the iron and aluminium industries.

b. The 'vitamin' elements of metallurgy. A whole range of non-ferrous metals are used in the steel industry, in conjunction with iron, to produce various kinds of steels. Alloying constituents are of course also added to other important metals such as aluminium. These added metals which produce specific properties, are sometimes called the 'vitamin' metals, and concern has been growing in the last few years that shortages of some of them could interfere with the rapid growth of metal industries based on plentiful resources in the crust, i.e. materials such as aluminium, iron, magnesium and titanium. It is to be hoped that extensive research will be devoted to find substitutes for the 'vitamins' based on the use of the more plentiful minerals as additives, to replace the scarcer ones. Perhaps the most important 'vitamins' which come into the category of being geochemically scarce (table 5.1) and hence potentially scarce commercially, are tungsten, niobium, molybdenum and nickel. The crustal abundance figures for these four metals are nickel, 0.007% (similar order to copper); niobium, 0.002%; and molybdenum and tungsten around 0.0001%. We will briefly review the resource aspect of each of these four elements, in decreasing order of crustal abundance.

Nickel. Roughly half the world's supply of nickel currently comes from Canada and about a fifth from the USSR. The only other major producing country is New Caledonia. Identified reserves of nickel ores are more plentiful in proportion to consumption than is the case with the ores of copper, lead, zinc and tin, but nevertheless there is no room for complacency since with present growth rates for nickel consumption, we have an

Exponential Index for reserve life of barely 50 years. However, interest is developing in a new type of nickel deposit, low in concentration, but extensive in distribution, which is likely to become of increasing importance to commerce in the future. This is found in the weathered zone formed over many mafic rocks in tropical regions. Here, nature has carried out a sort of beneficiation process lasting very many years, yielding an iron-rich laterite in which the concentration of nickel reaches 1% in places, but may be only 0.6%. These reserves of 'nickeliferous laterites', measured in hundreds of millions of tons, have been located in Cuba, the Philippines, Greece, Borneo and many other parts of the world. They are already being worked in Cuba, and constitute the world's greatest forward reserve of nickel. But is has to be born in mind that increases in production costs can be expected as the world begins extensively using these low-concentration deposits of nickel; a price rise of 10% per year has been predicted, and also their regional location in developing countries may possibly lead to supply difficulties for the developed countries. However, a new source of nickel has been identified; this is in the nodular deposits found on the bed of the ocean (see section 5.8.2 below).

Niobium. This element is used almost exclusively in high-strength low-alloy steels, stainless steels, and in superalloys based on nickel and cobalt. Overall growth in demand is now predicted to be $4\frac{1}{2}\%$ per year, but it is possible that future technological development could favour an even higher rate. To some extent, niobium is interchangeable with vanadium in high-strength low-alloy steels and with titanium in stainless steels, but the question of a substitute in superalloys is not so straightforward. It has been authoritatively stated recently that the known world reserves of niobium are 'nearly 13 billion pounds, at present prices, and more than sufficient to meet projected world demand to the year 2000'. These reserves are located mainly in Brazil, Canada and Russia–although only three countries, they are fairly well spread across the geographical and political spectrum. Clearly, we seem to have an assured supply for the iron and steel industry for some decades, but a careful watch will have to be kept on this element because of its expected high exponential growth in demand.

Molybdenum. Although this element was used first only as an alloying constituent in steels, it has been applied in nuclear and space applications and in the production of electrical and electronic components in recent years. About one fourth of the supply is now used in producing the materials for manufacturing industrial machines and machine tools. Growth rate throughout the world is likely to be between 4 and 5% per year for the rest of this century, possibly higher, depending upon the rate at which molybdenum displaces tungsten from alloy production. World production is currently around 150 million pounds per annum, two-thirds of this coming from the USA and most of the remainder from Canada, Chile and the USSR. The main ore worked is molybdenite (MoS_2) and because it has a highly erratic distribution, it is rather difficult to estimate the world's reserves. A significant fraction, about 25%, is derived as a by-product from porphyry copper mining; but the molybdenum content of porphyry copper ores is low–in the range 0.01 to 0.04%. However, the tonnages of copper ore processed are very large and the costs of concentrating molybdenite by flotation are apparently low enough to make a profitable recovery of this element possible at present. But the availability of this source of molybdenum is clearly going to be unreliable if copper becomes restricted. The same argument applies to molybdenum which is obtained from tungsten and uranium workings. However, there are extensive deposits of what are called the 'disseminated molybdenums' (so-called because the ore mineral is finely disseminated through a huge volume of rock) and these are found in places up the whole of the western USA and

Canada, for example. It seems that the world's main reserves of molybdenum lie within the western hemisphere, and much of it in the USA itself, but the US Bureau of Mines states that molybdenum deposits being mined by the year 2000 will be much deeper and will require more beneficiation, an indication of steadily growing scarcity. However, the increased extractive difficulties could well be offset by greater efficiencies resulting from the application of new mining and concentration technology and the price might therefore not rise considerably over this period of time. Present world reserves are estimated to be over 11×10^9 lb (5×10^6 tonnes), the Static Index for reserve life is roughly 70 years and the Exponential Index about 25 years. However, molybdenum is still a relatively new material, as is niobium, and we probably still know little about the distribution and ultimate workable resources of these two elements in the crust. We do, nevertheless, have the geochemist's figure of crustal abundance, which is only about one part per million. We are faced with an extremely high exponential growth rate in world demand, and obviously the resource supply situation must be further studied and kept under close scrutiny. The US Bureau of Mines point out that the supply would be greatly enhanced by the development of manufacturing processes that reduce scrap losses. Much of the scrap is lost because it is unsuitable for remelting into high-quality material and large losses are incurred during remelting.

Tungsten. Although tungsten is well known as a ferroalloy, or 'vitamin' material, nevertheless 33% of the metal is now actually used in the production of tungsten carbide (WC) which is itself used extensively in machine tools, rock drills and armour-piercing shells. It also has growing applications in electronics and shows promise for application in computers.

Tungsten deposits are very regional, with half the world's production coming from the east-Asian belt starting in the north from Korea, through China, to Malaya. However, there are still substantial reserves in Spain, the Congo and Australia. The main tungsten ore of commerce is wolframite ($FeWO_4$), and supplies of this mineral are said to cover the next 15–20 years, but it is difficult to obtain data and opinions of reserves of wolframite beyond that time scale. However, scheelite ($CaWO_4$) is a widespread ore of tungsten, although it is much more variable in quality and tungsten content than wolframite. In its poorer grades, it is cheaper than wolframite, though of lower tungsten content. The difficulty lies in the fact that scheelite is not used for making the best quality tungsten products, e.g. metal, and high-grade tungsten carbide. Work has been carried out in recent years aimed at using the cheaper grades of scheelite and this has met with partial success. Unfortunately, the present price of wolframite does not offer much incentive for effort in that direction. It can be expected that tungsten will be come increasingly scarce, the rate at which the price rises depending upon whether advances in mining and extraction technology can offset the loss of the richer deposits.

Recycling of tungsten metallic scrap has been receiving increasing attention and certain firms are now becoming concerned with the recovery of this metal from the chemical industry, which uses a good deal of tungsten in the form of catalysts. Indisputably, there is a need to find and develop substitutes for tungsten, because in the long term this is going to be the best solution to eventual scarcity. But there are many problems to be solved, and it is no use just turning to its associated minerals as substitutes if these are also going to become scarce around the same time. Conceivably, aluminium, silicon, and nitrogen, which are abundant elements, may be used to prepare substitute materials for the traditional roles of tungsten, such as in alloy steels, cutting materials, wear-resistant parts and high-temperature components. Ceramics will no doubt play a role here. But in the chemicals field, especially where the metal is used as a catalyst, it is more difficult to envisage promising solutions.

5.5.6 Cadmium, bismuth, selenium and indium

These metals of importance to advanced technology, together with other minor metals produced as by-products of base metal production, are likely to have scarcity problems in the future. As has been pointed out already, by-products from major base metal production operations are highly susceptible to becoming commercially scarce if the base metal reserves become seriously depleted and no individual, economic, resources of the by-product mineral can be found. The problem is a serious one because materials of this kind are used for highly sophisticated technological purposes, i.e. in electronics, ceramics, organic chemicals etc., which are increasing rapidly in demand. However, the price of the metal is usually a small fraction of the price of the final article and consumers will probably be prepared to pay highly increased prices in the future. These metals will no doubt lend themselves to both producer and merchant manipulation. Bismuth may be a special case as a very large deposit has recently been developed in Australia and this should slow down future price rises for this material in the medium term. Cadmium is important for batteries, and in the future, scarcity problems could restrict the growth and usage patterns of batteries based on this element.

5.6 RESOURCES FOR SYNTHETIC MATERIALS

5.6.1 Future demand for synthetics

Synthetic materials are here taken to mean plastics, man-made fibres and synthetic rubbers, with plastics being by far the major material in this group. The rapidly growing importance of synthetics merits the devotion of an entire section to this group, because it is currently based on organic chemicals, and the essential element of organic chemicals is *carbon*. Our main sources of carbon at the present stage of technology are essentially the fossil fuels, natural gas, crude oil and coal. But crude oil is also the main resource-base for organic chemicals; thus we see that energy resource and material resource availability problems interact when we turn to synthetics.

Just how important have organic-based materials now become? About 45% of manufactured materials on a *volume* basis, are currently organic-based. But even more important, owing to the relatively high exponential growth in demand for synthetics compared with, say, the older materials based on metallic minerals, we might expect synthetics to become the major group of materials in a decade or two. Some experts predict that the total *volume* of plastics which will be used in the 'Orwellian year' of 1984 will exceed that of steel products. An authority at the Rubber and Plastics Research Association (RAPRA) has stated that although the ratio of synthetic materials to metals in volume terms is at present about 1:2, sometime in the last 15 years of this century we might well find this ratio is reversed. Trend curves based on future demand forecasts have been prepared by the Economic-Technical Section of the Netherlands Organisation for Applied Scientific Research. The predictions for iron and steel products, for the two non-ferrous metals aluminium and copper, and for synthetics (which includes all plastics, man-made fibres and synthetic rubbers) show that synthetics are likely to have the highest growth rate. It is important to emphasise that this trend does not mean a decline in demand for metals and alloys; it is simply that the growth in synthetics is expected to take

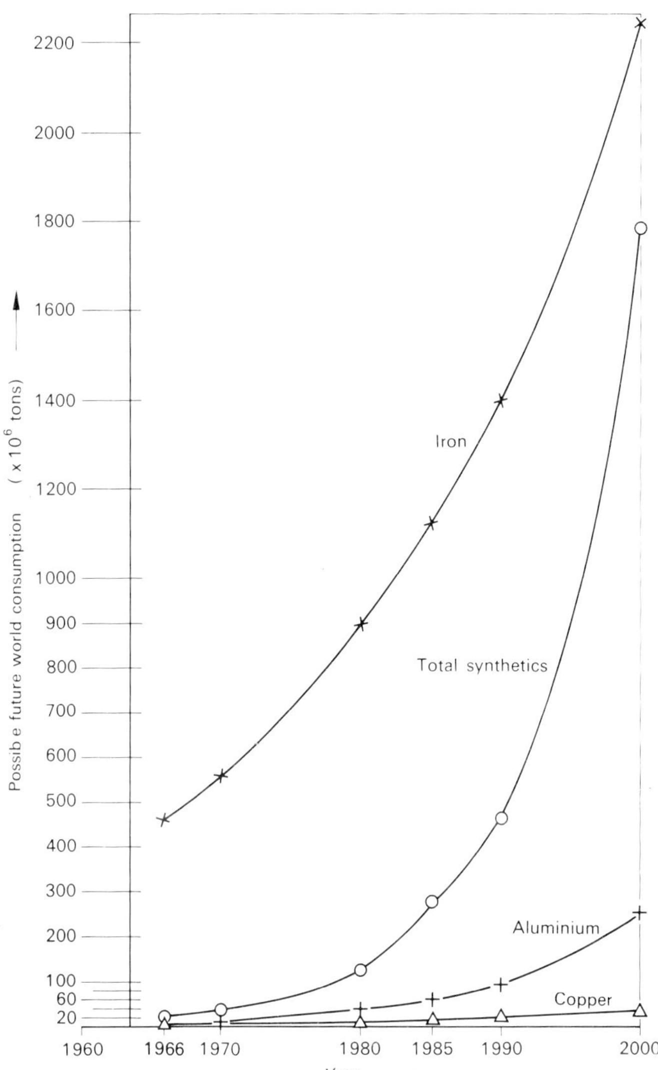

Fig. 5.8   Predicted world demand for plastics compared with key metals.

place at a much faster rate than metals. Of course, one cannot predict the future with any certainty, and the forecasts could turn out to be wrong–but in certain quarters it is believed that synthetics may grow even faster. Naturally, this all depends on the availability in the future of an adequate resource-base for synthetics.

5.6.2 Resources for synthetics

a. Crude oil. Our *major* supplier of organic chemicals for the synthetics industries has become the petrochemicals industry, which in turn depends upon crude oil as its raw material resource-base.

Hence we now begin to find materials technologists who are concerned about resources for the future taking an interest in the so-called 'energy crisis'. An increasing failure to discover new sources of mineral oil on a scale, and at a rate, to match the rate of growth in world demand is well illustrated in a forecast by Drury (see the chapter, *Resources and Environment*, in the book which the Open University uses for its Foundation Course in the Earth Sciences, *Understanding the Earth*, 2nd Edition, Artemis Press, 1972). Of course, in comparison with the total demand for oil for energy generation purposes, the present demand for the production of synthetics is very small indeed. But if scarcity leads to rapidly increasing prices for crude oil, the price of all petrochemicals is affected, as happened in 1974, for example.

Drury's curve is only a *prediction* of the life-cycle of crude-oil resources, and therefore is open to challenge. However, many authorities would tend to agree with the trend in growth of demand implied in the diagram, but what of the accuracy of the estimate of total ultimate resources, quoted at $2\,100 \times 10^9$ barrels (300×10^9 tonnes)? One can best check on this by consulting the oil industry experts themselves. Elwyn Jones, Chief Economist of the Esso Petroleum Company states that as a working hypothesis, meaning presumably for forward planning purposes, one might take a figure of $2\,000 \times 10^9$ barrels for the ultimate total resources of crude oil (seven barrels equal about 1 ton of oil).

Other oil experts could be quoted to support this view. Thus, there is some reason to believe that on present tendencies, the future life-cycle of crude oil might approximate to the pattern presented by Drury, i.e. similar to the forecast based on M. K. Hubbert's dynamic model, referred to in section 5.4.2(*c*) above.

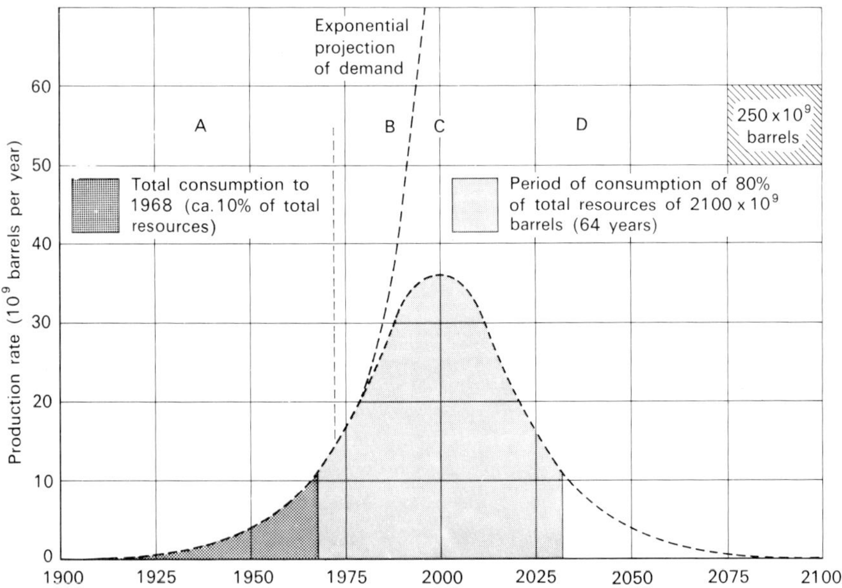

Fig. 5.9 Complete cycle of world crude-oil production: (A) exponential growing demands are supplied, (B) discovery and extraction become increasingly difficult, production is unable to meet demands, (C) maximum attainable production rate, (D) declining production, ultimately decaying exponentially.

What happens if we combine the predictions for demand in synthetics by the year 2000, with the predictions for the amount of crude oil likely to remain in the Earth by that date? Using the Elwyn Jones forecast, it seems that we might only have about 400×10^9 barrels of oil left by AD 2000. If demand for synthetics reached the figure of $1\,700 \times 10^6$ tons per annum by this date, then this might require an annual input of very roughly $1\,500 \times 10^6$ tons of crude oil or about 10×10^9 barrels. Thus, if for the sake of argument we were only using crude oil for synthetics production, having ceased burning it for energy generation by the year 2000, we could then only expect a Static Index reserve life of about 40 years. However, if the high exponential growth in demand for plastics etc. is assumed to continue after the year A.D. 2000, then the Exponential Index for the oil resource base could be as low as 12 years. Thus, if we were to go on using petroleum as a key source of energy up to the end of the century, based on predictions of demand growth, not enough of this resource would be left to support the huge plastics industry expected by the end of the century–an unsatisfactory outlook if petroleum were to be our sole resource-base for synthetics. Of course, there is a degree of uncertainty about the likely extent of future oil discoveries, and if we are very lucky, total oil resources in the world might turn out to grossly exceed $2\,000 \times 10^9$ barrels, although hardly anyone anticipates a figure in excess of $3\,000 \times 10^9$ barrels. But in any case, as has been pointed out in section 5.4.2 above, increasing resources when exponential growth rate in consumption is very high adds very little to the reserve life of a non-renewable resource. One should rather slow down the rate of increase in demand by substitute development or look for an alternative mineral source, if this is possible.

b. Coal. This mineral, being by far the most abundant and widespread of fossilized materials, presents a possible alternative to oil. Coal measures are confined to sedimentary basins, and particularly those with fresh-water sediments; geologists consider that all the *major* coal basins of the world have now been discovered. P. Averitt of the US Geological Survey estimated in 1969 that the world's reserves of recoverable coal were 9.5×10^{12} tons. Thus petroleum resources at 2000 billion barrels, i.e. about 0.3×10^{12} tons, are only 3% of the figure for coal. Although the estimated tonnage figure of coal reserves is enormous, this mineral varies tremendously in variety and quality, and also is very unevenly distributed over the Earth's surface. Averitt considered it a possibility, though this is only conjecture, that the coal resources may in fact be nearly doubled if one considers the large areas of the world unmapped and unexplored. North America, Europe and Asia are said to contain the majority of the world's coal. As a resource base for synthetics in the future, apart from the fact that we use coal as an energy source, causing it to be used up at an increasingly high rate, there are two main snags:

1.   It would be very variable as a raw material for a world coal-chemicals system. Various types of plant and processing units would have to be employed to deal with materials ranging from little more than peats, through lignites and sub-bitumous materials to bituminous coal and anthracites–because the coal resources figure covers all these variations.

2.   It is basically by no means as satisfactory a source for synthetics as petroleum on chemical grounds; essentially because of its lack of hydrogen. Petroleum has a much higher hydrogen content than coal and gives a higher yield of 'lights' (more volatile hydrocarbons). In principle, one could hydrogenate the coal–if one had plenty of cheap

hydrogen–or alternatively, extract the 'light' fraction from coal, leaving a carbon residue. A good deal of the technology is already known, but the economics is unfavourable compared with using crude oil for synthetics, on present-day prices.

    c. Tar sands and oil shales. These are two further potential resources of organic materials in the Earth's crust. The tar sands are impregnated with what is essentially a heavy crude oil that is too viscous to permit recovery by natural flowage into wells. There is no proper world inventory of reserves available as yet since tar sands have hardly been exploited. The best known of such deposits, and maybe the world's largest, are in northern Alberta, especially along the Athabasca River. One estimate puts these particular reserves at the equivalent of 300×10^9 barrels of crude oil. The tar is essentially a mixture of asphaltic hydrocarbons which act as a cementing agent for the grains of host sand. One snag is that there is an overburden ranging from the surface outcrop to about 2 000 feet (~ 700 metres) thick, and this has first to be removed before the tar sand is reached. To recover the hydrocarbons, the rock itself must be mined, broken up and heated with steam to obtain a tarry extract which, being in the same family chemically as crude oil, can in principle be processed in refineries of existing types. During the past, small-scale efforts to exploit these sands have failed, but development work proceeds by a number of oil companies and economic exploitation may well take place, as crude oil prices continue to rise. Although our reserves of tar sands are probably not very large in terms of their use as an energy source, this could be significant for chemical production for the manufacture of synthetics.

    The oil shales have an organic content, largely hydrocarbons, and when processed can yield up to one to two barrels of oil equivalent per ton of rock. However, these hydrocarbons are completely solid, not viscous as in the case of the tar sands. Also, they are chemically different from crude oil, containing many impurities. Thus, special refinery problems arise here and heavy capital investment would be necessary before an industry could be established to supply materials for synthetics manufacture using oil shales as the raw material. The total quantities of oil likely to exist in carbonaceous shales throughout the world have been estimated at approximately 30×10^{13} tons, which is roughly about thirty times the world's recoverable reserves of coal. This would appear to be a huge potential resource base for synthetics, but only a proportion of these resources is likely to be economically extractable. About 27×10^9 tons of oil is recoverable using known technology, and under foreseeable conditions; this is only one ten-thousandth of the estimated total reserves. The low concentration of oil in the shale seems severely to restrict the possibility of economically winning more than a tiny fraction of this resource in reality. However, the synthetics industries may eventually be able to absorb the necessary higher costs which would result from changing over to such raw material resource bases as tar sands and oil shales, whereas the energy-producing industries may not be able to in face of competition from coal and nuclear power.

    Finally, it must be pointed out that in this survey of resources of carbon for synthetics, no account has been taken of the oxidised forms of carbon in the Earth. In fact, the greatest quantities of the element carbon exist in the form of carbonates, mainly of magnesium and calcium. Of course large amounts of energy would be needed to get the carbon out in useful form. But carbonates could become useful in preparing carbides as intermediates for organic chemicals, by first carrying out a calcination process, perhaps collecting the carbon dioxide for use in synthesis gas. Carbides could also be prepared using coal as the source of carbon atoms. The carbides can be reacted chemically with water to produce a range of hydrocarbons and organics.

5.7 MEASURES TO CONSERVE NON-RENEWABLE MATERIAL RESOURCES

It has been shown in sections 5.5 and 5.6 that man could be facing scarcity problems in the future with regard to raw material resources for important sectors of materials technology, bearing in mind the general principles enunciated in section 5.4. What conservation measures might be considered for implementation to alleviate or avoid these problems?

To answer this question it will be useful to refer to an over-simplified flow diagram for the life-cycle of an engineering material. A raw mineral is extracted from its natural source, a mine or an oil-well for example, and is then *processed* through several stages to yield a metal or alloy, or perhaps a polymeric material. It is then *fabricated* into a geometric form, by, for example, casting, machining, moulding or extrusion. This product then enters the *construction* stage, which yields a useful article such as a motor car radiator, a metal bridge or a plastic vessel. During these manufacturing processes, waste material is produced which, in principle at any rate, is usually *recycled* back to earlier stages of production, with little delay. This material is sometimes called *primary* scrap. After the article has come to the end of its useful life, it is scrapped and becomes waste, which means some degree of pollution, or alternatively, it is returned as *secondary* scrap for reprocessing.

Now the points where conservation principles can possibly be applied in this system are as follows:

1.   Select a raw material with a wide resource base.
2.   Minimize the production of scrap during processing, fabrication and construction.
3.   Ensure the efficient collection of scrap for recycling, even after the useful article comes to the end of its life.
4.   Improve design to minimize material needs, if the resource base is likely to become scarce.

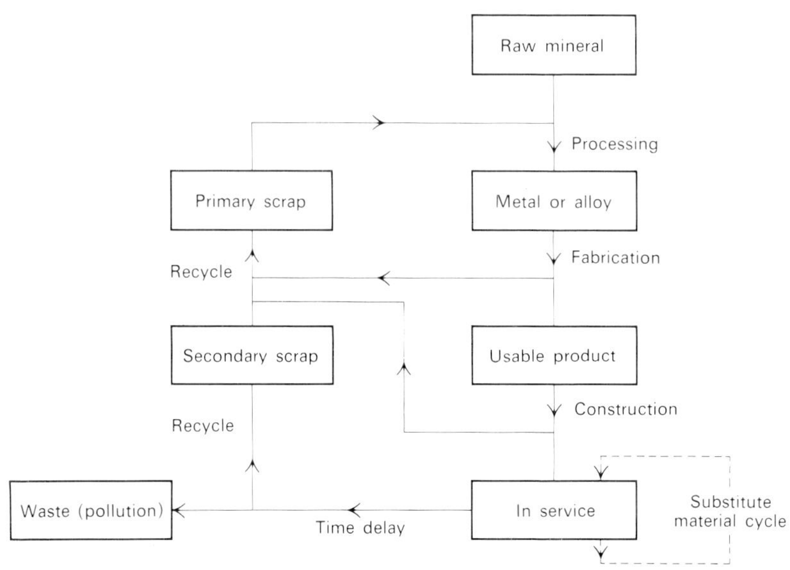

Fig. 5.10   The life-cycle of an engineering material.

5. Improve design to lengthen product's life.

6. Improve design to facilitate ease of recycle or re-use at end of product's life.

7. Develop and introduce substitute materials for the manufacture of the product, working from a far more plentiful resource base.

It is, of course, easy to list these approaches, but in practice they may be difficult to achieve, if economics, fashion or tradition lead in the opposite direction. The extent of secondary scrap recycled in the case of metals such as copper and lead varies considerably, depending on market prices. The modern, mass-produced motor car has a high degree of built-in obsolescence, thus shortening rather than lengthening its service life; and it is hardly designed for ease of recovery and recycling of its component non-ferrous metals. In the distant future, growing scarcity may force government intervention to improve conservation during the life-cycle of engineering materials.

It is necessary to bear other factors in mind in analysing a material's life cycle. It almost goes without saying that pollution of the environment must be minimized in all industrial operations, and energy requirements must be kept as low as possible. Thus, the materials flow system needs analysis and designing to minimize the *net* amount of pollution emanating from the whole cycle and to minimize the use of energy *overall*. Again, this is easier to say than to put into practice, because for very many materials and commodities preparation of a detailed life-cycle diagram would mean showing dozens of stages, involving several industries. In a free society, it is difficult to envisage means of optimizing resource conservation, and minimizing the pollution generation and energy consumption, without growing restrictions on business freedom. But this may have to be endured in the future in certain sectors of industry where, for example, non-ferrous metals and synthetics are involved.

Recycling may seem easy when one thinks of recovering lead from motor car batteries, but it is not so easy to envisage recovery of this metal from motor car exhaust gases. No one believes it would be easy, or cheap, to recover tin from tinplate owing to the thin layer of the coating on the steel sheet; similar considerations would apply to recovering zinc from galvanized steel. But recycling begins to look even more complex when one studies the life-cycle of materials, and takes into account exponential growth in demand and the lifetime of the commodity. On the latter point, if one considers buildings currently in course of construction, it may well be a hundred years before the metals incorporated will be available for recycling and re-use.

Substitution is of course a process which goes on continually in response to market forces and over the next 30 to 50 years we shall probably witness some revolutionary changes in engineering materials in response to scarcity. In broad terms, as those non-ferrous metals referred to in section 5.5 become scarce, there may well be widespread extension of the use of lightweight elements such as aluminium, titanium and magnesium and their alloys because there are relatively plentiful resources of the appropriate minerals. Further markets will be found for iron and steel products and alloys based on iron; it is clear that reserves of iron are extensive and may be viewed as being sufficient for a hundred years at the very least. Plastic materials for fabrication are experiencing a high growth rate and are substituting for metals in various applications. However, long-term supplies may have to be based on a raw material other than petroleum, because of the factors discussed in section 5.6. In the very long term, we may move towards large engineering industries based on non-metallic materials such as silicon and aluminium compounds and ceramics, for ductile metals and alloys will eventually become scarce and hence expensive. Even carbon resources may become scarce, putting restrictions on the availability of many organic compounds. However, the substitutional processes will in many important cases only become practically feasible after much research and development work has been carried out to enable the necessary advances in technology to be made.

With regard to the problem of minimizing the production of scrap or waste during the processing and fabrication stages, it is quite probable that a good deal of improvement will be effected here in the future. Even if all the primary scrap generated were to be fully recycled, inevitably some would be lost during the process. Furthermore, the very process of recycling often uses up other valuable resources such as energy and water, for example.

Finally, we come to the question of looking for more plentiful supplies of essential raw materials. The discussion of non-renewable metallic minerals in sections 5.4 and 5.5 was based on our conventional resources; that is, the ore bodies in the Earth's continental crust. But various authors and experts have postulated that as these historical sources approach exhaustion, we shall be able to develop other types of resources which are currently uneconomic, but which can be exploited commercially when metals prices have risen to appropriate levels. Thus, it is said in certain quarters that we need not run out of our key industrial minerals for many years to come, provided we are prepared to pay for them. This means finding the necessary capital and energy resources to do so. The possible new sources of minerals comprise ocean bed deposits, sea water itself, the minerals deep down in the Earth's crust, and the minerals present in trace quantities in surface rocks. In the final section of this chapter, we briefly examine some of these possibilities for the future.

5.8 POTENTIAL FUTURE RESOURCES OF MINERALS

5.8.1 *The structure of the Earth*

It will be useful to summarize, briefly, the structure of our planet by way of introduction. It is important to appreciate the concept of the Earth as a core surrounded by a series of concentric shells of varying size and characteristics; thus it is highly inhomogeneous. The core has a radius of 1790 miles (2860 km) and consists predominantly of iron and nickel. It is surrounded by the mantle, which consists mainly of dense oxide minerals rich in iron and magnesium. Outside the mantle is the crust, which only accounts for 0.375% of the total mass. The crust embraces of the *continental crust*, which projects above the oceans and also includes a narrow sea-covered fringe around each continent, the *continental shelf*. Below the ocean proper is the *oceanic crust*. At the Earth's surface are the oceans, lakes and rivers, plus water trapped in holes and fractures in soil and near-surface rocks; this is termed the *hydrosphere*, which accounts for 0.025% of the Earth's mass. The total mass of the Earth is 6.5×10^{21} tonnes. Surrounding everything is the gaseous envelope known as the atmosphere, which accounts for only 0.0001% of the mass. It is only from the three outermost zones – crust (continental and oceanic), hydrosphere and atmosphere, which together account for only 0.4% of the total mass of the Earth – that we must contemplate extracting our resources by any conceivable form of technology. It is very difficult indeed to envisage a way of winning materials from the mantle, let alone the core. In any case, we have abundant sources of iron, nickel and magnesium in the crust and hydrosphere and these are thought to be the

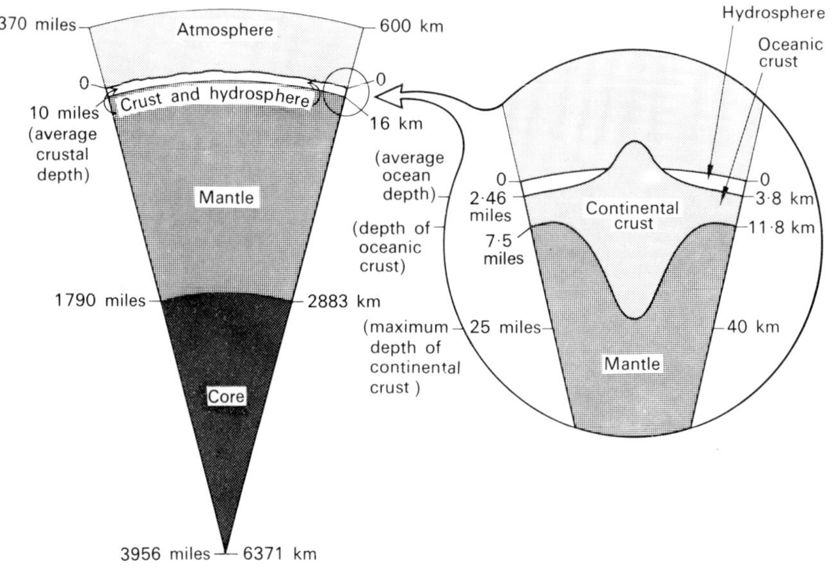

Fig. 5.11 Concentric structure of the Earth's zones.

main metals of the mantle and core; thus, the mantle and core may be disregarded for the forseeable future.

5.8.2 Ocean resources

The composition of the *oceanic crust* is not known with absolute certainty but it is highly likely to be basaltic and of relatively constant composition, not having suffered normal erosion. There is no evidence of any sort of large-scale mineral concentrations in the oceanic crust, although there is evidence of certain mineral accumulations or deposits on the sea floor – in particular, the extensive accumulations of nodular growths of mainly manganese oxides on the floors of the ocean deeps. These nodules are derived mainly from erosive processes on the land surface followed by a series of complex stages during which the manganese migrates slowly to the ocean bottom. According to one estimate there are probably 1.7×10^{12} tonnes which could be dredged from the bottom of the Pacific Ocean alone, although to extract anywhere near this quantity might cause great disruption of natural processes in the ocean, even if the technological problems are solved; nevertheless, some scientists and engineers, particularly in the USA, believe the exploitation date of this resource is not far off. The composition of the nodules varies from place to place, but according to Goldberg in 1954, the composition of the manganese nodules in the Pacific Ocean was thought to be on average as follows: 15–23% manganese, 10–12% iron, 0.5–1.5% copper, 0.6–1.2% titanium, 0.2–1.0% nickel, 0.2–0.7% cobalt, 0.2–1.0% phosphorus, 0.3–1.2% aluminium.

We have abundant resources of iron, aluminium and titanium, and manganese ore resources are fairly estensive in the continental crust. However, the copper concentration in the nodules seems comparable with that of ore bodies in the continental crust as worked in recent years, and as stated earlier, copper reserves are likely to become low in a few decades or so. Copper resources in the nodular deposits might be around 1.7×10^{10} tons, which is about fifty times our currently known crustal reserves of copper.

These nodules would therefore be useful for extracting copper, and the nickel and cobalt could well become potentially valuable at some point in the near future. Before leaving the subject of oceanic deposits, it should be mentioned that there are extensive deep sea deposits of mud in which metalliferous minerals have become concentrated by deposition from brines. If we examine the abundance of the geochemically 'scarce' minerals of table 5.1, it is apparent that the important ones are in very low concentration in these muds. For example, copper is between 270 and 740 p.p.m., lead is between 60 and 150 p.p.m., vanadium 125 to 450, and zinc about 370. It might well be imagined there are great technical problems to be faced, similar to those of the winning of nodular deposits, and these low concentrations of base metals make the deep sea mud deposits of less interest than the nodular deposits.

Turning to *sea water* itself as a source of scarce minerals, table 5.2 lists the average estimated concentration of important minerals in parts per million, and the quantity in tons in one cubic mile. The concentrations vary, however, from place to place, and with depth. The oceans, which cover about 71% of the Earth's surface to an average depth of 2.46 miles, act as a collection reservoir for many of the soluble materials formed in the crust, such as those released from rocks and soils by weathering, and from volcanic

Table 5.2 Average concentration of selected minerals in sea water, and total quantity in a cubic mile

	Concentration p.p.m.	Amount present in 1 cubic mile of sea water tons
Chlorine	19 000	89.5×10^6
Sodium	10 700	49.5×10^6
Magnesium	1 300	6.4×10^6
Sulphur	880	4.2×10^6
Calcium	400	1.9×10^6
Potassium	380	1.8×10^6
Bromine	65	0.3×10^6
Carbon	28	130 000
Boron	4.5	23 000
Aluminium	1.0	5 000
Lithium	0.17	800
Iodine	0.05	260
Manganese	0.01	47
Zinc	0.01	47
Tin	0.003	14
Copper	0.003	14
Uranium	0.002	10
Chromium	0.001	5
Molybdenum	0.0005	2.5
Gallium	0.0005	2.5
Thorium	0.0005	2.5
Lead	0.0003	1.5
Cadmium	0.0003	1.5
Silver	0.0003	1.5
Vanadium	0.0003	1.5
Nickel	0.0001	0.5
Mercury	0.00003	0.1
Gold	0.000006	40 lb

gases. It is common knowledge that sodium and chlorine are very abundant in sea water and together with magnesium, sulphur, calcium and potassium account for 99.5% of all dissolved solids in the ocean. At present, only sodium, chlorine, magnesium, and bromine are recovered commercially: sodium chloride often by solar evaporation in shallow ponds (bitterns), bromine by adding aniline or similar type of organic compound or chlorine gas, and magnesium by adding calcium hydroxide to precipitate magnesium hydroxide. It is abundantly clear from table 5.3 that the geochemically scarce minerals described in section 5.5 are only present in extremely low, trace concentrations in sea water, and economic extraction of them does not seem feasible in the foreseeable future because of the low concentration. Even in the case of the valuable and costly element, uranium, the time has not yet come when the market price would make sea-water extraction viable commercially. Zinc has only five times the concentration of uranium, tin and copper are of the same order, whereas lead, cadmium, silver and mercury are in lower concentration.

One sometimes hears the view expressed that we have enormous reserves of minerals in sea water, and certainly the volume of the seas is about 330×10^6 cubic miles. The potential resources are enormous for all minerals in table 5.3 higher than say, boron. But this excludes all the heavy metals. A more extensive use of minerals from the sea would really be confined to light elements and their compounds. In particular, any great expansion in demand for magnesium in the future would probably mean an expansion in the number of sea water processing plants around the world.

Before leaving the subject of minerals from the ocean, it is interesting to note that nature does a certain amount of concentrating and extracting scarce minerals for us. We have referred to the preferential concentration of several minerals in nodules and ocean-bed muds, and it will be seen that copper, manganese, nickel are orders of magnitude higher than in sea water, The concentration of minerals from sea water is brought about by various marine organisms, notably algae, fish bones and tunicates. This biogenic activity might conceivably be exploited by man at some time in the future but much research needs to be done to understand the processes involved.

5.8.3 Resources in the continental crust

All the geochemically scarce minerals listed in table 5.1 are present throughout the Earth's crust, but normally only at the approximate 'background concentration' as given in the table, except for ore deposits, which are economically viable deposits, as discussed in section 5.4. The continental crust forms about 0.29% of the mass of the planet, an enormous quantity of rock, and this when multiplied by any of the heavy metal concentrations in table 5.1 gives a huge figure of potential resources. Such figures are sometimes quoted to support the optimisitic case for man's reserves of metallic minerals being 'almost infinite'. But we have a situation which is perhaps parallel to that of minerals in sea water. The low abundance would mean processing enormous quantities of rock to extract small amounts of the minerals, and the energy and capital investments could be formidable. The likelihood of extensive environmental disturbance would be very great, and would be on a scale which would dwarf that due to any mining operations of ore deposits to date.

Nevertheless, mining of whole rock does take place on a large scale for the extraction of non-metallic minerals of commercial importance. Clay, limestone and dolomite, as well as siliceous rocks, are well-known examples. However, most of the rock is useful product; the waste material is much smaller than would result from the mining of whole rock to extract trace metallic elements. But there are environmental and pollution problems associated with any very large-scale project involving open-pit mining. Large holes are inevitably left in the ground, and there are always problems due to overburden removal and disposal during operations. If any washing or concentration process is involved, then the wet 'tailings' are an environmental problem to be dealt with. As individual site operations grow in size to exploit the economics of scale, so do the environmental problems increase, especially in the general vicinity of urban population or national park areas. This discussion on environmental/pollution issues may seem out of place in a chapter on material resources, but it is going to be one of the salient problems of the future which we shall have to face as we process crustal rocks, or oceanic deposits, or sea water, where man is only interested in extracting and marketing a small fraction of the total mass he is disturbing.

Finally, it is interesting to note that the crust is, on average, about 10 miles (16 Km) deep; reaching down to perhaps 25 miles under the continents but only about 7.5 miles under the bottom of the oceans. At these depths the interface with the mantle occurs, the position being inferred from seismic measurements. Now mining operations have so far only gone down a mile or two, at most, into the crust. In the deepest mines, there are difficulties in coping with the high temperatures involved – men can only operate down there for short periods, and refrigerated water has to be pumped down to make conditions at all tolerable. In future years, it may be envisaged that men will become increasingly reluctant to undergo such working conditions and completely new kinds of technology will have to evolve to enable remote control of unmanned operations at depths. It is hard to imagine mining of the Earth's crust being carried out at very much greater depths than those of existing deep mines, without much further research, investment and extensive break-throughs in new technology at present unforeseen. So it is reasonable to assume that the resources of only the *surface* layer of the Earth's crust will be available to us for very many years to come, and our conservation policies should be developed on this assumption.

One source of minerals on the Earth's surface we have not alluded to are the waste and spoil heaps of derelict mining operations. An example would be the extraction of fluorspar from old lead mine tips. In certain cases, metallic minerals are reworked from old mine waste heaps because high prices make this an economic proposition at certain times in the trade cycle. Increasing scarcity, and concomitant price rises on a more permanent basis, are bound to make the working of old mine wastes an increasingly attractive economic proposition. It is unfortunate that the high exponential growth in demand for minerals described early in this chapter will prevent such operations from supplying more than a small fraction of our future requirements. Furthermore, the mineral extraction industries of today are only interested in operating on a very large scale; the spoil heaps of the past are too widely scattered, in general, to offer an economic resource base owing to the high costs of transporting bulk materials to large central processing plants.

5.8.4 Concluding notes

It would seem that we will have to continue to rely on the mining of ore bodies as our main supply of non-renewable minerals which are 'geochemically scarce' and as prices rise with the passage of time, poorer grades of ore will progressively become profitable, provided they are identified at sites with a huge local deposit. Otherwise the capital for development will not be forthcoming – no mining company management will contemplate setting up a site with all the necessary capital facilities and infrastructure development if the local resources are liable to be exhausted within a few years.

As discussed in section 5.4, it is impossible to predict with any certainty how long the world's reserves of ore minerals will last. Clearly there will be concern about the adequacy of some of these, as discussed in section 5.5, well within a 50-year time-scale. This is not too far ahead to begin to take note of these facts, and to move towards the development of sound resource planning and control measures: in other words, establish mineral resource conservation policies on an international as well as a national basis.

5.9 LITERATURE

Resources and Man. A Study and Recommendations by the Committee on Resources and Man of the Division of Earth Sciences, US National Academy of Sciences – National Research Council, San Francisco, Freeman, 1969.

B. J. SKINNER, *Earth Resources.* New Jersey, Prentice-Hall, 1969.

J. L. MERO, *The Mineral Resources of the Sea.* New York, Elsevier, 1965.

P. T. FLAWN, *Mineral Resources – Geology, Engineering, Economics, Politics.* New York, Rand McNally, 1966.

E. W. ZIMMERMANN, *World Resources and Industries,* 3rd edn. Harper and Row, 1972.

Mineral Facts and Problems, US Bureau of Mines, Bulletin 650, 1970, London.

D. L. MEADOWS *et al., The Limits to Growth,* Earth Island Publications, 1972.

H. BROWN, J. BONNER and J. WEIR, *The Next Hundred Years,* New York, Viking Press, 1957.

Understanding the Earth, 2nd edn. Artemis Press, 1972.

ADDITIONAL LITERATURE

Because of the relative scarcity as yet of general publications in this rapidly developing subject, a number of useful articles in learned journals are listed below:

Important documents/symposia:

Elements of a National Materials Policy, Nat. Acad. Engineering MMAB 294, US Nat. Acad. of Sciences, Washington D.C., 1972.

Symposium on Scrap Recovery, Institute of Metals' Spring Meeting, London, 21–22 March, 1973 (see especially J. A. Clay, 'Future developments – a primary view of the secondary non-ferrous scrap recovery industry', Paper 17). Metal Society, 1, Carlton Terrace, S.W.1.

'The Limits to Growth Controversy', special issue of *Futures,* **5**, No. 1, Feb. 1973.

Useful papers and articles:

H. BROOKS, 'Materials in a steady state world', *Met. Trans.,* **3**, April 1972, p. 759.

F. ROBERTS, 'The next 50 years in engineering materials', *Chemical Engineering,* No. 260, April 1972, p. 146; *Atom,* Aug., 1972.

Proceedings of a conference on the Conservation of Materials, March 26–27, 1974, Harewell.

G. J. S. GOVETT and M. H. GOVETT, 'Mineral resource supplies and the limits of economic growth', *Earth-Sci. Revs.*, **8**, 1972, 275–290.

R. U. AYRES and A. V. KNEESE, *Economic and ecological effects of a stationary economy*, Resources for the Future Inc., USA, (reprint No. 99, Dec. 1971).

Y. SEKINE, 'On the concept of concentration of ore-forming elements and the relationship of frequency in the earth's crust', *Int. Geol. Rev.*, **5**, 505–15.

S. G. LASKY, 'How tonnage-grade relationships help predict ore reserves'; *Eng. Mining Journal*, 151(4), 1950b. 81–5.

W. R. HIBBARD, JR. 'Mineral Resources: Challenge or threat?'; *Science*, 160, 1968, 143–150.

HUBBERT M. KING, *Energy Resources: A Report to the Committee on Natural Resources*, Nat. Research Council Pub. No. 100-D, Nat. Acad. of Sciences, Washington D.C., 1962.

P. WEISS, *Renewable Resources*, Nat. Research Council Pub. 1000-A, US Nat. Acad. Sciences, Washington D.C., 1962.

F. ROBERTS, 'Future resources of engineering materials', *The Chartered Mechanical Engineer*, **20**, No. 4, April 1973.

V. WHALLEY, *RAPRA Bulletin*, **26**, No. 11, Nov. 1972, p. 332.

C. HANSON, 'The ocean as a commercial source of minerals', *The Chemical Engineer*, No. 264, p. 295, August 1972.

V. E. McKELVEY, 'Relation of the reserves of the elements to their crustal abundance'; *Am. J. of Sci.*, 258A, 1960, 234–41.

General Index

Bold figures indicate the number of the volume

Abalone, **5** 594, 595
Abalyn, **5** 143
Abel heat test, **4** 711
Aberdeen Angus breed, **8** 156, 158
Abhesive, **5** 65
Abies balsamea, **5** 152
Abies sibirica, **5** 23
Abies spp., **6** 102
Abietic acid, **5** 136, 142, 279
Ablative material, cork as, **5** 124
Abrade, definition of, **2** 118
Abrading, **2** 118-26
 purpose of, **2** 162-3
Abrasion resistance of glass, **2** 389-90
Abrasion tests
 for coke, **2** 603-4
 for vulcanizates, **5** 514
Abrasive
 definition of, **2** 118
 pumice as, **2** 66
 silicon carbide as, **2** 143-6
 white alumina, **2** 139
Abrasive cloth and paper, glue for, **5** 59, 77
Abrasive cutting, **2** 168-9
Abrasive grit, **2** 124, 125, 163-5
Abrasive material, **3** 570-1
 properties required of, **2** 119-25
 temperature and, **2** 125-6
Abrasive processes, **2** 118-19, 177
Abrasives
 diamond, **2** 132-5
 Federation of European Producers of, **2** 141
 ferrous, **2** 148-9
 health hazard of, **2** 132
 history of, **2** 115-18

indentation testing of, **2** 121
manufactures, **2** 135-47
metallic, **2** 147-50
for metal working, **3** 571
natural, **2** 126-32
shot blasting, **2** 147-9
special heavy duty, **2** 142-3
statistics for, **2** 181-2
technical advice on, **2** 180-1
uses of, **2** 162-81
Abrasive temper, **2** 117
Abrasive wheel sawing, **3** 570
 diamond for, **2** 489, 494-5
Abrasive Wheels Regulations 1970, **6** 67
Abscisic acid, **7** 210
Absolute joule, **2** 705
Absolute pressure, **8** 103-4, 105
Absolute reference electrode, **3** 843
Absolute temperature, **1** 41
Absolutes, **5** 9
Absorb, definition of, **2** 185
Absorbance, **2** 346
Absorbent pads, **5** 823
Absorption
 of alexandrite, **2** 512-13
 of aquamarine, **2** 517
 coefficient, **3** 76
 of corundum gems, **2** 510-11
 of diamond, **2** 466-8, 485
 of emerald, **2** 517-18
 of jade, **2** 530
 of sun's radiation, **2** 735
 for testing glass, **2** 399
 of water, *see* water absorption
ABS plastics, **6** 500, 529, 530-1, 533, 537, 563, 617-18
ABS resins, **4** 180
ABS rubber, **5** 82
Acacia, **5** 30, 404

Acacia gum, in tablets, **5** 765, 769
Acaricides, **4** 250; **7** 127-32, 143
 bridged diphenyl, **7** 105, 128-30
 carbamates, **7** 105, 130-1
Accelerated hammer, **3** 483
Accelerators, **1** 127; **6** 505, 506
 for concrete mixture, **2** 110-11
 guanidine, **5** 475, 476
 sulphenamide, **5** 475, 476
 of vulcanization, **5** 475, 476, 523-5
Accroides, **5** 132, 133, 151
Accumulator, **3** 264
 hydraulic, **3** 484, 487
Accumulator press system, **3** 475
Acenaphthene, **4** 96, 187
Acepromazine, **5** 798
Acetal, **4** 373
Acetaldehyde, **4** 142, 151, 353, 359-60
 oxidation of, **4** 424, 462, 481
 production of, **4** 156, 277, 279, 360
 reactions and uses of, **4** 283, 299, 301-2, 359-60, 361, 363
Acetal resins, **6** 500
Acetals, **4** 268, 355, 370-5
 definition and classification of, **4** 370-1
 preparation of, **4** 371-3
 properties of, **4** 371
 world production of, **4** 374-5
Acetamide, **4** 399, 525; **5** 796
Acetanilide, **4** 399, 524, 526, 555
Acetate fibres, **4** 463
Acetate rayon, **6** 282, 292, 301,

Novolak resins, **4** 324, 563; **5** 94, 95; **6** 587, 589-91
Nozzle, pressure, **1** 32-3
NPK fertilizers, **7** 77-9, 81, 82, 384
NR, *see* Rubber, natural
Nubuck, **5** 398
Nuclear batteries, **2** 782-5
Nuclear engineering, use of lithium in, **1** 482
Nuclear fission, **2** 727-31; *see also* Fission
Nuclear fuels, **2** 708
Nuclear fusion, **1** 655-6
Nuclear industry, molybdenum in, **3** 334
Nuclear magnetic resonance (n.m.r.), **8** 28, 29-30, 110 and structure of glass, **2** 335-6
Nuclear moderators, **2** 419, 434
Nuclear power, reserves of, **2** 711
Nuclear power stations, costs of, **2** 712
Nuclear reactions
electrical energy from, **2** 782-5
heat energy from, **2** 725-37
Nuclear reactors, *see* Reactors
Nuclear technology, use of isotopes in, **1** 628-9
Nuclear weapons, **1** 655-6
Nucleating agent, **3** 404-5
Nucleation, **3** 404-5
Nucleation-growth process, **3** 24-5
Nucleic acids, **7** 228, 811
Nucleoproteins, **8** 142, 144
Nuhydrazone, **5** 645-7 *passim*
Numerical control
advantages of, **3** 576-7
essential features of, **3** 577
types of system, **3** 577-9
Numerically controlled machine tools, **3** 576-9
Nusselt number, **2** 742
Nutmeg *(Myristica fragrans)*, **5** 245; **7** 653
Nutmeg butter, **5** 220, 224
Nut oils, **5** 286
Nutria, **5** 553, 571
Nutrient solution culture, **7** 42
Nuts, **7** 403-10
Nuts, self-locking, **3** 587
Nyldrin, **5** 790
Nylon
dyeing of, **4** 672, 675, 680-1, 683, 707
production of, **4** 89, 348, 434-5, 514-15, 524, 528-30,

558
structure of, **4** 670
Nylon 6, **6** 283, 321, 324-5, 338, 507, 570-3, 577, 618
Nylon 6/6, **6** 279, 283, 285, 338-9, 570, 577
characteristics of, **6** 323
manufacture of, **6** 321-2, 501, 574-5, 618
structure of, **6** 571, 573-4
Nylon 6/10, **6** 323-4, 575, 577, 618
Nylon 11, **5** 189; **6** 325, 338-9, 575, 577, 618
Nylon 12, **6** 575-6
Nylon fibres, **6** 250, 257, 278, 279, 299, 321-5, 442-3, 570-6, 577
bleaching of, **6** 486
dyeing of, **6** 488
flammability of, **6** 495
as moulding material, **6** 500
production of, **6** 618
Nylon salt, **6** 322
Nystatin, **5** 626, 627

Oak, European *(Quercus robur or petraea)*, **6** 6, 7, 29-30, 35-6, 49, 106, 113, 130
veneer, **6** 81
Oak bark, for tanning, **5** 403, 405, 407
Oak galls, **5** 404
Oats *(Avena* spp.), **7** 1, 4, 37, 45, 240-1, 252, 274-7
botany of, **7** 232-3, 235, 275-6
cultivation of, **7** 275
diseases of, **7** 46, 49, 275
statistics for, **7** 253-8
types of, **7** 275-6
uses of, **7** 276-7, 870
Obeche *(Triplochiton scleroxylon)*, **6** 4, 29-30, 38, 43, 47, 87, 89, 113
blue stain in, **6** 110
Oberth, H., **4** 737
Oblique cutting, **3** 527-8
Obstetrics, **5** 810
Oceanic crust, **8** 395, 396
Ocean resources, **8** 396-8
Ocelot, **5** 552
Octadecanoic acid, **5** 226, 248
Octadecyl stearate, **5** 264
Octagon work, **2** 475
Octahedral coordination, **2** 335
Octane number, **4** 62-3
Octanoic acid, **5** 244
Octan-2-ol, **4** 292; **5** 267
Octyl alcohol, **5** 260, 261

Octyl caprate, **5** 265
Octylphenol, **4** 149; **5** 310
Odorants, for rubber mixes, **5** 526
Odoroside, **5** 747, 748
Oersted, H. C., **3** 243-4
Oestradiol, **5** 821
Oestrogen hormones, **5** 821
Ohlsson, **4** 722
Ohm's law, **2** 738
Oil
for cutting fluid, **3** 538
for ground laying, **2** 312
for infiltration of bearings, **3** 771
in production of tinplate, **3** 705
removal of, from metal, **3** 677-80
self-curing drying, **3** 377
Oil-bearing seeds, **7** 56; *see also* Oil seeds
Oil blacks, *see* Furnace blacks
Oil burner, **2** 719
Oil cake, **8** 71, 72-4, 76
Oil colours, **5** 376
Oil of cucumbers, **5** 5
Oil and dusting, **2** 312
Oil fields, **4** 4
Oil furnace blacks, *see* Furnace blacks
Oil furnace process, **2** 680, 682
Oil furnaces, **2** 725
Oiling of furskins, **5** 559-60
Oil-modified alkyd resins, **5** 214
Oil paints and varnishes, **5** 320-2
Oil palm *(Elaeis guineensis)*, **8** 6, 11, 38, 41-2
harvesting of, **8** 4, 69-70
Oil reserves, **4** 20-2
Oil reservoir, **4** 4
Oils and fats, **5** Ch. 8, 181-2; **8** Ch. 1
analysis and testing of, **8** 16-34
animal and vegetable, **5** 219-23, 225-7
biochemistry of, **8** 11-12
bleaching of, **2** 188, 193, 196
classification of, **8** 1-2
colour of, **8** 15, 26
cooking, *see* Cooking fats and oils
crude, **8** 2
definition of, **5** 187; **8** 1
edible and inedible, **8** 2, 4, 14, 135
extraction of, *see* Animal fats,

Stannous oxide, **1** 556
Stannous sulphate, **1** 557; **3** 700
Staphylococcus aureus, **4** 319; **5** 603
Staphylomycin, **5** 633
Staple length
  of auchenia, **6** 255-6
  of cashmere, **6** 253
  of cotton, **6** 210, 211, 215
  of mohair, **6** 252
  of natural fibres, **6** 264-6
  of wool, **6** 244-6, 249
Star anise oil, **5** 17
Starch, **2** 566; **5** 3, 69, 70, 702; **6** 113, 180; **7** 198-9, Ch. 7
  acid hydrolysis of, **7** 195, 543, 546, 608-10, 613, 870
  appearance of, **7** 511
  in cereals, **7** 200, 231, 271, 276, 441-2, 518, 520, 523-36
  chemical structure of, **7** 512-13
  components of, **7** 513-19
  cross-linked, **7** 552-3
  enzyme hydrolysis of, **7** 195, 515-17, 543, 595, 610-13, 815-16, 870
  in flour, **7** 459, 463, 470
  formation of, **7** 38
  fructose manufacture from, **7** 624-5
  granules, **7** 520-1, 524-5, 532
  and gums, **5** 28-9, 44, 110-11
  hydrolysis of, **7** 195, 514, 559, 604-5, 861
  isolation of, **7** 511
  modified, **7** 542-7
  oxydised, **7** 547-50
  pre-gelatinized, **7** 553-4
  properties of, **7** 521, 529-31, 533, 536, 540-1
  in root crops, **7** 288, 295-6, 298, 302-3, 315-16, 536-42
  statistics for, **7** 555-8
  uses of, **7** 521-2
Starch acetates, **7** 553
Starch ethers and esters, **7** 550-3
Starch phosphates, **7** 552
Star facet, **2** 478
Stark Brothers, **7** 352
Star ruby, **2** 537
Star of South Africa, **2** 443, 479
Stas-Otto process, **5** 710-11
Static binding, **6** 20
Static fatigue, in glass, **2** 343
Static friction, coefficient of, **2** 425

Stationery inks, **5** 379-80
Statistical theory for coal sampling, **2** 573-4
Staudinger, Hermann, **6** 278, 500, 529, 540, 596
Stauffer, **7** 133, 134
Steam, **2** 628, 745
  applications of, **2** 235-6, 597, 748-51
Steam distillation, **2** 749-50; **5** 6, 282
Steam engine, reciprocating, **2** 755
Steam generators, **2** 745-9
Steam-hydrocarbon process, **5** 237
Steam-iron process, **5** 237
Steam processes, **2** 750
Steam sterilization, **2** 751
Steam turbine oils, **4** 69
Steam turbines, **2** 753-5
Steaoropten, **5** 24
Steapsin, **5** 701
Stearates, **5** 320
Stearic acid, **5** 187, 248, 251, 260; **6** 505; **7** 202; **8** 7, 101, 108
  in animal fats, **8** 62, 66
  composition and characteristics of, **5** 226, 246, 247
  in hydrogenation, **8** 99
  for lubrication, **3** 751, 752
  preparation of, **5** 232, 234-6
  and soap, **5** 276, 279, 292, 294
  in vulcanization, **5** 475, 517, 521
Stearin, **8** 38, 167
Stearin pitch, **5** 168, 396
Stearyl alcohol, **4** 292; **5** 261-3 *passim*
Stearn, C. H., **6** 277
Steatite, **3** 45-8, 322-4
Steatite bodies, **2** 322-4
Steel, **2** 111-12, 225, 235-6
  and abrasives, **2** 142, 147
  glass-lined, **2** 405
  grinding of, **2** 123, 126, 166, 179
  primers for, **5** 362-3
  properties of, **2** 121, 123, 738, 740
Steel furnaces, **2** 268, 276
Steel grit, **2** 148
Steel industry, use of oxygen in, **1** 77
Steel-making, **2** 75, 84; **3** 151-91
  chemistry of, **3** 153

  direct, **3** 189-90
  graphite for, **2** 430, 433-4, 674
  instrumentation in, **3** 190-1
  in oxygen furnaces, **3** 165-72
  spray, **3** 189
Steel powders, **3** 747-8
Steels, **3** 191-209
  age-hardening of, **3** 219
  alloy, **3** 200, 203-4
  anaerobic corrosion of, **3** 812
  annealing of, **3** 34-5, 216
  austenitic, fatigue in, **3** 60-2
  carbon, **3** 21-3, 28, 199-202
  carbon removal from, **3** 189
  carburizing of, **3** 31
  castings of, **3** 434, 440
  classification of, **3** 197-200
  coating of, **3** 684. 706-10, 723
  cold extrusion of, **3** 479
  corrosion of, **3** 809, 812, 821-2
  corrosion rates of, **3** 820
  corrosion resistant, machining of, **3** 536
  and creep resistance, **3** 230
  deoxidation of, **3** 183-5
  descaling of, **3** 674
  effect of elements in, **3** 178
  electroplating of, **3** 702-3
  enamelling of, **3** 718
  etching of, **3** 680
  examples of use of, **3** 183
  ferritic, fatigue in, **3** 60-2
  grainsize of, **3** 195-6
  hardenability of, **3** 221-31
  hardening of, **3** 26-32
  hardness of, **3** 219-32
  heat resisting, **3** 536, 620-1, 829
  heat treatment of, **3** 209-19
  high carbon, **3** 197-8, 201-2
  high-speed, **3** 537, 774
  high-strength low-alloy, **3** 203-4
  impact properties of, **3** 55-6
  impingement attack and, **3** 806
  influence of working of, **3** 196-7
  isothermal treatment of, **3** 28-31
  low-alloy, **3** 822-3, 827-8
  low-carbon, **3** 50, 200-1
  for making rolls, **3** 471-2
  maraging, **3** 207-8
  medium carbon, **3** 201
  medium-carbon alloy, **3** 204

Recent Developments in Materials and Technology

Technology does not stand still. For all the care taken by the Editorial Board to present *Materials and Technology* in a way which would not rapidly be dated, new materials and new techniques have continued to emerge during the preparation of the Encyclopaedia; inevitably some have come too late for mention in the text. In order to remedy this situation, it was decided to include a chapter in this last Volume, in which the original authors would be given the chance to draw attention to noteworthy new developments not covered in their original chapters. Many authors were glad of this opportunity and their contributions will be found in the following pages. Others, in some of the more static technologies, felt that their original texts were still sufficiently accurate to stand without correction. In some instances, for various reasons, contact with the original author was not possible and here the Editorial Board had to decide either to let the original contribution stand or to endeavour to find someone competent to advise on the relevant Chapter and to provide up-dating information where required. For this reason, some of the original Chapters are not mentioned in the up-dating Appendix which follows. Where the up-dating material has not been supplied by the original author, this is indicated by a star(★).

Some basic decisions had also to be taken. One important one was not to attempt a comprehensive up-dating of statistical information. Statistics inevitably become quickly obsolete and new figures would soon be subject to the same criticism as those they replaced. Moreover, the intention when statistics were originally included was to provide information on trends and not to replace the standard reference books of statistical data. For this reason, changes in statistics are only included in the up-dating text incidentally, when they are quoted to illustrate some important result or effect of technological change.

One other change concerns the reading list appended to each Chapter. Wherever possible, important new reference books have been quoted. However, publication of a new book on a subject does not necessarily mean that previous works are superseded. For example, the definitive work on thermal conductivity as a test method is still that by Ingen–Hausz (1789) (*J. Phys. Radium* **34** (68) 380.) Nevertheless, if important new works have been published recently, attention is drawn to them.

Another important decision was not to change information as a result of technological change in a specific country. For example, the introduction of North-Sea gas into the UK has resulted in the obsolescence of a number of methods of producing town gas by carbonization. Nevertheless, these methods may still be of vital interest in countries not fortunate enough to have access to natural gas and they have therefore been retained.

The arrangement and order of subjects is the same as that of the original volumes. No changes in inorganic chemicals (Volume 1) fall within the guide lines set by the Editorial Board and outlined above, so the up-dating material starts with Volume 2, *Non Metallic Minerals and Rocks.*

VOLUME 2

Chapter 3. Abrasives

The latest advances in abrasives have been in the upgrading of products already available. Thus particular abrasive composites have been altered in their composition and make-up so that they better perform the task they were designed for when judged on a cost performance basis.

In the preparation of abrasive grains new methods of making, crushing, selection and measurement are continuously investigated with a view to improvement. New types of grain are also being made and evaluated.

High-speed grinding is moving forward, with various research groups finding the best methods of operation. Grinding-wheel companies are developing extra strong vitrified wheels to meet their high-speed requirements.

Resinoid snagging wheels are being improved mainly by the use of much tougher grains, amongst which the zirconia alumina grains are outstanding. The shape, size and toughness of these grains is controlled so as to optimize them for each application. Very large reinforced cut-off wheels – 4 ft (1.2 m) dia. are coming into more general use. Ceramically-coated grains are also finding their way into applications where water resistance is required.

Coated abrasives are being upgraded by improved backing, bonds and special selection of grain for the various duties and close control over coating conditions. As in the other areas, products are optimized for various tasks where proper evaluations can be made.

Efforts by the suppliers of both natural and man-made diamond to improve performance, have led to shape and friability controlled grains, and metal cladding, the latter for various reasons giving improved performance when used with resin bonds.

In the field of abrasives made by high pressure, high-temperature methods, cubic boron nitride (Borazan) is finding increasing use in grinding the more conventional metals. It will be recalled that the main use for diamond, its companion abrasive, is grinding very hard materials like ceramics, and carbides.

There has been a great upsurge in grinding research on a world-wide basis in the last decade and volumes of literature dealing with the various concepts are now available. Universities, Research Associations, and private companies are all contributing to this continuing research.

ADDITIONAL LITERATURE

L. COES JR. 'Abrasives,' *Applied Mineralogical Series No. 1.* Vienna Springer-Verlag, 1971.

M. C. SHAW, Ed. 'New Developments in Grinding,' Proceedings of the International Grinding Conference, 18–20 April 1970, Pittsburgh Penna, Pittsburgh Carnegie Press, 1972.

International Machine Tool Design and Research, Published annually, Pergamon.

Chapter 4. Adsorptive Materials *

IUPAC RECOMMENDED TERMINOLOGY

The classification of pore sizes given has been superseded by the following IUPAC (International Union of Pure & Applied Chemistry) recommendation (*Manual of Symbols and Terminology for Physicochemical Quantities and Units*; Appendix II, Definition, Terminology and Symbols in Colloid and Surface Chemistry; Part I. IUPAC, 1972):

 (i) Pores with widths exceeding about 50 nm are called *macropores*

(ii) Pores with widths not exceeding about 2 nm are called *micropores*

(iii) Pores of intermediate size (2–50 nm) are called *mesopores.*

ADDITIONAL LITERATURE

S. J. GREGG and K. S. W. SING. *Adsorption, Surface Area and Porosity*, London, Academic Press, 1967.

Chapter 6. Glass

AIR POLLUTION

Increasing attention has been given in recent years to the problem of air pollution caused by industrial emissions, and, particularly in the USA and Japan, effort is being directed to the measurement and control of material emitted into the atmosphere during melting of glass. These emissions may consist of one or more of the following:

1.   Sulphur dioxide produced by oxidation of the sulphur present in the fuel oil.

2.   Fine particles, generally in the size range 0.1–$1.0\,\mu$m, of non-toxic solid condensate, e.g. sodium sulphate from the melting of soda-lime glass, or sodium borate and boric oxide from borosilicate glass.

3.   Particles or vapour of toxic material, e.g. lead oxide from the melting of lead-containing glasses, and hydrofluoric acid from fluoride opal glasses.

Abatement equipment currently in use includes bag filters for dust or fume removal, electrostatic precipitators for removal of solid particles and liquid droplets, and scrubbing for cleaning of waste gas.

AUTOMATIC INSPECTION OF GLASS CONTAINERS

With increasing speeds of glass container manufacture (machines now in use will produce at over 300 bottles per minute) inspection of containers for faults has become an automated process, and in modern production lines the containers, after annealing, are passed through several inspection machines which automatically reject those faulty in the following respects:

(a) Incorrect diameter or height tolerance
(b) Cracks or imperfections in the sealing surface
(c) Cracks in the base
(d) Incorrect bore tolerance
(e) Foreign particles in the glass, or particles of glass inside the container
(f) Incorrect thickness distribution in glass sidewall
(g) Low strength in the glass sidewall

Most of these inspection machines operate on electro-optical principles, whereby a beam of light is transmitted or reflected in an abnormal manner when a faulty container is placed in the light path, and the deviant light is used to activate a bottle rejection device.

PLASTIC COATING OF GLASS CONTAINERS

The coating of glass bottles with a relatively thick layer of a thermoplastic material has been employed for some years, but in the very limited area of glass aerosol bottles. Recent developments in the technology of applying plastic coatings to glass have led to a substantial usage of coated glass containers, particularly for large-size carbonated soft-drink bottles. The protection against impact and surface damage afforded by such coatings means that thinner glass containers can be used, and the cost savings thereby engendered offset to a considerable extent the additional cost of applying the plastic coating.

Two types of coating process are presently in use:

1. The container is wrapped with a pre-formed decorated foam polystyrene sleeve, and subsequently the sleeve is heat-shrunk round the sidewall and bottom of the container.

2. The container is heated, electrostatically sprayed with a powdered thermoplastic polymer, and reheated to flow the plastic.

Plastic-coated containers produced by process (2) are resistant to shattering, in that when breakage occurs the glass fragments are substantially retained within the usually unbroken plastic film.

Chapter 8. Diamonds and Synthetic Gemstones

In 1970 a team of workers in the Laboratories of the General Electric Co. in Schenectady, NY, USA, succeeded in manufacturing for the first time diamonds of gem quality in sizes up to rather more than 1 carat in weight.

The best crystals were classified as equal in quality to white natural rough having very small flaws and inclusions. The crystals were tabular in habit, consisting of truncated octahedra with modified cube faces. Not only white stones but crystals of several colours, including blue and yellow, were produced. A distinguishing feature of these synthetic diamonds was their electrical conductivity–a property associated in nature only with blue diamonds of the 'Type IIb' variety.

The process involved sustained heating, under enormous pressures, of small synthetic diamonds in a bath of nickel or iron catalyst in a chamber in which the temperature was at its highest in the centre, the dissolved carbon crystallizing out on small diamond seed crystals placed at the cooler ends. The growth rate is about 100 hours per carat, and the cost of production far greater than the recovery of natural diamonds. One of the objectives of the project was to produce semi-conducting diamonds, and this was successfully accomplished.

In the field of synthetic gemstones the firm of Pierre Gilson, in the Pas de Calais, has achieved two more successes to add to the problems faced by the jewellery trade. The first of these is a synthetic turquoise which has the composition and crystal structure of natural turquoise, and nearly the same density and refractive index. In distinguishing these, a characteristic 'bland' appearance and a uniform granular structure seen under magnification are helpful features.

The second, and surprising success, is a synthetic opal resembling natural black opal very closely. Workers in Australia had already succeeded in producing tiny uniform silica spheres which the electron microscope had shown to be the basis for the play of colour in opal. But to fix these spheres into a permanent form which could be cut and polished as a gemstone proved to be an extremely difficult problem, which Gilson seems to have solved.

Apart from these synthetic stones, the precious stone trade has been plagued in recent years by ingenious composite stones or 'doublets' designed not only to deceive the eye but also to pass muster under the more simple forms of testing, especially when the specimens are mounted in jewellery. Examples may be quoted; doublets consisting of a crown of natural poor quality Australian sapphire cleverly cemented at the girdle to a base of synthetic sapphire or of synthetic ruby to provide the required colour; and a colourless Yttrium aluminate crown cemented to a base of strontium titanate. The latter provides a diamond-like stone in which the upper surface has a hardness equal to topaz while the dispersive quality of the base provides the necessary 'fire'.

The world production of gem diamonds and industrial diamonds was virtually unchanged in the years 1969, 1970 and 1971.

ADDITIONAL LITERATURE

E. BRUTON. *Diamonds*, London, N.A.G. Press Ltd., 1970.

H. BANK. *From the world of gemstones*, Innsbruck, Pinguin-Verlag, 1973.

G. F. HERBERT SMITH (revised F. C. PHILLIPS) *Gemstones*. London, Chapman and Hall, 1972. This is the revised edition of a previously recommended book.

Webster's *Gems*. A new edition was scheduled for publication in 1974.

Chapter 9. Solid Mineral Fuels[*]

Because of changes in the cost and availability of other forms of energy, more attention is now being paid to coal which in terms of known reserves vastly exceeds all other forms of fossil fuel. In particular, total gasification is once more arousing interest, and it seems likely that gasifiers such as the Koppers-Totzek may again become important, unless some more modern form of gasifier is developed to supersede them.

Underground gasification is another process which is being re-examined particularly in the USA. In the production of coke and gas from coal, the horizontal retort method has been largely superseded, except in one or two out-dated plants.

Coal gas is not at present used as an industrial and domestic fuel in the UK and many other European countries. Its place has been taken by natural gas, for example methane from the North Sea, or by reformed light petroleum spirit. However, the use of the latter is tending to decrease because of increased raw material costs. Gas from coal may be used as a fuel in the future, but it is unlikely to be made by a carbonization process, as understood in the past. Experiments were being conducted in 1974 on the manufacture of substitute natural gas from coal, using a conventional Lurgi plant.

Producer gas and water gas are no longer made in significant quantities, although once again, these processes may reappear in the future, perhaps with new designs of production equipment.

Readers interested in the general problem of fuel resources and relative importance are referred to Chapter 5, Volume 8, where this subject is discussed in detail.

VOLUME 3

Chapter 1. Metals in General

CONTINUOUS COOLING TRANSFORMATION (CCT) DIAGRAMS

Most industrial heat treatment processes involve continuous cooling rather than isothermal transformation and the application of isothermal (TTT) diagrams for such industrial processes has obvious disadvantages. A further aid to practical heat treatment is now found in the use of continuous cooling transformation (CCT) diagrams, a typical example of which is shown in Fig. A.1. The two axes of this diagram represent the temperature and diameter of an oil-quenched bar, and for any given bar diameter, the temperature at which transformation begins (0%) and ends (100%) as well as intermediate degrees of transformation (10%, 50%, 90%) may be ascertained. On the bottom axes there are three scales which enable the progress of transformation on cooling from an austenitizing temperature to be followed for the bar axis, the

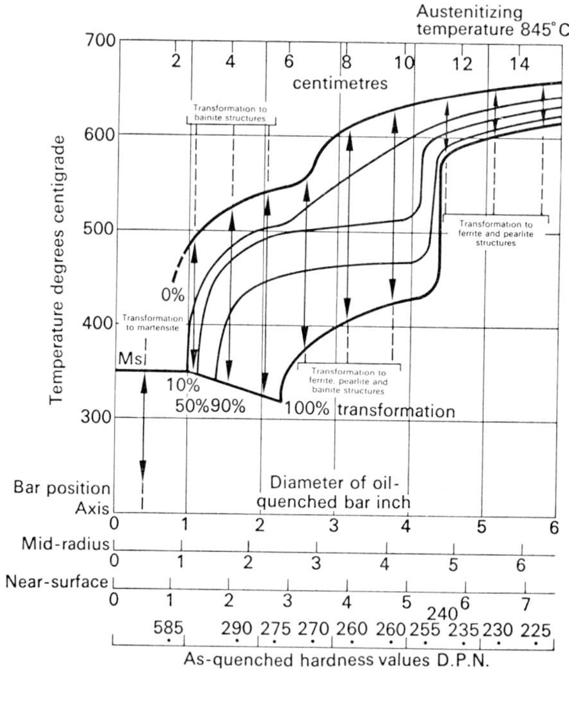

Structures present in as-quenched bars

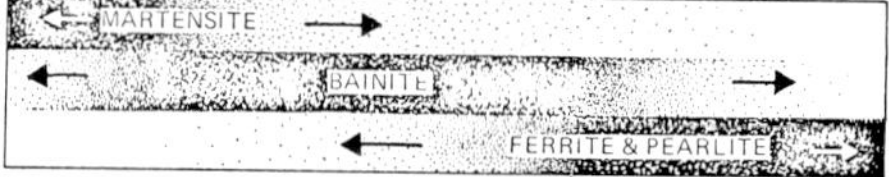

Fig. A1   Continuous Cooling Transformation diagram for BS En 111 steel.

mid-radius and the near surface position. The structures and hardness values which may be obtained are also usually included with CCT diagrams.

Thus a $2\frac{3}{4}$ in (69 mm) diameter bar at its axis will begin to transform at 590°C to a ferrite-pearlite structure. At lower temperatures the austenite that remains will transform to bainite and transformation will be complete at 375°C. In contrast the near-surface will transform over a temperature range from 550° to 320°C giving a predominantly bainitic structure and the as-quenched hardness will be approximately 275 DPN. The information obtained from CCT diagrams may be a useful guide as to the suitability of a particular steel for an application requiring specific properties and structure.

MARAGING STEELS

Special iron-based alloys capable of giving yield strengths up to 135 tonf/in^2 (210 kgf/mm^2) with excellent fracture toughness have been developed and are referred to as Maraging steels. The steels are so named by virtue of their martensitic structure in the annealed state plus the response to ageing that is achieved. Typical compositions are given in Table A.1.

The heat treatment necessary to develop the ultra-high strength properties depends on the alloy being processed. The 25 Ni steel would normally be in the soft austenitic state but after a suitable conditioning treatment at 650°C may be encouraged to transform to martensite on cooling to room temperature. Refrigeration of the steel at -70°C (i.e. a continuation of the quenching process) will further assist the transformation. Subsequent reheating of the steel to 400 to 500°C allows ageing to take place and strengths of 70 tonf/in^2 (110 kgf/mm^2) to develop. The 20 Ni steel does not require the conditioning treatment although the refrigeration part of the process is recommended. In both these steels the ageing effect is due to the precipitation of titanium-aluminium compounds assisted by the presence of niobium.

An 18% Nickel maraging steel is available without niobium but containing both cobalt and molybdenum. Both these elements assist in the ageing process the extent of which gives rise to the designation 18 Ni (280) 18 Ni (250) and 18 Ni (200) corresponding to yield strengths of 280, 250 and 200 ksi.

The 25%, 20% and 18% nickel maraging steels are normally produced in the wrought form but a similar range of steels has been formulated for the production of cast shapes.

Table A.1. Composition of nickel-maraging steels

Designa-tion	C	Mn	Si	S	P	Ni	Co	Mo	Ti	Al	Nb
	max	max	max	max	max						
25 Ni	0.03	0.10	0.10	0.01	0.01	25/26	—	—	1.5	0.2	0.4
20 Ni (280)	0.03	0.10	0.10	0.01	0.01	19/20	—	—	1.5	0.2	0.4
18 Ni (250)	0.03	0.10	0.10	0.01	0.01	18/19	9	5	0.6	0.10	
18 Ni (200)	0.03	0.10	0.10	0.01	0.01	17/19	8	5	0.4	0.10	

(From *ASM Metals Handbook*, Vol. 2, American Society of Metals, Ohio)

Table A.2. Summary of typical heat-treatment atmospheres; composition, uses and costs

Ref.	Atmosphere	Source	Approx. composition %						Typical uses	Advantages	Disadvantages	Approx. Cost per $1000\ m^3$ pence
			CO	CO_2	H_2	H_2O	N_2	CH_4				
1a	Exothermic, rich	Part burnt fuel gas	9–12	5–7	11–15	2–3	60–70	1–2	Bright anneal, brazing and soldering mild steel. Normalizing: Cu brazing and sintering.	Reducing, cheap	Just toxic & combustible Can cause sooting and decarb.	$1–1\frac{3}{4}$
1b	Exothermic, lean	Part burnt fuel gas	0–3	10–13	0–4	2–3	80–85	—	Bright and clean anneal Cu, Ni, Al, and brasses. Cu brazing.	Non-combustible cheap.	Sometimes traces O_2	$\frac{3}{4}–1\frac{1}{2}$
1c	Exo., rich, stripped	As 1a, but CO_2, and H_2O removed	10–13	—	12–15	—	70–75	1–2	Substitute for 2 for most purposes.	Fairly cheap, Non decarb.	Can cause sooting	1–2
1d	Exo., lean, stripped	As 1a, but CO_2, and H_2O removed	0–3	—	0–4	—	93–100	—	Bright anneal and hardening carbon alloy steels. Carrier gas for carbon restn.	Fairly cheap, Non-decarb:	Sometimes traces O_2. Can cause sooting.	1–2
1e	Exo., stripped modified.	As 1a, but CO_2, and H_2O removed	20–30	—	3–12	—	60–75	—	Low-C and mild steel long cycle bright anneal. Bright hardening. Carrier for carburizing.	Cheap, non-sooting. Reducing.	Just combustible; toxic.	$1\frac{1}{2}–2\frac{1}{2}$
2	Endothermic	Catalytic reaction. fuel gas/air.	20–25	—	30–45	<0.1	30–50	1.0	Bright anneal, clean hardening, brazing carbon and alloy steels. Carburizing carrier gas.	Strongly reducing Non-decarb:	Toxic; explosive. Can cause sooting. Good maintenance and control required.	$1\frac{1}{2}–3\frac{1}{2}$
3a	Cracked Ammonia	Catalytic cracking of ammonia.	—	—	75	<0.1	25	—	Bright and clean anneal, sintering and brazing of stainless and other high alloy steels, certain copper and nickel alloys, clean anneal of brass.	Pure, strongly reducing, cheaper than H_2.	Explosive, can cause nitriding.	$5\frac{1}{4}–7$
3b	Part burnt ammonia.	Catalytic oxidation of ammonia.	—	—	1–25	<1	Bal.	—	As for 3a.	Pure, cheaper and less explosive than 3a.		5–7
3c	Fully burnt ammonia.	As for 3b.	—	—	0–0.5	—	Bal.	—	Bright, anneal, hardening and sintering of special non-ferrous metals and steels. Anneal silicon steels.	Non-combustible. Purer than 1d.		$4\frac{1}{2}–5\frac{1}{2}$

4	Hydrogen	Trailer or bottles	—	—	99.9	—	—	—	As for 3a, but no nitriding effects.	Pure, low capital and maintenance	Costly and explosive.	$8\frac{3}{4}–17\frac{1}{2}$
5	Nitrogen	Bulk tank	—	—	—	—	99.9	—	As for 1d and 3c but more exacting applications.	Non-explosive, inert most metals, low capital and maintenance.	Fairly expensive.	$4\frac{1}{2}–7$
6	Argon	Bottles	—	—	—	—	—	—	Special bright treatments on reactive metals, titanium, Nimonics, special steels.	Inert and non-combustible. Low capital and maintenance.	Very expensive.	$35\frac{1}{2}–106$
7	Steam	Electric boiler	—	—	—	99.9	—	—	Tempering and blueing steels. Bright anneal copper.	Very cheap, non-combustible.	Condensation difficulties Risk of H_2 embrittlement.	$\frac{1}{2}–\frac{3}{4}$
8	Carbon Dioxide	Bulk tank	—	99.9	—	—	—	—	As for 7, also anneal magnesium alloys.	No risk H_2 embrittlement. Non-combustible.	Fairly expensive.	

*Dewpoint of wet gases may be reduced if required from room temperature to less than $-50°C$ by means of suitable drying agents.

CONTROLLED ATMOSPHERE

Modern technology has developed a wide range of gaseous atmospheres which may be used in heat-treatment furnaces, either to prevent surface oxidation of the metal being treated or to cause certain desirable chemical reactions to take place (as in the carburizing of steel). The atmospheres are commonly manufactured in generators close to the furnace and may be produced from town gas, natural gas, propane, butane ammonia, or from other gases in bulk or in bottles.

A summary of these atmospheres, their source, composition, use and approximate cost is given in Table A.2.

ADDITIONAL LITERATURE

Transformation Characteristics of Direct-Hardening Nickel Steels, London, International Nickel Co. Ltd.
ASM Metals Handbook, Ohio, American Society of Metals.

Chapter 3. Iron and Steel*

The scale of manufacturing operations has continued to increase. Ore ships of 170000 tonnes capacity and more* are now in use and in the UK the new terminals at Port Talbot and Redcar can take 100000 t ships and 130000 t ships respectively, while that being constructed at Hunterston on the Clyde will eventually be able to take 250000 t ships.

World expansion of rich ore production means that ores containing less than 55 to 60% of iron when shipped are declining in importance.

The blast furnace remains the main producer of iron. The world's largest furnaces now have hearth diameters of about 14 m (46 ft) producing 8000 to 10000 t per day; even larger furnaces are being built. Such high production rates require special forms of furnace-top instead of the traditional double bell, conveyor belt rather than hoist charging, and the provision of multiple tap holes. High top pressures of up to 2.5 bars are used. Coke consumption per tonne of iron continues to fall, partly because of continued improvement in burden preparation and the use of higher blast temperatures of up to 1200°C, and partly because some of the coke can be replaced by oil or natural gas injected at the tuyeres. Best world figures are now in the region of 400 kg. coke plus 40 kg. oil per tonne of iron.

Other iron-producing processes (p. 149) have not yet made serious inroads on the supremacy of the blast furnace because the latter remains the cheapest way of making iron on a really large scale. However, the usefulness of direct reduction for making iron in places where large-scale production is not required, or where specially cheap reducing gas is available, is increasingly recognized; world capacity for direct reduction

* Some of the giant ships designed to carry both oil and ore cargoes are able to carry up to 300000 t of ore, but present port limitations prevent their full loading as far as ore is concerned.

is now in the region of 5 m. t per year and is growing. Some of the processes listed on pp. 149–151 have dropped out and others have appeared. An example of a recent process now being operated on a commercial production scale at several works abroad is the Midrex process in which ore pellets are reduced in a shaft furnace with hot reducing gas moving upwards in counterflow to the descending ore. At present the various commercial processes are moving up or down in favour according to the local economics of the time, and obviously much depends on the relative prices of oil, gas, and solid reducing agents.

In the field of steel-making the air-blown Bessemer converter is almost obsolete. While much steel is still being made in open-hearth furnaces, this process is becoming obsolescent as well, though new methods of using oxygen such as the SIP process mentioned below may help to delay obsolescence in some situations. Practically all new steel-making plants use either the LD or similar processes† if molten iron is the main raw material, or the electric arc furnace if scrap or prereduced iron is the main material. Because most iron is now made from rich ore low in phosphorus the LDAC, Kaldo, and Rotor processes, which were primarily designed to use high-phosphorus iron, are declining in importance. Converters are increasing in size and modern units hold 250 to 300 t of steel.

Recent developments may herald a return to bottom blowing. As mentioned on p. 165, pure oxygen cannot be used in the traditional Bessemer converter because the intense heat generated destroys the bottom. The VLN process was one solution but was found uneconomic compared with the LD process and the only VLN plant in the UK, at Port Talbot, was replaced by a new LD shop in 1969. However, there is now a bottom-blown process in which double (concentric annular) tuyeres are used. Oxygen is blown through the central tube but is surrounded by a relatively small quantity of hydrocarbon gas (e.g. propane) blown through the outer annulus. Thermal decomposition of this gas cools and protects the tuyere region, but extra heat is thereby available in the body of the vessel, allowing more scrap to be used. This is one of the advantages claimed for the process, others are that capital costs are lower and there is better control of quality. This process is known as the OBM or LWS process on the Continent, and the Q-BOP process in the USA. The double tuyere can be built in to the bottom of tilting open-hearth furnaces to allow blowing of oxygen/hydrocarbon below the slag line. This is known as the SIP (submerged injection) process and because it speeds up production dramatically, is the subject of much interest by open-hearth operators anxious to prolong the economic life of their plants. There is one installation operating in Canada and others are planned.

In the arc furnace field the main tendencies are the increased use of big furnaces, higher powered transformers, and the use of prereduced pellets in arc furnaces, referred to on p. 190, is now well established.

There has been continued growth in vacuum degassing (p. 178). For example the whole ladle need not be placed in the vacuum, a small vacuum chamber can be positioned above the main ladle and portions of steel sucked up into it successively so as to treat the whole after a number of such operations. Such cheaper methods make vacuum degassing economic for a wider range of steels. Secondary refining processes are being increasingly used, particularly when the time required to make special qualities occupies an expensive steel furnace too long. Finishing the steel in a specially

† The terms BOP, BOS or BOF (basic oxygen-process, -steel-making, or -furnace, respectively) are now frequently used to describe oxygen-blown converter processes generally.

adapted ladle, sometimes fitted with electrodes for supplementary heating and with provision for gas injection and/or vacuum treatment, allows quicker release of the main furnace for the next cast. An example of this practice is the ASEA-SKF process. In new processes for making stainless steel the electric furnace is used for melting the scrap only, the steel being finished in a small 'converter' blown with argon and oxygen (the AOD process) or with steam and oxygen (the CLU process).

Continuous casting (p. 185) has made rapid progress and large plants are now in use. The most notable development is sequence casting (the current name for the 'continuous-continuous' casting referred to on p. 188) which is now well established; in October 1972 it was reported that 89 successive heats had been sequence-cast at a Japanese plant.

The spray steel-making process (p. 189) has been abandoned for the present because it was doubtful whether it could ever compete economically with the low costs of modern LD converters producing steel at the rate of 500 t/h or more. Commercial continuous steel-making still eludes us though good progress is said to have been made with a pilot plant in France. In this process molten iron is continuously fed to the base of an LD-type vessel, top blown with oxygen, and the steel produced overflows into a second vessel where slag is removed, then into a third where the final composition adjustments are made.

In the field of heat treatment mention should be made of 'controlled rolling', meaning rolling at a temperature controlled at a lower level than in normal rolling, thus giving a heat-treatment as well as a deforming effect. While this imposes a greater load on the rolling mill, and adds to cost, improved metallurgical and mechanical properties are given to the product. Some of the high-strength carbon steels, e.g. those used for high-duty pipelines, are now made in this way.

Chapter 4. Non-ferrous Metals

BERYLLIUM

The bertrandite deposits in the western USA have been developed and now provide about one-half of the world's beryllium, other than in the USSR. Electrolytic flake beryllium is being produced in the USA in both high-purity (electro-refined) and ultra-purity (double refined) grades. In France, beryllium metal is no longer manufactured, while production of beryllium-copper has been increased.

Important advances using high-purity beryllium have led to properties which are significantly improved over those reported previously. Improved ductilities are also available in conventional sheet products, as shown in the following table:

Typical mechanical properties of beryllium mill products

	Ultimate tensile strength p.s.i.	Yield strength p.s.i.	Elongation %
Powder metallurgy products			
Hot-pressed regular-purity block	48 000	35 000	2
Isopressed high-purity block (high ductility)	60 000	37 000	6
Isopressed high-purity block (high strength)	85 000	65 000	4
Sheet	77 000	55 000	15

MAGNESIUM*

Extraction. Most magnesium is now produced by electrolysis, and most of this comes from sea water. The important plants are in the USA and Norway. In a recent commercial development, brine is dehydrated to produce a mixture of anhydrous magnesium and alkaline earth chlorides suitable for electrolysis. This process produces chlorine as a saleable by-product and should thus lead to a reduction in the extraction cost of magnesium. A later development of the thermal process operates at higher temperatures and by incorporating bauxite in the charge, utilizes a fluid reaction mixture. The magnesium vapour is condensed directly to the liquid stage and then refined and cast into ingots.

Uses. Anodes are used for the protection of ships' hulls, piers and jetties; for buried pipe work e.g. gas, oil and water pipes and for domestic water storage systems. The diversity of uses of magnesium is further illustrated by its application as electrodes in the construction of primary electric batteries, for dosing cattle to avoid 'grass staggers', for the production of rigid diving suits, and for the desulphurization of blast furnace iron.

ALUMINIUM

In 1973, the Aluminum Company of America announced that it had applied for patents on the Alcoa smelting process, a new electrolytic method of producing primary aluminium. The new process is expected to reduce by as much as 30% the electricity required by the most efficient units of the Hall process presently used worldwide.

Alcoa said the new process uses a system free of undesirable emissions and affords a superior employee working environment. In addition to requiring less energy, the process is more tolerant of power interruptions than is the Hall process, and can accept power reductions during daily periods of peak demand by the public. Its total operating costs are expected to be lower. It also will permit plants to be located on smaller sites, with greater location flexibility. Alcoa stated that the smelting process has been tested in a full-scale developmental unit. It is now ready for the next step, which is to evaluate the process in a first unit of a commercial plant, which will take three to four years. While they have great hopes for the new process, it should be clearly understood that this does not signal any over-night obsolescence of established facilities for smelting aluminium.

Alcoa explained that, as in the Hall process, the new method employs alumina, the oxide of aluminium, which currently is refined from bauxite ore. Alumina obtained from ores other than bauxite will be equally suitable for the new process.

In the Alcoa smelting process, alumina is combined with chlorine in a reactor unit which chemically converts the oxide to aluminium chloride. The chloride is electrolytically processed in a completely enclosed cell which separates the compound into molten aluminium and chlorine. The latter is continuously recycled back to the reactor in a closed loop.

In the traditional Hall electrolytic process, alumina is dissolved in molten cryolite, a fluoride chemical which has become increasingly scarce and costly. Alcoa's new process not only disposes of the need for cryolite, but also the expense of containing fluoride emissions.

GERMANIUM

Coal is probably now the most important source of this element.

TIN

The world production and consumption of tin in 1972 was close to 200 000 tonnes. The growth rate over the last 10 years for production averaged about $3\frac{1}{2}\%$ p.a., while the consumption has increased only by 1.2% p.a. The largest consumer (United States) has decreased its consumption but world usage has increased because the expanding packaging industry has led to increased consumption of tin as tinplate. Countries not mentioned previously whose consumption exceeds 2 000 tonnes p.a. are Poland (4 215), Spain (3 206), Czechoslovakia (3 500), and Rumania (2 860)–these figures being for 1972.

Changes in the canning industry have led to increased usage of differentially coated tinplate with the thicker layer on the inside for corrosion protection, while there is a trend to use pure tin as the solder for the side seam to reduce lead pick-up of the contents. A high proportion of beer and beverage cans in the USA are now made from lacquered chromium-plated steel with a glued seam, instead of from tinplate. Tinplate has regained some markets by introduction of the drawn and wall-ironed can where the base and body are integral and elimination of the side seam allows complete decoration of the exterior.

Laboratory research has led to some usage of tin powder as an aid to the production of sintered iron components for the automotive and other industries, while a stronger tin-rich bearing metal chromium has become available.

The last few years have also seen increased use of a plated, bright tin coating, partly as a substitute for cadmium plating.

LEAD

World consumption of refined lead in 1973 was estimated at $3\frac{1}{2}$ million tons, excluding the USSR and allied countries which are estimated to consume another 1 million tons. About a quarter of the world total is recovered from scrap, although in the UK and USA the proportion is nearer a half. Japan should be added to the list of major importing countries.

BISMUTH

The residues from electrolytic copper refining cells and slags from non-ferrous smelting generally provide enough bismuth to meet current demands.

COBALT

High-temperature materials. The demand for enhanced corrosion resistance because of the range of fuels consumed and environmental conditions (e.g. marine) of operation of gas-turbines, has stimulated greater interest in the application of cobalt–high-chromium base alloys. The incorporation in this base of elements such as tantalum, aluminium and rare earth metals has led to alloys with significantly improved corrosion properties. One of these alloys is also being used successfully as a coating alloy.

A simpler cobalt–chromium–iron alloy is finding increased usage in metallurgical furnace furniture, chemical processing plant and cement works.

Magnetic materials. A figure of merit which can be used for the comparison of permanent magnets is that of energy product. The conventional cast Alnico-type magnets typically have an energy product range of 2 to 10 MGOe. The development of cobalt–rare-earth compound magnets, typically Co_5Sm, yield materials with energy products of the order of 20 MGOe. These very powerful magnet materials will obviously find many applications where the advantages of power availability and small section size are important.

Wear-resistant alloys. Composite electrodeposits of chromium carbide dispersed in a cobalt matrix provide a surface coating technique which has found successful applications at working temperatures, up to, and in excess of 750°C. Other new compositions available both in massive form and as powders for spraying are demonstrating exceptional low friction and wear properties under conditions of boundary lubrication.

Tool steels and cutting alloys. Progress is being made with new iron-cobalt base compositions processed by powder metallurgical techniques which offer substantial advantages over conventional tool steels for cutting. For die applications particularly where advanced equipment is unable to achieve economical operation because of die-life limitation, cobalt base alloys are demonstrating substantial improvements.

Vanadium pentoxide is used as a catalyst in the chemical industry, for example in sulphuric acid manufacture and in polymerisation reactions.

NICKEL

The largest single market for nickel is for special steels and, since 1969, changes in steel-making processes have enabled the producers to use less-pure grades of nickel such as nickel oxide sinters and ferro nickels. This major change in demand has affected the market for refined nickel and has caused the traditional suppliers in Canada and New Caledonia to increase their range of products. Other sources of supply of these grades of nickel have been stimulated; for example, the Dominican Republic and Greece are now major producers and there are plans for production in Australia, Guatemala, Indonesia and in other areas of New Caledonia.

CHROMIUM

In the two-stage Perrin process for the production of low carbon ferrochromium, a charge of chromite, silica and coke is fed to a submerged arc furnace, producing an iron-chromium-silicon alloy. This is cast and broken into pieces; it is then used to reduce a chromite-lime slag prepared in a similar arc furnace. In this second stage, silicon in the alloy acts as the reducing agent to give a low silicon, low carbon ferrochromium suitable for the manufacture of stainless steel.

The exothermic reaction for reducing chromite to chromium using silicon is:

$$Cr_2O_3 + 3\,Si = 4\,Cr + 3\,SiO_2$$

MOLYBDENUM

Production. The molybdenum oxide plant of the Climax Molybdenum Company is no longer used. Western-World production of molybdenum in concentrate form in 1972 was 163 million lb.

Uses. Molybdenum is also used in inhibiting pigments and corrosion inhibitors.

ADDITIONAL LITERATURE

E. F. EMLEY. Principles of Magnesium Technology, London, Pergamon, 1956.
Information on the use of molybdenum base alloys can be obtained from a paper by J. E. Briggs and R. Q. Barr (1971) in *High Temperatures – High Pressures*, Vol. 3, pp. 363–409.

Chapter 5. Metal Casting

LOW-PRESSURE DIE CASTING

In addition to the well-established gravity die casting and pressure die casting processes, a third process, low-pressure die casting, has grown in popularity in recent years.

In many ways, the low-pressure die casting technique is applicable for castings which fall in between the typical gravity die cast and the typical pressure die cast product. At present the application of the process is largely restricted to aluminium-base alloys.

The following advantages are claimed for the process:

1. Higher productivity and metal yield compared to similar castings made by the gravity die casting process.

2. When the die has been designed properly to ensure progressive solidification towards the ingate, proper feeding of the shrinkage areas can take place during solidification, resulting in denser castings, free from shrinkage cavity.

3. Lower capital cost of plant, compared to pressure die casting; the less complicated equipment is also easier to maintain.

4. Lower die costs, compared to equivalent castings made by the pressure die casting method. Dies for low-pressure die castings are usually made of cast iron.

5. Sand cores (usually resin-bonded) can be used for low-pressure die casting but not for pressure die casting.

6. Low-pressure die cast products can be heat-treated to improve the mechanical properties, whereas this is not generally practicable for the alloys used in pressure die castings.

The molten alloy is held in a crucible made of silicon carbide or cast iron. The crucible is contained within a hermetically-sealed pressure vessel. Electrical heating elements are also provided.

An air pressure of 3 to 10 lbf/in^2 (20–68 KNm^2), introduced into the pressure vessel, forces the molten metal upwards through a vertically mounted hollow tube (stalk), into a die which is clamped to the top of the furnace vessel. The air pressure is released only after the casting (in the die) has solidified. The molten metal in the stalk falls back to its level in the crucible. (A schematic diagram of the set-up is shown in Fig. A.2)

The solid casting is now removed from the die, and after closing, the cycle is repeated. Manually-operated, semi-automatic or fully-automatic plants are in use.

The important point to note is that the cooling conditions in the die must be so arranged that when the casting solidifies, it does so in a sequence starting at the extremities of the die cavity and progressively towards the ingate region at the top of the stalk. The ingate region is much larger than those used for other methods of casting, since this (ingate) is also the reservoir for feed metal to compensate for solidification shrinkage. For this reason, heating and/or cooling arrangements may need to be incorporated in the die assembly to ensure progressive (directional) solidification, as illustrated in Fig. A.3.

One of the disadvantages of the earlier machines used for low-pressure die casting, was that the molten metal could not be treated (for degassing, grain refinement, etc) in

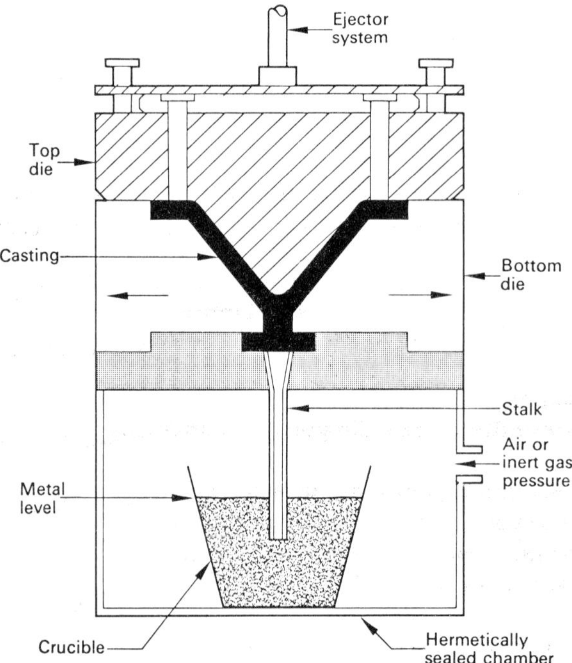

Fig. A2   General arrangements for low-pressure die casting.

the crucible within the chamber of the machine. In the newer machines, it is possible to treat the metal in the crucible within the chamber. Instead of air pressure, it is also possible to use an inert gas pressure (e.g. argon) to push the molten metal upwards into the die. This is useful in preventing oxidation and deterioration of molten metals, like magnesium alloys used to cast road wheels for motor cars and components for the aircraft industry.

THE V-PROCESS

It has been known since 1972 that the cast metals industry in Japan has developed a casting process, called the V-process, which eliminates bonding agents and other additives from the sand used in making the moulds. This is achieved by supporting dry unbonded sand in the mould box with the help of vacuum suction. The vacuum system remains in operation until molten metal has been poured into the mould and the casting has solidified. After solidification, the vacuum is released and the free-flowing sand drops out of the mould box together with the clean casting. The sequence of operation may be summarized as follows (see Fig. A.4):

(a) An assembled pattern plate, containing a large number of vent holes, is placed on a suction box, and the surface covered with a thin (0.05 to 0.10 mm) sheet of plastic.

(b) Heat is applied to the plastic sheet so that it softens and drapes over the pattern surface as a tightly adhering film (due to suction).

(c) A moulding box, also containing vent holes, is now set on the plastic-film

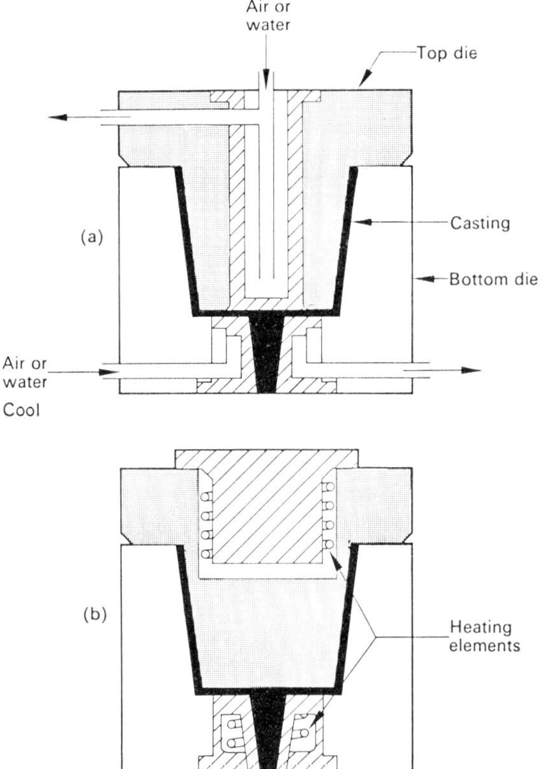

Fig. A3 Progressive solidification may be aided by inserts for cooling (a), or heating arrangements (b) in the die.

covered pattern, and filled with dry free-flowing unbonded sand. The assembly is vibrated during filling and a plastic sheet placed over the sand-filled box.

(d) Vacuum suction is applied to the sand-filled box through the vent holes and the vacuum maintained (until after pouring and solidification of the casting).

(e) The vacuum in the pattern is now released and the mould stripped from the pattern (maintaining vacuum in the mould-box).

(f) The other half (cope or drag) of the mould is prepared in a similar manner, cores, etc., set, and the two mould halves closed to complete the mould cavity (maintaining vacuum in each half).

(g) The molten metal is now poured into the mould cavity and the casting allowed to solidify.

(h) The vacuum is released after solidification of the casting and the sand and castings are allowed to drop out of the mould-box.

It is worth recording that the concept of using unbonded moulding media is by no means new. Efforts have been made in the past, for example, to support the moulding material in the flask within a magnetic field. However, all such systems necessitate the use of moulding media with special properties (e.g. magnetic properties in this case), and are, therefore, limited in application. The vacuum support system is not restricted in this way.

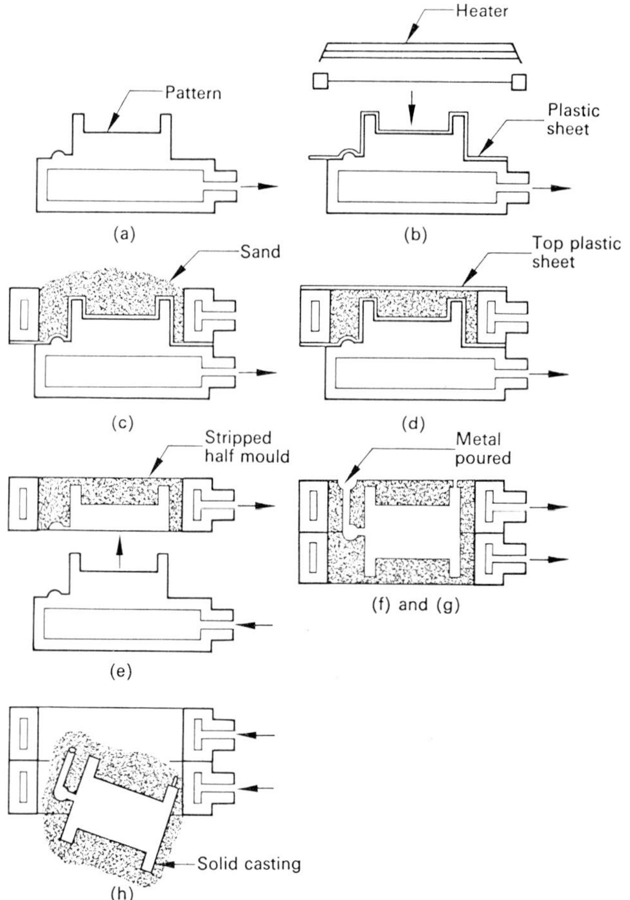

Fig. A4 V-process.

The implications of this development are far-reaching and may result in many improvements. For example, it can lead to the virtual elimination of the large, and often complex, sand plants used in foundries at present. Also plant and tackle for vibratory or other knock-out systems may not be required. Since dry sand (or other granular refractory material) is free-flowing, moulding machines used for packing and compaction of the sand would become redundant. The bonding agents and other additives present in conventional 'sand mixes' can give rise to fusion and other metal/mould interface reactions. Their elimination may, thus, lead to a reduction in fettling and cleaning operations. Furthermore, remarkable improvements may take place in the in-plant environment due to the possible elimination of:

Additives to the sand, such as coal dust;

Noise in foundries from sand plants, moulding machines and knock-out systems;

Impact/vibration problems from machinery such as jolters;

Contamination from organic and inorganic chemicals and clay substances;

Fumes and mould gases resulting from binders and additives when the metal is poured.

Although it is known that at least three Japanese foundries have used the V-process for the commercial production of castings, the published account of the process is limited to superficial descriptions of the concept, and authoritative information on the scientific, technological, economical and sociological aspects of the process is not yet fully documented. Both ferrous and non-ferrous metals can be successfully cast by the V-process.

LITERATURE

Publications of the American Foundrymens' Society. Des Moines, Illinois, U.S.A.
Cast Metals Handbook.
Foundry Sand Handbook.
Industrial Engineering in the Foundry.
Moulding Methods and Materials.
Shell Process Foundry Practice.
Symposium on Solidification.
The Cupola and its Operation.
Design of Ferrous Castings,
Copper-base Alloys Foundry Practice.
Design of Die Castings.
Investment Casting Handbook.
Pattern Makers Manual.
Engineering Manual for Control of In-Plant Environment in Foundries.
Gating and Feeding of Light-Metal Castings.
Gating of Die Castings.
Principles of Metal Casting.
Advanced Cost Accounting Methods for Gray Iron Castings.
Basic Cost Principles with Charts for Accounts (Non-Ferrous Founders Society).

A. M. KOROLKOV. *Casting Properties of Metals and Alloys* (Translated from Russian), New York, Consultants Bureau.

G. LAMBERT (Ed.). *Typical Microstructures of Cast Metals.* Institute of British Foundrymen, London, 1966.

R. A. FLINN. *Fundamentals of Metal Casting,* Addison-Wesley, University of Michigan, 1963.

A. B. CHELYUSTKIN. *Application of Computing Techniques to Automatic Control Systems in Metallurgical Plant* (translated from Russian by D. P. Barrett), Oxford, Pergamon, 1964.

J. P. VIDOSIC. *Computer Applications in Metallurgical Engineering,* New York, Ronald Press.

BENJAMIN W. NIEBEL and ALAN B. DRAPER. *Product Design and Process Engineering,* New York, McGraw-Hill, 1974.

Chapter 6. Metal Deformation

THERMO-MECHANICAL TREATMENT

Conventionally, properties of metals and alloys are developed by control of the individual processes of hot working, cold working and heat treatment. In recent years thermo-mechanical treatments generally involving deformation between the normal hot- and cold-working ranges have been applied to steels. A representation of the various treatments is shown in Fig. A.5. (*a–d*). Low-temperature mechanical treatment (LMT) or ausforming, shown in (*a*), involves heating of the steel into the austenitic range. The steel is then cooled to a temperature allowing metastable austenite to exist for a limited period where it is mechanically worked to effect deformation before final quenching. Because of the need for a bay of austenite stability the treatment can normally only be applied to alloy steels where substantial improvements in strength properties are claimed.

The deformation of stable austenite by high temperature mechanical treatment (HTMT) is shown in (*b*). Improvements in strength and ductility are claimed for certain alloys. It is thought that the benefits are attributable to the rate of recrystallization of the deformed austenite and/or the precipitation of alloy carbides during the mechanical treatment.

The mechanical treatment of steel during the ferrite-pearlite transformation is referred to as isoforming (*c*). Some slight increase in strength may be obtained but the major benefit is that of increased toughness.

A fourth treatment involving the warm working of steel is shown in (*d*). This may be applied to quenched and tempered steels, to steels previously subjected to the LTMT process or to ferrite-pearlite structures. The warm working temperature is generally between 150°C and 300°C and increased strength may be obtained, normally, however, with a loss of toughness and ductility.

SHAPE MEASUREMENT AND CONTROL

Variations in the roll separating force during the rolling of flat products will lead to a greater or lesser amount of roll bending than has been allowed for in the camber on the rolls. When this takes place the sheet or strip would tend to have an over-rolled middle or edge and would be described as having bad shape.

Improved control over the rolling process is now possible with the development of shape meters. A British design is shown in Fig. A.6. This is a 'Vidimon' visual reading meter and basically it comprises a series of separate air-bearing rotors on a shaft, forming a composite roll which is maintained in contact with the tensioned strip. Any variation in force applied to any roller segment will affect the pressure distribution in the air film and the changes in air pressure are sensed by pneumatic points on each segment and a signal transmitted for oscilloscope display. (Fig. A.7.)

A shape meter developed in Japan specifically for hot rolling is shown in Fig. A.8. The meter operates on a non-contact principle, employing an optical system in which the image of a specially designed cylinder-type incandescent lamp, reflected from the surface of rolled steel plate, is monitored by a special TV camera to detect any variation from flatness. The results are fed into the mill control system so that the necessary correction may be rapidly made.

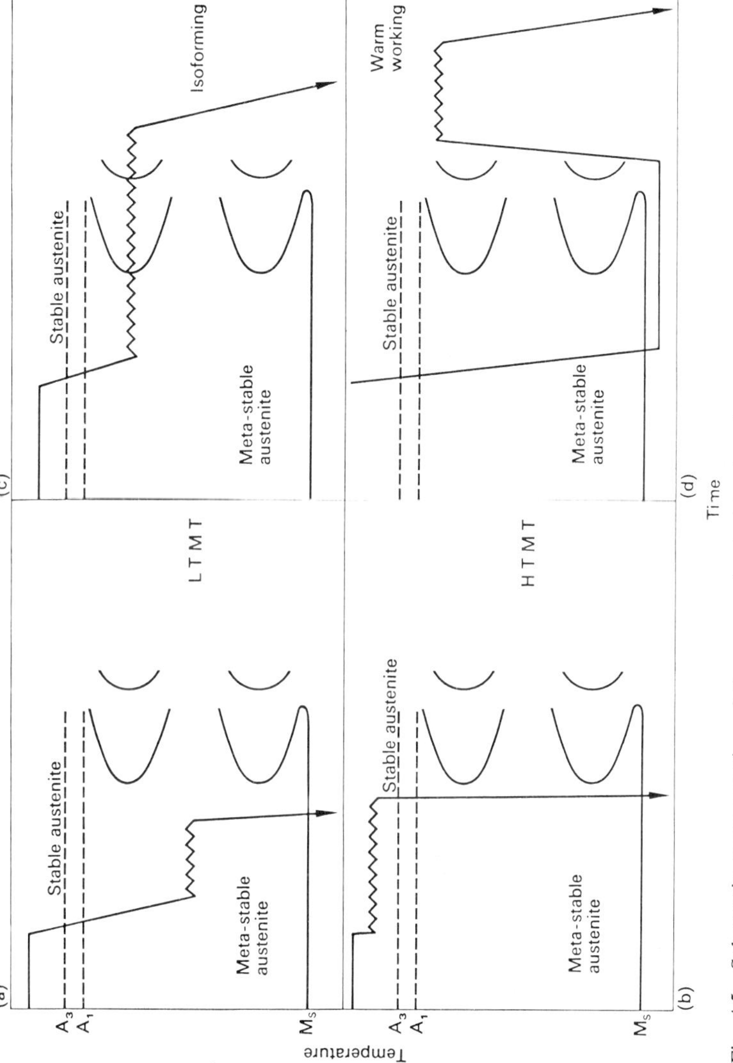

Fig. A5   Schematic representation of thermo-mechanical treatments.

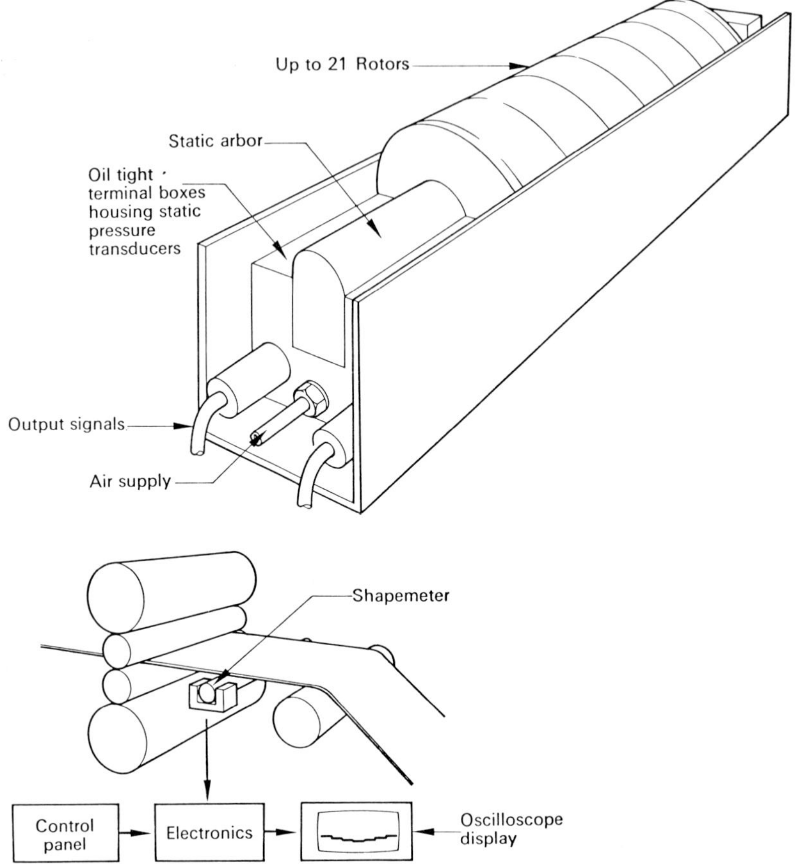

Fig. A6   General view and installation of 'Vidimon' shape meter.

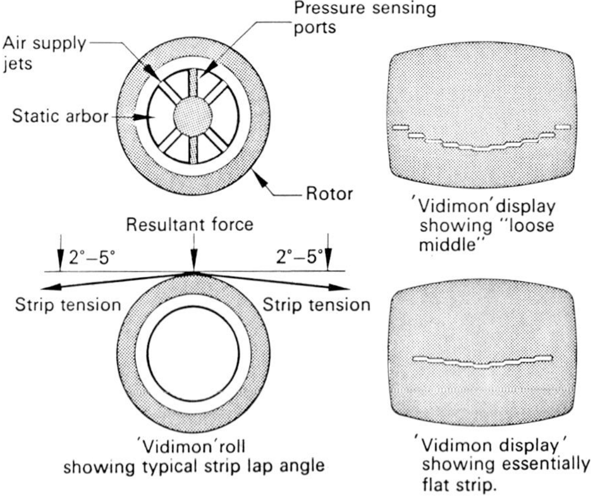

Fig. A7   Basic concept of 'Vidimon' visual shape meter.

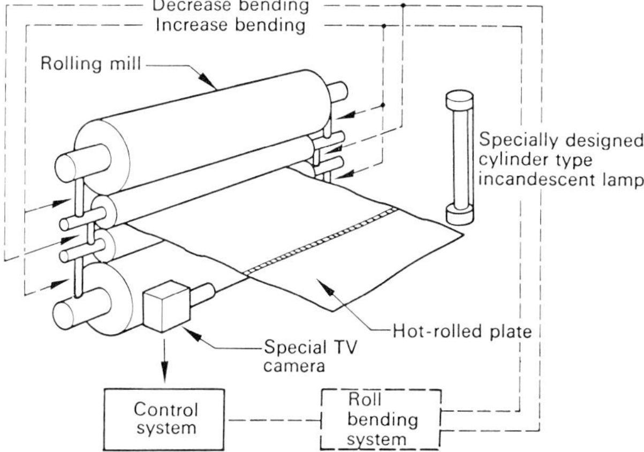

Fig. A8 Diagrammatic arrangement of a non-contact shape meter developed for hot rolling control.

Chapter 7. Working of Metals by Cutting

ELECTRICAL DISCHARGE MACHINING (EDM)

This process, commonly known as *spark erosion*, depends for its operation on the erosion of the surface of the workpiece caused by the rapid discharge of small electric arcs between a conducting tool and the work. Local temperatures as high as 10000°C are created, melting surface particles which are then carried away by the electrolyte surrounding the working area.

The tool can be made of any conducting material, copper and brass being the most easily formed, though where longer life is required and higher temperatures are being generated tungsten alloys and sometimes graphite tools are used. By connecting the work to the positive pole and the tool to the negative pole of the spark generator, it is found that very little wear occurs at the tool in comparison with the volume of metal removed from the work.

Two alternative methods have been used to generate the high-frequency spark required:

(*a*) relaxation circuit, by charging and discharging a condenser; or

(*b*) pulse generator circuit using solid state transistor devices.

The latter gives a better control of rate of spark discharge (which may lie between 1000 and 100000/sec according to application) and also largely eliminates tool wear. Intensity of spark is assisted by immersion of the process in a dielectric fluid, paraffin being normally used for this purpose.

Metal removal rates are relatively low, 0.2 in³/min (50 mm³/s) being typical, though higher rates can be achieved in a 'roughing' cut, this being followed by a finishing cut to produce the best accuracy and surface finish. Tolerances as good as 0.001 in (0.025 mm) and surface finish of 10 μin (0.25 μm) can be met.

The capital cost of EDM machines is moderate, and they are widely employed for shaping dies, moulds, and small press tools. Because no force has to be applied, they also provide a method of drilling very fine holes in any conducting materials.

ELECTRO-CHEMICAL MACHINING (ECM)

This process operates on the electrolysis principle, material being removed from the workpiece by a reverse electro-plating method (Fig. A.9). A low voltage, heavy d.c. current is passed between the work (connected to the anode) and the tool (connected to the cathode). Both are immersed in an electrolyte which serves to remove metal from the anode (in the form of charged 'ions'), and carry it to the cathode. To prevent these particles being deposited and causing build-up on the cathode tool, the electrolyte is pumped at high speed through the space between tool and work.

Copper is the most frequently used tool material, though brass and sometimes stainless steel are used. Salt solutions are required as electrolyte, 10% sodium chloride

Fig. A9   An electrical discharge machine with a 30 amp generator.

brine being the cheapest available and satisfactory unless corrosion is a major problem.

Metal removal rate is proportional to current density, and can reach 1 in^3/min (250 mm^3/s) in machines capable of passing 10 000 amps.

Accuracy of form is to about 0.002 in (0.05 mm) and surface finish as good as 6 μin (0.15 μm) can be achieved.

ECM machines are more expensive than EDM machines, but metal removal rate is higher and tool costs lower due to the absence of tool wear.

Applications include the shaping of gas turbine blades and nozzles, and forming and die-sinking operations.

ELECTROLYTIC GRINDING

Similar in principle to ECM, but using an abrasive grinding wheel as a cathode. The wheel is rotated at high speed as in conventional grinding processes, but very little actual contact occurs between the wheel and the work. Electrolyte is sprayed continuously between the two, acting as a coolant and particle remover. Up to 90% of metal is removed by electrolytic action, the remainder by abrasive action. Quicker than conventional grinding, this process avoids the formation of heat and stress cracking in the work surface. Used mainly for grinding tungsten carbide tools and other hard materials.

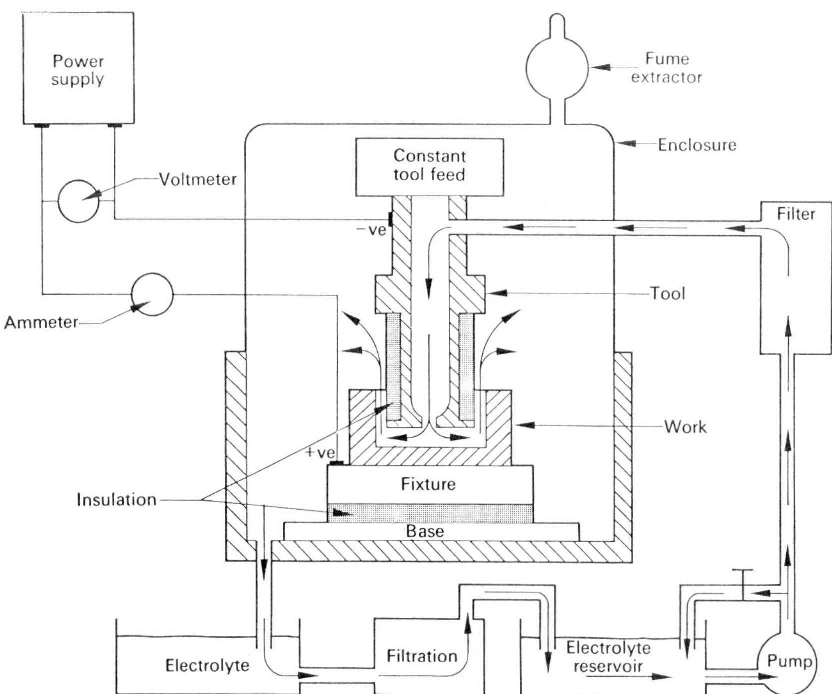

Fig. A10 Schematic diagram of ECM process.

the size of a compacting press can be drastically reduced, with a concomitant reduction in costs. The high-speed operation overcomes some of the problems of non-uniform compaction inherent in powder pressing, and may contribute to better bonding of powder particles in the green state, judging by the measurements of electrical resistivity carried out on actual samples. A large compact can be produced successfully on a small machine, by this process.

Composite materials. The increasing use of composites containing metal artefacts has demonstrated a new use for metal powder. The incorporation of powder allows more homogeneous mixing of various components of the composite, especially when hot pressing is used for compaction. The properties of fibre reinforced materials using metal 'whiskers' as fibres have improved by the use of powder techniques, and may have considerable development potential.

Chapter 11. Corrosion of Metals*

Two methods for measurement of rate of corrosion in the field have recently come into prominence. The measurement of polarization resistance, already used as a laboratory technique, has been developed so that it can be carried out on site with an instrument known as the Waverley Corrosion Monitor which operates with a pulsed square-wave current. It has been used for the monitoring of inhibitor injection in seawater pipe systems, for the monitoring of boiler corrosion caused by acid cleaning and for monitoring cables laid on the sea bed. The method, however, only provides information on general corrosion and will not provide quantitative measurements of pitting attack. The Magna Corrosometer is a portable instrument in which a probe element of the metal under investigation is exposed to the corrosive environment and its electrical resistance measured with reference to a control element of the same metal. These instruments are used to monitor pipelines carrying liquids or gases and tanks.

The ASTM have issued a series of standards relating to corrosion research:

G1-67 Preparing, clearing and evaluating test specimens.

G3-68 Conditions applicable to electrochemical measurements.

G4-68 Conducting plant corrosion tests.

G5-71 A standard reference method for making potentiostatic and potentiodynamic anodic polarization measurements.

G16-71 Applying statistics to analysis of corrosion data.

ADDITIONAL LITERATURE

'Report of the Committee on Corrosion and Protection', Dept. of Trade and Industry (UK), London HMSO, 1971.

J. O'M. BOCKRIS and A. M. N. REDDY. *Modern Electrochemistry* (2 vol.), London, Macdonald, 1970.

Code of Practice for Cathodic Protection BS: CP1021: 1973, London, British Standards Institution, 1973.

T. P. HOAR. 'The production and breakdown of the passivity of metals', *Proc. 3rd International Congress on Metallic Corrosion*, Vol. 1 English language edition, Moscow, 1966.

Chapter 3. Tar and Pitch

Coke ovens. In the past, the dimensions of a single coke oven chamber have generally been about $4\frac{1}{2}$ m in height, $10\frac{1}{2}$ to 12 m in length and 400 and 500 mm in width, and very many such ovens are still in operation. However, there is a 'new generation' of large-capacity ovens with heights up to 7 m and lengths up to 16 m. The common long walls separating one chamber from its neighbours enclose vertical flues in which gas is burnt for heating the coal charge (up to 20 t per chamber in conventional ovens and 35 t in large-capacity ovens).

Naphthalene. Purification of hot-pressed or phthalic-anhydride-grade naphthalene to material suitable for chemical synthesis or mothball production is achieved either by chemical means or by multistage fractional crystallization. This latter method, employing Schildknecht rotating helix columns or the Brodie crystallizer developed by Union Carbide (Australia), is a recent development but is now employed by the largest UK producer of pure naphthalene. The chemical purification method generally involves contacting the molten naphthalene with concentrated sulphuric acid (frequently and desirably preceded by a wash with formaldehyde), followed by neutralization, water washing and distillation. For the highest-purity material, the distilled naphthalene is sublimed.

Briquetting pitch. Until fairly recently, one of the major outlets for coal-tar pitch, particularly in the European export market, was for briquetting coal fines. Powdered medium–soft pitch (8–12%) is mixed with the pulverized coal and heated to about 90°C in a pug mill. The plastic mass from the mill is shaped into rectangular briquettes by a ram press or into ovoids on a continuous roll press. Legislation to prevent atmospheric pollution and the increasing use of diesel and electric locomotives by railways have reduced the importance of this outlet very considerably and little pitch is now sold in the UK or exported for the production of uncarbonized coal briquettes. In 1973, however, 75 000 t were used in the UK for the production of 'Phurnacite' smokeless fuel. This takes the form of pitch-bound ovoids which have been subjected to partial carbonization to drive off volatile matter which would otherwise cause smoke on burning.

Electrodes. The increased demand for electrodes, particularly by the aluminium industry, raised the output of electrode binding pitch from 30 000 t in 1969 to 66 000 t in 1973.

Road tar. The two main uses for tar-based binders are for the maintenance of roads by surface dressing and for the production of precoated macadams and asphalts used as wearing courses and base courses in new road construction. The use of straight road tar of 38° to 46°C e.v.t. for surface dressing declined from 100 000 t in 1969 to 40 000 t in 1973 but the use of tar/bitumen blends (40% tar/60% bitumen) for this purpose increased to 42 000 t. An increasing use for pitch is in pitch/bitumen blends (20–25%

pitch), which are now widely used as binders for non-skid hot-rolled asphalt surfacings on motorways and other heavily trafficked roads; 30 000 t of pitch were used for this purpose in 1973. Tarmacadam wearing courses and base courses are constructed as either open-, medium- or dense-textured carpets by selecting suitable size gradings for the stone aggregate and varying the amount of filler added. The binder viscosity for dense-tar surfacing is normally in the 55° to 60°C e.v.t. range and for open-textured material in 34° to 42°C e.v.t. range. In 1973, 135 000 tonnes of road tar were used for macadams.

Chapter 4. Natural Gas

Estimates of 'proved' natural gas reserves are subject to many uncertainties, not the least of which is the optimum gas pressure in a field down to which it is economic to operate the field. Furthermore, in countries with major oil fields many non-associated gas fields have not been evaluated since sufficient natural gas is available to these countries from gas associated with the oil fields. However, such figures as have been published suggest that total proved reserves at the end of 1970 were about 41×10^{12} cubic metres.

Attempts have also been made to make estimates of undiscovered or predicted reserves to arrive at the total world hydrocarbon reserves, and the United States Geological Survey has given a figure for the theoretical total recoverable natural gas existing in the world today of the order of ten times the present proved reserves. Current consumption of natural gas throughout the world is about 1×10^{12} cubic metres per year, and this may well rise to over 1.8×10^{12} cubic metres by the 1980s, which would probably then represent about 25% of the total world primary energy.

ADDITIONAL LITERATURE

D. SCOTT WILSON. *North Sea Heritage. The Story of Britain: Natural Gas*, London, The Gas Council, 1972.

E. N. TIRATSOO. *Natural Gas: A study*, 2nd edn., Scientific Press 1972.

Chapter 5. Hydrocarbons★

UNSATURATED HYDROCARBONS

Alkenes are now made almost exclusively by the thermal cracking of hydrocarbon feedstocks. Current practice in Western Europe and Japan favours the use of light naphtha which until recently has been freely available at low cost. Severe conditions are used to maximize the yield of ethylene which is the most valuable product. In a typical installation a 2:1 light naphtha–steam mixture at one atmosphere pressure is passed through a vertical tubular reactor maintained at 800° to 900°C. The steam acts as a diluent so that a low partial pressure of reactant–favourable to high yields of products–may be combined with a high flow rate; the residence time in the reactor must be no more than 0.5 s if decomposition to carbon and hydrogen is to be minimized. The hot product stream is rapidly cooled and quenched with water and oil to stop secondary reactions. Distillation of the volatile products at atmospheric pressure separates liquid hydrocarbons from the gaseous compounds which are then dried, compressed and fractionated at high pressure and low temperature. The product mixture is complex: 32–35% by weight of ethylene, 13–17% propylene, 4–6% butadiene, 3–5% mixed butenes, 18–20% each of residual gas (mainly hydrogen and methane) and of C_5–C_{10} hydrocarbons, and 1–3% fuel oil–and the economics of the process turn to a considerable extent on finding markets for all the materials formed. Very large crackers are favoured in order to realize the very substantial economies of scale which are possible: most of those recently commissioned or currently under construction have annual ethylene capacities of 300 000 to 500 000 tons.

In the USA light naphtha is comparatively expensive because it is much in demand as a gasoline constituent. Most ethylene is therefore made by the steam cracking of ethane obtained from natural gas; this is a comparatively straightforward process which yields 75 to 85% ethylene and only small amounts of higher alkenes. Propylene, the butenes and butadiene are made by cracking or dehydrogenating natural gas propane and butane or are obtained as by-products of catalytic cracking. The recent dramatic rise in the price of crude oil, and therefore naphtha, is bound to stimulate European interest in cracking processes based on natural gas, although North Sea gas is unfortunately very largely composed of methane and contains only about 4% by volume of higher hydrocarbons. Another alternative to naphtha is the use of heavy feedstocks such as gas oil, fuel oil or even crude oil itself. Such materials are considerably cheaper than naphtha but give relatively low yields of ethylene compared to other products. Very severe cracking conditions are necessary and the capital costs suggested appear dismayingly high.

Ethylene. This continues to be the most important alkene and indeed the most important single organic compound. Excluding the Eastern bloc, world production exceeded 18 m. tons in 1971 and was then expected to approach 30 m. tons by 1975. Polyethylene remains the major outlet for ethylene but the manufacture of vinyl chloride by the oxychlorination route is of rapidly increasing importance. Vinyl acetate is now produced on a substantial scale by the reaction of ethylene, oxygen and acetic acid at 175° to 200°C over a palladium catalyst. The development of these processes has enabled ethylene to displace the more expensive acetylene in the manufacture of both vinyl chloride and vinyl acetate. The 1970 production figures and usage patterns for ethylene in three industrial areas were as follows:

	USA	UK	*Other* *Western Europe*
Production ('000 tons)	7850	982	4820
Usage pattern			
Polyethylene	36%	40%	52%
Ethylene oxide	21	18	14
Ethylbenzene	9	4	9
Ethyl alcohol	8	8	2
Ethylene dichloride etc.	14	10?	17
Other	12	20	6

Propylene. Only slightly less important to the chemical industry than ethylene. World production in 1971 was about 9 m. tons, of which the USA accounted for 33%, Japan for 27%, and the UK for 6%, and was expected to rise to about 16 m. tons by 1975. The major chemical outlets for propylene in the USA in 1970 were *iso*propanol (21%), oxo alcohols (15%), polypropylene, propylene oxide and acrylonitrile (14% each), and cumene (8%). The UK consumption pattern is rather different: of 508000 tons produced in 1971 33% was converted to acrylonitrile, 20% to *iso*propanol, 18% to polypropylene and 14% to cumene. In both areas acrylonitrile and polypropylene are outlets of growing importance while *iso*propanol is declining.

Butadiene. In Europe this is now very largely obtained as a by-product of naphtha cracking but in the USA the majority comes from the dehydrogenation of mixed *n*-butenes over calcium nickel phosphate or in the presence of oxidizing systems. Since 80 to 90% of all butadiene is used in the manufacture of synthetic rubbers it is not surprising that production is increasing steadily rather than dramatically. World production in 1971 was about 5 m. tons of which Japan and the USA accounted for 2.1 and 1.5 m. tons respectively. British production was 188000 tons.

Acetylene. This has declined in both relative and absolute importance over the past ten years. Although chemically versatile it is excessively expensive. The carbide route is particularly costly because of its heavy electricity consumption: under British conditions in 1972 power costs were as much as £45 to 50 per ton of acetylene. The processes based on the pyrolysis of hydrocarbon feedstocks have proved less attractive than expected. They appear best suited to outputs of no more than 30000 to 50000 tons per year; capital costs are therefore high while the severe conditions used lead to heavy expenditure on utilities and maintenance. Acetylene derived from hydrocarbons cost two to three times as much as ethylene in 1972, while that produced via carbide cost more than four times as much. With the development of routes from ethylene to acetaldehyde, vinyl chloride and vinyl acetate, and from propylene to acrylonitrile it is hardly surprising that acetylene has been displaced from many of its former uses. World production reached a peak of 1.95 m. tons in 1969 but had fallen to 1.2 m. tons by 1971; in the latter year only 10% was made by the carbide route which appears to be on the verge of extinction. It remains to be seen whether the recent rise in crude oil prices makes natural-gas-based acetylene a more attractive starting material in the future.

AROMATIC HYDROCARBONS

The catalytic reforming of light naphtha now provides the majority of the aromatic hydrocarbons required by the chemical industry. The C_6—C_8 naphtha fraction is passed over a platinum-alumina catalyst at 400° to 500°C and 25 to 35 atm. pressure to give a 45 to 55% yield of aromatics. Platinum-alumina catalysts are unstable at the low pressures which favour cyclodehydrogenation and aromatization but the recently introduced platinum–rhenium–alumina catalysts function satisfactorily at 10 to 20 atm. pressure and increase the yield of aromatics to 75%. The aromatic fraction is extracted with di-ethylene glycol–water, N-methylpyrrolidone–water or sulpholane and then fractionally distilled to give benzene, toluene and the xylenes.

The main problem with reforming processes is that they yield mainly toluene and xylenes whereas the main market demand is for benzene. Hydrodealkylation processes to convert toluene and other alkylbenzenes to benzene have therefore been developed. The feedstock and hydrogen–usually obtained from the reforming process itself–are passed through a reactor at 550° to 750°C and 20 to 60 atm. to give high yields of benzene and alkane. Both purely thermal and metal oxide catalysed processes are used. Alternatively catalytic transalkylation reactions may be employed to convert, say, toluene to benzene and xylene. By these means the economic viability of reforming processes may be maximized.

Benzene. This remains the most important aromatic compound which in 1971 was produced on a scale of 8.5 m. tons. About 1.2 m. tons was obtained from coal and the remainder from petroleum. Even in countries where the coal industry continues to be important, coal-based benzene is of declining significance: thus only 20% of the 550 000 tons of benzene produced in the UK in 1971 came from coal. The main outlets for benzene in North America and Western Europe in 1971 were styrene (42%), cumene (20%), *cyclo*hexane (15–23%) and other (17–23%).

Toluene. Production in 1971 was about 4.4 m. tons of which 96% came from petroleum. In the USA it is very largely used as a solvent and gasoline constituent or is dehydroalkylated to benzene; in Europe, however, dehydroalkylation is less important and about 30% of the 750 000 tons produced in 1971 was converted to phenol, caprolactam or toluene di-isocyanates.

Xylenes. These continue to be of importance mainly as solvents and gasoline ingredients and only a minority of production is separated into individual isomers. World production of *o*-xylene in 1970 was about 600 000 tons, almost all of which was oxidized to phthalic anhydride. Production of *p*-xylene is higher at 1.3 m. tons but future trends depend on the demand for polyester fibres. *m*-xylene has few uses–about 50 000 tons is oxidized to *iso*phthalic acid–and is generally isomerized to the *p*-isomer.

Naphthalene. Still remains a coal-based chemical to a considerable extent: over half the rather static 1970 production of about 750 000 tons was a by-product of coal carbonization. Approximately 75% is oxidized to phthalic anhydride.

ADDITIONAL LITERATURE

A. L. WADDAMS. *Chemicals from Petroleum*, 3rd. edn., London: John Murray, 1973.

B.G.REUBEN and M.L.BURSTALL. *The Chemical Economy*, London, Longman, 1973.

Chapter 11. Ketones, Ketens and Quinones

Within the last few years many discoveries relating to the properties of ketones, ketens and quinones have been made. Research into the chemistry of ketones has been especially stimulated by the necessity to find new routes to synthesize novel prostaglandins–hormones which have very wide physiological activity in minute amounts.

KETONES

Preparation

1. Primary amines are converted into ketones with N-bromo-succinimide, e.g.

2. Oximes rapidly yield ketones on treatment with thallium (III) nitrate in methanol at room temperature, e.g.

3. Thallium (III) nitrate can also be used to oxidize some olefins to ketones.

VOLUME 5

Chapter 1. Some Essential Oils

Many formerly readily available essential oils have become scarce and some of the oils which were traditionally produced in Europe are not available at all. This change is a direct result of the fact that the production of essential oils is labour intensive and hence under the present economic climate the cultivation of essential oil-bearing crops is not profitable. This means that the perfumery industry must find the synthetic equivalents of such oils and some progress has been achieved in this direction.

Chapter 2. Gums

Recent increases in the production of xanthan gum to between 2000 and 3000 tons per annum have been noted. Most of this material seems to be used in oil-well drilling and similar processes.

Chapter 4. Adhesion and Adhesives

Perhaps the most interesting adhesive of recent years is the so-called anaerobic adhesive; developed in the 1960s this has now become established. As the name to some extent implies, these adhesives function under certain anaerobic conditions: the molecules of the monomer liquid contain double bonds which react with free radicals to form a polymer chain, but the reaction can only take place when oxygen is eliminated from the area of the glueline because oxygen in the air combines with the free radicals forming a stable unreactive compound. Thus the adhesive does not cure outside the glueline.

The present anaerobic adhesives comprise: (*a*) a high proportion of a monomeric acrylate ester; (*b*) a lesser proportion of a liquid polyester adhesive; and (*c*) a peroxy-type polymerization catalyst.

As a class these adhesives should be included under 'acrylates'. Their most important use is in the adhesive-locking of threaded nuts.

As aircraft speeds continue to increase, so the need of adhesives capable of withstanding higher and higher temperatures becomes more and more urgent. In the field of organic adhesives research on aromatic-heterocyclic polymers continues; and two polymers from this general class, polyimides and polybenzothiazoles, are used with some success on a limited scale.

Although inorganic compounds in general hold only a minor place as adhesives, ceramic substances have a limited usefulness. They are primarily of interest in bonding stainless steel, and they do have the highest heat resistance of *any* adhesive system. Like the aromatic-heterocyclic polymers they are brittle and have a low breaking strain and therefore low peel strength.

For very many years the use of 'sulphite waste liquor' from pulp mills has interested formulators of adhesives. The material is cheap and available in large quantities. It can be dried and powdered but remains soluble in water. However, under certain conditions it can be used to make chipboard of acceptable quality, but not as easily as with amino or phenolic resins. If wood chips are first coated with the waste liquor, then mixed with a strong mineral acid such as sulphuric acid and subjected to a relatively long heat and pressure cycle, preferably followed by a post-curing cycle, the water-soluble lignin is converted to a water-insoluble product and makes the manufacture of wood chipboard feasible.

Largely as a result of the development of the fluidized-bed process there is a revival of interest in the application of adhesives in powder form.

Concerning the application of adhesives, the increased interest being shown in gluing horses' shoes may lead to the non-blacksmithing farrier. The use of an adhesive in place of nails was first done with an epoxy-type adhesive many years ago.

Papers on the theory of adhesion, fewer than in the 1960s, have not thrown any fresh light on why or how things stick together, but the known parameters are always under closer examination.

Related to the subject of adhesion is that of the treatment of the surfaces of adherends. Instead of the conventional chemical treatments, adherends can be placed in a plasma environment where the surfaces are attacked by the reactive ionized gas and rendered wettable. The method, which is non-messy, can be applied with greatest advantage to plastic films such as polyethylene.

Allied to the treatment of adherends is the use of adhesion promoters. The alkoxy silanes continue to be first choice and are used with a variety of adhesives primarily for bonding metals.

Chapter 8. Non-edible Fatty Oils and Fats

There have been no major technical developments in the field of non-edible fatty oils and fats since the publication of Chapter 8. There have been, however, appreciable price fluctuations in some of the products since the last figures quoted and this has affected the extent of utilization of the different materials, particularly in the paint field.

Linseed oil, the main paint-drying oil, has increased moderately in price and at the end of 1972 was about £125 per ton, compared to £100 in 1968. The main competitor to linseed oil is soya bean oil and the price and availability of this oil depends on the extent to which it is required for edible purposes. Although there have been continued prophecies of replacement of oil in paints by synthetic materials, this change, mainly to water-based synthetic polymer emulsion paints is taking place very slowly in Gt. Britain. The conventional decorative gloss paint is still oil-based and likely to remain so for the next decade. In USA substitution of oil-based paints is taking place rather more rapidly.

Tung oil continues to fluctuate widely in price but the paint industry has little faith in the continuity of low price supplies and use this oil only when the unique properties cannot be matched in any other way. From the price of £105 per ton in 1968 there was a rise to £250 per ton by the end of 1969, falling to £130 per ton by 1972.

The most extreme rise was shown by castor oil, which, after remaining steadily at about £185 per ton from 1969 to 1971 rose to £405 per ton by the end of 1972. Castor oil as a basis of non-drying plasticizing alkyd resins is usually replaceable by coconut oil, and the best, although not entirely satisfactory substitutes for dehydrated castor oil as a drying oil are the isomerized oils.

Tall oil is a useful raw material for the paint industry but supply has not generally been reliable. It is hoped that a new source from Bulgaria to supplement that from Scandinavia and America may improve matters.

Chapter 11. Varnishes, Paints and Other Painting Compositions

Several new types of paint and methods of application have been developed recently. Closer regard for environmental pollution and toxic hazards have become major factors affecting both composition and application method.

New types of building paints include thixotropic matt emulsion paints, glossy emulsion paints, and solvent-borne paints with easy brush clean-up. The thixotropic emulsion paints are of conventional formulation with zirconium or titanium chelates added to give thixotropy. Glossy emulsion paints can be formulated on conventional acrylic latices, contain no extender pigments, but include thickeners selected to give good flow-out of brushmarks. Such gloss finishes are almost solvent-free, have easy brush clean-up, and excellent resistance to yellowing, but have lower gloss and inferior exterior durability, flow, and lapping properties than conventional solvent-borne alkyd paints. Recently developed air-drying solvent-borne gloss paints can be washed out of brushes with a mixture of water and liquid washing-up detergent instead of petroleum solvent.

Closer consideration of pollution effects has resulted in an increased tendency, particularly in car and industrial finishes, to introduce new finishes and methods of application which release the least amounts of solvent, etc, into the atmosphere, and into sewage systems. In addition, paint compositions have been changed to reduce still further the low level of toxic ingredients in air-drying paints. There is thus a trend toward water-borne finishes, to high solids solvent-borne finishes, and to methods of application (such as electrodeposition, powder coating, and radiation curing) which largely eliminate atmospheric pollution by solvents. In air-drying alkyd building paints, toxic lead drier catalysts are being largely replaced by combinations of zirconium and cobalt driers.

The development of non-aqueous dispersions (NAD) in recent years has led to the introduction of high-solids car finishes based on thermoplastic and thermosetting acrylic resins. Conventional acrylic finishes are based on high molecular weight polymers, which have low solubility even in polar solvents. With NAD, the polymers are formed as resin suspensions in non-aqueous solvents, giving a much higher solids content. Fewer coats of paint are required, spraying is easier and more efficient, and atmospheric pollution by solvents is greatly reduced during spraying and stoving.

Powder coating involves coating a metal object with a powdered mixture of resin and pigment, by electrostatic spraying or dipping into a fluidized bed of the mixture, followed by heating in an oven to fuse and cure it. Such coatings can be based on polyvinyl chloride, epoxy, acrylic, cellulose acetobutyrate, or polyamide resins.

In radiation curing, clear or translucent solutions of polymerizable resins (unsaturated polyesters, acrylics, urethanes) are dissolved in reactive monomers (styrene, vinyl toluene, acrylic), and the coated objects (preferably flat) are exposed to radiation using an electron beam, or ultra violet light (with a photo-initiator such as benzil in the coating).

ADDITIONAL LITERATURE

H. WARSON. *The Applications of Synthetic Resin Emulsions*, London, Ernest Benn Ltd, 1972.
T. C. PATTON. *Pigment Handbook*, New York-London, J. Wiley and Sons, 1973.
Series on Coatings Technology, Federation of Societies for Paint Technology, Philadelphia, 1964–72.

Chapter 12. Leather

In view of the continued demand for leather hide and skins, producing countries are now more fully aware of the need to improve the quality of their cattle, sheep and goats and to reduce needless loss of quality through faulty flaying and poor preservation. The Food and Agriculture Organization of the United Nations has continued to give practical aid towards improving methods in many industrial developing countries with a fair measure of success to the benefit of producer and tanner. In recent years a number of new tanneries have been built, equipped with modern machinery, in the developing African and Asian countries enabling a larger proportion of hides and skins of local origin to be utilized to meet local demands and further leather exports in the finished or partially treated condition. Through research and development a wide range of new chemicals dyes and oils, etc. are now available for the tanning and finishing of leather. These include new synthetic tannins which react with chromium compounds yielding leather with improved physical properties in relation to water and perspiration etc. In the field of tannery machinery there has been considerable progress to meet the demand for rapid through-put units combined with labour saving processing.

The use of processing vessels, based on the concrete mixer principle, machinery and ancillary equipment covering the whole range of leather manufacture has undergone numerous changes. These include reinforced glass fibre or stainless-steel processors with a wide field for speed and mixing actions, suitable for all types of hides and skins including woolled sheepskins and furs. New hydraulic high-capacity fleshing machines, requiring only simple maintenance and capable of operation by semi-skilled labour, are now available.

A new feed-out shaving machine in which striking out has been eliminated by combining shaving with setting-out in one operation is gaining in popularity.

Punch card control systems have been introduced for both tanning and dyeing drums and give full control of running times, water and chemical additions, temperature and rinsing operations, are increasingly being used in tanneries.

An ironing machine giving an improved ironing effect with better heating and electronic temperature control is now available. Further improvements on this machine include a through feed fitted with an endless stainless-steel band and automatic servo-hydraulic steerer with heating by means of infra-red elements. In the field of leather measurement an electronically operated machine with variable speeds working on the same principle as the pin wheel type ensures greater accuracy, is easier to use when dealing with stretchy leathers and may be used in conjunction with a batch counter and adding machine print out. Leather driers have also received attention, the new vacuum driers providing a larger working area and the whole operation fully automatic.

Process rationalization with the aim of combining process stages wherever practicable and the presentation of chemicals and auxiliaries to minimize environmental pollution has received considerable attention in recent years.

Efforts are being made by the supply industries both in this and in process development to obtain new or improved properties in leather or for the minimization of waste in use and are continuing to back the public acceptance of leather as a high fashion material.

The demand for clothing leathers, combining softness and light-weight texture continues and the boom for leatherwear in world markets shows no signs of abating.

For shoe leathers, those of the full grain softee type ensuring the unspoilt character of natural leather combined with a smooth waxy feel and soft handle are now in increasing demand.

Chapter 13. Artificial Leather and Linoleum

POLYURETHANE-COATED FABRICS

During the last three years there has been a considerable increase in the use of polyurethane coated fabrics. It is difficult to quote exact figures but many millions of yards are used annually for consumer goods such as clothing, footwear, upholstery and handbags. The main reason for their growth is due to the considerable improved technical and aesthetic properties of polyurethane coated fabrics.

Technically, these materials are superior to vinyl-coated fabrics in a number of important ways. Firstly, polyurethanes can form soft, flexible films without the aid of plasticizers. This means that there are no oils to be washed out by solvents and therefore the coatings can be formulated to be dry-cleanable. The absence of plasticizers also means that there are no migration problems where the coated fabrics are used in conjunction with other materials, for example with adhesives or leathers. Secondly, polyurethanes are much more resistant than PVC to stiffening at low temperatures. Thus a garment made from polyurethane will still feel supple on a cold day in winter, whereas a PVC garment would feel stiff and uncomfortable. Also the PU-coated fabric is more resistant to flex cracking at low temperatures, an important property for garments and footwear, and for car upholstery. Thirdly, polyurethane is very resistant to abrasion in thin films, enabling lightweight, but serviceable coatings to be produced. Fourthly, in thin films polyurethane has a high permeability to water vapour, yet is waterproof. Advantage can be taken of this, in conjunction with manufacturing and structural techniques, to produce PU-coated fabrics with a measure of permeability.

Again, aesthetically, PU-coated fabrics offer many advantages over vinyls. The softness and suppleness of PU coatings result in excellent drape properties for clothing, properties which can be enhanced by the selection of flexible base fabrics, such as warp knitted nylon or circular knitted cotton. By selection of polymer and compounding ingredients, PU's can be formulated to give a very dry handle, even in high-gloss films, or they can be given the somewhat greasy handle characteristic of certain leathers. By virtue of their construction they have a pleasant, characteristic leather 'break', and this same construction imparts a softness and fullness to the touch.

This remarkable combination of aesthetic and technical properties accounts for the growth in usage of PU-coated fabrics. To date, generally, this has not been growth at the expense of vinyl-coated fabrics, though there are certain fields in which this could happen in the future. The polyurethanes have provided an additional and novel coated fabric which has opened up new markets and has provided further opportunities in those markets approaching saturation.

However, with all pros there are cons, and in the case of polyurethane, there are two distinct disadvantages with respect to vinyl. Firstly, it is much more expensive, the actual cost of raw material in the coating being from 10 to as much as 15 times more expensive than a vinyl coating of equivalent thickness. Secondly polyurethane must be applied from a solvent-based coating compound. Vinyls are applied from a paste which is virtually 100% solids. For example, if a vinyl paste coating having a thickness of ten-thousandths of an inch is applied to a substrate, after the heating operation a coating thickness of approximately ten-thousandths of an inch is obtained. With polyurethanes, where the polymer is dissolved in solvent, the solution may have a solids content of perhaps 25%. Thus on spreading a polyurethane solution of ten-thousandths of an inch onto a substrate after drying off the solvent, one is left with only 2.5 thousandths of an inch coating thickness.

CUSHIONED VINYL FLOORING

To a large degree, the domestic printed flooring as described in the original chapter has been replaced by material containing a cellular layer of PVC. In almost all cases this product has been given a novel appearance by the application of the 'chemical embossing' principle. Several qualities have been developed but they vary mainly in thickness. They all have a softer and more flexible feel than their predecessors, and a wide and interesting range of designs. The great novelty of these floorcoverings is that the material has an embossed appearance, which is accurately in register with the rest of the design so that, for example, the mortar line between tiles can be considerably lower than the surface of the tiles themselves.

Usually the backing is an asbestos felt base in which the asbestos fibres are bound by a synthetic rubber. A pigmented (usually white) PVC plastisol containing a blowing agent is spread on to the asbestos and the coated felt passed through an oven at a temperature sufficient to gel, but not fully cure, the resin and below the decomposition temperature of the blowing agent. At this stage the material is capable of being rolled up, transported, handled and so on, and it is then printed in a photogravure printing machine with several colours; frequently trichromatic printing is used to produce a sophisticated effect. After printing, a further coat of clear plastisol is spread on top of the print and, in tandem with this, the product passed through an oven at a considerably higher temperature than previously.

The oven is divided into a series of zones of increasing temperature and in this process the resin in both case sheet and wear layer fuses, while at the same time the blowing agent gives off gas, thus converting the base layer into a cellular one. Usually the asbestos is of the order of 0.76 mm in thickness and the cellular layer about 1.01 mm and the clear layer about 0.17 mm although thicker versions are produced for heavier duty.

It is necessary to strike the right balance between the rate of the evolution of gas from the blowing agent, and the melt viscosity of the PVC base layer in which this is occurring, in order to obtain a good cellular structure and avoid the bubbles escaping into the clear wear layer. This is done partly by temperature control in the oven, partly by choosing the correct grades of PVC resin and partly by having in the base sheet an additive known as a kicker or accelerator. There are various compounds available for this purpose, such as salts of zinc, lead or cadmium, and their effect is to cause a drop in the temperature at which the blowing agent evolves gas, and thus give a control over this feature.

The embossed appearance is produced by one of two means, both of which interfere with the blowing agent/accelerator system. In the process developed by Congoleum-Nairn an inhibitor is added to the printing ink when it is desired to prevent foaming. It is thought that, in the final fusion, the inhibitor penetrates and reacts with the accelerator in the base layer so that at this point a much higher temperature would be needed to cause gas evolution. The result is that it is possible to produce an embossed design in which foamed and non-foamed areas are accurately kept in register, a process which prior to this development was not thought to be possible.

An alternative method is described in the Armstrong patent, in which a monomer which can be polymerized is added to the base sheet. At the point where it is desired to prevent foaming a peroxide is added to the printing ink, which is then followed by the normal transparent wear layer, heating and fusing. It is claimed that the peroxide penetrates the base layer and causes polymerization of the monomer to occur at these points. The resultant increase in viscosity resists the effect of the evolving gas, so that once again an emboss is produced in register with the design.

These cellular floorcoverings have become very popular during the last few years as they show a considerable improvement in quietness and warmth over the earlier hard printed vinyl floorcoverings. They are also very easy to maintain, and the chemical embossing process has opened up a new dimension in design.

Although the process as described sounds simple, it is basically very sophisticated as regards technique. In the main, plasticizers of the butyl benzyl phthalate type are preferred because of their relatively good stain resistance, and the correct choice of resin is extremely important, particularly as regards the balance of k values between the base sheet and the wear layer. The ovens in which the heating is done also demand a very precise and accurate control of temperature in order to avoid variation in thickness from side to side and other faults.

Chapter 14. Rubber

The pattern of the end usage of rubber has not changed to any extent since the end of the nineteenth century. The pneumatic tyre still accounts for about 60% of the world rubber consumption, this being 10 m. tons in 1973.

The other traditional end uses, such as conveyor belting, hosing or footwear, absorb about 1% and even less in the case of other non-tyre uses (1 000 to 10 000 tons a year).

The trend today is for innovations aimed at new or improved production methods, safety devices for cars or the replacement of traditional materials. The producers will supply special compounds as used in the plastics industry, probably to an extent of 50 000 tons in 1975. An example is ready-to-mould-and-cure EPDM compounds for the car industry suitable for window sections, shock-absorbing bumper systems, etc. Natural rubber and SBR will be replaced in many non-tyre products to be used at normal temperatures (20–30°C) if the third generation of rubbers of the thermoplastic type (TPR) will develop favourably into easy processable materials needing no vulcanization. The borderline between plastics and rubber will fade away, when a big part of the non-tyre market will be opened to the plastic processors using block polymers of styrene and butadiene or styrene and isoprene, aromatic polyesters, thermoplastic poly-urethanes or ionomer resins.

The marketing of SMR (Standard Malayan Rubber) will be improved, the target being 1 m. tons for 1975, about 70% of the Malayan natural rubber production. Malayan producers have developed a tailor-made SMR for lorry and earthmover tyres, consisting of latex and sheet rubber, field coagulum and plasticizer.

The consumption of natural rubber and its counterpart SBR depends not only on technology (production, compounding, processing) but also on world economy. Factors affecting it include costs of oil supply from the Middle East shortage of energy, the development of non-tyred vehicles (hovercraft) and also the possible development of a less expensive method of collecting latex and preparation of raw rubber by means of new dewatering and drying machinery. It is not likely, however, that these costs and the freight charges overseas will decrease.

ADDITIONAL LITERATURE

P. W. ALLEN. *Natural rubber and the synthetics*, London, Crosly.

Chapter 20. Alkaloids and Herbs

ALKALOIDS

Important changes in the form of trade statistics have been made in the UK following the introduction on the 1 January 1970 of the 'Tariff and Overseas Trade Classification', due to Common Market entry. The statistics are now more detailed; also re-exports are not separately distinguished but are included with exports of UK produce. With respect to alkaloids the following information is now available in *Annual Statements of the Trade of the United Kingdom* up to and including 1971 and in abbreviated form in the monthly publication *Overseas Trade Statistics of the United Kingdom* (HMSO) for 1970–73:

(*a*) annual imports and exports to various countries of: Opium alkaloids, cocaine, caffeine, quinine and other vegetable alkaloids, their salts and derivatives,

(*b*) detailed information of the weight, cost and countries involved in the import of all alkaloids and in particular caffeine, theobromine, emetine, nicotine, quinine,

(c) similarly detailed information for exports (only caffeine, emetine and quinine were dealt with in detail before 1970),

(*d*) more detailed figures for import and export of cinchona bark, ipecacuanha and opium.

ADDITIONAL LITERATURE

J. E. SAXTON. *The Alkaloids*, Vols. 1 (1971), 2 (1972), and 3 (1973), London, The Chemical Society.

Chapter 21. Pharmaceutical Products

TABLETS

Direct compression bases

Anhydrous lactose which consists of 75% β-lactose has proved a useful material in many formulations due to its low equilibrium moisture content at those relative humidities commonly encountered in normal storage. Several physically *modified starches* have become available the main varieties of interest being a maize starch–Starex 1500 and a rich starch.

These show excellent binding characteristics and good disintegration properties but require glidants to give adequate flow properties. Their relatively high moisture contents may be a problem in formulations susceptible to hydrolysis.

Film coating

A new range of water-based film coating materials has been introduced and this consists of aqueous dispersions of polymethylmethacrylate resins, known as Eudragit dispersions. The advantage of these preparations is that being water-based they are not susceptible to the inflammable and toxic hazards associated with film coating solutions using organic solvents. A variety of Eudragit dispersions can be used to modify or delay the release of drugs from tablet formulations.

NEW PHARMACEUTICAL PREPARATIONS

The Alza Corporation, USA, have developed some unique preparations which are designed to be placed in body cavities and release their drug contents over long periods of time, e.g. 1 week to 1 month. The advantage of these preparations over conventional dosage forms such as tablets and capsules is that they release the drug at a constant rate and may therefore give a more reproducible response. Furthermore they can be removed from the patient at any desired time during therapy.

The two preparations which have undergone the most extensive clinical tests are the 'Ocusert' and the 'Progestasert'.

The Ocusert is placed in the conjunctival sac of the eye and consists of a small piece of modified gelatin containing the drug in a depot or reservoir distributed throughout the device. In this way the drug, e.g. Pilocarpine is delivered at a constant rate, e.g. 20–40 microgrammes per hour for a week. The Progestasert is an intrauterine device which releases progesterone to the endometrium at a constant rate. It is known as the 'once a year' contraceptive method.

ADDITIONAL LITERATURE

J. T. CARSTENEN. *Theory of Pharmaceutical Systems*, Vols. 1 and 2, Academic Press, 1972.
M. J. GROVES. *Parenteral Products*, London, Heinemann, 1973.
E. SHOLTEN. and K. RIDGWAY. *Physical Pharmaceutics*, Oxford, 1974.

Chapter 22. Cosmetics

ADDITIONAL LITERATURE

M. S. BOLSAM and E. PEGARIN. *Cosmetics – Science and Technology*, 2nd edn, Vols. I and II, Wiley-Interscience, 1972.
J. STEPHAN JELLINCK. *Formulation and Function of Cosmetics*, Wiley-Interscience, 1970.

MONTFORD A. JOHNSEN. *The Aerosol Handbook*, 1st edn, Caldwell, N.J., Wayne E. Dorland Company, 1972.

JOHN J. SCIARRA and L. STOLLER, Eds. *The Science and Technology of Aerosol Packaging*, Wiley-Interscience, 1974.

'The Hygienic Manufacture and Preservation of Toiletries and Cosmetics', A monograph prepared by a Working Party for the Society of Cosmetic Chemists of Great Britain, *Journal of the Society of Cosmetic Chemists*, 1970, Vol. 21, pp. 719–800.

Code of Good Practice. Toilet Preparations Federation in co-operation with the Society of Cosmetic Chemists of Great Britain.